NANOMEDICINE FOR THE TREATMENT OF DISEASE

From Concept to Application

NANOMEDICINE FOR THE TREATMENT OF DISEASE

From Concept to Application

Edited by
Sarwar Beg
Mahfoozur Rahman
Md. Abul Barkat
Farhan J. Ahmad

AAP APPLE ACADEMIC PRESS

Apple Academic Press Inc.	Apple Academic Press Inc.
3333 Mistwell Crescent	1265 Goldenrod Circle NE
Oakville, ON L6L 0A2	Palm Bay, Florida 32905
Canada USA	USA

© 2020 by Apple Academic Press, Inc.

First issued in paperback 2021

Exclusive worldwide distribution by CRC Press, a member of Taylor & Francis Group

No claim to original U.S. Government works

ISBN 13: 978-1-77463-443-1 (pbk)
ISBN 13: 978-1-77188-767-0 (hbk)

CIP data on file with Canada Library and Archives

Library of Congress Cataloging-in-Publication Data

Names: Beg, Sarwar, 1987- editor. | Rahman, Mahfoozur, 1984- editor. | Barkat, Md. Abul, 1987- editor. | Ahmad, Farhan J., editor.

Title: Nanomedicine for the treatment of disease : from concept to application / editors, Sarwar Beg, Mahfoozur Rahman, Md. Abul Barkat, Farhan J. Ahmad.

Description: Toronto ; New Jersey : Apple Academic Press, 2019. | Includes bibliographical references and index.

Identifiers: LCCN 2019015309 (print) | LCCN 2019016142 (ebook) | ISBN 9780429425714 (ebook) | ISBN 9781771887670 (hardcover : alk. paper)

Subjects: | MESH: Nanoparticles | Drug Carriers | Drug Delivery Systems | Technology, Pharmaceutical

Classification: LCC RS420 (ebook) | LCC RS420 (print) | NLM QV 785 | DDC 615.1/9--dc23

LC record available at https://lccn.loc.gov/2019015309

Apple Academic Press also publishes its books in a variety of electronic formats. Some content that appears in print may not be available in electronic format. For information about Apple Academic Press products, visit our website at **www.appleacademicpress.com** and the CRC Press website at **www.crcpress.com**

About the Editors

Sarwar Beg, PhD

Dr Sarwar Beg is currently serving as an Assistant Professor of Pharmaceutics & Biopharmaceutics at School of Pharmaceutical Education and Research, Jamia Hamdard (Hamdard University), New Delhi, India. Prior to joining Jamia Hamdard, Dr Sarwar has worked with Jubilant Generics Limited, Noida, India, as Research Scientist, where he was solely responsible for implementation of Quality by Design (QbD) in formulation development and analytical development of generic products. He has nearly a decade of experience in systematic development and characterization of novel and nanostructured drug delivery systems using QbD paradigms including Design of Experiments (DoE), Quality Risk Management (QRM), Multivariate Chemometric Approaches, Advanced Biopharmaceutics and Pharmacokinetic. Besides, Dr Sarwar has acquired know-how of applying advanced release kinetic modeling, pharmacokinetic modeling and in vitro/in vivo correlation (IVIVC) for efficient development of drug products. Till date, he has authored over 135 publications including research and review papers in high impact peer-reviewed journals, 08 journal special issues, 08 books, 38 book chapters and 3 Indian patent applications. He has very good citation record with Google Scholar H-index of 23, i10-index of 56 and 2100 citations to his credit. He is serving as the Regional Editor-Asia of the journal Current Nanomedicine (Bentham Science) and is an editorial board member of several other journals. Dr Sarwar has also participated and presented his research work at several conferences held in India, China, Bangladesh, Canada and USA, bagging several "Best Paper" awards. He has also been awarded with "Innovative Pharma Researcher Award 2016" by SIPRA Lab (Hyderabad), "Eudragit® Award 2014" in South-Asia by M/s Evonik (Germany), "Budding QbD Scientist Award 2014" and "Budding ADME Scientist Award 2013" by M/s Select Biosciences (UK) and "Novartis Biocamp Award 2012" (Hyderabad). In 2017, Dr Sarwar was felicitated by Honorable Union Health Minister of India and Managing Director, Sun Pharmaceutical Industries, with prestigious "Sun Pharma Science Foundation" Award.

Mahfoozur Rahman, MPharm

Mahfoozur Rahman is an Assistant Professor in the Department of Pharmaceutical Sciences, Shalom Institute of Health and Allied Sciences, Sam Higginbottom University of Agriculture, Technology and Sciences (SHUATS), Allahabad, India, where is also currently working on his PhD in Pharmaceutical Sciences. He completed his master's in Pharmacy at Jamia Hamdard, New Delhi, India. His major areas of research interest include development and characterization of nanosized drug delivery systems for inflammatory disorders, such as psoriasis, arthritis, cancer, etc. To date he has published over 90 publications in peer-reviewed journals with H-index of 18, i10-index of 35 and 25 book chapters with various publishers. He is a member of the IPGA (Indian Pharmacy Graduates' Association) and an editorial board member and reviewer of various internationals journals.

Md. Abul Barkat, PhD

Md. Abul Barkat, PhD, is currently working as an Assistant Professor at the School of Pharmacy, K.R. Mangalam University, Gurgaon, India. He has received PhD in Pharmaceutical Sciences (Pharmaceutics) from Integral University Lucknow, India, in the year 2016. He completed his masters in Pharmaceutical Sciences from Hamdard University, New Delhi, India. He has more than five years of research experience in the field of pharmaceutics. His major areas of research interest include herbal and silver-based development and characterization of diverse nanostructured systems, controlled release drug delivery systems, bioenhanced drug delivery systems, nanomaterials, and nanocomposites. To date, he has authored over 20 publications in various peer-reviewed journals, two book chapters, and two books, and also has an US and Indian patent application to his credit. Dr. Barkat has also participated and presented his research work at conferences in India, Canada, and Spain.

Prof. Farhan Jalees Ahmad

Farhan Jalees Ahmad, PhD, is currently a Professor, School of Pharmaceutical Education & Research, and Dean, Interdisciplinary Sciences, Jamia Hamdard, New Delhi, India is an internationally known researcher in the area of Pharmaceutical Sciences. Having obtained his degree in M.Pharma and Ph.D (Medicine) from Jamia Hamdard, he continues to teach and leads a very productive research group which has been funded extensively by National and international funding agencies.

Dr. Ahmad has a 26 years of rich experience in Research and Teaching. He has experience of industrial research through his fruitful association with Ranbaxy Research Laboratories for six long years as Scientist. His domain of research therein included development, scale-up, technology transfer and launching of pharmaceutical products, both for domestic and international markets.

He is working in the area of Nanomedicine for the last 15 years and has published a number of research papers on nanomedicine in peer reviewed journals. Two of his nanoproducts are approved by Drug Controller of India for Phase-III clinical studies. He has organized many International conferences and more than 40 National symposiums, workshops and seminars. Besides, he is also working on amalgamation of herbal medicinal plants with modern therapeutics in order to deliver a scientifically acceptable therapy for various diseases.

He has been granted projects to a tune of rupees 4.5 crores from National agencies like DBT, CCRUM, AYUSH, UGC, DST and International agencies like FIP and OPCW etc.He is also involved in nationally conceived projects under the aegis of institutes like INMAS, GB Pant Hospital, RBTB Hospital and IIT, Delhi.

He has earned many accolades including Young Scientist Fast Track from DST, Scientist of the Year-2005 from NESA, UGC Research Award 2011, Bharat Jyoti Award 2011, Pharma Ratan 2017, ABAP Senior Scientist Award 2017 and also prizes in scientific poster session in National Pharmacy Symposium. He has a US patent, Two PCT and 24 Indian patents to his name. He has published more than 300 research and review papers, 12 Book chapters, 9 books with a total citation of 9000, H Index of 44 and i-10 index of 200and attended many national and international conferences for presentation of research papers. He has guided around 35 M.Pharm students and about 33 PhD scholars. Besides this, he is offering consultancy to small pharmaceutical setups for pharmaceutical product development and scale-up, and troubleshooting specific problems.

He has been an expert, chairman, co-chairman, invited speaker, co-ordinator and examiner of various seminars, workshops in and outside the university.

He played a pivotal role in developing and maintaining the state-of-the-art Nanomedicine Lab and Central Instrumentation Facility at Faculty of Pharmacy. He is an active member of bodies like IPA, APTI, NESA, MAAS, IPGA, IHPA, CRS and ICA, and also holds their life memberships. He has served IPA, Delhi Branch in the capacity of Member Executive Committee, Honorary Secretary and as President for two terms. At the moment He is serving as General Secretary for SPDS, North chapter.

Contents

Contributors

Muhammad Afzal
Department of Pharmacology, College of Pharmacy, Aljouf University, Sakaka, KSA

Farhan J. Ahmad
Nanomedicine Research Lab, Department of Pharmaceutics, Faculty of Pharmacy, Jamia Hamdard,
New Delhi–110062, India

Kainat Alam
Christian College of Nursing, Faculty of Health Sciences, Sam Higginbottom Institute of Agriculture,
Technology, and Sciences, Allahabad, Uttar Pradesh, India

Kumar Anand
Department of Pharmaceutical Technology, Jadavpur University, Kolkata,West Bengal, India.
Tel.: +91-8017136385 E-mail: sanmoykarmakar@gmail.com

Firoz Anwar
Department of Biochemistry, Faculty of Science, King Abdul Aziz University, Jeddah,
Kingdom of Saudi Arabia

Mohd Aqil
Department of Pharmaceutics, School of Pharmaceutical Education and Research, Jamia Hamdard,
New Delhi–110062, India

Md. Abul Barkat
School of Medical & Allied Sciences, Department of Pharmaceutics, K.R. Mangalam University,
Gurgaon, India

Sarwar Beg
Product Nanomedicine Research Lab, Department of Pharmaceutics, School of Pharmaceutical
Education and Research, Jamia Hamdard (Hamdard University), New Delhi, India

Shreyasi Chakraborty
Department of Pharmaceutical Technology, Jadavpur University, Kolkata–700032, India

Mohini Chaurasia
Amity Institute of Pharmacy, Amity University, Lucknow–226010, India

Sundeep Chaurasia
Formulation Research & Development, Mankind Research Centre (A Division of Mankind Pharma Ltd.),
191-E, Sector-4/II, IMT Manesar-122 051, Gurgaon, Haryana, India

Manish K. Chourasia
Senior Scientist, Pharmaceutics and Pharmacokinetics Division, CSIR-Central Drug Research Institute,
Sector–10, Jankipuram Extension, Lucknow–226031, U.P., India, E-mail: manish_chourasia@cdri.res.in

Moumita Dhara
Department of Pharmaceutical Technology, Jadavpur University, Kolkata–700032, India

Surbhi Dubey
Drug Delivery Research Laboratory, Department of Pharmaceutical Sciences,
Dr. Harisingh Gour Vishwavidyalaya, Sagar, M.P., 470003, India

Debasmita Dutta
Department of Pharmaceutical Technology, Jadavpur University, Kolkata–700032, India

Lopamudra Dutta
Department of Pharmaceutical Technology, Jadavpur University, Kolkata–700032, India

Iman Ehsan
Department of Pharmaceutical Technology, Jadavpur University, Kolkata–700032, India

Clara Fernandes
Assistant Professor, Shobhaben Pratapbhai Patel School of Pharmacy & Technology Management, SVKM's NMIMS, V.L. Mehta Road, Vile Parle (W), Mumbai–400056, India, E-mail: clara_fern@yahoo.co.in

Laxmikant Gautam
Drug Delivery Research Laboratory, Department of Pharmaceutical Sciences, Dr. Harisingh Gour Vishwavidyalaya, Sagar, M.P., 470003, India

Pallab Ghosh
Bengal School of Technology, Chinsurah, Hooghly, West Bengal, India, Tel.: +91-8981088093, E-mail: adil503@yahoo.co.in

Sadaf Jamal Gilani
Department of Pharmaceutical Chemistry, Glocal School of Pharmacy, The Glocal University, Saharanpur, Uttar Pradesh–247121, India

Ramesh K. Goyal
Delhi Pharmaceutical Science & Research University, Pushp Vihar Sector–3, MB Road, New Delhi–110017, India

Arijit Guha
Bengal School of Technology, Chinsurah, Hooghly, West Bengal, India

Madhu Gupta
Delhi Pharmaceutical Science & Research University, Pushp Vihar Sector–3, MB Road, New Delhi–110017, India, Tel.: +91-9205916054, E-mail: madhugupta98@gmail.com

Abdul Hafeez
Glocal School of Pharmacy, Glocal University, Saharanpur, U.P., India

Harshita
School of Medical & Allied Sciences, Department of Pharmaceutics, K.R. Mangalam University, Gurgaon, India

Ashique Al Hoque
Department of Pharmaceutical Technology, Jadavpur University, Kolkata–700032, India

Chowdhury Mobaswar Hossain
Bengal School of Technology, Chinsurah, Hooghly, West Bengal, India

Syed Sarim Imam
Associate Professor, Department of Pharmaceutics, Glocal School of Pharmacy, The Glocal University, Saharanpur, Uttar Pradesh–247121, India, Tel.: +91-9536572892, E-mail: sarimimam@gmail.com

Arun K. Iyer
PhD, Assistant Professor, Use-Inspired Biomaterials and Integrated Nanodelivery Systems Laboratory, Department of Pharmaceutical Sciences, Eugene Applebaum College of Pharmacy and Health Sciences, Wayne State University, Detroit, MI 48201 | Molecular Imaging Program, Barbara Ann Karmanos Cancer Institute, School of Medicine, United States, Tel.: 313-577-5875, E-mail: arun.iyer@wayne.edu

Anamika Jain
Drug Delivery Research Laboratory, Department of Pharmaceutical Sciences, Dr. Harisingh Gour
Vishwavidyalaya, Sagar, M.P., 470003, India

Sanmoy Karmakar
Department of Pharmaceutical Technology, Jadavpur University, Kolkata, West Bengal, India

Sushil K. Kashaw
Department of Pharmaceutical Sciences, Dr. Harisingh Gour University, Sagar, India

Imran Kazmi
Glocal School of Pharmacy, Glocal University, Mirzapur Pole, Saharanpur, Uttar Pradesh, India

Lalit Kumar
Faculty of Pharmaceutical Sciences, Department of Pharmaceutics, PCTE Group of Institutes,
Ludhiana, Punjab 142021, India.and I.K. Gujral Punjab Technical University, Jalandhar,
Punjab–144601, India, Tel.: +91-7837-344360, Fax: +91-1887-221276,
E-mail: lkpharma27@gmail.com

Vikas Kumar
Natural Product Drug Discovery Laboratory, Department of Pharmaceutical Sciences, Faculty of Health
Sciences, Shalom Institute of Health and Allied Sciences (SIHAS), Sam Higginbottom Institute of
Agriculture, Technology, and Sciences (SHUATS), Allahabad, Uttar Pradesh, India,
Tel.: +91-8627985598, E-mail: mahfoozkaifi@gmail.com

Leena Kumari
Department of Pharmaceutical Technology, Jadavpur University, Kolkata–700032, India

Bikash Medhi
Department of Pharmacology, Postgraduate Institute of Medical Education and Research (PGIMER),
Chandigarh–160012, India

Jaya Gopal Meher
Pharmaceutics and Pharmacokinetics Division, CSIR-Central Drug Research Institute, Sector–10,
Jankipuram Extension, Lucknow–226031, U.P., India

Yu Mi
Department of Radiation Oncology, University of North Carolina at Chapel Hill, Chapel Hill,
North Carolina 27599, United States, E-mail: yumi1@med.unc.edu

Brahmeshwar Mishra
Professor of Pharmaceutics, Department of Pharmaceutical Engineering & Technology, Indian Institute
of Technology (Banaras Hindu University), Varanasi–221005, U.P., India, Tel.: +91 542 6702748,
Fax: +91 542 2368428, E-mail: bmishra.phe@iitbhu.ac.in

Nishi Mody
Drug Delivery Research Laboratory, Department of Pharmaceutical Sciences, Dr. Harisingh Gour
Vishwavidyalaya, Sagar, M.P., 470003, India

Biswajit Mukherjee
Professor and Head, Department of Pharmaceutical Technology, Jadavpur University,
Kolkata–700032, India, Tel.: & Fax: +91-33-24146677, E-mail: biswajit55@yahoo.com,
Tel.: +91-33-2457-2588/2414-6677/2457-2274

Paramita Paul
Department of Pharmaceutical Technology, Jadavpur University, Kolkata–700032, India

Priyanka Prabhu
Shobhaben Pratapbhai Patel School of Pharmacy & Technology Management, SVKM's NMIMS,
Vile Parle West, Mumbai, India

Deo Nandan Prasad
I. K. Gujral Punjab Technical University, Jalandhar, Punjab–144601, India; and Department of Pharmaceutical Chemistry, Shivalik College of Pharmacy, Nangal, Punjab–140126, India

Mahfoozur Rahman
Assistant Professor & Researcher, Department of Pharmaceutical Science, Faculty of Health Science, Shalom Institute of Health and Allied Sciences SHUATS-State University (Formerly Allahabad Agriculture Institute) Naini, Allahabad-211007, India. Tel.: +91-8627985598, E-mail: mahfoozkaifi@gmail.com

Md. Rizwanullah
Department of Pharmaceutics, School of Pharmaceutical Education and Research, Jamia Hamdard, New Delhi–110062, India

Somdatta Roy
Department of Pharmaceutical Technology, Jadavpur University, Kolkata–700032, India

Rehan Abdur Rub
Phmaceutics Research Lab. SPER, Jamia Hamdard, New Delhi-62, India

Prashant Sahu
Department of Pharmaceutical Sciences, Dr. HarisinghGour University, Sagar, India

Samaresh Sau
Use-Inspired Biomaterials and Integrated Nanodelivery Systems Laboratory, Department of Pharmaceutical Sciences, Eugene Applebaum College of Pharmacy and Health Sciences, Wayne State University, Detroit, MI, United States

Soma Sengupta
Department of Pharmaceutical Technology, Jadavpur University, Kolkata–700032, India

Md. Adil Shaharyar
Bengal School of Technology, Chinsurah, Hooghly, West Bengal, India, and
Department of Pharmaceutical Technology, Jadavpur University, Kolkata,West Bengal, India. Tel.: +91-9748902723, Email: adil503@yahoo.co.in, adil503@yahoo.co.in and

Mohd Farooq Shaikh
Neuropharmacology Research Laboratory, Jeffrey Cheah School of Medicine and Health Sciences, Monash University Malaysia, Selangor Darul Ehsan, Malaysia

Rajeev Sharma
Drug Delivery Research Laboratory, Department of Pharmaceutical Sciences, Dr. Harisingh Gour Vishwavidyalaya, Sagar, M.P., 470003, India

Vikas Sharma
Shri Rawatpura Sarkar Institute of Pharmacy, NH–75, Jhansi Road, Datia, MP, India, E-mail: vikassharma15@gmail.com

Saritha Shetty
Shobhaben Pratapbhai Patel School of Pharmacy & Technology Management, SVKM's NMIMS, Vile Parle West, Mumbai, India, E-mail: saritha.shetty@nmims.edu

Anita Singh
Department of Pharmaceutical Sciences, Bhimtal Campus, Kumaon University, Nainital–263136, India

Pankaj K. Singh
Pharmaceutics and Pharmacokinetics Division, CSIR-Central Drug Research Institute, Sector–10, Jankipuram Extension, Lucknow–226031, U.P., India

Yuvraj Singh
Pharmaceutics and Pharmacokinetics Division, CSIR-Central Drug Research Institute, Sector–10, Jankipuram Extension, Lucknow–226031, U.P., India

Shringika Soni
Department of Pharmacology, Postgraduate Institute of Medical Education and Research (PGIMER), Chandigarh-160012, India

Divya Suares
Shobhaben Pratapbhai Patel School of Pharmacy & Technology Management, SVKM's NMIMS, V.L. Mehta Road, Vile Parle (W), Mumbai–400056, India

Puneet Utreja
I. K. Gujral Punjab Technical University, Jalandhar, Punjab–144601, India; and
Faculty of Pharmaceutical Sciences, Department of Pharmaceutics, PCTE Group of Institutes, Ludhiana, Punjab 142021, India.

Shivani Verma
Department of Pharmaceutics, Rayat-Bahra College of Pharmacy, Hoshiarpur, Punjab 146001, India; and I.K. Gujral Punjab Technical University, Jalandhar, Punjab–144601, India

Nikhar Vishwakarma
Drug Delivery Research Laboratory, Department of Pharmaceutical Sciences, Dr. Harisingh Gour Vishwavidyalaya, Sagar, M.P., 470003, India

S.P. Vyas
Drug Delivery Research Laboratory, Department of Pharmaceutical Sciences, Dr. Harisingh Gour Vishwavidyalaya, Sagar, M.P., 470003, India, E-mails: vyas_sp@rediffmail.com, spvyas54@gmail.com

Youli Xia
Department of Bioinformatics & Computational Biology Curriculum, University of North Carolina at Chapel Hill, Chapel Hill, North Carolina 27599, United States

Feifei Yang
Department of Radiation Oncology, University of North Carolina at Chapel Hill, Chapel Hill, North Carolina 27599, United States | Institute of Medicinal Plant Development (IMPLAD), Chinese Academy of Medical Sciences & Peking Union Medical College, Haidian District, Beijing, P.R. China

Abbreviations

μm	micron meter
AA	allergic asthma
AAV1	adenoviral vectors of serotype 1
AB	amlodipine besylate
ABC	ATP-binding cassette
AB-NE	amlodipine besylate loaded nanoemulsion
ACE2	angiotensin-converting enzyme 2
AD	Alzheimer's disease
AD	atopic dermatitis
ADR	adriamycin
AGT	angiotensinogen
AIDS	human immunodeficiency syndrome
ALA	aminolevulinic acid
AONS	antisense oligonucleotides
APC	antigen-presenting cell
ApoE	apolipoprotein-E
AR	allergic rhinitis
ARVs	antiretroviral drugs
ASO	antisense oligonucleotide
AT2R	angiotensin II receptor type 2
ATF	amino-terminal fragments
Au-NP	gold nanoparticles
AZTTP	azidothymidine 5′-triphosphate
Aβ	amyloid-β
BASCs	brown adipose-derived stem cells
BBB	blood-brain barrier
BCC	basal cell
Bcl-2	B-cell lymphoma 2
BCNT	boron capture neutron therapy
BCRP	breast cancer resistance protein
BM	basement membrane
BMHP1	bone marrow homing motif–1
BPDMA	benzoporphyrin derivative monoacid ring A
BSA	bovine serum albumin

CAR	carvedilol
CAR	chimeric antigen receptor
CAR-NLCs	carvedilol loaded nanostructured lipid carriers
CC	candesartan cilexetil
CC-SLNs	candesartan cilexetil loaded solid lipid nanoparticles
CD13	aminopeptidase-N
CD2	cluster differentiation 2
CD47	cluster of differentiation 47
CDX	candid candoxin
CED	convection-enhanced delivery
CED	convection enhanced diffusion
CEH	cholesteryl ester hydrolase
CEL	carboxyl ester lipase
CEM	cholecyst-derived extracellular matrix
CFTR	cystic fibrosis transmembrane regulator
CFTR	cystic fibrosis transport regulator
cGMP	cyclic guanosine monophosphate
CINOD	cyclooxygenase inhibiting nitric oxide donor
CINs	chitosan nanoparticles
CLIC	clathrin-independent carriers
CLSM	confocal laser scanning microscopy
CMTKP	carboxymethyl tamarind kernel polysaccharide
CNS	central nervous system
CNT	carbon nanotube
COL	chitosan oligosaccharide lactate
COPD	chronic obstructive pulmonary diseases
COX	cyclooxygenase
CPs	choroid plexuses
CQ	clioquinol
CRs	complement receptors
CS	celiac disease
CS	chitosan
CS	cromolyn sodium
CSF	cerebrospinal fluid
CSF–2	colony stimulating factor–2
CS-PLHNs	core-shell PLHNs
CT	computed tomography
CVD	cardiovascular disease
CVD	chemical vapor deposition
DBCO	dibenzocyclooctyl

DOTAP	dioleoyl–3-trimethylammonium-propane
DOX	doxorubicin
DOX-M	doxorubicin magnetic conjugate
DPI	dry powder inhalation
DVT	deep vein thrombosis
EAs	edge activators
ECM	extracellular matrix
EGCG	epigallocatechin–3-gallate
EGFR	epidermal growth factor receptor
ELP	elastin-like polypeptide
EMA	endomysium antibodies
ENT1	equilibrative nucleoside transporter–1
EPR	enhanced permeability and retention
ETA	endothelin A receptor
FCS	fetal bovine serum
FD	felodipine
FDA	Food and Drug Administration
FDG-PET	fluorodeoxyglucose-positron emission tomography
FGFR	fibroblast growth factor receptor
FITC	fluorescein–5-isothiocyanate
GBM	glioblastoma multiform
GCGR	glucagon receptor
GCK	glucokinase
GDNF	glial-derived neurotrophic factor
GEEC	GPI-anchored protein-enriched early endosomal compartment
GI	gastrointestinal
GIST	gastrointestinal stromal tumor
GIT	gastrointestinal tract
GLP-1	glucagon-like peptide-1
GLUT	glucose transporter
GNPs	gold nanoparticles
GPx	glutathione peroxidase
GWAS	genome-wide association study
GXV	gene expression vaccine
HA	hyaluronan
HA	hyaluronic acid
HAART	highly active antiretroviral therapy
HAD	HIV-associated dementia
HAND	HIV-associated neurocognitive disorders

HA-NPs	hyaluronic acid nanoparticles
HBP	high blood pressure
HD	Huntington disease
HDACI	histone deacetylase inhibitor
HDL	high-density lipoproteins
HETCAM	Hen's egg test chorioallantoic membrane
HGC-NPs	hydrophobically modified glycol chitosan nanoparticles
HIV	human immunodeficiency virus
HIVE	HIV encephalitis
HNC	head and neck cancer
HNF	hepatocyte nuclear factor
HNLC	high shear homogenization technique
HO	heme oxygenase
HOMO	highest occupied molecular orbital
HRE	hypoxia response element
HST	hirsutenone
HT	hypertension
IA	intra-arterial
ICAM1	intercellular adhesion molecule
ICV	intracerebroventricular
IFN	chitosan/interferon
IFNγ	interferon gamma
IgE	immunoglobulin E
IGF-1	insulin growth factor-1
IGF-II	insulin-like growth factor-II
IGRT	image-guided radiation therapy
IL	interleukin
Ins	insulin
IPAH	idiopathic pulmonary arterial hypertension
IPF1	insulin promoter factor-1
IPR	intellectual property rights
ISFIs	in-situ forming injectable implants
KLF11	kruppel-like factor 11
LAC	lacidipine
LAT1	large neutral amino acid carrier-1
LBL	layer-by-layer
LDL	low-density lipoproteins
LDLR	low-density lipoprotein receptor
LLC	Lewis lung carcinoma
LMWH	low molecular weight heparin

LPS	lipopolysaccharide
LRP2	lipoprotein-related protein-2
LUMO	lowest unoccupied molecular orbital
LUV	large unilamellar vesicles
MA	myristic acid
MAP	multiple antigen peptide
MARCKS	myristoylated, alanine-rich C-kinase substrate
MCMD	minor cognitive motor disorder
MCT1	monocarboxylate transporter
MDI	metered dose inhalators
MDM	monocyte-derived macrophages
MDR	multidrug resistance
MHC	major histocompatibility complex
MI	myocardial infarction
miRNA	microRNA
MLs	magnetoliposomes
MLV	multilamellar vesicles
mm	millimeter
MMASPM	methylmethacrylate-sulfopropyl methacrylate
MNLC	microemulsion technique
MP	methylprednisolone
MPLs	magneto-plasmonic liposomes
MPP	1-methyl–4-phenylpyridinium
MPS	mononuclear phagocyte system
MRI	magnetic resonance imaging
MRP	multi-resistance drug associated protein
MRs	mannose receptors
MS	multiple sclerosis
MSCs	mesenchymal stem cells
MSN	mesoporous silica nanoparticles
MTB	mycobacterium tuberculosis
MTC	modified trimethyl chitosan-cysteine
MTR-TB	multi-drug resistant tuberculosis
MTX	methotrexate
MWCNTs	multiwalled carbon nanotubes
MYH9	myosin heavy chain 9 gene
NALT	nasal-associated lymphoid tissues
NCs	nanocrystals
ND	nisoldipine
ND-SLNs	nisoldipine loaded solid lipid nanoparticles

NE	nanoemulsions
NEP	natriuretic peptides
NERD	erosive reflux disease
NEUROD1	neurogenic differentiation 1
NHE 3	sodium/hydrogen exchanger 3
NIH	National Institute of Health
NIR	near-infrared
NIRF	near-infrared fluorescence
NLC	nanostructured lipid carrier
NLS	nuclear localization signal
nm	nanometer
NNI	nanotechnology initiative
NP	nanoparticles
NPRA	natriuretic peptide receptor an
NPs	nanoparticles
NR1	N-methyl D-aspartate receptor 1
NS	nanosuspensions
NSAID	non-steroidal anti-inflammatory drugs
NSCLC	non-small-cell lung cancer
OAT	organic anion transporters
OATP	organic anion transporting polypeptide family
OATs	organic anion transporters
OCT	organic cation transporters
OM	olmesartan medoxomil
OM-NE	olmesartan medoxomil loaded nanoemulsion
OM-NLCs	olmesartan medoxomil loaded nanostructured lipid carriers
PAA	polyaspartic acid
PAH	pulmonary arterial hypertension
PAMAM	poly(amidoamine)
PAX4	paired box gene 4
PBA	phenylboronic acid
PBCA	poly(n-butyl cyanoacrylate)
PBMC	peripheral blood mononuclear cell
PBR	peripheral benzodiazepine receptor
PC	phosphatidylcholine
PCA	poly(cyanoacrylate)
PCDC	pressure-controlled colon delivery capsules
PCL	poly (ε-caprolactone)
PCP	pneumocystis pneumonia
PD	Parkinson's disease

PDE-5	phosphodiesterase-5
PDT	photodynamic therapy
PECA	poly (ethyl cyanoacrylate)
PEG	polyethylene glycol
PEI	polyethyleneimine
PEO	poly(ethylene oxide)
PEPE	polythene-co-polyester
PET	positron emission tomography
PFC	perfluorocarbon
PG	poly(DL-glycolide)
PGF	placental growth factor
P-gp	P-glycoprotein
PLA	poly(lactic acid)
PLG	poly(lactide-co-glycolide)
PLGA	polylactic-co-glycolic acid
PLHNs	polymeric-lipid hybrid nanoparticles
PLLA	poly(l-lactic acid)
PMMA	poly (methyl methacrylate)
PNIPAM	poly(N-isopropyl acrylamide)
PNP	polymeric nanoparticle
POH	perillyl alcohol
PPG	poly(propylene glycol)
PPI	polylysine
PPI-GATG	polypropylenimine-gallic acid-triethylene glycol
PPO	poly(propylene oxide)
pre-mRNA	precursor mRNA
PSGL–1	P-selectin glycoprotein ligand–1
PUD	peptic ulcer disease
PUFA	polyunsaturated fatty acids
PVP	polyvinylpyrrolidone
QD	quantum dots
RAAS	renin angiotensinogen aldosterone system
RES	reticuloendothelial cells
RGD	arginine-glycine-aspartic acid
RISC	RNA-induced silencing complex
RNAi	RNA interferences
ROS	reactive oxygen species
RPE	retinal pigment epithelium
RSV	respiratory syncytial virus
RVE	retinal vascular endothelium

RVLM	rostral ventrolateral medulla oblongata
SA	stearic acid
SAPNS	self-assembly peptide nanofiber scaffolds system
SC	stratum corneum
SCC	squamous cell carcinomas
SCF	supercritical fluid technology
SCI	spinal cord injury
SD	Sprague-Dawley
SEM	scanning electron microscopy
SGK1	serum- and glucocorticoid-inducible kinase 1
shRNA	short hairpin RNA
SiNP	silica nanoparticles
siRNA	small interfering RNA
SIRPα	signal regulatory protein-α
SLC	solute carrier
SLN	solid lipid nanoparticles
SNPs	single nucleotide polymorphisms
SOD	superoxide dismutase
SPECT	single photon emission computed tomography
SPIO	superparamagnetic iron oxide
SPIONs	superparamagnetic iron oxide nanoparticles
SRs	scavenger receptors
STAT4	signal transducer and activator of transcription 4
SUV	small unilamellar vesicles
SWCNTs	single-walled carbon nanotubes
SWNT	small walled carbon nanotube
TAT	trans-activating transcriptor
TATp	trans-activating transcriptional peptide
TB	tuberculosis
TCF2	transcription factor–2
TEM	transmission electron microscopy
TfRMAb	transferrin receptor monoclonal antibody
TGF	transforming growth factor
TH	tyrosine hydroxylase
TH2	T-helper cells
THF	tetrahydrofolate
TLRs	toll-like receptors
TM	trimyristin
TMC	trimethyl chitosan
TNF	tumor necrosis factor

TNF-α	tumor necrosis factor-α
TPGS	d-α-tocopheryl polyethylene glycol 1000 succinate monoester
TRAIL	tumor necrosis factor–related apoptosis-inducing ligand
TRE	tretinoin
UPA	urokinase plasminogen activator
VAL	valsartan
VAL-PL	valsartan loaded proliposomes
VCAM1	vascular cell adhesion molecule
VEGF	vascular endothelial growth factor
VIP	vasoactive intestinal peptide
WHO	World Health Organization
XDR-TB	extremely drug-resistant TB

Preface

"Nano" is not a buzzword today in the era of the twenty-first century. With the advent of nanotechnology, the world has witnessed a significant momentum in exploring the precepts and perspectives of this evolutionary technology for diverse applications in various fields. Importantly, nanotechnology applications in the healthcare sector have benefited many people by serving unmet medical needs. In this regard, nanotherapeutic medicines, also referred as nanomedicines, are the substances that have remarkable applications in disease treatment in the forms of nanodrug delivery devices, medical diagnostics in the form of nanoelectronic biosensors, and many more for the purpose.

The present book is a compilation of the application of nanomedicines with a particular focus on their use in the treatment of diseases. Notwithstanding the benefits of nanotherapeutic devices, the healthcare sector has tremendously benefited in terms of reducing the mortality rate beyond the expectations. A total of 17 chapters encompassed in this book are contributed by eminent scientists, researchers, and nanotechnologists across the globe from countries including the US, China, and India, with the primary goal of highlighting the key advancements, challenges, and opportunities in the area of application of nanomedicines for disease treatment. The book, therefore, carries a lot of potential as a repertoire of knowledge and package of information for pharmaceutical scientists, nanoscientists, and nanobiotechnologists to provide holistic information on the subject of interest.

A succinct account on key highlights of each of the chapters included in the book has been discussed in the below-mentioned text.

Chapter 1 on the topic of *Nanomedicine and Diseases: An Updated Overview* provides a general outline on the recent developments and huge opportunities in the field of nanomedicines for treatment of life-threatening diseases along with critical challenges associated with the translation of nanomedicines into the market for clinical benefits.

Chapter 2 on the topic *Multitasking Role of Dendrimers: Drug Delivery and Disease Targeting* discusses the application of dendritic architecture in the treatment of diseases with the help of their excellent drug loading, surface tunability, and functionalization ability.

Chapter 3 on the topic *Nanomedicines for the Treatment of Respiratory Diseases* has been compiled with the goal of highlighting emerging therapeutic applications of nanomedicine for the treatment of respiratory diseases. The chapter will also serve as a guide for clinicians and researchers in working on the development of novel nanotherapeutic strategies improving the health effects for successful utilization in the form of next-generation nanomedicines.

Chapter 4 on the topic *Nanomedicine for the Treatment of Neurological Disorders* discusses the precepts and perspectives on the application of nanotherapeutic carriers for targeting drugs to the brain for treatment and management of several CNS disorders.

Chapter 5 *Nanomedicinal Genetic Manipulation: Promising Strategy to Treat Some Genetic Diseases* discusses several strategies used for manipulation of the expression of the lethal gene(s) specifically responsible for the diseases. With the usage of nanomedicines, the gene defects are corrected by delivering genes from external sources.

Chapter 6 on *Nanomedicine: Could It Be a Boon for Pulmonary Fungal Infections?* has focused on the current treatment approaches with nanoformulations for the treatment of pulmonary fungal infections via different routes of administration, and various factors that need to be considered for designing of nanoformulations and their nanotoxicological aspects.

Chapter 7 on *Nanotechnological Approaches for Management of NeuroAIDS* has highlighted the barriers to the delivery and advancements in nanomedicine to design and develop specific size therapeutic cargo for targeting HIV brain reservoirs. Moreover, this chapter also provides insight about the nanoenabled multidisciplinary research to formulate efficient nanomedicine for the management of neuroAIDS.

Chapter 8 on *Nanomedicines for the Treatment of Cardiovascular Disorders* has discussed the application of nanopharmaceutical formulations for the management of several cardiovascular diseases. Since several literature reports have demonstrated the useful aspects of nanomedicines in cardiac problems, the present chapter provides an in-depth review of the literature.

Chapter 9 on *Nanomedicines for the Treatment of Gastric and Colonic Diseases* focuses on various gastric disorders and their pathophysiology and recent developments in nanomedicines for the detection and treatment of these ailments.

Chapter 10 on *Angiogenesis Treatment with CD13 Targeting Nanomedicines* covers the discussion on structure and function of CD13 protein in the development of malignant tumor cells, along with recent advancements on various CD13-assisted nanomedicines, such as liposomes, polymeric nanoparticles, quantum dots, drug conjugates, and many more for their utility in targeted drug delivery, gene therapy and imaging applications.

Chapter 11 on *Liposomal Nanomedicines as State-of-the-Art in the Treatment of Skin Disorders* discusses the emergence of liposomal systems for the treatment of various skin disorders. Several literature reports have been published in the past few decades on liposomal topical drug delivery for psoriasis, skin cancer, acne, atopic dermatitis, antifungal, and anti-inflammatory drugs. A number of liposomal formulations are on the market, and many are under clinical trial, and many more will receive approval in the near future.

Chapter 12 on *Conventional to Novel Targeted Approaches for Brain Tumor: The Role of Nanomedicines for Effective Treatment* has discussed the current treatment opportunities with the usage of nanomedicinal carriers for targeting drugs into the brain for treatment of cancer. Since several research reports have demonstrated the success in treating brain disorders with nanomedicines, the present chapter focuses on drug delivery applications of nanocarriers to the brain tumor.

Chapter 13 on *Macrophage Targeting: A Promising Strategy for Delivery of Chemotherapeutics in Leishmaniasis and Other Visceral Diseases* deals with macrophage targeting for treatment of leishmaniasis and other visceral diseases. The chapter also gives an overview of diseases, origin, and role of macrophages, phagocytosis, and receptor targeting as well as on macrophage targeting.

Chapter 14 on *Nanomedicine for the Diagnosis and Treatment of Cancer* has discussed current opportunities with nanomedicines for treatment of diverse verities of cancer. In this chapter, we describe different types of nanomedicines, such as quantum dots and different nanoparticulate systems, including silver nanoparticles, gold nanoparticles, etc. At present different types of cancer therapies are being used in the treatment of cancer, such as radiation therapy, chemotherapy, targeted therapy, hormone therapy, stem cell transplants, etc. It is further discussed how the field of nanomedicine can be applied to diagnose, monitor, and treat cancer cell annexation.

Chapter 15 on *Nanomedicine Therapeutic Approaches to Overcome Hypertension* focuses on various nanoformulations available for different administration routes for improving solubility, dissolution, and consequently systemic availability of antihypertensive drugs.

Chapter 16 on *Nanomedicine for the Treatment of Ocular Diseases* has discussed the application of nanocarriers for treatment of various ocular pathogenic conditions. The chapter highlights the incredible potential of nanotechnology for treatment of ocular diseases and focuses on utilization of myriad nanocarriers to deliver a variety of therapeutics to different regions within the eye at a desired rate of release. To conclude, nanotechnology will play a pivotal role in the treatment of ocular conditions by providing highly efficacious therapies with a concomitant decrease in dosing frequency and side effects.

Chapter 17 on the application of *Nanomedicine for Radiation Therapy* depicts the concept of nanomedicine for radiation therapy and summarizes its applications in cancer treatment and diagnosis for the purpose.

Foreword

The word 'nano' is no longer a buzzword but has actually received more respect and credibility over the last couple of decades. Application of this branch of science to medicine and health care is gaining momentum given the immense therapeutic benefits. The edited book, entitled *Nanomedicine for the Treatment of Disease: From Concept to Application,* is a very interesting compilation of the translational components of the science and technology of nanomedicine and its applications to human health. I would, therefore, like to start by complimenting the Editors for having taken the initiative to get the best names working in these areas to contribute their chapters for this book.

With a wonderful introduction, the very first chapter of this book looks at issues of drug delivery and drug targets as applied to respiratory diseases, neurological disorders, and gastric and intestinal disorders. Skin and treatment of cancer and ocular diseases, angiogenesis, cardiovascular disorder, and hypertension are also covered. Applications of nanomedicine to address issues of bacterial and parasitic disease will further make this book a valuable one. What makes this book very user-friendly is the language and the treatment of the technical content, while at the same time not compromising on the recent advancements, opportunities, and the challenges.

In my opinion, this book would be of immense use not only to the basic scientists but also physician-scientists and clinicians who may like to get a better understanding of nano-based therapies. Overall, this compilation represents a wonderful treatise based on expert experience with a clear bearing on applications in human health.

I would like to once again congratulate the Editors for coming out with this wonderful book.

—**Professor Dr. Seyed Ehtesham Hasnain,**
Vice Chancellor, Jamia Hamdard (Deemed to be University),
and Professor & Head, Jamia Hamdard-Institute of Molecular Medicine,
Hamdard Nagar, New Delhi, 110062, India
&
Invited Professor
Kusuma School of Biological Sciences,
Indian Institute of Technology, Delhi (IIT,D),
Hauz Khas, New Delhi, 110 016, India

CHAPTER 1

Nanomedicine and Diseases: An Updated Overview

SARWAR BEG[1*], MD. ABUL BARKAT[2], HARSHITA[2],
MAHFOOZUR RAHMAN[3], and FARHAN J. AHMAD[4]

[1]*Product Nanomedicine Research Lab, Department of Pharmaceutics,
School of Pharmaceutical Education and Research,
Jamia Hamdard (Hamdard University), New Delhi, India*

[2]*School of Medical & Allied Sciences, Department of Pharmaceutics,
K.R. Mangalam University, Gurgaon, India*

[3]*Department of Pharma Sciences, Shalom Institute of Health & Allied
Sciences, Sam Higginbottom University of Agriculture,
Technology & Sciences, Allahabad–211007, UP, India*

[4]*Nanomedicine Research Lab, Department of Pharmaceutics,
Faculty of Pharmacy, Jamia Hamdard, New Delhi–110062, India*

Corresponding author: Sarwar Beg

ABSTRACT

Nanomedicine is a branch of medicine that applies the knowledge and tools of nanotechnology to the prevention and treatment of disease. Nanomedicine involves the use of nanoscale materials, such as biocompatible nanoparticles (NPs) and nanorobots, for diagnosis, delivery, sensing, or actuation purposes in a living organism. In the recent years, the successful introduction of several novel nanomedicine products into clinical trials and even onto the commercial market has shown successful outcomes of fundamental research into clinics. This chapter is intended to examine several nanomedicines for disease treatments and/or diagnostics-related applications, to analyze the trend of nanomedicine development, future opportunities, and challenges of this fast-growing area.

1.1   INTRODUCTION TO NANOMEDICINES FOR THE TREATMENT OF DISEASES

The European Science Foundation's has defined *nanomedicine* as "the nano-sized tools used for diagnosis, prevention, and treatment of diseases, and to gain an increased understanding of the complex pathophysiology associated with them. The ultimate goal is to improve the quality of life" (Astruc, 2016). It involves three nanotechnology areas such as diagnosis, imaging agents and drug delivery with the help of nanoparticles (NPs) in 1–1000 nm range, bioships (from both "top-down" and "bottom-up" sources) and polymer therapeutics (Peer et al., 2007; Duncan, 2006). A relevant more recent terminology is that of "theranostics" involving both diagnostics and therapy with the same nanopharmaceutics (Picard and Bergeron, 2002; Bardhan et al., 2011). In fact, nanomedicine can be traced back to the use of colloidal gold in ancient times (Anconi, 1618; Macker, 1766), but Metchnikov and Ehrlich (Nobel Prize for Medicine in 1908) are the modern pioneers of nanomedicine for their works on phagocytosis (Cooper, 2008) and cell-specific diagnostic therapy (Ehrlich, 1913), respectively. The seminal works on NPs for nanomedicine were increasingly developed in the last 30 years of 20th century, which include liposomes (Gregoriadis et al., 1971; Bangham, 1972), DNA-drug complexes (Cornu et al., 1974), polymer-drug conjugates (Ringsdorf, 1975), antibody-drug conjugates (Hurwitz et al., 1975), polymeric nanocapsules (Kreuter and Speiser, 1976; Couvreur et al., 1977; Duncan and Kopecek, 1984), polymer-protein conjugates, albumin-drug conjugates (Davis, 2002; Trouet et al., 1982), block-copolymer micelles (Gros et al., 1981), anti-arthritis gold NPs (Dequeker et al., 1984), and anti-microbial silver NPs (Russell and Hugo, 1994). These nanomedicines have various size ranges that are often not strictly within the standard definition of the nanoworld, i.e., 1–100 nm (Duncan and Gaspar, 2011). Clinical toxicities including side effects have been broadly studied and sometimes point toward patient individualization.

1.2   NANOPOLY ARCHITECTURES IN DRUG DELIVERY AND TARGETING

Dendrimers are three-dimensional, immensely branched, well-organized nanoscopic macromolecules (typically 5000–500,000 g/mol), possess low polydispersity index and have displayed an essential role in the emerging field of nanomedicine (Buhleier et al., 1978). These are polymeric architectures that are known for their defined structures, versatility in drug delivery and high functionality whose properties resemble with biomolecules. These nanostructured macromolecules have shown their potential abilities in entrapping and/or conjugating the high molecular weight hydrophilic/hydrophobic entities by host-guest interactions and covalent bonding (prodrug approach) respectively. In recent years, improved pharmacokinetics, biodistribution, and controlled release of the drug to the specific targeted site have been achieved with polymer-based drug delivery (Allen and Cullis, 2004). Unlike traditional polymers, dendrimers have received considerable attention in biological applications due to their high water solubility (Soto-Castro et al., 2012), biocompatibility (Duncan and Izzo, 2005), polyvalency (Patton et al., 2006), and precise molecular weight (Patton et al., 2006). These features make them an ideal carrier for drug delivery and targeting applications (Kesharwani et al., 2013).

1.3   NANOMEDICINES FOR THE TREATMENT OF RESPIRATORY DISEASES

Treatment of respiratory diseases and infections has proved to be a challenging task, with the incidence of these ailments increasing worldwide (HuldaSwai et al., 2009). Respiratory diseases include a wide spectrum of illnesses affecting individuals in all age groups from the fetal period to the elderly. By increased life expectancy, hope for a comfortable life for the elderly becomes especially important, and nanomedicine may help. The lung is a very suitable target for drug delivery due to easy, non-invasive, and safe administration via inhalation aerosols. Direct delivery to the site of action for the treatment of lung disease and injuries, and because of availability of lavage surface areas for local drug action and systemic absorption of drugs (Moslem Bahadori and Forozan Mohammadi, 2012). In this regard, nanomedicine researchers by considering three basic principles in this subject, respiratory disease, namely: (a) diagnosis and imaging based on nanotechnology (b) targeted drug delivery (c) reconstructive surgery were able to benefit from nanomedicine technology in some chronic pulmonary diseases (Pison et al., 2006). Since nanocarrier systems can be easily transferred to the airways (Dames et al., 2007), many respiratory diseases including chronic obstructive pulmonary disease, cystic fibrosis and some other genetic disorders, tuberculosis, and infectious diseases, cancer, and pediatric diseases and we are reviewing some of them.

1.4 NANOMEDICINE FOR TREATMENT OF NEUROLOGICAL DISORDERS

The neurological disease mainly includes Alzheimer's disease (AD), Parkinson (PD), Huntington disease (HD) and stroke, epilepsy, etc. Of these, the neurodegenerative disease is a class of chronic progressive neurological diseases caused by the loss of neurons and/or their myelin (Chun-Xiao Wang and Xue Xue, 2017). Conventional drug delivery strategies are unable to restore cytoarchitecture and connection pattern in the central nervous system (CNS) disorders. Nanotechnologies overcome these problems due to its nanoscale quantum effect, small, and high surface area to volume ratio (Soni et al., 2016; Dinda and Pattnaik, 2013). Basically, nanotechnology is a convergence of science and engineering, which needs one-dimensional designing and characterization at the nanometric scale. NPs used in CNS drug delivery should have following promising features: (a) they should biodegradable, non-toxic, and biocompatible; (b) their physical properties should easily manipulate according to mode of delivery; (c) different NPs with modified chemical properties should achieve organ- or cell-specific drug delivery; and (d) the formulation should be cost-effective. Drugs need to chemically modify and transported to the brain via loading with different nanomaterial-based vehicles by breaching the blood-brain barrier (BBB). The different nanoformulation carrier has used for targeted drug delivery, some of them are NPs, lipid-based vehicle, carbon nanostructure-based vehicle and polymer-based vehicle (Soni et al., 2016).

1.5 NANOMEDICINES FOR GENETIC MANIPULATION

Effective diagnostic and therapeutic interventions of genetic diseases have been particularly pending on technological development. It was not until the last two decades of the past century that molecular biology and recombinant DNA have provided diagnostic and therapeutic tools for these diseases. The application of nanotechnology principles and tools to the field of medicine, known as nanomedicine, is also a relatively young field that has vastly expanded within the last couple of decades to render new diagnostic and treatment approaches. However, delivery of active agents assisted by these technologies is still a relatively unexplored strategy in the case of genetic diseases. Several properties of nanomedicine designs, mainly pertaining their biocompatible size and a high degree of manipulation that allow adaptation to different biomedical applications, have caused this field to be considered a new technological revolution. Nanotechnology has opened new possibilities for ex vivo detection methods (e.g., applicable to mutation screening) as well as biomarkers of disease, with several technologies being also applicable for in vivo imaging (Cai et al., 2003; Corstjens et al., 2005; Bailey et al., 2009; Cheung et al., 2010; Choi, and Frangioni, 2010; Hsu and Muro, 2011). In a broad definition, gene therapy encompasses the modulation of the expression of genes affected in genetic conditions. It can be achieved at the level of providing codifying gene sequences that can enable the transcription and translation of functional proteins otherwise affected by these defects, or other regulatory sequences that can up-regulate or down-regulate said expression at any stage during transcription or translation. Several nanoconstructs with varied chemical nature, architectural design, and functional properties have been successfully translated into the clinics, mainly for applications other than treatment of genetic conditions. Although, their application in this field has been only modestly explored, these systems rather represent general platforms, offering a unique opportunity to develop alternative and complementary therapeutic interventions applicable to the treatment of genetic diseases (Hsu and Muro, 2011).

1.6 NANOMEDICINES FOR PULMONARY FUNGAL INFECTIONS

Local administration of therapeutics by inhalation for the treatment of lung diseases has the ability to deliver drugs, nucleic acids, and peptides specifically to the site of their action. And therefore, enhance the efficacy of the treatment, limit the penetration of nebulized therapeutic agent(s) into the bloodstream and consequently decrease adverse systemic side effects of the treatment (Perlroth et al., 2007; Alangaden, 2011; Limper et al., 2010). Nanotechnology allows for a further enhancement of the treatment efficiency. The present review analyzes the modern therapeutic approaches of inhaled nanoscale-based pharmaceutics for the detection and treatment of various lung diseases. Analysis of modern achievements of various lung diseases by inhalation of nanomedicines clearly shows advantages of direct local delivery of nanopharmaceutics specifically to the diseased cells in the lungs. In addition, most probably such advanced multifunctional treatment will include the delivery of several drugs with different mechanisms of action, enhance the efficacy of treatment of lung diseases and limit adverse side effects on healthy tissues (Kuzmov and Minko, 2015).

1.7   NANOMEDICINES FOR MANAGEMENT OF NEURO-AIDS

Currently, there is no cure and no preventive vaccine for HIV/AIDS. Combinational antiretroviral therapy has dramatically improved the treatment, but it has to be taken for a lifetime. It has major side effects and is ineffective in patients, in whom the virus develops resistance. Nanotechnology is an emerging multidisciplinary field that is revolutionizing medicine in the 21st century. It has a vast potential to radically advance the treatment and prevention of HIV/AIDS (Gallo, 2002; Gallo and Montagnier, 2003; Montagnier, 2002). Nanotechnology can impact the treatment and prevention of HIV/AIDS with various innovative approaches. Treatment options may be improved using nanotechnology platforms for delivery of antiretroviral drugs. Controlled and sustained release of the drugs could improve patient adherence to drug regimens, increasing treatment effectiveness. Targeted NPs utilizing ligands such as mannose, galactose, tuftsin, and fMLF peptides have been used to target macrophages, major HIV viral reservoirs. In the future, targeted co-delivery of two or more antiviral drugs in a NP system could radically improve the treatment of viral reservoirs (Mamo et al., 2010).

1.8   NANOMEDICINES FOR THE TREATMENT OF CARDIOVASCULAR DISORDERS

Cardiovascular disease (CVD) encompasses all pathologies of the heart or circulatory system, including coronary heart disease, peripheral vascular disease, and stroke. CVDs are the primary morbidity and mortality cause in the world and seen among nearly 25% of the adult population who are over 20 years of age, although it differs in continents and regions. The primary conditions underlying the great majority of CVDs are dyslipidemia, atherosclerosis, and hypertension. Nanomedicine is the application of nanotechnology in monitoring, diagnosing, preventing, repairing or curing diseases and damaged tissues in biological systems and it is gaining importance for the treatment of CVD. In the near future, it is considered that nanomedicine approach will enable the establishment of patient-specific "personalized medicine." It is also considered that gene therapies for cardiovascular applications will have a potential use in the field of cardiovascular applications in upcoming years (Gündogdu et al., 2014; Arayne et al., 2007).

1.9 NANOMEDICINES FOR THE TREATMENT OF GASTRIC AND COLONIC DISEASES

Gastrointestinal (GI) diseases are the diseases affecting the GI tract, from the esophagus to the rectum, and the accessory digestive organs such as liver, gall bladder, and pancreas. GI diseases include acute, chronic, recurrent or functional disorders while covering a broad range of diseases, including the most common acute and chronic inflammatory bowel disease (Bellmann et al., 20115). Nanotechnology enables the scientists to figure out the barriers of conventional approaches, and now it is possible to deliver the hydrophobic drugs; specific targeting of drugs to particular regions of GI tract; transcytosis of drugs across the intestinal barriers and intracellular delivery of drugs (Ugazio et al., 2002). Nanomedicine is a promising therapeutic representative in the GI tract. Their physiochemical properties such as high surface to volume ratio, less cytotoxic to healthy tissues, ease in hydrophobic drug delivery, stability, and biodegradability are important to be considered. Different types of NPs have prospective uses in the field of gastroenterology as they overcome the conventional system of treatment in many GI diseases due to their less toxic, target specific, efficient, reliable, and practical approach and they can be used in the diagnosis and personalized therapies (Riasat et al., 2016).

1.10 ANGIOGENESIS TREATMENT WITH CD13 TARGETING NANOMEDICINES

Aminopeptidase-N (CD13), a membrane-bound enzyme, is associated with angiogenesis and has been used for tumor vascular targeting via the NGR/CD13 ligand-receptor system. One successful example was a liposomal formulation of doxorubicin bearing the NGR peptide at the outer surface, which led to dramatic vascular damage and neuroblastoma eradication. Dual-targeting liposomes modified with APRPG and GNGRG peptides were further developed by Murase et al. (XXXX) to enhance the potential of active endothelial targeting. These peptides were observed to cooperatively facilitate the association of liposomes to proliferating endothelial cells cooperatively. Moreover, doxorubicin encapsulated in the dual-targeting liposomes was demonstrated to significantly suppress angiogenesis and tumor growth (Yanping Ding et al., 2013).

1.11   LIPOSOMAL NANOMEDICINES FOR TREATMENT OF SKIN DISORDERS

The treatment of skin disease mainly implies the application of a drug to skin with an impaired epidermal barrier and linked with the penetration profile of the drug substance as well as the carrier into the skin. To elucidate this, the effect of skin barrier damage on the penetration profile of various bioactive applied as solid lipid nanoparticles (SLNs) or nanostructured lipid nanoparticles (NLCs) composed of different lipids, varying in polarity, was studied. It is appreciated that the topical delivery of drugs in skin diseases by lipidic NPs serves as an excellent tool in the dermatological field (Prow et al., 2011; Madhu et al., 2012; Bulbake et al., 2017).

1.12   NANOMEDICINE FOR EFFECTIVE TREATMENT OF BRAIN TUMORS

In the past, a wealth of studies has been performed to reveal the best way of delivering drugs to the CNS, mainly for the treatment of brain tumors. Targeted delivery by nanocarriers can increase the amount of drug delivered to the brain but, from the data available, the percentages of the injected drug dose found in the brain after targeting with nanocarriers is <1%, and the main proportion is localized in the liver (Ambruosi et al., 2006; Vergoni et al., 2009; Van Rooy et al., 2010). The CNS, one of the most delicate microenvironments of the body, is protected by the blood-brain barrier (BBB) regulating its homeostasis due to its highly complex structure, also protecting it from injuries and diseases. It is also important to point out that the tailoring of NPs to enhance drug delivery to the brain does not necessarily imply their ability to cross the BBB themselves (Haque et al., 2012; Martin-Banderas, 2011). Currently, several types of NPs are available for biomedical use with different features and applications facilitating the delivery of neuroactive molecules such as drugs, growth factors and genes, and cells to the brain. NPs offer clinical advantages for drug delivery such as decreased drug dose, reduced side effects, increased drug half-life, and the possibility to enhance drug crossing across the BBB (Massimo, 2013).

1.13 NANOMEDICINES FOR LEISHMANIASIS AND VISCERAL DISEASES

Leishmaniasis is a group of diseases caused by trypanosomatids from the genus *Leishmania*. Transmission occurs in 88 tropical and subtropical countries where the sandfly vector is present, meaning that approximately 350 million people are at risk of contracting the disease (Chowdhary et al., 2016; WHO, 1990; Piscopo et al., 2007). It is one of the most neglected tropical diseases, with a major impact among the poorest. The numbers of leishmaniasis cases are increasing worldwide. Some reasons are the lack of vaccines, difficulties in controlling vectors and the increasing number of parasites resistance to chemotherapy. The available therapeutic modalities for leishmaniasis are overwhelmed with resistance to leishmaniasis therapy. Mechanisms of classical drug resistance are often related with the lower drug uptake, increased efflux, the faster drug metabolism, drug target modifications and over-expression of drug transporters. The high prevalence of leishmaniasis and the appearance of resistance to classical drugs reveal the demand to develop and explore novel, less toxic, low cost and more promising therapeutic modalities. The review describes the mechanisms of classical drug resistance and potential drug targets in *leishmania* infection. Moreover, current drug-delivery systems and future perspectives towards leishmaniasis treatment are also covered. Nanotechnology presents an extraordinary opportunity in the rational delivery of drugs to the infection site following systemic administration. Examples of nanotechnology progress in the pharmaceutical product include liposomes, niosomes, nanodisks, emulsions, SLNs, polymeric NPs, and polymeric drug conjugates (Yasinzai, 2013; Juliana, 2015).

1.14 NANOMEDICINE FOR THE DIAGNOSIS AND TREATMENT OF CANCER

NPs have been proposed for improved systems in medical imaging for disease diagnosis. Much of the potential for nanomaterials as diagnostic agents comes from their ability to enhance contrast in spectroscopy (Sabzichi et al., 2016). In particular, superparamagnetic iron oxide has been shown to enhance magnetic resonance imaging and as a result, can aid in the detection of liver metastases (Rappeport et al., 2007). Angiogenesis is a key hallmark of cancer, and thus its detection would be of importance. NPs can be used for targeting sites of angiogenesis and enhancing diagnostic imaging. For example, cyanoacrylate microbubbles can be conjugated to ligands specific to biomarkers, such as vascular endothelial growth factor receptor and αvβ3integrin, which are more abundant with increased angiogenesis (Palmowski et al., 2008).

Nanotechnology is at the forefront of both targeted drug delivery and intrinsic therapies. For instance, NPs can already be injected into the tumor and then be activated to produce heat and destroy cancer cells locally either by magnetic fields, X-rays or light. Meanwhile, the encapsulation of existing chemotherapy drugs or genes allows much more localized delivery both reducing significantly the quantity of drugs absorbed by the patient for equal impact and the side effects on healthy tissues in the body. Coupling both modes of action has also been achieved with gold nanorods carrying chemotherapy drugs and locally excited in the tumor by infrared light. The induced heat both releases the encapsulated drug and helps destroying the cancer cells, resulting in a combined effect of enhanced delivery and intrinsic therapy (Sebastian, 2017).

1.15 NANOMEDICINE TO OVERCOME HYPERTENSION

Nanotechnology is a promising approach in resolving several constraints of antihypertensive. Targeted NP can effectively take antihypertensive to its site of action whether it is kidney, heart, or smooth muscle (Alam et al., 2017). Chronotherapeutics in conjunction with nanotechnology can effectively regulate the high blood pressure which can not only just modify the release pattern of the drug but can also increase the bioavailability of the drug. Gene silencing technology is an innovative therapeutic tool which could definitely play a major role in future to treat hypertension. The challenges in gene delivery like cellular uptake and pharmacokinetics could be overcome by the use of suitable nanocarriers. However, the oral drug delivery system of siRNA is still under its infancy for hypertension. But, several researches on different disease state using siRNA technology are developing very fast from preclinical to clinical trial level. Ultimately, the success of the treatment depends upon the versatility of the nanoparticulate system which can entrap a wide variety of molecule including peptides and proteins and its targeting potential apart from its stability in the external environment and in physiological condition (Alam et al., 2017).

1.16 NANOMEDICINE FOR THE TREATMENT OF OCULAR DISEASES

Nanotechnology has been proven to be a powerful and effective tool for treatment and detection of ocular diseases by fabricating nanosystems (Yang et al., 2002; Finger et al., 2005; Townsend et al., 2008; Weng et al., 2017). Several nanosystems with different payloads have shown great potential in ocular delivery either *in vitro* or *in vivo*. It is reported that NPs seem to grow in size and aggregate inside the tissues after intravitreous injection or other administration routes (Marmor and Negi, 1985; Amrite, 2005; Cheruvu et al., 2008). This phenomenon could decrease the delivery efficiency and affect drug distribution. Further studies need to improve our understanding of the fundamentals of NPs and facilitate the development of proper delivery routes for the application (Weng et al., 2017).

1.17 CONCLUSIONS

Despite the tremendous scope and applications of nanotechnological tools, such systems are still considered to be at the very nascent stage of growth due to multiple challenges associated with their manufacturing, scalability, stability, and toxicity. Hence, more in-depth research is still required for establishing proof-of-concept and exploring the real-time application of the nanomedicines for human health.

KEYWORDS

- **biocompatible nanoparticles**
- **fundamental research**
- **nanomedicine**
- **nanorobots**
- **nanotechnology**

REFERENCES

Alam, T., Khan, S., Gaba, B., Haider, M. F., Baboota, S., & Ali, J., (2017). Nanocarriers as treatment modalities for hypertension. *Drug Deliv., 24*(1), 358–369.

Alangaden, G. J., (2011). Nosocomial fungal infections: Epidemiology, infection control, and prevention. *Infect Dis. Clin. North Am., 25*(1), 201–225.

Allen, T. M., & Cullis, P. R., (2004). Drug delivery systems: Entering the mainstream. *Science, 303*, 1818–1822.

Ambruosi, A., et al., (2006). Biodistribution of polysorbate 80-coated doxorubicin loaded [14C]-poly(butyl cyanoacrylate) nanoparticles after intravenous administration to glioblastoma-bearing rats. *J. Drug Target, 14*, 97–105.

Amrite, A. C., & Kompella, U. B., (2005). Size-dependent disposition of nanoparticles and microparticles following subconjunctival administration. *J. Pharm. Pharmacol., 57*, 1555–1563.

Anconi, F. (1618). Panacae aurea-auro potabile, bibliopolio frobeniano. Hamburg, Germany.

Araújo, J., Gonzalez, E., Egea, M. A., Garcia, M. L., & Souto, E. B., (2009). Nanomedicines for ocular NSAIDs: Safety on drug delivery. *Nanomedicine, 5*(4), 394–401.

Arayne, M. S., Sultana, N., & Qureshi, F., (2007). Nanoparticles in delivery of cardiovascular drugs, *Pak. J. Pharm. Sci., 20*(4), 340–348.

Astruc, D., (2016). Introduction to nanomedicine. *Molecules, 21*(1), 4.

Bailey, V. J., Puleo, C. M., Ho, Y. P., Yeh, H. C., & Wang, T. H., (2009). Quantum dots in molecular detection of disease. *Conf. Proc. IEEE Eng. Med. Biol. Soc., 4089–4092*.

Bangham, A. D., (1972). Lipid bilayers and biomembranes. *Annu. Rev. Biochem., 41*, 753–776.

Bardhan, R., Lal, S., Joshi, A., & Halas, N. J., (2011). Theranosctic shells: From probe design to imaging and treatment of cancer. *Acc. Chem. Res., 44*, 936–946.

Bellmann, S., Carlander, D., Fasano, A., Momcilovic, D., Scimeca, J. A., et al., (2015). Mammalian gastrointestinal tract parameters are modulating the integrity, surface properties, and absorption of food-relevant nanomaterials. *Wiley Interdiscip Rev. Nanomed. Nanobiotechnol., 7*, 609–622.

Buhleier, E., Wehner, W., & VÖGtle, F., (1978). *Synthesis "Cascade"-and "Nonskid-Chain-Like" Syntheses of Molecular Cavity Topologies, 2*, 155–158.

Bulbake, U., Doppalapudi, S., Kommineni, N., & Khan, W., (2017). Liposomal formulations in clinical use: An updated review. In: Yallapu, M. M., (ed.), *Pharmaceutics, 9*(2), 12. doi: 10.3390/pharmaceutics9020012.

Cai, H., Zhu, N., Jiang, Y., He, P., & Fang, Y., (2003). Cu @ Au alloy nanoparticle as oligonucleotides labels for electrochemical stripping detection of DNA hybridization. *Biosens. Bioelectron., 18*(11), pp. 1311–1319, 0956–5663(Print), 0956–5663 (Linking).

Cheruvu, N., Amrite, A. C., & Kompella, U. B., (2008). Effect of eye pigmentation on transscleral drug delivery. *Invest Ophthalmol. Vis. Sci., 49*, 333–341.

Cheung, W., Pontoriero, F., Taratula, O., Chen, A. M., & He, H., (2010). DNA and carbon nanotubes as medicine. *Adv. Drug Deliv. Rev., 62*(6), 633–649.

Choi, H. S., & Frangioni, J. V., (2010). Nanoparticles for biomedical imaging: Fundamentals of clinical translation. *Mol. Imaging, 9*(6), 291–310.

Chowdhary, S. J., Chowdhary, A., & Kashaw, S., (2016). Macrophage targeting: A strategy for leishmaniasis specific delivery. *International Journal of Pharmacy and Pharmaceutical Sciences, 8*(2), 16–26.

Chun-Xiao, W., & Xue, X., (2017). Nanomedicine: A new approach for treatment neuropsychiatric diseases. *Advances in Materials, 6*(3), 24–30. doi: 10.11648/j.am.20170603.12.

Cooper, E. L., (2008). From Darwin and Metchnikoff to burnet and beyond. *Contrib. Microbiol., 15*, 1–11.

Cornu, G., Michaux, J. L., Sokal, G., & Trouet, A., (1974). Daunorubicin-DNA: Further clinical trials in acute non-lymphoblastic leukemia. *Eur. J. Cancer, 10*, 695–700.

Corstjens, P. L., Li, S., Zuiderwijk, M., Kardos, K., Abrams, W. R., Niedbala, R. S., & Tanke, H. J., (2005). Infrared up-converting phosphors for bioassays. *IEE Proc. Nanobiotechnol., 152*(2), pp. 64–72, 1478–1581 (Print), 1478–1581 (Linking).

Couvreur, P., Tulkens, P., Roland, M., Trouet, A., & Speiser, P., (1977). Nanocapsules: A new type of lysosomotropic carrier. *FEBS Lett., 84*, 323–326.

Dames, P., Gleich, B., Flemmer, A., Hajek, K., Seidl, N., Wiekhorst, F., et al., (2007). Targeted delivery of magnetic aerosol droplets to the lung. *Nat. Nanotechnol., 2*(8), 495–499.

Davis, F. F., (2002). The origin of penology. *Adv. Drug Deliv. Rev., 54*, 457–458.

Dequeker, J., Verdickt, W., Gevers, G., & Vanschoubroek, K., (1984). Long-term experience with oral gold in rheumatoid arthritis and psoriatic arthritis. *Clin. Rheumatol., 3 (Suppl. 1)*, 67–74.

Dinda, S. C., & Pattnaik, G., (2013). Nanobiotechnology-based drug delivery in brain targeting. *Curr. Pharm. Biotechnol., 14*(15), 1264–1274.

Duncan, R., & Gaspar, R., (2011). Nanomedicines under the microscope. *Mol. Pharm., 8*, 2101–2141.

Duncan, R., & Izzo, L., (2005). Dendrimer biocompatibility and toxicity. *Adv. Drug Deliv. Rev., 57*, 2215–2237.

Duncan, R., & Kopecek, J., (1984). Soluble synthetic polymers as potential drug carriers. *Adv. Polym. Sci., 57*, 51–101.

Duncan, R., (2006). Polymer conjugates as anticancer nanomedicines. *Nat. Rev. Cancer, 6*, 688–708.

Ehrlich, P., (1913). Address in pathology on chemotherapy. Delivered before the 17th International Congress of Medicine. *Br. Med. J., 16*, 353–359.

Finger, P. T., Kurli, M., Reddy, S., Tena, L. B., & Pavlick, A. C., (2005). Whole body PET/CT for initial staging of choroidal melanoma. *The British Journal of Ophthalmology, 89*(10), 1270–1274. doi: 10.1136/bjo.2005.069823.

Gallo, R. C., & Montagnier, L., (2003). The discovery of HIV as the cause of AIDS. *N. Engl. J. Med., 349*(24), 2283–2285.

Gallo, R. C., (2002). Historical essay. *The Early Years of HIV/AIDS Science, 298*(5599), 1728–1730.

Gregoriadis, G., Leathwood, P. D., & Ryman, B. E., (1971). Enzyme entrapment in liposomes. *FEBS Lett., 14*, 95–99.

Gros, L., Ringsdorf, H., & Schupp, H., (1981). Polymeric antitumor agents on a molecular and cellular level. *Angew. Chem. Int. Ed., 20*, 301–323.

Gündogdu, E., Senyigit, Z., & Ilem, O. D., (2014). Nanomedicine for diagnosis and treatment of cardiovascular disease: Current status and future perspective. *Cardiovascular Disease, 1*, 187–200.

Haque, S., Md, S., Alam, M. I., Sahni, J. K., Ali, J., & Baboota, S., (2012). "Nanostructure-based drug delivery systems for brain targeting." *Drug Development and Industrial Pharmacy, 38*(4), pp. 387–411.

Hsu, J., & Muro, S., (2011). Nanomedicine and drug delivery strategies for treatment of genetic diseases. In: Plaseska-Karanfilska, D., (ed.), *Human Genetic Diseases* (pp. 241–266). Rijeka, Croatia: InTech.

Hulda, S., Boitumelo, S., & Lonji, K., (2009). Paul Chelule, Kevin Kisich and bob sievers. Nanomedicine for respiratory diseases. *Nanomed. Nanobiotechnol., 1*, 255–263.

Hurwitz, E., Levy, R., Maron, R., Wilchek, M., Arnon, R., & Sela, M., (1975). The covalent binding of daunomycin and adriamycin to antibodies, with retention of both drug and antibody activities. *Cancer Res., 35*, 1175–1181.

Juliana, P., Bezerra, De M., Carlos, E., Sampaio, G., Antônio, L. De O., Almeida, P., et al., (2015). "Advances in development of new treatment for leishmaniasis." *BioMed. Research International*, p. 11. Article ID 815023. doi: 10.1155/2015/815023.

Kesharwani, P., Jain, K., & Jain, N. K., (2013). Dendrimer as nanocarrier for drug delivery. *Prog. Polym. Sci.*

Kreuter, J., & Speiser, P. P., (1976). In vitro studies of poly(methyl methacrylate) adjuvants. *J. Pharm. Sci., 65*, 1624–1627.

Kuzmov, A., & Minko, T., (2015). Nanotechnology approaches for inhalation treatment of lung diseases. *J. Control. Release, 219*, 500–518.

Limper, A. H., Knox, K. S., Sarosi, G. A., Ampel, N. M., Bennett, J. E., Catanzaro, A., et al., (2010). An official American thoracic society statement: Treatment of fungal infections in adult pulmonary and critical care patients. *American Journal of Respiratory and Critical Care Medicine, 183*(1), 96.

Macker, P. J. (1766). Dictionnaire de Chymie, Lacombe: Paris, France.

Madhu, G., Udita, A. S., & Vyas, P., (2012). Nanocarrier-based topical drug delivery for the treatment of skin diseases. *Expert Opin. Drug Deliv. Jul., 9*(7), 783–804.

Mamo, T., Moseman, E. A., Kolishetti, N., Salvador-Morales, C., Shi, J., Kuritzkes, D. R., et al., (2010). Emerging nanotechnology approaches for HIV/AIDS treatment and prevention. *Nanomedicine, 5*(2), 269–285.

Marmor, M. F., Negi, A., & Maurice, D. M., (1985). Kinetics of macromolecules injected into the subretinal space. *Exp. Eye Res., 40*, 687–696.

Martin-Banderas, L., Holgado, M. A., Venero, J. L., Alvarez-Fuentes, J., & Fernàdez-Aréalo, M., (2011). "Nanostructures for drug delivery to the brain." *Current Medicinal Chemistry, 148*(34), pp. 5303–5321.

Massimo, M., (2013). "Nanoparticles for Brain Drug Delivery." *ISRN Biochemistry*, p. 18. Article ID 238428. doi:10.1155/2013/238428.

Montagnier, L., (2002). Historical essay: A history of HIV discovery. *Science, 298*(5599), 1727–1728.

Moslem, B., & Forozan, M., (2012). Nanomedicine for respiratory diseases, *Tanaffos., 11*(4), 18–22.

Palmowski, M., Huppert, J., Ladewig, G., Hauff, P., Reinhardt, M., et al., (2008). Molecular profiling of angiogenesis with targeted ultrasound imaging: Early assessment of antiangiogenic therapy effects. *Mol. Cancer Ther., 7*, 101–109.

Patton, D. L., Cosgrove, S. Y. T., McCarthy, T. D., & Hillier, S. L., (2006). Preclinical safety and efficacy assessments of dendrimer-based (SPL7013) microbicide gel formulations in a nonhuman primate model. *Antimicrob. Agents Chemother., 50*, 1696–1700.

Peer, D., Karp, J. M., Hong, S., Farokhzad, O. C., Margalit, R., & Langer, R., (2007). Nanocarriers as an emerging platform for cancer therapy. *Nat. Nanotechnol., 2*, 751–760.

Perlroth, J., Choi, B., & Spellberg, B., (2007). Nosocomial fungal infections: Epidemiology, diagnosis, and treatment. *Med. Mycol., 45*(4), 321–346.

Picard, F. J., & Bergeron, M. J., (2002). Rapid theranostic in infectious diseases. *Drug Discov. Today, 7*, 1092–1101.

Piscopo, T. V., & Mallia, A. C., (2007). Leishmaniasis. *Postgrad. Med. J., 83*, 649–657.

Pison, U., Welte, T., Giersig, M., & Groneberg, D. A., (2006). Nanomedicine for respiratory diseases. *Eur. J. Pharmacol., 533*(1–3), 341–350.

Prow, T. W., Grice, J. E., Lin, L. L., et al., (2011). Nanoparticles and microparticles for skin drug delivery. *Adv. Drug Deliv. Rev., 63*, 470–491.

Rappeport, E. D., Loft, A., Berthelsen, A. K., Von der Recke, P., Larsen, P. N., et al., (2007). Contrast-enhanced FDG-PET/CT vs. SPIO-enhanced MRI vs. FDG-PET vs. CT in patients with liver metastases from colorectal cancer: A prospective study with intraoperative confirmation. *Acta Radiol., 48*, 369–378.

Riasat, R., Guangjun, N., Riasat, Z., & Aslam, I., (2016). Effects of nanoparticles on gastrointestinal disorders and therapy. *J. Clin. Toxicol., 6*, 313. doi: 10.4172/2161–0495.1000313.

Ringsdorf, H., (1975). Structure and properties of pharmacologically active polymers. *J. Polym. Sci. Polym. Symp., 51*, 135–153.

Russell, A. D., & Hugo, W. B., (1994). Antimicrobial activity and action of silver. *Prog. Med. Chem., 31*, 351–370.

Sabzichi, M., Samadi, N., Mohammadian, J., Hamishehkar, H., Akbarzadeh, M., et al., (2016). Sustained release of melatonin: A novel approach in elevating efficacy of tamoxifen in breast cancer treatment. *Colloids Surf B Biointerfaces, 145*, 64–71.

Sebastian, R., (2017). Nanomedicine - the future of cancer treatment: A review. *J. Cancer Prev. Curr. Res., 8*(1), 00265.

Soni, S., Ruhela, R. K., & Medhi, B., (2016). Nanomedicine in central nervous system (CNS) disorders: A present and future prospective. *Advanced Pharmaceutical Bulletin, 6*(3), 319–335. doi:10.15171/apb.2016.044.

Soto-Castro, D., Cruz-Morales, J. A., Ramírez, A. M. T., & Guadarrama, P., (2012). Solubilization and anticancer-activity enhancement of Methotrexate by novel dendrimeric nanodevices synthesized in one-step reaction. *Bioorg. Chem., 41*(2), 13–21.

Townsend, K. A., Wollstein, G., & Schuman, J. S., (2008). Clinical application of MRI in ophthalmology. *NMR in Biomedicine, 21*(9), 997–1002. doi: 10.1002/nbm.1247.

Trouet, A., Masquelier, M., Baurain, R., & Deprez-de, C. D., (1982). A covalent linkage between daunorubicin and proteins that is stable in serum and reversible by lysosomal hydrolases, as required for a lysosomotropic drug-carrier conjugate: *In vitro* and *in vivo* studies. *Proc. Natl. Acad. Sci. USA, 79*, 626–629.

Ugazio, E., Cavalli, R., & Gasco, M. R., (2002). Incorporation of cyclosporin A in solid lipid nanoparticles (SLN). *Int. J. Pharm., 241*, 341–344.

Van Rooy, I., et al., (2010). Comparison of five different targeting ligands to enhance accumulation of liposomes into the brain. *J. Control. Rel., 150*, 30–36.

Vergoni, A. V., et al., (2009). Nanoparticles as drug delivery agents specific for CNS: *In vivo* biodistribution. *Nanomedicine NBM, 5*, 369–377.

Weng, Y., Liu, J., Jin, S., Guo, W., Liang, X., & Hu, Z., (2017). Nanotechnology-based strategies for treatment of ocular disease. *Acta Pharmaceutica Sinica B., 7*(3), 281–291. doi: 10.1016/j.apsb.2016.09.001.

WHO Control of the Leishmaniases, (1990). Report of a WHO expert committee, World Health Organ. *Tech. Rep. Ser., 793*, 1–158.

Yang, M. S., Hu, Y. J., Lin, K. C., & Lin, C. C., (2002). Segmentation techniques for tissue differentiation in MRI of ophthalmology using fuzzy clustering algorithms. *Magn. Reson. Imaging, 20*(2), 173–179.

Yanping, D., Suping, L., & Guangjun, N., (2013). Nanotechnological strategies for therapeutic targeting of tumor vasculature. *Nanomedicine, 8*(7), 1209–1222.

Yasinzai, M., Khan, M., Nadhman, A., & Shahnaz, G., (2013). Drug resistance in leishmaniasis: Current drug-delivery systems and future perspectives. *Future Med. Chem. Oct., 5*(15), 1877–1888.

CHAPTER 2

Multitasking Role of Dendrimers: Drug Delivery and Disease Targeting

SAMARESH SAU[1,#], PRASHANT SAHU[2,#], SUSHIL K. KASHAW[2], and ARUN K. IYER[1,3,*]

[1]*Use-Inspired Biomaterials and Integrated Nanodelivery Systems Laboratory, Department of Pharmaceutical Sciences, Eugene Applebaum College of Pharmacy and Health Sciences, Wayne State University, Detroit, MI, United States*

[2]*Department of Pharmaceutical Sciences, Dr. HarisinghGour University, Sagar, India*

[3]*Molecular Imaging Program, Barbara Ann Karmanos Cancer Institute, Wayne State University, School of Medicine, Detroit, MI, United States*

*Corresponding author. E-mail: arun.iyer@wayne.edu

#Equal contribution: Disclosures: There is no conflict of interest and disclosures associated with the manuscript.

ABSTRACT

Dendrimers are nanometer-scaled materials contained with three major components: a central core, an interior dendritic structure (the branches), and an exterior surface with variety of functional surface groups. Presence of diverse components in a dendrimer molecule provide a unique feature of shapes, sizes and they are potent candidate for applications in both biological and materials sciences. The variety of attached surface groups improved the solubility and bioavailability, and sustained drug releasing characteristics. Dendrimers have wide variety of applications as highlighted in this chapter including drug delivery, energy harvesting, gene delivery, catalysis, photo thermal activity, rheology modification. The three-dimensional macromolecular structure of dendrimer is broadly recognized to generate polydisperse products with different molecular weights that are helpful in manipulating biological properties, self-assembling, electrostatic interactions, chemical stability, low cytotoxicity, disease targetability and solubility. These diverse features make dendrimers a unique material for the medical field, and this chapter discusses their applications.

2.1 INTRODUCTION

Dendrimers are the polymeric macromolecular with three-dimensional geometry arising from a central core and surrounded by successive addition of branching layers (Wu et al., 2015). Dendrimers exhibit a high degree of molecular symmetry, narrow particle size, ease of chemical functionalization, high drug loading efficiency. Along with that, the physicochemical characteristics of dendrimer can be manipulated for biodegradable backbone designing that will be helpful for several applications including drug delivery, targeted therapy, imaging, and disease diagnosis (Basij et al., 2018; Sau et al., 2018). Dendrimers can also serve as a pro-drug delivery vehicle for tumor site-specific receptor-mediated drug and imaging agent delivery (Luong et al., 2016). In this chapter, we briefly discuss about the synthesis, properties, and application of dendrimer for cancer treatment paying attention to the role of dendrimer as a versatile carrier for drug formulations. Dendrimers are large, circular polymeric complex that has very well-defined chemical structures and they possess three unique structural components, including (a) exterior part (terminal functionality useful for chemical modification) connected with interior generation; (b) interior layer (varies with different generations), composed of repeating units, connected with the initiator core; and (c) initiator core (Figure 2.1). There are two major dendrimer synthesis strategy has been utilized, such as divergent method and convergent growth process. In a divergent approach, monomeric modules are coupled in a branch-upon-branch on certain dendritic rules (Abbasi et al., 2014). The advantages of divergent dendrimer synthesis approach are included fast synthesis, cheap reagents, exponential growth, and the possibility to prepare large dendrimers (Nanjwade et al., 2009; Bae et al., 2011). But the major disadvantages of this strategy are difficulties in purification thousands of by-products with variable molecular weight, polarity, charge, hydrophilicity, and chirality and major limitation is low yield due to the steric hindrance of higher generation dendrimers, which prevent couplings of desired building blocks and cause major defects the surface of dendrimer (Iyer et al., 2006). In convergent growth process, several dendrons are reacted with a multifunctional core to obtain a final product. Clearly, the reaction must be very 'clean' and high yielding for the construction of large targets to be feasible. Dendrimers are commonly synthesized with an interactive sequence of reactions, and each additional interaction leads to higher generation dendrimer. Each of the new layers creates a new 'generation.' Due to the presence of hydrophobic interior, dendrimers are suitable for accommodating many clinically active hydrophobic drugs, through hydrophobic interactive forces (Kesharwani et al., 2015; Hu, Hu, and Cheng, 2016).

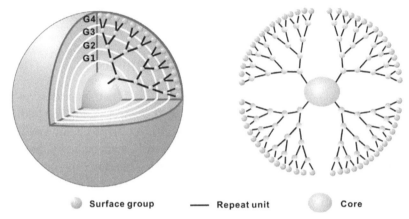

FIGURE 2.1   Structure outline of dendrimer and dendrimer is consisting of the exterior, interior layer, and interior parts. The branching of dendrimer represents the development of different generation (G1–G4). Source: Used with permission from Hu, Hu, and Cheng (2016). © 2016 Elsevier.

2.2   *'CLICK'* CHEMISTRY-BASED DENDRIMER SYNTHESIS

In the field of drug delivery, use of ease chemical synthesis is a very important task to avoid toxic side effect in human and preclinical use (Tatiparti et al., 2018). Among the many types of chemical conjugation reactions between two chemical functional groups, such as (i) DCC-DMAP for coupling of carboxylic acid and alcohol; (ii) EDC-NHS for coupling between carboxylic and amine functional group; (iii) copper catalyzed and copper free strain promoted *'click'* reaction between alkyne and azide are widely used for fabrication of nanoparticles with targeting ligands and drugs and imaging agents (Almansour et al., 2017; Alsaab et al., 2017; Sahu et al., 2017; Sau et al., 2017a; Therapy Targeted Cancer, 2017). Among all these coupling reactions, *'click'* reaction become very popular for high yielding, more stereo-specificity, faster reaction rate (Hein, Liu, and Wang, 2008; Kolb, Finn, and Sharpless, 2001). As the copper-catalyzed *'click'* reaction produces toxic byproducts, thus clinically it becomes a non-viable method of synthesis. To overcome the challenges, currently copper-free, strain promoted click reactions become a very popular option. For this click reaction, dibenzo-cyclooctyl (DBCO), a cyclic strained alkyne and Azide (-N$_3$) are used, and coupling can be occurred in aqueous condition (Hein, Liu, and Wang, 2008). Currently, it has been using extensively for various types of nanoparticle surface functionalization. This is also applied to PAMAM dendrimer, where amine groups are because it has several amine groups to couple with DDCO-COOH (Figure 2.2).

2.3   DENDRIMERS AS DRUG DELIVERY CARRIERS

Oral drug delivery is the most anticipated route for drug transporters and has been employed for the delivery of many drugs (Luong et al., 2017; Mukherjee et al., 2016; Sahu et al., 2017; Sau et al., 2017a, 2017b). Conversely, the oral bioavailability of drug transporter is frequently inadequate by their huge mass and molecular weight (Braun et al., 2005). Therefore, the deviation in the contrivance and net perviousness of drug transporters have been accredited to the alteration in their structural structures such as volume, molecular weight, molecular structure, and surface charge (Kim, Oh, and Crooks, 2004). Dendrimers as drug delivery agents are of great interest due to their extremely manageable structure and size, and the incurable functional collections of dendrimers display advanced chemical reactivity associated with other polymers. The functional groups of dendrimers have been associated with various biologically vigorous particles (Duan, Ji, and Nie, 2015). In this repute, PAMAM dendrimers have been widely examined for oral drug delivery, because they are a personal of water-soluble polymers considered by unique tree-like bifurcating style and a compressed spherical shape in solution (Bhise et al., 2017; Gawde et al., 2017; Mukherjee et al., 2016; Sau et al., 2014, 2017; Tatiparti et al., 2017). Their prospective as a drug carrier ascends from a large number of supports and surface amine groups that can be employed to immobilize drugs, enzymes, antibodies, or other bioactive representatives. Original studies have displayed that PAMAM dendrimers can pervade across epithelial barriers by a grouping of transcellular and paracellular routes and can briefly open tight junctions. This improves their conveyance via the paracellular pathway suggesting their possible as oral drug carriers (Twibanire and Grindley, 2012). Nevertheless, incomplete data are accessible for a regular association of the assembly of these polymers with their conveyance across biological barricades, precisely the intestinal epithelia. In this sense, a universal trend practical is that cationic PAMAM dendrimers are

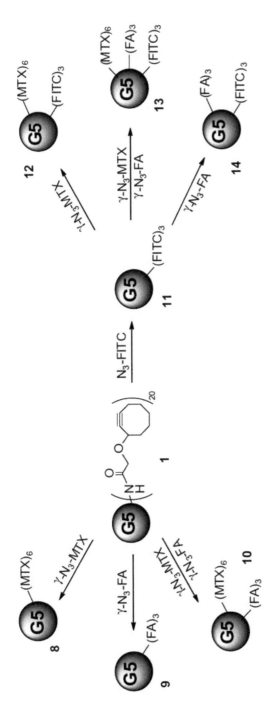

FIGURE 2.2 Schematic presentation of various folate and methotrexate conjugated G5-PAMAM dendrimer synthesis using copper free *'click'* chemistry. Source: Used with permission from Huang et al. (2012). © 2012 Elsevier.

more poisonous than their an ionic pledge parts, larger dendrimers are more toxic associated with smaller dendrimers of comparable surface functionality, and covering cationic remains with non-charged clusters recovers tolerability of PAMAM dendrimers and their approval by the epithelial cells (Sharma and Kakkar, 2015). The consequence of PAMAM dendrimers size, charge, and absorption on uptake and conveyance across the adult rat intestine in vitro use the rat intestinal sac organization. They studied cationic PAMAM dendrimers (generations 3 and 4) and anionic PAMAM dendrimers (generations 2.5, 3.5, and 5.5) (Figure 2.3). They displayed that I-labeled dendrimers optional that conveyance across the intestinal membrane was charge reliant, and they established that cationic dendrimers tissue acceptance was advanced than serosal conveyance (Maiti and Bagchi, 2006). Finally, 2.5G and 3.5G dendrimers were expected to be elated through third-phase endocytosis. In contrast, 5.5G and cationic dendrimers were experiential to be taken up by precise or generic adsorptive endocytosis.

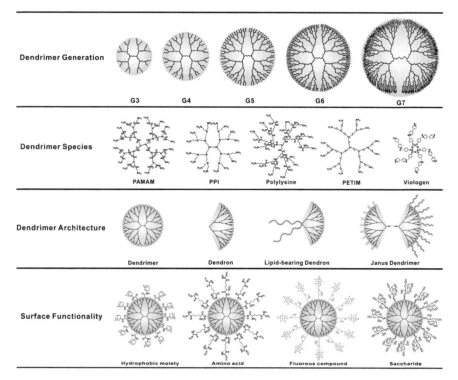

FIGURE 2.3 Summary of dendrimer with a different generation, composition, and surface functionality. They can be tailored for cancer drug delivery and targeting purpose. Source: Used with permission from Hu, Hu, and Cheng (2016). © 2016 Elsevier.

2.4   DIAGNOSIS AND THERAPY USING DENDRIMERS

Dendrimers have been castoff to transmit an assortment of tiny particle medications. These particles have an amount of compensations for usage as transporters or supports for judgment and treatment. For example, the amphiphilic stuff and the inner hollows of dendrimers can be castoff to condense hydrophobic or hydrophilic remedies contingent on dendrimers mechanisms (Peterson et al., 2001). In accumulation, dendrimers are mono-molecular polymer micelles, and this stuff evades the unpredictability of drug preparations by means of outdated amphiphilic polymers. On the other hand, the cohort size rectilinear association of dendrimers incomes that they can be designated for an appropriate scope for precise biomedical requests (Figure 2.4). For example, medium-sized dendrimers of about 5nm are castoff for MRI disparity mediators in the analysis of lymphatic organizations. The extremely divided, multivalent landscape of dendrimers makes the ideal contenders for tissue engineering submissions, and they are used as cross-linking managers, modulators of surface charge and surface chemistry, and prime mechanisms in skeletons that impressionist natural extracellular condition (Pan et al., 2006).

2.5   DENDRIMERS IN ONCOLOGY

Early readings of dendrimers as latent delivery schemes absorbed on their use as unimolecular micelles and "dendritic boxes" for the non-covalent encapsulation of drug moieties. For example, in primary training, DNA was conjugated with PAMAM dendrimers for gene delivery presentations, and hydrophobic medications and dye particles were combined into several dendrimer nucleuses (Jung et al., 2007). An improvement of consuming dendritic unimolecular micelles rather than orthodox polymeric micelles is that the micellar edifice is continued at totally deliberations because the hydrophobic sections are covalently associated. However, this method agonizes from a general disadvantage; in that, it is problematic to regulator the announcement of particles from the dendrimer staple. In some cases, harsh circumstances are compulsory, while in others the summarized drug is not well booked and the particles are unconfined moderately fast (Farokhzad and Langer, 2009).

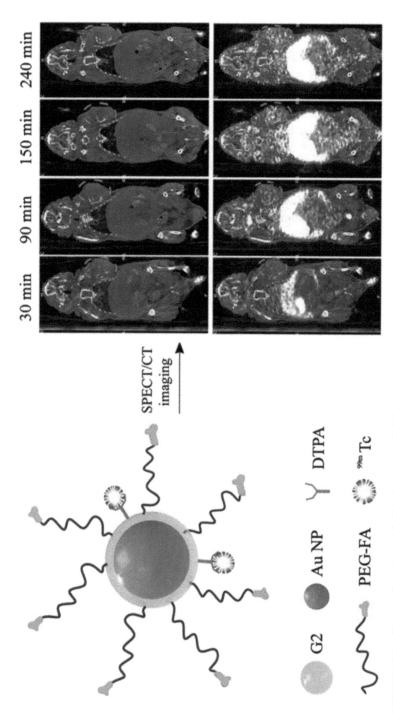

FIGURE 2.4 (See color insert.) Development of folate receptor targeting dendrimer-entrapped gold nanoparticles [{(Au0)6-G2-DTPA(99mTc)–PEG-FA}] for targeted SPECT/CT dual-mode imaging of tumors. This delivery system showed promise for efficient, low-cost tumor diagnosis. Source: Used with permission from Li et al. (2016). © 2016 American Chemical Society.

2.6   DENDRIMERS IN MRI IMAGING

Magnetic resonance imaging (MRI) is a central tool in contemporary treatment, providing high-quality three-dimensional pictures without the use of destructive ionizing emission. The signal strength in MRI stems mostly from the reduction rate of *in vivo* water protons and is improved by the supervision of a difference mediator previous to the image. Such managers comprise a paramagnetic metal particle that diminutions the lessening times of nearby water protons (Liu et al., 2013). Dissimilar clusters of difference managers are recognized for clinical submission: gadolinium chelates, superparamagnetic iron oxide atoms, and hepatobiliary difference managers. However, the gadolinium chelates establish the major collection of MRI contrast managers and are measured to be safe (Nikam, Ratnaparkhiand, and Chaudhari, 2014). The vital possessions of MRI distinction mediators comprise moral biocompatibility, low harmfulness, and high lessening. Low molecular weight MRI contrast negotiates or sprolix speedily from blood vessels into the interstitial cosmos and is expelled from the body very speedily (Noriega-Luna et al., 2014).

2.7   DENDRIMERS IN TISSUE REGENERATION

The arena of tissue engineering has developed importantly over the ancient few periods. The area of this ground is the renewal of innate tissue by complementing the natural curative procedure of the body or the formation of complete organs for movement (Figure 2.5). The first step building material theory is the range of a suitable skeleton sensible. Platforms may be as humble as a two-dimensional exterior upon which cells can grow, or erudite frameworks may capture one or more cell natures in three magnitudes (Criscione et al., 2009). Most platforms are arranged as porous edifices that consent for the transmission of nutrients to and unused merchandise left from the cells. The objective of tissue engineering claims is for the condensed cells to progressed extracellular matrix (ECM) and finally substitute the shell overall. Consequently, the framework must biodegrade at those accompaniments the biosynthesis of novel ECM (Gao, Chan, and Farokhzad, 2010).

2.8   DENDRIMERS IN CELL REPAIR

Integral extracellular backgrounds (ECMs) have established possible as biomaterials in numerous tissue engineering and clinical solicitations

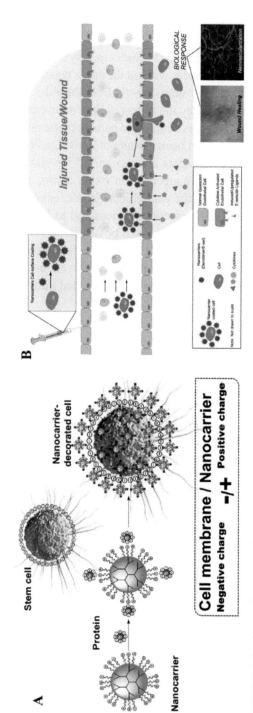

FIGURE 2.5 (See color insert.) (A) Schematic design of dendrimer coated stem cells through electrostatic interaction. (B) The data indicates a wound healing response in the mice model, and their biological response study indicates the nanoparticle can generate neo-vascularization Source: Used with permission from Li et al. (2016). © 2016 American Chemical Society.

(Daniell, 2012). These ECM frameworks deliver a natural three-dimensional provision to assistance the original mechanical supplies essential to sustenance injured or expunged tissue. In totaling, ECM delivers vivacious biological prompts for cellular appreciation which is important for an original cellular supplement, succeeding cellular differentiation, in-growth of vascular networks, and secretion of new ECM requisite for eventual scaffold remodeling and tissue regeneration. Dendrimer works as a linker to the scaffold and as a carrier of bioactive molecules (Garg et al., 2011). The dense functional groups terminated on the surface of dendrimers can react with intrinsic functional groups in protein-based scaffolds to form covalent binding. Conveniently in this way, support stability can also be personalized by regulating the degree of cross-linking, which has the advantage of encompassing there in vivo existence. Obtainability of numerous functional groups successfully increases the number of locations accessible for conjugation with exogenous bioactive moieties (Online n.d.). For example, the group 1 polyamidoamine (G1 PAMAM) dendrimer recycled as a typical dendrimer and combined into the cholecyst-derived extracellular matrix (CEM), a unique intact ECM derived from the perimuscular subserosal connective tissue of porcine cholecysts established in the lab. They pragmatic the combination of diverse feed absorptions of PAMAM dendrimer in CEM by the EDC/NHS cross-linking scheme causing covalent binding of PAMAM on CEM. Diverse grades of cross-linking, enhanced constancy of CEM to enzymatic deprivation, augmented amine functional assemblies beneficial in securing bioactive proxies, conservation of flexible asset but augmented elasticity of skeleton, and protection of the aptitude of DENCEM to provision cells in-vitro was pragmatic (Rai et al., 2016).

2.9 DENDRIMERS IN VACCINE DELIVERY

Vaccination has established to be a very economical method of monitoring transferable ailments produced by microbial pathogens and has been recognized in its contemporary form since the original effort in the late 18th century when Jenner announced vaccinia (cowpox virus) as the first dependable vaccine (Emole, 2012). Consequently, many other positive inoculations have been established, empirically, on the basis of lessened or killed microbes or their contaminants. However, efficient vaccines are still required for HIV1, tuberculosis, malaria, and assembly of the respirator and contagious intestinal ailments (Jassal, Nautiyal, and Kundlas, 2015).

Vaccination has been a controlling device for eliminating smallpox and declining previously extensive ailments like polio, measles, and rinderpest. Uniform so, a wide variety of contagious ailments still plentifully exist around the globe (malaria, tuberculosis, and bacterial and viral diarrhea being the most widespread). Such ailments are usually produced by composite pathogens where a more balanced method in vaccine project is desired (Majoros et al., 2009).

2.10   DENDRIMERS IN ANTIGENS DELIVERY

The ultimate objective of human immunodeficiency syndrome (AIDS) exploration is the development of a vaccine, since any permutation of medications has not instigated a comprehensive remedial of AIDS. In this logic, vaccine purpose of the multiple antigen peptides (MAP) which was produced by combining the V3loop (third variable region of envelope glycoprotein gp120 of HIV) structure to lysine dendrimers (Kojima et al., 2011). They perceived that the MAP produced relatively high and sustained levels of HIV–1-specific counteracting antibodies in animals. In the same situation, the amalgamation of the first glycopeptide dendrimer-type AIDS vaccine prototypical entailing of a V3 loop peptide-succinyl-maltose-prolinepoly (lysine) dendrimer platform (Abbasi et al., 2014). To upsurge the antigenic effectiveness of V3 loop or associated peptides, they considered glycopeptide dendrimers type AIDS vaccine with improved V3 peptide structures. This vaccine prototypical is altered from the formerly conveyed dendrimer vaccine, because it comprises a sugar molecule. By presenting sugar into the dendrimer, improvement in water solubility and reductions in the cell noxiousness and the antigenicity of dendrimer itself can be completed (Tripathy and Das, 2013).

2.11   DENDRIMERS IN ELECTRODE GENERATION

The expanse of electrochemical biosensors has momentous expansion in the previous few years, with the expansion of enzyme biosensors, immune beams, and DNA beams. Strategies are being effectively used in medical chemistry, food industry, and environmental grounds. To accomplish high specificity, high compassion, rapid reply, and suppleness of use, it is clear that the research endures to an emphasis on new meeting plans (Li et al., 2015). The use of layer-by-layer (LBL) meeting technique for the enterprise of electrochemical biosensors has involved extensive consideration. Freshly, other approaches have been useful in the edifice of LBL squeaky films over hydrogen-bonding influences, charge-transfer connections, biological connections, and so onward. It would be conceivable to the concept many dissimilar kinds of multilayer thin films over different LBL assemblage plans (Abbasi et al., 2014). The LBL method has been originating as an extremely valuable way for accumulating a number of organic and inorganic materials with varied nature, counting proteins, DNA, viruses, dendrimers, nanogels, and nanoparticles (Baig et al., 2015). The prepared multilayer constructions have fascinated great concentration because of their extensive assortment of submissions in the grounds of biosensors, nonlinear optical strategies, surface modification, etc. (Nigam, Chandra, and Bahadur, 2015).

2.12 CONCLUSION

There are limited studies addressing dendrimer toxicity and bio-distribution, particularly on parenteral dosing. But some literature has reported that due to high cationic nature PAMAM dendrimers are non-specifically up-taken by liver and spleen. Thus, the use of biodegradable natural amino acid based dendrimers such as poly-lysine showed reduces toxicity in animal model studies. There are several potential applications of a dendrimer that are currently being investigated with different surface functionality, targeting ligand that will help cross the plasma membrane, immune barrier, and other biological barriers. However, more research is required to unravel the molecular mechanisms in modulating dendrimer uptake involved through endocytosis or binding with scavenger receptors in cancer cells.

KEYWORDS

- **dendrimer as cancer therapy**
- **dendrimer**
- **regenerative medicine**
- **tumor imaging**
- **types of synthesis**

REFERENCES

Abbasi, E., et al., (2014). "Dendrimers: Synthesis, applications, and properties." *Nanoscale Research Letters, 9*(1), 247.

Almansour, A. I., et al., (2017). "Design, synthesis and antiproliferative activity of decarbonyl luotonin analogs." *European Journal of Medicinal Chemistry, 138*, 932–941.

Alsaab, H. O., et al., (2017). "PD-1 and PD-L1 checkpoint signaling inhibition for cancer immunotherapy: Mechanism, combinations, and clinical outcome." *Frontiers in Pharmacology, 8*, 1–15. http://journal.frontiersin.org/article/10.3389/fphar.2017.00561/full.

Bae, J. W., et al., (2011). "Dendron-mediated self-assembly of highly PEGylated block copolymers: A modular nanocarrier platform." *Chemical Communications, 47*(37), 10302–10304.

Baig, T., et al., (2015). "A review about dendrimers : Synthesis, types, characterization, and applications." *International Journal of Advances in Pharmacy, Biology and Chemistry, 4*(1), 44–59.

Basij, M., et al., (2018). *"Combined Phased-Array Ultrasound and Photoacoustic Endoscope for Gynecologic Cancer Imaging Applications."* In, 105800V–10580–7. https://doi.org/10.1117/12.2296563.

Bhise, K., Sushil, K. K., Samaresh, S., & Arun, K I., (2017). "Nanostructured lipid carriers employing polyphenols as promising anticancer agents: Quality by design (QbD) approach." *International Journal of Pharmaceutics Volume 526,* Issues 1–2, 30 June 2017, Pages 506–515

Braun, C. S., et al., (2005). "Structure/function relationships of polyamidoamine/DNA dendrimers as gene delivery vehicles." *Journal of Pharmaceutical Sciences, 94*(2), 423–436.

Criscione, J. M., et al., (2009). "Self-assembly of pH-responsive fluorinated dendrimer-based particulates for drug delivery and noninvasive imaging." *Biomaterials, 30*(23–24), 3946–3955.

Daniell, H., (2012). *"NIH Public Access,"* 76(October 2009), 211–220.

Duan, X. D., Chang, J. J., & Lin, N., (2015). "Formulation and development of dendrimer-based transdermal patches of meloxicam for the management of arthritis." *Tropical Journal of Pharmaceutical Research, 14*(4), 583–590.

Emole, J., (2012). "Cancer diagnosis and treatment : An overview for the general practitioner." *Intechopen (ISBN: 978–953–51–0539–8, InTech),* 175–186.

Farokhzad, O. C., & Robert, L., (2009). "Impact of nanotechnology on drug delivery." *ACS Nano, 3*(1), 16–20.

Gao, W., Chan, J. M., & Farokhzad, O. C., (2010). "pH-responsive nanoparticles for drug delivery." *Molecular Pharmaceutics, 7*(6), 1913–1920.

Garg, T., Onkar, S., Saahil, A., & Murthy, R. S. R., (2011). "Dendrimer: a novel scaffold for drug delivery." *International Journal of Pharmaceutical Sciences Review and Research, 7*(2), 211–220.

Gawde, K. A., et al., (2017). "Synthesis and characterization of folate decorated albumin bio-conjugate nanoparticles loaded with a synthetic curcumin difluorinated analog." *Journal of Colloid and Interface Science, 496*, 290–299.

Hein, C. D., Xin-Ming, L., & Dong, W., (2008). "Click chemistry, a powerful tool for pharmaceutical sciences." *Pharmaceutical Research, 25*(10), 2216–2230.

Hu, J., Ke, H., & Yiyun, C., (2016). "Tailoring the dendrimer core for efficient gene delivery." *Acta Biomaterialia, 35*, 1–11.

Huang, B., et al., (2012). "The facile synthesis of multifunctional PAMAM dendrimer conjugates through copper-free click chemistry." *Bioorganic and Medicinal Chemistry Letters, 22*(9), 3152–3156.

Iyer, A. K., Greish, K., Jun, F., & Hiroshi, M., (2006). "Exploiting the enhanced permeability and retention effect for tumor targeting." *Drug Discovery Today, 11*(17), 812–818.

Jassal, M., Ujjwal, N., & Jyotsana, K., (2015). *"Indian Journal of Pharmaceutical and Biological Research (IJPBR) A Review : Gastroretentive Drug Delivery System (GRDDS),"* *3*(1), 82–92.

Jung, J., et al., (2007). "pH-sensitive polymer nanospheres for use as a potential drug delivery vehicle." *Biomacromolecules, 8*(11), 3401–3407.

Kesharwani, P., et al., (2015). "PAMAM dendrimers as promising nanocarriers for RNAi therapeutics." *Materials Today, 18*(10), 565–572.

Kim, Y., Sang-keun, O., & Richard, M. C., (2004). "Preparation and characterization of 1–2 Nm preparation and characterization of 1–2n Nm dendrimer-encapsulated gold nanoparticles having very narrow size distributions." *Chem. Mater., 16*(18), 167–172.

Kojima, C., et al., (2011). "Dendrimer-based MRI contrast agents: The effects of PEGylation on relaxivity and pharmacokinetics." *Nanomedicine: Nanotechnology, Biology, and Medicine, 7*(6), 1001–1008.

Kolb, H. C., Finn, M. G., & Barry, S. K., (2001). "Click chemistry: Diverse chemical function from a few good reactions." *Angew. Chem. Int. Ed., 40*, 2004–2021.

Li, J., et al., (2015). "Recent advances in targeted nanoparticles drug delivery to melanoma." *Nanomedicine: Nanotechnology, Biology, and Medicine, 11*(3), 769–794.

Li, X., et al., (2016). "99mTc-labeled multifunctional low-generation dendrimer-entrapped gold nanoparticles for targeted SPECT/CT dual-mode imaging of tumors." *ACS Applied Materials and Interfaces, 8*(31), 19883–19891.

Liu, X., et al., (2013). "Enhanced retention and cellular uptake of nanoparticles in tumors by controlling their aggregation behavior." *ACS Nano, 7*(7), 6244–6257.

Liu Z-J, Daftarian P, Kovalski L, Wang B, Tian R, Castilla DM, et al. (2016) Directing and Potentiating Stem Cell-Mediated Angiogenesis and Tissue Repair by Cell Surface E-Selectin Coating. PLoS ONE 11(4): e0154053. https://doi.org/10.1371/journal.pone.0154053.

Luong, D., et al., (2016). "PEGylated PAMAM dendrimers: Enhancing efficacy and mitigating toxicity for effective anticancer drug and gene delivery." *Acta Biomaterialia*, 1–16.

Luong, D., et al., (2017). "Folic acid conjugated polymeric micelles loaded with a curcumin difluorinated analog for targeting cervical and ovarian cancers." *Colloids Surf B Biointerfaces.* 2017 Sep 1;157:490-502. doi: 10.1016/j.colsurfb.2017.06.025. Epub 2017 Jun 21.

Maiti, P. K., & Biman, B., (2006). "Structure and dynamics of DNA-dendrimer complexation: Role of counterions, water, and base pair sequence." *Nano Letters, 6*(11), 2478–2485.

Majoros, I. J., Christopher, R. W., Andrew, B., & James, R. B., (2009). "Methotrexate delivery via the folate-targeted dendrimer-based nanotherapeutic platform." *Wiley Interdisciplinary Reviews: Nanomedicine and Nanobiotechnology, 1*(5), 502–510.

Mukherjee, S., et al., (2016). "Green synthesis and characterization of monodispersed gold nanoparticles: Toxicity study, delivery of doxorubicin and its bio-distribution in mouse model." *J Biomed Nanotechnol. 2016 Jan;12(1):165-81.* Green Synthesis and Characterization of Monodispersed Gold Nanoparticles: Toxicity Study, Delivery of Doxorubicin and Its Bio-Distribution in Mouse Model.

Nanjwade, B. K., et al., (2009). "Dendrimers: Emerging polymers for drug-delivery systems." *European Journal of Pharmaceutical Sciences, 38*(3), 185–196.

Nigam, S., Sudeshna, C., & Dhirendra, B., (2015). *"Dendrimers Based Electrochemical Biosensors." 2*(1), 21–36.

Nikam, A. P., Mukesh, P. R., & Shilpa, P. C., (2014). *"Review Article Nanoparticles–An Overview," 3*(5), 1121–1127. Available online at http//www.ijrdpl.com

Noriega-Luna, B., et al., (2014). "Applications of dendrimers in drug delivery agents, diagnosis, therapy, and detection." *Journal of Nanomaterials.* Volume 2014, Article ID 507273, 19 pages http://dx.doi.org/10.1155/2014/507273

Pan, B., et al., (2006). "Cellular uptake enhancement of polyamidoamine dendrimer modified single walled carbon nanotubes." *ICBPE 2006 - Proceedings of the 2006 International Conference on Biomedical and Pharmaceutical Engineering, 1*, 541–544.

Peterson, J., Arkadi, E., Veiko, A., & Margus, L., (2001). "Synthesis and case analysis of pamam dendrimers with an ethylenediamine core." *Proc. Estonian Acad. Sci. Chem., 50*(3), 156–166.

Rai, A. K., Ruchi, T., Priyanka, M., & Pooja, Y., (2016). "Dendrimers : A potential carrier for targeted drug delivery system." *Pharmaceutical and Biological Evaluations, 3*(3), 275–287.

Sahu, P., Sushil, K. K., Samaresh, S., & Arun, K. I., (2017). *"Stimuli-Responsive Bio-Hybrid Nanogels: An Emerging Platform in Medicinal Arena," 1*(3), 6–8.

Sahu, P., Sushil, K. K., Sanyog, J., et al., (2017). "Assessment of penetration potential of pH-responsive double walled biodegradable nanogels coated with eucalyptus oil for the controlled delivery of 5-fluorouracil: *In vitro* and *ex vivo* studies." *Journal of Controlled Release, 253*, 122–136. http://dx.doi.org/10.1016/j.jconrel.2017.03.023.

Sau, S., & Hashem, O. A., et al., (2017a). "Advances in antibody-drug conjugates: A new era of targeted cancer therapy." *Drug Discovery Today,* http://dx.doi.org/10.1016/j.drudis.2017.05.011.

Sau, S., et al., (2014). "Cancer cell-selective promoter recognition accompanies antitumor effect by glucocorticoid receptor-targeted gold nanoparticle." *Nanoscale, 6*(12), 6745–6754. http://www.ncbi.nlm.nih.gov/pubmed/24824564.

Sau, S., et al., (2018). "A tumor multicomponent targeting chemoimmune drug delivery system for reprograming the tumor microenvironment and personalized cancer therapy." *Drug Discovery Today,* http://linkinghub.elsevier.com/retrieve/pii/S1359644617304208.

Prashant Sahu, Sushil K. Kashaw, Samaresh Sau, Varun Kushwah, Sanyog Jain, Ram K.Agrawala, Arun K.Iyer. pH Responsive 5-Fluorouracil Loaded Biocompatible Nanogels For Topical Chemotherapy of Aggressive Melanoma. Colloids and Surfaces B: Biointerfaces Volume 174, 1 February 2019, Pages 232-245

Sau, S., Hashem, O. A., et al., (2017b). "Advances in antibody: Drug conjugates: A new era of targeted cancer therapy." *Drug Discovery Today*. 2017 Oct;22(10):1547-1556. doi: 10.1016/j.drudis.2017.05.011 http://linkinghub.elsevier.com/retrieve/pii/S1359644617300788.

Sau, S., Sujan, K. M., et al., (2017). "Combination of cationic dexamethasone derivative and STAT3 inhibitor (WP1066) for aggressive melanoma: A strategy for repurposing a phase I clinical trial drug." *Molecular and Cellular Biochemistry*, 1–18.

Sharma, A., & Ashok, K., (2015). "Designing dendrimer and miktoarm polymer based multi-tasking nanocarriers for efficient medical therapy." *Molecules, 20*(9), 16987–16915.

Tatiparti, K., Samaresh, S., Kaustubh, G., & Arun, I., (2018). "Copper-free 'click' chemistry-based synthesis and characterization of carbonic anhydrase-IX anchored albumin-paclitaxel nanoparticles for targeting tumor hypoxia." *International Journal of Molecular Sciences, 19*(3), 838. http://www.mdpi.com/1422–0067/19/3/838.

Tatiparti, K., Samaresh, S., Sushil, K., & Arun, I., (2017). "siRNA delivery strategies: A comprehensive review of recent developments." *Nanomaterials, 7*(4), 77. http://www.mdpi.com/2079–4991/7/4/77.

Hashem Alsaab, Samaresh Sau, Rami M. Alzhrani, Vino T. Cheriyan, Lisa A.Polin, Ulka Vaishampayam, Arun K.Rishi, Arun K.Iyer. Biomaterials, Volume 183, November 2018, Pages 280-294

Tripathy, S., & Malay, K. D., (2013). "Dendrimers and their applications as novel drug delivery carriers." *Journal of Applied Pharmaceutical Science, 3*(9), 142–149.

Twibanire, J. D. K., & Bruce, G. T., (2012). "Polyester dendrimers." *Polymers, 4*(1), 794–879.

Wu, L. P., et al., (2015). "Dendrimers in medicine: Therapeutic concepts and pharmaceutical challenges." *Bioconjugate Chemistry, 26*(7), 1198–1211.

CHAPTER 3

Nanomedicines for the Treatment of Respiratory Diseases

BRAHMESHWAR MISHRA[1*], and SUNDEEP CHAURASIA[2]

[1]*Professor of Pharmaceutics, Department of Pharmaceutical Engineering & Technology, Indian Institute of Technology (Banaras Hindu University), Varanasi–221005, U.P., India, Tel.: +91 5426702748, Fax: +91 5422368428, E-mail: bmishra.phe@iitbhu.ac.in*

[2]*Formulation Research & Development, Mankind Research Centre (A Division of Mankind Pharma Ltd.), 191-E, Sector-4/II, IMT Manesar-122 051, Gurgaon, Haryana, India*

Corresponding author. E-mail: bmishra.phe@iitbhu.ac.in

ABSTRACT

Nanotechnology has revolutionized medicine over the past decade. The unique physicochemical characteristics of engineered nanomedicine enable novel therapeutic applications, particularly for the treatment of respiratory disorders. The research over the past decade has provided insights into biological properties and applications of nanomedicine in respiratory diseases. This book chapter provides a comprehensive review of the respiratory applications of the nanomedicine and aims to enlighten the readers about novel nano-based therapeutic strategies for treating respiratory diseases. This chapter discusses strategies to treat various airways diseases through nanomedicine applications. In summary, the book is focused on emerging cutting-edge applications of nanotechnology in respiratory medicine and aims to synchronize the efforts of respiratory biologists, nano-chemists, and clinicians to develop novel nano-based therapeutic systems for the treatment of airway diseases. This chapter has been compiled with the goal to serve both academic institutions and industry for education, training, and research. It is written to educate graduate and postgraduate students on emerging therapeutic applications of nanomedicine for the treatment of respiratory diseases. It will also serve as a guide for both clinicians and researchers in developing novel therapeutics while closely monitoring the health effects of next-generation nanomedicine. Overall, this is a Wikipedia of respiratory nanomedicine that discusses the scope of both current and future nanotechnologies for respiratory applications.

3.1   INTRODUCTION

The incidence of respiratory diseases and infections is increasing worldwide. Based on the physiological conditions, respiratory diseases are classified as obstructive or restrictive. Obstructive diseases generally prevent the flow rate entry and exit of the lungs, whereas a restrictive disease shows the reduction in the functional volume of the lungs (Respiratory Tract Infection, 2009). The very common obstructive diseases are chronic obstructive pulmonary diseases (COPD), asthma, occupational lung diseases, respiratory allergies, and pulmonary hypertension. Presently, 80 million people suffer from moderate-to-severe COPD while millions of others have mild COPD, allergic rhinitis and other often under-diagnosed chronic respiratory diseases and 300 million people suffer from asthma (WHO, 2002; Enarson, 1999). Respiratory tract infections are caused by either viral or bacterial, in the lower or upper position of the respiratory tract. The World Health Organization has classified chronic respiratory diseases as one of the key diseases afflicting the human and as a result efforts have been dedicated to their prevention, diagnosis, and treatment (WHO, 2002).

Poor curative outcomes have been reported for a number of treatment regimens for both respiratory diseases and infections, including asthma and tuberculosis (TB), respectively. These can be attributed mainly to patient non-compliance to the prescribed medication, which in several cases is a result of inadequate modes of drug administration (Gonda et al., 1998). The management of respiratory infections is relatively delicate as pre-examination of the aetiological agents of the disease appears are very challenging (Ruiz et al., 1999). A rise in drug-resistant strains of infectious organisms such as *Streptococcus pneumoniae* and *Mycobacterium tuberculosis* has created challenges in the treatment of these infections (Butler et al., 1996; Kunin, 1993; Petrini & Hoffner, 1999). *S. pneumoniae* is well known to be resistant to most of the antibiotics used to treat pneumonia, including penicillin, macrolides, doxycycline, and trimethoprim-sulfamethoxazole, and resistance to second- and third-generation cephalosporins has also increased (Butler et al., 1996). Multi-drug resistant *M. tuberculosis* is resistant to two of the first-line anti-TB drugs, that is, isoniazid, and rifampicin (Petrini & Hoffner, 1999). Recently, cases of extremely drug-resistant TB (XDR-TB) have been reported in South Africa and rest of the world, with high prevalence in HIV-positive individuals. XDR-TB is resistant to both first- and second-line anti-TB drugs (Wright, 2006). Another factors that play a major role in increasing the rates of infection include allergy and toxic exposures, including tobacco smoke and ambient pollution (Enarson, 1999).

Respiratory diseases are increasing pathologic entities in the old age human. The high expectations for good quality of life and changing lifestyle of our society, more efficient and affordable healthcare. Although our understanding of the functioning of the human body at the molecular and nanoscale has improved tremendously. Our diagnostic and therapeutic options for the effective treatment of severe and chronic respiratory diseases have increased only slowly over the past. Diseases like cancer, diabetes, lung, and cardiovascular problems, inflammatory, infectious diseases and neurological disorders are serious challenges to be dealt with. Multidisciplinary approaches that bring together material and medical scientists will speed up the development of new diagnostic and therapeutic solutions.

Over the past decades, colloidal drug delivery systems and especially nanoparticles or nanomedicine have received great attention. Nanomedicine can be administered through different routes of administration like parenteral, oral, intraocular, transdermal or pulmonary inhalation. Aerosol therapy using particulate drug carrier systems is becoming a popular strategy to deliver therapeutic or diagnostic compounds either locally or systemically (Ely et al., 2007) as shown by the development of inhalable insulin (Quattrin, 2004). This is due to the large alveolar surface area comfortable for drug absorption, the low thickness of the epithelial barrier, extensive vascularization and relatively low proteolytic activity in the alveolar space compared to other routes of administration and the avoidance of the first-pass metabolism (Patton & Platz, 1992; Clark, 2002; Courrier et al., 2002; Gill et al., 2007). In general, delivery of nanomedicine to the lungs is a challenging as well as most attractive concept because it can cause retention of the nanoparticles in the lungs accompanied with *in vitro* sustained drug release if large porous nanoparticle matrices are used (Edwards et al., 1997). Moreover, studies have been shown that nanomedicine uptake by alveolar macrophages can be reduced if the nanoparticles size are less than 260 nm (Edwards et al., 1997; Tsapis et al., 2002; Niven et al., 1995). Combination of both effects might improve local respiratory drug therapy. However, the mean particle size of medically used nanomedicine is too small to be suitable for direct lung delivery. A pre-requisite for deep drug delivery through lung is the design of proper nanocarrier systems (Ely et al., 2007). Successful respiratory delivery of inhaled particles depends mainly on the mean particle size, particle density, and mass median aerodynamic diameter (Taylor & Kellaway, 2001).

Treatment of respiratory diseases and infections still poses a great challenge. Much progress has been done with regard to varying the mode of delivery of therapeutic drugs for respiratory diseases and infections. The respirable fraction of an inhalable powder is generally the fraction of particles with an aerodynamic diameter ranging between 1 and 5 µm. this size range guarantees a maximum deposition in the deep lung (Taylor & Kellaway, 2001). Currently, interest in the use of nanotechnology-based delivery systems has gained momentum. Nanotechnology offers a broad range of opportunities for improving the diagnosis and therapy for respiratory diseases and infections, in particular, nanotechnology-based drug delivery systems, which represents an area of particular promise for the treatment of lung diseases. In this chapter, we focused mainly research performed during the last three decades in the area of nanoparticle delivery with special focus on nanomedicine targeting to the lungs. Whereas, direct respiratory delivery of dry powder formulations containing nanomedicine is rather a new approach, this chapter will also review nanomedicine delivery to the lungs for the treatment of respiratory disorders through different routes of administration.

3.2   CLASSIFICATION OF RESPIRATORY DISEASES

3.2.1   *OBSTRUCTIVE RESPIRATORY DISEASES*

The obstructive airway diseases represent one of the major global causes of disability and death. The various obstructive respiratory diseases and their description are given in the following subsections.

3.2.1.1   *BRONCHIAL ASTHMA*

Bronchial asthma is a chronic inflammatory disorder of the airways in which many cells and cellular elements play a major role: in particular, mast cells, eosinophils, T-lymphocytes, macrophages, neutrophils, and epithelial cells. In susceptible individuals, this inflammation causes recurrent episodes of wheezing, breathlessness, chest tightness, and coughing, mainly at nighttime or in the early morning. These episodes are usually associated with widespread but variable airflow obstruction that is often reversible either spontaneously or with treatment. The inflammation also causes an associated increase in the existing bronchial hyperresponsiveness to a variety of stimuli. Reversibility of airflow limitation may be incomplete in some patients with bronchial asthma (Jiang et al., 2014). The Centers for Disease Control and Prevention estimate that 1 in 11 children and 1 in 12 adults have bronchial asthma in the United States of America (Jiang et al., 2014). According to the World Health Organization, bronchial asthma affects 235 million people worldwide (WHO, 2016).

3.2.1.2 *CHRONIC OBSTRUCTIVE PULMONARY DISEASES*

The obstructive airway diseases, COPD represents one of the major global causes of disability and death. In this respect, COPD is estimated to become the third most common cause of death by 2020. The structural and pathophysiologic findings in COPD diseases appear to be easily differentiated in the extremes of clinical presentation. However, a significant overlap may exist in individual patients regarding features such as airway wall thickening on computer tomography or reversibility and airway hyperresponsiveness in lung function tests. In chronic COPD, airway inflammation is characterized by an increased number of T-lymphocytes, particularly CD8+T cells, macrophages, and neutrophils. With the progression of the disease severity macrophage and neutrophils numbers increase. Although there may be a partial overlap between bronchial asthma and COPD in some patients, the differences in functional, structural, and pharmacological features clearly demonstrate the consensus that COPD is different disease along all their stages of severity.

3.2.1.3 *RESPIRATORY ALLERGIES*

It is also known as allergic rhinitis (AR), is mainly evoked by an IgE-mediated response and shares several features with allergic asthma (AA). AR is often associated with sinusitis or other co-morbidities such as conjunctivitis (Cazzola et al., 2013; Jarvis et al., 2012; Kariyawasam & Rotiroti, 2013; Bonini et al., 2007) and precedes AA. AR and AA not only have a common biochemical inception but, to some extent, also have common treatment.

The interdependence between morbidities of the upper and lower airways is now known under the concept of "united airways," and the need for the concomitant treatment of these diseases is recognized. The IgE-mediated response is not a unique mechanism of allergic reaction onset; other less known mechanisms exists. In fact, five years ago, the ARIA group of experts wrote (Bousquet et al., 2008) "allergen-specific IgE, synthesized in response to allergens in the environment, becomes fixed to FcεRI on the membranes of mast cells and basophils, this aggregation results in the production of mediators (histamine, leukotrienes, and others) that produce the allergic response. However, a direct non-IgE-dependent mechanism also exists, and the relative importance of non-IgE to IgE-mediated mechanisms is undetermined.

3.2.1.4   OCCUPATIONAL LUNG DISEASES

Occupational lung diseases are a broad group of diagnoses caused by the inhalation of dusts, chemicals, or proteins. "Pneumoconiosis" is the term used for the diseases associated with inhaling mineral dusts. The severity of the disease is related to the material inhaled and the intensity as well as the duration of the exposure. Even individuals who do not work in the industry can develop occupational disease through indirect exposure. Although these diseases have been documented as far back as ancient Greece and Rome, the incidence of the disease increased dramatically with the development of modern industry.

In most cases, these diseases are man-made, resulting from inorganic dust exposure during mining, processing, or manufacturing. In New York and New Jersey in the 1970s, asbestosis could be diagnosed in over 70 percent of asbestos insulation workers with greater than 20 years of exposure (Rom and Markowitz, 2007). After the introduction of regulatory agencies and prom-ulgation of dust regulations and their enforcement, these high prevalence rates and others dropped dramatically. For instance, the prevalence of coal workers pneumoconiosis dropped to 5 percent among miners with greater than 25 years of exposure (Rom and Markowitz, 2007). The pneumoco-nioses primarily affect those exposed at work, but environmental exposure can make others sick as well. Asbestos insulators expose their wives and children by bringing home their asbestos-covered clothing, and asbestos factories and mines expose residents of nearby neighborhoods.

Different exposures result in different diseases. With silica exposure, the classic and most common disease is chronic silicosis, which develops decades after exposure and is characterized by the silicotic nodule, predominantly in the upper lobes of the lungs, and "eggshell" calcification of the lymph nodes. These findings do not always have clinical symptoms. Higher intensity exposure can result in accelerated or acute silicosis, in which symptoms develop much earlier. Acute silicosis is the least frequent, but it also has the highest mortality rate. The accelerated and chronic forms of silicosis can become complicated silicosis or progressive massive fibrosis, in which the silicotic nodules coalesce into larger lesions in the upper lobes of the lung, and the patients develop increasing breathing with difficulty.

3.2.1.5   *PULMONARY HYPERTENSION*

Pulmonary hypertension is high blood pressure in the arteries going to the lung. In healthy individuals, the blood pressure in these arteries is much lower than in the rest of the body. In a healthy individual, the blood pressure of the arteries going to the rest of the body is around 120/80 millimeters of mercury (mm Hg), and pulmonary artery blood pressure is about 25/10 mm Hg. If the pulmonary arterial pressure exceeds about 40/20 mm Hg or the average pressure exceeds 25 mm Hg, then pulmonary hypertension is present. If pulmonary hypertension persists or becomes very high, the right ventricle of the heart, which supplies blood to the pulmonary arteries, is unable to pump effectively, and the person experiences symptoms that include shortness of breath, loss of energy, and edema, which is a sign of right heart failure. Many diseases and conditions increase the pulmonary artery pressure.

The exact prevalence of all types of pulmonary hypertension in the United States and the world is not known. The number of patients in the United States is certainly in the hundreds of thousands, with many more who are undiagnosed. About 200,000 hospitalizations occur annually in the United States with pulmonary hypertension as a primary or secondary diagnosis (Hyduk et al., 2005). About 15,000 deaths per year are ascribed to pulmonary hypertension, although this is certainly a low estimate.[28] Most medical references to heart failure are for left heart failure, which in the United States has a prevalence of about 4.9 million and an annual incidence of 378 per 100,000 (Hunt et al., 2001; Roger et al., 2004). Pulmonary hypertension, which causes right heart failure, affects all races and socio-economic levels.

The most common cause of pulmonary hypertension in the developing world is schistosomiasis, a parasitic infection in which the parasite's eggs can lodge in and obstruct the pulmonary arteries. Another risk factor for pulmonary hypertension is high altitude. More than 140 million persons worldwide and up to 1 million in the United States live 10,000 feet or more above sea level (Penaloza and Arias-Stella, 2007). In African Americans, sickle cell anemia is an important cause of pulmonary hypertension.

A specific type of pulmonary hypertension in which the disease process occurs in the pulmonary arteries themselves is called pulmonary arterial hypertension (PAH). This condition generally affects young and otherwise healthy individuals and strikes women twice as frequently as men. The average age of diagnosis is 36 years, and three-year survival after diagnosis is only about 50 percent. Each year, between 10 and 15 people per million populations are diagnosed with the disease. With improved treatments and survival, the number of U.S. patients living with the disease has increased to between 10,000 and 20,000 (Badesch et al., 2010).

Because so many disorders can result in severe pulmonary hypertension and treatments may vary dramatically, it is important for a thorough evaluation to occur when pulmonary hypertension is detected or suspected. For instance, pulmonary hypertension related to blood clots in the pulmonary arteries (pulmonary embolism and thromboembolic pulmonary hypertension) requires anticoagulation and, in some cases, surgical removal of the clots. Because about 250,000 cases of pulmonary embolism occur each year in the United States, thousands of patients are annually at risk of residual pulmonary hypertension from this disorder (Silverstein et al., 1998). The actual number is not easily determined because most cases of pulmonary embolism go undiagnosed.

3.2.2   CYSTIC FIBROSIS

Cystic fibrosis, a frequent inherited, autosomal recessive disorder. It is caused by a dysfunction of the epithelial chloride channel CFTR (cystic fibrosis transmembrane regulator) (Rosenstein and Zeitlin, 1998). So far, more than 500 mutations of the CFTR gene are known that are associated with cystic fibrosis (Stern, 1997). Apart from gastrointestinal manifestations such as pancreatic insufficiency, the major cause of morbidity results from airway disease (Rosenstein and Zeitlin, 1998). The hypersecretory-induced airway changes in cystic fibrosis are characterized by submucosal gland and goblet cell hyper- and metaplasia, leading to mucus over-production and distortion of the mucociliary clearance. As a result, airway plugging by mucus leads to chronic inflammatory changes and bacterial colonization (Groneberg et al., 2002; Ramsey, 1996).

The major component of airway mucus is made up of large, oligomeric, gel-forming mucin glycoproteins with molar masses ranging between 10 and 40 million Da (Davies et al., 1996; Gupta and Jentoft, 1992; Thornton et al., 1990, 1991). A number of different mucin proteins which are products of different genes have been identified in the respiratory tract so far. The mucins are produced primarily by two different airway cell types: goblet cells and glandular cells. While MUC2 and MUC5AC expression has been localized to cells of the surface epithelium (Groneberg et al., 2002; Davies et al., 1996; Chung et al., 2004; Li et al., 1997; Wickstrom et al., 1998), MUC5B and MUC7 mucins are expressed primarily in glandular cells (Wickstrom et al., 1998; Groneberg et al., 2003; Sharma et al., 1998), but can also be induced in epithelial cells under states of inflammation (Groneberg et al., 2002). Out of the different respiratory mucins, MUC5AC and MUC5B have been identified as major gel-forming mucins whereas MUC2 contributes only to a lesser extent to the matrix (Davies et al., 1996).

3.2.3 PULMONARY TUBERCULOSIS

Tuberculosis (TB) is an infectious disease that occurs worldwide. It is caused by *Mycobacterium tuberculosis* complex. Most diseases are caused by *M. tuberculosis* (M. TB). TB is currently the second most deadly infectious disease in the world (WHO, 2016). The Czech Republic is among the countries that have effectively controlled TB, and the incidence of this disease is still declining. In 2014, a total of 512 cases of TB of all forms and localizations as well as their recurrence (4.9 patients/100,000) were reported to the TB registry. TB is clinically manifested by fatigue, loss of appetite, weight loss, decreased physical performance, sub febrile night sweats, and a dry or productive cough, possibly with hemoptysis, dyspnea, and pleural pain. However, the clinical course of TB in some cases can be without symptoms (Kolek et al., 2014).

Infection occurs when a person inhales droplet nuclei containing tubercle bacilli that reach the alveoli of the lungs. These tubercle bacilli are ingested by alveolar macrophages; the majority of these bacilli are destroyed or inhibited. A small number may multiply intracellularly and are released when the macrophages die. If alive, these bacilli may spread by way of lymphatic channels or through the bloodstream to more distant tissues and organs (including areas of the body in which TB disease is most likely to develop: regional lymph nodes, apex of the lung, kidneys, brain, and bone). This process of dissemination primes the immune system for a systemic response (Kolek et al., 2014).

3.2.4　LUNG CANCER

Lung cancer is the most common cancer worldwide (Parkin et al., 2002) and the second most common cancer in the LSS cohort of the atomic bomb survivors in Hiroshima and Nagasaki. Evidence suggests that environmental exposures such as cigarette smoking and radiation have increased the risks of lung cancer (Cruz et al., 2011). Statistical analysis suggests that more than 90% of lung cancer is caused by these extrinsic factors (Preston et al., 2007). In addition, lung cancer incidence among atomic bomb survivors is strongly associated with radiation, with an estimated excess relative risk per Gy of 0.81 and excess 20 absolute risks per Gy of 7.5 per 10,000 people per year (Preston et al., 2007). Therefore, approaches are imperatively needed to explore how radiation affects the development of lung cancer. A pilot study suggests that mutation frequencies of certain genes (e.g., the TP53 tumor suppressor gene) and methylation levels (e.g., the retrotransposon LINE1) may be associated with radiation exposure. It is widely accepted that the formation of nearly all sorts of tumors is largely owing to the dynamic changes in the genome. There are three types of genes that are responsible for tumorigenesis, which are oncogenes, tumor-suppressor genes and stability genes (Vogelstein and Kinzler, 2004). In the early 1950s, a multistage model was introduced as an essential tool to understand tumorigenesis (Armitage and Doll, 1954). This model describes the tumorigenesis as a process of an infinite number of mutations turning a normal cell into malignant or cancerous cells. It shows that the logarithm of incidence was a linear function of the logarithm of age. With the advances of molecular biology, clonal expansion was recognized as an essential stage in carcinogenesis.

3.3　CURRENT TREATMENT FOR RESPIRATORY DISEASES

To date, the most effective treatment for respiratory diseases resulting in airway inflammation has been oral or injectable corticosteroids administered generally to treat bronchial asthma and COPD. Many systemic side effects can occur as a result of the chronic use of corticosteroids. However, such advances have been made in this area, in that corticosteroids can be given by inhalation (Wright, 2006). This route of delivery has minimized systemic absorption of the drugs and many complications previously observed with injectable and oral dosages form (Todd et al., 2002). Although inhalation delivery of the drug has addressed these factors, the persistent challenge is that the lung is functionally and anatomically heterogeneous. Thus the dose and drug distribution in the lungs play an important role in reproducible delivery and thus successful therapy (Gonda et al., 1998). Viral respiratory infections such as influenza have no effective and safe antiviral compound and are not susceptible to antibiotic treatment. However, antibiotics are generally prescribed for secondary infections. Ribavirin, an antiviral compound with activity against a number of DNA and RNA viruses, has been used to treat viral respiratory infections such as influenza and respiratory syncytial virus (RSV) infection (Smith et al., 1980). At present oral ribavirin is used in Mexico against influenza, and the aerosol dosage form has been used to treat RSV-related diseases in children. The challenges with ribavirin are the associated adverse side effects, such as hemolytic anemia, which have been reported to be dose-dependent. An additional concern is that this compound has been identified as a teratogen in some animal species (Bani-Sadr et al., 2005). Bacterial respiratory infections, on the other hand, are treated with oral or injectable antibiotics. Although drugs against respiratory infections such as *S. pneumoniae* and *M. tuberculosis* are effective, these drugs generally have to be administered as combination therapy in high doses for long durations of treatment to maintain therapeutic levels, and they also have poor bioavailability (Butler et al., 1996; du Toit et al., 2006). Because of the high doses administered and the associated side effects, patient non-compliance as mentioned above has led to the inefficacy of the treatment regimen. The incorrect dosing of chemotherapy has also been reported to be linked to the emergence of multi-drug resistant strains (Lipsitch and Samore, 2002), and these challenges have posed a need to develop novel ways of delivering the therapeutic compounds (Prabakaran et al., 2004).

3.4 ADVANTAGES OF NANOMEDICINE OVER RESPIRATORY DISEASES

The burgeoning interest in colloidal carriers (nanocarrier systems) has led to increasing attention for respiratory diseases. The lung is an attractive target for nano-drug delivery due to non-invasive means to provide not only local lung effects but possibly high systemic bioavailability, avoidance of first-pass metabolism, more rapid onset of therapeutic action, and the availability of a huge surface area (Yang et al., 2008; Patton & Byron, 2007). Nanomedicine offers several advantages for the treatment of respiratory diseases. These advantages include the following: (i) the potential to achieve relatively uniform distribution of drug dose among the alveoli, (ii) an achievement of enhanced solubility of the drug than its own aqueous solubility, (iii) the sustained-release of drug which consequently reduces the dosing frequency, (iv) suitability for delivery of macromolecules, (v) decreased incidence of side effects, (vi) improved patient compliance, and (vii) the potential of drug internalization by cells (Sung et al., 2007; Bailey and Berkland, 2009).

Because of the size of the nanoparticles, which could be in the form of either a nanocapsule or a nanosphere, these particles are able to reach the 'hard' to targeted sites of the body. The ability to target the lung via inhalable micro- or nanoparticles has the potential of minimizing drug resistance, reducing side effects and also lowering the therapeutic dose which is usually administered orally (Pandey and Khuller, 2005). The entrapment of the drugs in polymeric particles allows drug release from the polymeric material in a pre-designed manner which may be either sustain over a long period, cyclic, or it may also be triggered by the cellular microenvironment or other external factors, such as pH, temperature, oxidative conditions or an external magnetic field (Schmoljohann, 2006). Functional groups on the surface of the particles allow chemical conjugation of various ligands, peptide, DNA, and sugars to enable actively targeted delivery of particles to cells where these molecules will be recognized (Torchilin, 2006). Furthermore, in order to make the particles stealth, polymers such as polyethylene glycol (PEG) and polyvinylpyrrolidone (PVP) have been attached onto their surface. These polymers have been reported to minimize opsonization, thus increasing the residence time of the particles in the blood circulation (Evora et al., 1998). In recent years, biodegradable polymers such as poly(DL-lactide), poly(lactic acid) (PLA), poly(DL-glycolide) (PG), poly(lactide-co-glycolide) (PLG), and poly(cyanoacrylate) (PCA), as well as chitosan, alginate, and gelatine to name a few that have attracted considerable attention as potential drug delivery devices in view of their applications in the control release of drugs, their ability to target particular organs/tissues as carriers of DNA in gene therapy and in the ability to deliver proteins, peptides, and genes (Kumaresh et al., 2000). Several polymers for respiratory drug delivery using nanocarrier systems are presented in Table 3.1.

3.5   NANOMEDICINE FOR OBSTRUCTIVE RESPIRATORY DISEASES

Among obstructive pulmonary diseases, bronchial asthma and COPD are among global health hazards in terms of mortality and morbidity. The basis of nanomedicine activities in diseases were discussed in a study by John *et al.* in Ann Arbor, MI (John et al., 2003). For bronchial asthma, experimental studies have already been conducted to assess the use of such nanosystems.

TABLE 3.1 Several Polymers for Respiratory Nanodrug Delivery Systems

S. No.	Name of polymers	Drugs and references	Size of the nanoparticles
I	**Chitosan**		
	Chitosan	Plasmid DNA (Plumley et al., 2009)	91–164 nm
		Small interfering RNA (Almeida and Souto, 2007)	40–600 nm
	Chitosan/tripolyphosphate	Insulin (Koping-Hoggard et al., 2004)	300–388 nm
	Trisaccharide-substituted chitosan	Plasmid DNA (Howard et al., 2006)	77–90 nm
	Urocanic acid–modified chitosan	Programmed cell death protein (Grenha et al., 2005)	NA
II	**Gelatin**		
	Gelatin type A	Fluorescein amine (Grenha et al., 2005)	277.8 nm
	Gelatin type B	Sulforhodamine 101 acid chloride (Issa et al., 2006)	242±14 nm
	PEGylated gelatin	Plasmid DNA (Brzoska et al., 2004)	100–500 nm
III	**Alginate**		
	Sodium alginate	Rifampicin, isoniazid, pyrazinamide (Kimura et al., 2009)	235.5 nm
IV	**Polyalkylcyanoacrylate**		
	Polybutylcyanoacrylate	Insulin (Kaipel et al., 2008)	254.7 nm
		Doxorubicin (Sham et al., 2004)	173±43 nm
		Ciprofloxacin (Kaul and Amiji, 2005)	156–259 nm
V	**PEI**		
	PEI	Chimeric oligonucleotide (Hwang et al., 2007)	30–100 nm
		Plasmid DNA (Beck-Broichsitter et al., 2009)	50–100 nm
	PEI-alt-PEG	Small interfering RNA (Shahiwala and Misra, 2005)	NA
	Glucosylated PEI	Programmed cell death protein 4 (Gautam et al., 2000)	NA
	Galactose-PEG-PEI	Plasmid DNA (Lynch et al., 2007)	105–210 nm
	Cell-penetrating peptides-PEG-PEI	Plasmid DNA (Chen et al., 2008)	113–296 nm

TABLE 3.1 *(Continued)*

S. No.	Name of polymers	Drugs and references	Size of the nanoparticles
VI	**Poly-L-lysine**		
	PEGylated poly-l-lysine	Plasmid DNA (Nguyen et al., 2008; Ziady et al., 2003)	211±29 nm
VII	**Proticle**		
	Protamine-oligonucleotide	Vasoactive intestinal peptide (Park et al., 2008)	177–318 nm
VIII	**PLGA**		
	PLGA	Rifampicin, isoniazid, pyrazinamide (Gong et al., 2005)	570–680 nm
	PEG-PLGA	Nuclear factor κB decoy oligodeoxynucleotide (Cryan et al., 2005)	44 nm
	Chitosan-modified PLGA	Elcatonin (Bivas-Benita et al., 2004)	650 nm
	Chitosan/PLGA	Antisense oligonucleotide 2-O-methyl-RNA (Gautam et al., 2002)	135–175 nm
	Poly[vinyl 3-(diethylamino)propyl-carbamate-co-vinyl acetate-covinyl alcohol]-graft-PLGA	5(6)-carboxyfluorescein (Ely et al., 2007)	195.3±7.1 nm
IX	**Dendrimer**		
	G9 PAMAM	Plasmid DNA (Ziady et al., 2003)	NA
	G2/G3 PAMAM	Low molecular weight heparin (Zhang et al., 2001)	NA
	Pegylated G3 PAMAM	Low molecular weight heparin (Huang and Wang, 2006)	17.1±7.1 nm

Abbreviations: PEG, poly(ethylene glycol); PLGA, poly(lactic-co-glycolic acid); PEI, poly(ethylenimine); PAMAM, polyamidoamine; NA, not available.

Nanomedicine technology was applied to discover a potent nanoparticle P-selectin antagonist with strong anti-inflammatory effects in a murine model allergic asthma (John et al., 2003). The background of the study was to assess the role of P-selectin for the development and progression of peri-bronchial inflammation in allergic airway disease. Since selective P-selectin inhibitors may lead to an attenuation of the ongoing inflammatory processes present in allergic bronchial asthma, a panel of novel P-selectin inhibitors were synthesized using polyvalent polymeric nanoparticles (John et al., 2003). First, a construct that binds efficiently to P-selectin was generated by assembling a particle with the ligands acting as mimetics of the binding elements that mediate the adhesion of P-selectin to its ligand P-selectin glycoprotein ligand–1 (PSGL–1). Then, an *in vitro* assay was used to evaluate the different inhibitors by examining the interactions between P-selectin coated capillary tubes and circulating cells. It was shown that they preferentially bind to selectins expressed on activated endothelial cells (John et al., 2003). After these *in vitro* experiments, *in vivo* studies were conducted using a murine model of allergic asthma and a significant reduction of allergen-induced peri-bronchial inflammation airway and airway hyperreactivity present (John et al., 2003). This indicating the validity of the new compounds.

Chitosan/interferon (IFN)-γ pDNA nanoparticles was used in the further study. These nanoparticles were established to analyze the quantity of *in situ* IFN-γ production. The reason for this approach is that adenovirus-mediated IFN-γ gene transfer reduces airway hyperresponsiveness in mice but is limited so far by the frequency of gene delivery required. The nanomedicine were given to ovalbumin-sensitized mice to assess their efficacy to modulate ovalbumin-induced inflammation and airway hyperresponsiveness (Kumar et al., 2003). It was shown that the mice treated intranasally with 25µg of chitosan-IFN-γ nanoparticles (i.n.) had a lower airway hyperresponsiveness to methacholine challenge. They also had less histopathological signs of airway inflammation. The nanoparticles also led to an increased production of IFN-γ while Th2-cytokines, interleukin (IL)–4, IL–5, and ovalbumin-specific serum IgE were reduced in comparison to the control group (Kumar et al., 2003). It was also demonstrated that the nanomedicine inhibited the epithelial inflammation within 6 h of delivery which was paralleled by an induction of apoptosis of goblet cells. On the molecular level, the treatment with nanomedicine involved signal transducer and activator of transcription 4 (STAT4) signaling because STAT4-deficient mice did exhibit reduced airway hyperresponsiveness and inflammation (Kumar et al., 2003). Together, these studies using chitosan/interferon (IFN)-γ pDNA nanoparticles demonstrated that nanoparticles could be coupled to bioactive molecules that target inflammatory events in allergic airway inflammation and that the intranasal administration of this nanomedicine can effectively reduce established allergen-induced experimental bronchial asthma.

In contrast to these two studies focusing on the use of nanomedicine in the treatment of bronchial asthma, no studies have so far been conducted to assess the potential use of nanomedicine for the treatment of COPD. However, with new compounds being developed currently, experiments using this new technology in COPD will be conducted soon using animal models of COPD (Kumar et al., 2003; Chung et al., 2004).

In another study, weekly treatment with budesonide encapsulated in stealth liposomes was as effective as daily budesonide therapy in decreasing lung inflammation and lowering eosinophil peroxidase activity, peripheral blood eosinophils, and total serum IgE levels. In none of the other groups was there a significant decrease in the inflammatory parameters evaluated. We conclude that weekly therapy with budesonide encapsulated liposomes is as effective as daily budesonide in reducing markers of lung inflammation in experimental asthma. This novel strategy offers an effective alternative to standard daily budesonide therapy in asthma and has the potential to reduce toxicity and improve compliance (Konduri et al., 2003).

Drug delivery of protein and peptide-based drugs, which represent a growing and important therapeutic class, is hampered by these drugs and having very short half-lives. High susceptibility towards enzymatic degradation necessitates frequent drug administration followed by poor adherence to therapy. Among these drugs is vasoactive intestinal peptide (VIP), a potent systemic and pulmonary vasodilator, which is a promising drug for the treatment of idiopathic pulmonary arterial hypertension (IPAH). Encapsulation of VIP into the nanoparticle matrix of biodegradable protamine-oligonucleotide nanoparticles (proticles) protects the peptide against rapid enzymatic degradation. Additionally, the nanoparticle matrix will be able to sustain drug release. Proticles consist of 18mer non-sense oligonucleotides and protamine, a polycationic arginine-rich peptide. VIP encapsulation occurs during self-assembly of the components. In this study, size of the nanoparticle (hydrodynamic diameter), zeta potential of VIP-loaded proticles, encapsulation efficiency and VIP release was done to evaluate. Further, the pharmacological VIP response of "encapsulated VIP" is investigated using an *ex vivo* lung arterial model system. The results found that favorable encapsulation efficiency (up to 80%), VIP release (77–87%), and an appropriate nanoparticle size (177–251 nm). Investigations on rat pulmonary arteries showed a modified VIP response of proticle-associated

VIP. We noted differences in the profile of artery relaxation where VIP proticles lead to a 20–30% lower relaxation maximum than aqueous VIP solutions followed by prolonged vasodilatation. These above data indicate that proticles could be a feasible nano-drug delivery system for a respiratory VIP depot formulation (Wernig et al., 2008).

In another study, development of cromolyn sodium (CS) encapsulated polylactide-co-glycolide (PLGA) nanoparticles (CS-PNs) for enhancing intestinal permeation, and this CS is used as an allergic rhinitis. The CS-PNs were prepared by double emulsification solvent evaporation method (W1/O/W2). The "Quality by Design" approach using Box-Behnken experimental design was employed to enhance encapsulation of CS inside CS-PNs without compromising with particle size. The polymer concentration, surfactant concentration, and organic/aqueous phase ratio significantly affected the physicochemical properties of CS-PNs. The optimized CS-PNs were subjected to various solid-state and surface characterization studies using FTIR, DSC, XRD, TEM, and AFM, which pointed towards the encapsulation of CS inside the spherical shaped nanoparticles without any physical as well as chemical interactions. *Ex vivo* intestinal permeation study demonstrated ~4 fold improvements in CS permeation by forming CS-PNs as compared to pure CS. Further, *in vivo* intestinal uptake study performed using confocal microscopy, after oral administration confirmed the permeation potential of CS-PNs. Thus, the findings of the studies suggest that CS-PNs could provide a superior therapeutic carrier system of CS, with enhanced intestinal permeation (Patel et al., 2016).

In another study, CS, encapsulated core-shell polymeric-lipid hybrid nanoparticles (PLHNs) for enhancing its oral bioavailability, by improving its intestinal permeability through lymphatic uptake. The CS encapsulated PLGA-lecithin based core-shell PLHNs (CS-PLHNs) were engineered by a double emulsification solvent evaporation method and optimized using a response surface methodology based on "Quality by Design" approach. The Box-Behnken experimental design was imperatively enforced to enhance encapsulation of CS inside PLHNs without compromising particle size. Optimized CS-PLHNs exhibited a particle size of 227±3.8 nm and EE of 57.8±1.32% with unimodal size distribution. The physicochemical characterizations of CS-PLHNs suggested the encapsulation of CS in an amorphous form inside PLHNs without any interactions. The morphological studies pointed towards the existence of smooth, spherical core-shell architecture of CS-PLHNs, which extended-release up to 48 h by a controlled diffusion process. The optimized CS-PLHNs exhibited remarkable stability at different environmental conditions as well as in biological milieu. An ex vivo intestinal permeation study showed that the permeation of CS was significantly improved by encapsulating it inside PLHNs compared to that of pure CS, which was additionally confirmed by the in vivo intestinal uptake study using confocal microscopy. A pharmacokinetic study in rats further exhibited a 11.9-fold enhancement in the oral bioavailability of CS after its incorporation into PLHNs. In a nutshell, PLHNs can serve as a superior therapeutic carrier system by imparting lipophilicity to potentially obtain the high oral bioavailability of CS, which can further be extended towards numerous hydrophilic drug molecules (Patel et al., 2015).

Cromolyn sodium (CS), a mast cell stabilizer, is widely employed for the prevention and treatment of allergic conditions. However, high hydrophilicity and poor oral permeability hinder its oral clinical translation. Here, solid lipid nanoparticles (SLNs) have been developed for the purpose of oral bioavailability enhancement. The CS-SLNs were engineered by double emulsification method (W1/O/W2) and optimized by using Box-Behnken experimental design. The surface and solid-state characterizations revealed the presence of CS in an amorphous form without any interactions inside the spherical-shaped SLNs. The in-vitro release study showed an extended-release up to 24 hr by diffusion controlled process. *Ex vivo* and *in vivo* intestinal permeation study showed ~2.96-fold increase in permeability of CS by presentation as SLNs ($p<0.05$). Further, in-vivo pharmacokinetic study exhibited ~2.86-fold enhancements in oral bioavailability of CS by encapsulating inside SLNs, which clearly indicate that SLNs can serve as the potential therapeutic carrier system for oral delivery of CS (Patel et al., 2016).

3.6   NANOMEDICINE FOR CYSTIC FIBROSIS

In view of the high levels of CFTR expression in glandular cells which play an important role in the composition of mucus hydration and their attributed pathophysiological importance in the progression of cystic fibrosis (Engelhardt et al., 1993), numerous studies have focused on the identification of differences between cystic fibrosis and normal airway secretions (Boat and Cheng, 1980; Roussel et al., 1975; Zhang et al., 1995). Alterations such as increased sulfation and fucosylation and decreased sialytion of secreted mucins have been demonstrated. Biochemical studies also indicated a higher heterogeneity of cystic fibrosis mucin as compared to the normal mucus composition (Chace et al., 1985). However, a molecular treatment option targeting mucin gene expression has not been established so far, and only one study has led to promising results using a myristoylated, alanine-rich C-kinase substrate (MARCKS) protein (Singer et al., 2004). Since not only mucus hyper-secretion but also the deficient channel protein CFTR may be targeted by nanosystems, cystic fibrosis seems to be an ideal candidate for the therapeutic use of such systems.

In a recent study, a gelatine and DNA nanoparticle coacervation containing chloroquine and calcium has been developed as a gene delivery vehicle. In this vehicle, the cell ligand transferring is covalently bound to gelatine (Truong-Le et al., 1999). The coacervation conditions which resulted in the formation of distinct nanoparticles were studied, and it was demonstrated that the nanosystems formed within a narrow range of DNA concentrations and achieved incorporation of more than 98% of the DNA in the reaction (Truong-Le et al., 1999). It was further studied whether the cross-linking of gelatine to stabilize the particles affects electrophoretic mobility of the DNA, but no effect was found. The DNA in the nanosystems was also present to be partially resistant to digestion with concentrations of DNase I that result in extensive degradation of free DNA but is completely degraded by high concentrations of DNase. An optimum cell transfection by nanosphere DNA required the presence of calcium and nanospheres containing transferrin. The biological integrity of the nanosphere DNA was demonstrated with a model system utilizing DNA encoding the cystic fibrosis transport regulator (CFTR). In transfection studies of cultured human tracheal epithelial cells (9HTEo) with nanospheres containing the plasmid, a CFTR expression in over 50% of the cells was found (Truong-Le et al., 1999). It was also demonstrated that human bronchial epithelial cells (IB–3–1) defective in CFTR-mediated chloride transport regained an effective transport activity when transfected with nanospheres containing the CFTR transgene (Truong-Le et al., 1999). Therefore, nanosystems might be an attractive candidate to deliver new therapeutic compounds in cystic fibrosis.

In another embodiment, the preparation of mucus-penetrating nano-medicine for respiratory administration of ibuprofen in patients with cystic fibrosis is described. A fluorescent derivative of α,β-poly(N–2-hydroxyethyl)-D,L-aspartamide is synthesized by derivatization with rhoda-mine, polylactide, and poly(ethylene glycol), to obtain polyaspartamide-polylactide derivatives with different degrees of pegylation. Starting from these copolymers, fluorescent nanoparticles with different poly(ethylene glycol) content, empty, and loaded with ibuprofen, showed spherical shape, colloidal size, slightly negative ζ potential, and biocompatibility toward human bronchial epithelial cells. The high surface poly(ethylene glycol) density of fluorescent nanoparticles and poly(ethylene glycol) brush-like conformation assumed on their surface, conferred to pegylated nanoparticles the mucus-penetrating properties, properly demonstrated by assessing their ability to avoid interactions with mucus components and to penetrate cystic fibrosis artificial mucus. Finally, ibuprofen release profile and uptake capacity within human bronchial epithelial cells in the presence of cystic fibrosis artificial mucus showed how these mucus-penetrating nanoparticles could rapidly diffuse through the mucus barrier reaching the mucosal surface, where they could offer a sustained delivery of ibuprofen at the site of disease (Craparo et al., 2016).

In another study, encapsulate Amiloride Hydrochloride into nano-liposomes, incorporate it into dry powder inhaler, and to provide prolonged effective concentration in airways to enhance mucociliary clearance and prevent secondary infection in cystic fibrosis. Liposomes were prepared by thin film hydration technique, and then dispersion was passed through high-pressure homogenizer to achieve size of nanometer range. Nano-liposomes were separated by centrifugation and were characterized. They were dispersed in phosphate buffer saline pH 7.4 containing carriers (lactose/sucrose/mannitol), and glycine as anti-adherent. The resultant dispersion was spray dried. The spray-dried powders were characterized, and *in vitro*, drug release studies were performed using phosphate buffer saline pH 7.4. *In vitro* and *in vivo* drug respiratory deposition was carried out using Andersen Cascade Impactor and by estimating drug in bronchial alveolar lavage and lung homogenate after intratracheal instillation in rats, respectively. Nano-liposomes were found to have mean volume diameter of 198 ± 15 nm, and $57\pm1.9\%$ of drug entrapment. Mannitol based formulation was found to have low density, good flowability, particle size of $6.7\pm0.6\mu$ determined by Malvern MasterSizer, maximum fine particle fraction of $67.6\pm0.6\%$, mean mass aerodynamic diameter $2.3\pm0.1\mu$, and geometric standard deviation 2.4 ± 0.1. Developed formulations were found to have prolonged drug release following Higuchi's controlled release model and *in vivo* studies showed maximal retention time of drug of 12 hr within the lungs and slow clearance from the lungs. This study provides a practical approach for direct lung delivery of Amiloride Hydrochloride encapsulated in liposomes for controlled and prolonged retention at the site of action from dry powder inhaler. It can provide a promising alternative to the presently available nebulizers in terms of prolonged pharmacological effect, reducing systemic side effects such as potassium retention due to rapid clearance of the drug from lungs in patients suffering from cystic fibrosis (Chougule et al., 2006).

3.7 NANOMEDICINE FOR PULMONARY TUBERCULOSIS

One of the first studies addressed the issue of nanoparticle-encapsulated antitubercular drugs as potential oral drug delivery systems against murine tuberculosis (TB) (Pandey et al., 2003). One of the major problems with long duration TB chemotherapy is patient non-compliance. A reduction in the frequency of dosing using nanoparticle-encapsulated compounds might, therefore, lead to a significant improvement in the therapy. To assess this issue, poly (DL-lactide-co-glycolide) (PLG) nanoparticle-encapsulated formulations of the three front line anti-tubercular drugs rifampicin, isoniazid, and pyrazinamide were studied in mice. The drug encapsulation efficiencies were reported to be 66.3±5.8% for isoniazid, 68±5.6% for pyrazinamide, and 56.9±2.7% for rifampicin. After single oral administrations of the nanomedicine formulations the preparations were found in the circulation for over 9 days (isoniazid and pyrazinamide) and 6 days (rifampicin) while the therapeutic concentrations in the tissues were maintained for 9 to 11 days (Pandey et al., 2003). In further infection studies, the effects of oral administration of nanoparticle-encapsulated drugs to *Mycobacterium tuberculosis*-infected mice every 10th day was studied. No tubercle bacilli were detected after 5 oral doses (day 50 of treatment). This finding suggests that nanoparticle-based antituberculous therapy may be beneficial since it may lead to a reduction in dosing frequency (Pandey et al., 2003). Next to this study on oral administration, also the aerosolic bioavailability of antitubercular drugs was examined (Pandey et al., 2003) using a formulation of rifampicin, isoniazid, and pyrazinamid encapsulated in poly (DL-lactide-co-glycolide) nanoparticles. These nanosystems are suitable for nebulization and are prepared by the use of the multiple emulsion techniques.

Different drugs were examined for pharmacokinetic characteristics and chemotherapeutic potentials after nebulization in a model of *M. tuberculosis*-infected guinea pigs. It was shown that the aerosolized particles exhibit a mass median aerodynamic diameter of 1.88±0.11 μm that is known to be favorable for a bronchoalveolar deposition (Groneberg et al., 2003). A single nebulization to guinea pigs was found to lead to sustained therapeutic levels in the lungs for up to 11 days and in the plasma for 6–8 days. Nanoparticle-encapsulation also led to a significantly prolonged mean residence time and elimination half-life of the different drugs when compared to the orally administered conventional drugs formulations. A comparison to the oral route of administration resulted in an enhanced relative bioavailability for the aerosolically delivered nanoparticle-encapsulated preparations (rifampicin

12.7-fold, pyrazinamide 14.7-fold, and isoniazid 32.8-fold). In addition to these beneficial features, the absolute bioavailability in comparison to intravenous administration was increased (6.5-time for rifampicin, 13.4-time for pyrazinamide, and 19.1-time for isoniazid).

Nebulization of the nanoparticle-encoated drug in the *M. tuberculosis* infection model on every 10th day led to an absence of *M. tuberculosis* in the lungs of the infected animals after five treatment doses. By contrast, 46 daily doses of orally administered drug were needed to obtain similar therapeutic benefits. Together these studies demonstrated that an aerosolic administration of nanoparticle-encoated antitubercular drugs can be used experimentally to improve pharmacological management of pulmonary tuberculosis.

In another embodiment, subcutaneous delivery of a nanoparticle-based system using three anti-tuberculosis drugs: isoniazid, rifampin, and streptomycin. They studied *in vitro* the accumulation of these drugs in human monocytes and their anti-microbial activity against *Mycobacterium tuberculosis* residing in human monocyte-derived macrophages. Their results showed that nanoparticle encapsulation increased the intracellular accumulation of all three tested drugs, but only the anti-microbial activity of isoniazid and streptomycin was increased. Also in their study, they showed that the activity of encapsulated rifampin against intracellular bacteria was not higher than that of the free drug (Anisimova et al., 2000).

In another study, efficiency of oral encapsulated ethambutol in combination with PLG nanoparticles loaded with rifampicin, isoniazid, and pyrazinamide in a murine tuberculosis model. The study concluded that polymeric nanoparticles using a combination of 4-drugs have a significant potential to shorten the duration of tuberculosis chemotherapy besides reducing the dosing frequency (Pandey et al., 2006).

These studies showed that oral delivery of anti-tuberculosis drugs incorporated in nanoparticles might be a feasible alternative to conventional oral drug delivery to achieve better patient compliance. This is due to the decreased dosing frequency. However, it is still not known why certain drugs did not have an increased therapeutic effect when delivered to macrophages via nanoparticulate carriers even if their local concentration increased. The promising results in the tuberculosis management using nanoparticles might be due to different effects. It is possible that the constant drug plasma levels are more effective than fluctuating drug plasma levels after oral administration of free drugs, or even a small nanoparticle accumulation in the lungs might cause an effective increase in drug concentration locally. Combined with a constant controlled drug release such nearly undetectable drug concentration increases might be the key for an improved drug delivery.

More mechanistic studies are needed to gain a better understanding of the improved drug therapy in the treatment of tuberculosis.

In another study, HLA-A2 transgenic mouse model to investigate the effects of pulmonary delivery of a new DNA plasmid encoding eight HLA-A*0201-restricted T-cell epitopes from *M. tuberculosis* formulated in chemically modified or unmodified chitosan nanoparticles (CINs). They have shown that respiratory administration of the DNA plasmid incorporated in CINs induced increased levels of IFN-γ secretion. Maturation of dendritic cells was also observed when compared with respiratory delivery of plasmid in solution and the more frequently used intramuscular immunization route (Bivas-Bentia et al., 2004), in line with observations of other groups (Eyles et al., 1998; Lagranderie et al., 1993). The studies summarized above indicate the advancement of research in the area on nanoparticulate gene delivery systems. The success of polymer compacted DNA in the treatment of asthma has also been reported for CF, a disease where gene therapy has been explored over a decade ago.

3.8 NANOMEDICINE FOR RESPIRATORY VIRAL INFECTIONS

Nanomedicine may also be used to stimulate long-lasting and protective immune responses to respiratory viral infections. A number of studies already assessed this issue experimentally, i.e., by using polylactic acid nanoparticles coated with a hydrophilic polyethylene glycol coating and tetanus toxoid (Vila et al., 2004) or chitosan-DNA nanospheres containing a cocktail of plasmid DNAs encoding respiratory syncytial virus (RSV) antigens (Kumar et al., 2002). In mice, a significant reduction of viral antigen load and viral titers after an acute RSV infection were found. Treatment also resulted in the induction of RSV-specific nasal IgA antibodies, cytotoxic T-lymphocytes, IgG antibodies, and interferon-gamma production suggesting a potential use of these nanocarriers in the development of new RSV vaccination strategies (Kumar et al., 2002). Nanoparticles may also be used to induce immunity to para-influenza virus infections. Proteins of the bovine para-influenza type 3 virus (PI–3) were incorporated into the two nanoparticles PLGA and polymethylmethacrylate (PMMA)[128], and it was demonstrated that mice immunized with the bovine PI–3 protein-containing PLGA nanoparticles had higher levels of virus-specific antibodies (Shephard et al., 2003).

Respiratory syncytial virus has been recognized as the leading cause of severe bronchiolitis and pneumonia in infants worldwide and also as resulting in lower respiratory tract infections in immune deficient and elderly adults (Kong et al., 2007; Xie et al., 2007). Natural immunity to RSV is incomplete, and infection recurs throughout life (Xie et al., 2007). In another report, the prophylactic effects of short interfering non structural proteins (siNS1) construct in preventing RSV infection in rats, have illustrated that the siNS1 treatment reduced RSV titers significantly, and prevented the accompanying lung damage and airway hyper-reactivity when *p*DNAs expressing anti-NS1 RNA or an unrelated sequence were complexed with CINs and instilled intranasally 1 day prior to intranasal infection with RSV (Kong et al., 2007). Similar results in a murine model confirmed an effective prophylactic effect of a mucosal gene expression vaccine (GXV) made up of a cocktail of at least four different *p*DNAs encoding corresponding RSV antigens, coacervated with chitosan to formulate nanospheres (Shahiwala et al., 2007; Mohapatra et al., 2003). The intranasal administration with GXV resulted in significant induction of RSV specific antibodies, nasal IgA antibodies, cytotoxic T-lymphocytes and IFN-γ production in the lung and splenocytes (Mohapatra et al., 2003).

A similar effect in the reduction of allergic response was reported in ovalbulim-sensitized mice that were administered chitosan/pIFN-γ nanoparticles prior to an asthma-inducing challenge (Dang and Leong, 2006). Subsequent to prophylactic treatment with chitosan/pIFN-γ particles, splenocytes collected from treated mice exhibited increased secretion of IFN-γ and decreased secretion of IL–5 and IL–4 (Dang and Leong, 2006). One researcher found that monovalent influenza subunit vaccine-loaded *N*-trimethyl chitosan (TMC) nanoparticles were an effective carrier system for nasal delivery (Amidi et al., 2007). Furthermore, they observed that the immune responses elicited by the antigen-loaded TMC nanoparticles were likely attributed to cellular uptake of the nanoparticles in the nasal epithelium and nasal-associated lymphoid tissues (NALT) and subsequent access of the vaccine to sub-mucosal lymphoid tissues (Amidi et al., 2007).

3.9   NANOMEDICINE FOR LUNG CANCER

Cancer gene therapy for the treatment of lung cancer has recently been demonstrated to have beneficial effects in experimental and in preclinical trials (Gopalan et al., 2004). The lung cancer gene therapy is currently limited to treating localized tumors since there is a host-immunity response against the gene delivery vector and the transgene. In this respect, studies are currently performed to develop novel gene delivery vectors that are non-immunogenic (Gopalan et al., 2004). One attractive vehicle is the non-viral vector, N-[1- (2,3-dioleoyloxy) propyl]-N,N,N trimethylammonium chloride (DOTAP):cholesterol (DOTAP:Chol) nanoparticle that has been shown to be an effective systemic gene delivery vectors in preclinical studies (Gopalan et al., 2004). They used small molecule inhibitors against the signaling molecules such as naproxen and showed that these small molecules could suppress nanoparticle-mediated inflammation without affecting transgene expression. Their results might be of clinical significance both in terms of suppressing toxicity, as well as, increasing the therapeutic window.

In another study, the possibility of gelatin nanoparticles as plasmid DNA delivery system on Lewis lung carcinoma (LLC) bearing mice models (Kaul et al., 2005). They encapsulated reporter plasmid DNA encoding for β-galactosidase (pCMV-β) in gelatin and PEGylated gelatin nanoparticles. They showed that PEGylated gelatin nanoparticles are superior transfection reagents compared to gelatin nanoparticles and lipofectin. Also, the *in vivo* expression of β-galactosidase in tumor mass showed that PEGylated gelatin nanoparticles could transfect with 61% efficiency after i.v., administration relative to the intratumoral administration. They attributed the high transfection efficiency of PEGylated gelatin nanoparticles to the biocompatible, biodegradable, and long circulating nature of the carrier system. Nanoparticles do not complex DNA molecules and preserve their supercoiled structure which is critical for nuclear uptake and efficient transfection. Fink et al. (2006) showed that nanoparticles consisting of a single molecule of DNA condensed with polyethylene glycol-substituted lysine–30-meters efficiently transfected the lung epithelium following intrapulmonary administration into mice. One researcher developed a ligand-targeted and sterically stabilized nanoparticle formulation for the targeted delivery of antisense oligodeoxynucleotides and small interference RNA into lung cancer cells (Li and Huang, 2006). Their results showed that the ligand-targeted and sterically stabilized nanoparticles could provide a selective delivery of antisense oligodeoxynucleotides and siRNA into lung cancer cells which might be used for cancer therapy. These studies show that nanoparticles have the potential to be used as carrier for safe and effective gene delivery to treat certain respiratory diseases in the future.

In another study, a multicentre phase II clinical trial showed that Abraxane®, a novel cremophore free, albumin-bound paclitaxel nanoparticle formulation, showed encouraging results in regard to efficacy and safety in patients with non-small-cell lung cancer (NSCLC) (Green et al., 2006). A significant tumor response and prolonged disease control was documented in forty-three patients. This study shows the potential that nanoparticles may have as carriers of chemotherapeutics to treat tumors and lung cancer in particular. It should be mentioned that Abraxane® nanoparticles are suspected to dissolve shortly after administration; therefore, it is not known if the nanoparticles simply improve paclitaxel injectability and dissolution or if they contribute to the observed therapeutic efficacy in other ways (e.g., affecting drug biodistribution). More clinical research involving different nanoparticles and drugs is needed to assess if nanoparticles can improve drug therapy especially for cytotoxic drugs; however, preclinical studies suggest that nanoparticles might revolutionize chemotherapy.

Further, using magnetic nanoparticles, either for diagnostic or treatment purposes, was the center of interest during the last two decades. However, there were only two articles published on the specific delivery of nanoparticles to the lungs. In one of the first studies, Mykhaylyk et al. (2005) evaluated the pharmacokinetics of doxorubicin magnetic conjugate (DOX-M) nanoparticles in a mouse model. They investigated the efficiency of a non-uniform magnetic field on the clearance of the magnetic DOX-M. In this study, they injected DOX-M suspensions into the eye sinus vein of adult male mice, and applied a magnetic field centered over the left lung. They showed that a non-uniform magnetic field was a potent factor in modifying the DOX-M conjugate pharmacokinetics. The magnetic field application resulted in considerable enrichment of DOX-M in the lungs, and a depletion in the liver of the magnetic carrier compared to a reference without a magnetic field. They showed that the application of a magnetic field could significantly increase the bioavailability of DOX-M in the lungs. Although their work in mice showed some promising results, the outcome of the application of magnetic fields in humans for increasing the localization of a drug in the lungs containing magnetic nanoparticles has not yet been proven. Contrary to the results from the previous study, Wu et al. (2007) showed that an external magnetic field applied to rats after intravenous injection of dextran-coated Fe_3O_4 did not change the accumulation of the nanoparticles in the lungs. Generally, the use of magnetic nanocarriers has merit for diagnostic or treatment purposes. The delivery of magnetic nanoparticles to the lungs might be worth more detailed research to be used as effective drug delivery system or as a safe diagnostic tool.

The development of monoclonal antibodies and utilization of their targeting properties (Kohler et al., 1975) can be used for active targeting. This can be used to improve drug delivery by attaching antibodies to drug molecules or drug delivery systems (Garnett, 2001; Funaro et al., 2000; Trail and Bianchi, 1999; Trail et al., 2003; Lambert, 2005). In an attempt to target lung tumors using antibody modified nanoparticles, injected bovine serum albumin (BSA)-conjugated with Lewis lung carcinoma monoclonal antibodies to Lewis lung carcinoma-bearing mice (Akasaka et al., 1988). They showed that nanoparticles made from the BSA-conjugate with monoclonal antibodies were only slightly localized in the carcinoma tissue. 24 hr after injection the amount of the nanoparticles localized in the carcinoma tissue was rather low. This study, however, showed that the particle size was more important than the affinity of the monoclonal antibodies to the tumor cells. Although the study showed some promising data for targeted delivery of nanoparticles to lung tumors, there is still room for improvements.

Another method for the delivery of nanoparticles was spraying or nebu-lization of a nanoparticle suspension using a nebulizer. In this study, one researcher introduced a novel surfactant free biodegradable nanoparticle system for aerosol therapy (Daily et al., 2003). They formulated nanopar-ticle suspensions from a branched polyester, diethylaminopropyl amine-poly (vinyl alcohol)-grafted-poly (lactide-co-glycolide) (DEAPA-PVAL-g-PLGA), as well as with increasing amounts of carboxymethyl cellulose. They showed that this new polymer has high encapsulation efficiency for drug molecules by utilizing electrostatic interactions. They claimed that using these nanoparticles "alveolar deposition can be easily achieved by either jet or ultrasonic nebulization of the nanoparticle suspensions." They also showed that not only polymer hydrophilicity was necessary to maintain stability during nebulization, but also the formation of well-defined nanoparticles is an important factor. In their study they showed that formulations containing free DEAPE-PVAL-g-PLGA tend to aggregate and, therefore, only anionic formulations will be suitable for nebulization. Also, they showed that a critical amount of carboxymethyl cellulose is needed to prevent particles from agglomeration.

In another study, one researcher evaluated the role of lymphatic drainage in the uptake of inhaled solid lipid nanoparticles (SLN) (Videira et al., 2002). They studied the biodistribution of SLNs (200 nm) following the aerosolization of a ^{99m}Tc-SLN suspension in a group of rats. Their study showed an important and significant uptake of the radio-labeled SLN into the lymphatic system after inhalation, and a high rate of distribution in periaortic, axillary, and inguinal lymph nodes. Nanoparticle accumulation in the regional lymph nodes suggests that the translocation mechanism of SLN may involve phagocytosis by macrophages followed by migration to the lymphatic system (Videira et al., 2002). This study showed that inhala-tion could be an effective route to deliver radio-labeled SLN to the lungs, representing an alternative to the intraperitoneal route for targeting colloidal carriers to the lymphatics. This technology may provide the possibility of using radio-labeled SLN as a lymphoscintigraphic agent. Additionally, it can allow the delivery of cytotoxic drugs to lung cancers in different stages.

Nanoparticles have widely been studied in drug delivery research for targeting and controlled release. To deposit effectively deep the particles in the lungs, the PLGA nanoparticles loaded with the anticancer drug 6-{[2-(dimethylamino)ethyl]amino}–3-hydroxyl–7H-indeno[2,1-c] quinolin–7-one dihydrochloride (TAS–103) were prepared in the form of nanocomposite particles. The nanocomposite particles consist of the complex of drug-loaded nanoparticles and excipients. In this study, the anticancer effects of the nanocomposite particles against the lung cancer cell line A549. Also, the concentration of TAS–103 in blood and lungs were determined after administration of the nanocomposite particles by inhalation to rats. TAS–103-loaded PLGA nanoparticles were prepared with 5% and 10% of loading ratio by spray drying method with trehalose as an excipient. The 5% drug-loaded nanocomposite particles were more suitable for inhalable agent because of the sustained release of TAS–103 and higher FPF value. Cytotoxicity of nanocomposite particles against A549 cells was higher than that of free drug. When the nanocomposite particles were administered in rats by inhalation, drug concentration in lung was much higher than that in plasma. Furthermore, drug concentration in lungs administered by inhalation of nanocomposite particles was much higher than that after intravenous administration of free drug. From these results, the nanocomposite particle systems could be promising for treatment of lung cancer (Tomoda et al., 2009).

3.10   CONCLUSION AND FUTURE CHALLENGES

Respiratory drug delivery is becoming more and more important. This is due to the specific physiological environment of the lung as an absorption and treatment organ. Concepts of nanomedicine offer numerous novel therapeutic options in pharmacotherapy. Nanomedicine based on different materials and functions have been developed and applied in the therapeutic treatment of respiratory diseases in the last two decades. The utilization of polymer, lipid or metal-based nanoparticle systems in the field of targeted gene delivery has grown tremendously and showed promising *in vitro* or *in vivo* experimental results in therapeutic efficacy. In the treatment of respiratory diseases, nanoparticles carrying gene molecules showed high transfection efficiency and targeting ability on lung cancer tumors through systematic or localized administrations. In addition, therapeutic efficiency of gene therapy can be improved by active targeting on specific lung cancer tumors or metastases through modification or conjugation of targeting agents on the surface of the nanoparticles. However, translating these novel nanoparticle-mediated gene delivery techniques into clinical practice is a huge challenge. The challenges of applying nanoparticle-mediated gene delivery within the body, such as maintaining the stability of nanoparticles and gene molecules during delivery, controlling the bio-distribution and pharmacokinetics, penetrating biological barriers and minimizing the potential cytotoxicity of the nanoparticles, need to be considered and overcome before entering into clinical trials. To expand the application of nanoparticle systems in gene therapy in clinics, standards in the examination of nanoparticle safety and evaluation of therapeutic efficacy should be established to guide the direction of research and intervention in gene therapy using nanoparticles. However, the gap between the concepts of nanomedicine and the published experimental data and the clinical reality is huge. Future experimental and clinical studies have to reveal precisely the diagnostic and therapeutic potential of nanomedicine.

KEYWORDS

- **asthma**
- **lung infections**
- **nanomedicine**
- **nanotechnology**
- **respiratory diseases**

REFERENCES

Akasaka, Y., et al., (1988). Preparation and evaluation of bovine serum albumin nanospheres coated with monoclonal antibodies. *Drug Des. Deliv.*, *3*, 85–97.

Almeida, A. J., & Souto, E., (2007). Solid lipid nanoparticles as a drug delivery system for peptides and proteins. *Adv. Drug Deliv. Rev.*, *59*, 478–490.

Amidi, M., et al., (2007). *N*-Trimethyl chitosan (TMC) nanoparticles loaded with influenza subunit antigen for intranasal vaccination: Biological properties and immunogenicity in a mouse model. *Vaccine*, *25*, 144–153.

Anisimova, Y. V., et al., (2000). Nanoparticles as anti-tuberculosis drugs carriers: Effect on activity against Mycobacterium tuberculosis in human monocyte-derived macrophages. *J. Nanopart. Res.*, *2*, 165–171.

Armitage, P., & Doll, R., (1954). The age distribution of cancer and multistage theory of carcinogenesis. *Br. J. Cancer.*, *8*, 1–12.

Asthma. World Health Organization. Archived from the original on June 29, 2011. Retrieved 2016-03-29.

Badesch, D. B., et al., (2010). Pulmonary arterial hypertension: Baseline characteristics from the REVEAL Registry. *Chest*, *137*, 376–387.

Bailey, M. M., & Berkland, C. J., (2009). Nanoparticle formulations in pulmonary drug delivery. *Med. Res. Rev.*, *29*, 196–212.

Bani-Sadr, F., Carrat, F., & Pol, S., (2005). Risk factors for symptomatic mitochondrial toxicity in HIV/hepatitis C virus-co-infected patients during interferon plus ribavirin-based therapy. *J. Acquir. Immune Defic. Syndr.*, *40*, 47–52.

Beck-Broichsitter, et al., (2009). Pulmonary drug delivery with aerosolizable nanoparticles in an *ex vivo* lung model. *Int. J. Pharm.*, *367*, 169–178.

Bivas-Benita, M., et al., (2004). Cationic submicron emulsions for pulmonary DNA immunization. *J. Cont. Rel.*, *100*, 145–155.

Bivas-Benita, M., et al., (2004). Pulmonary delivery of chitosan-DNA nanoparticles enhances the immunogenicity of a DNA vaccine encoding HLA-A*0201-restricted T-cell epitopes of *Mycobacterium tuberculosis*. *Vaccine*, *22*, 1609–1615.

Boat, T. F., & Cheng, P. W., (1980). Biochemistry of airway mucus secretions. *Fed. Proc.*, *39*, 3067–3074.

Bonini, S., et al., (2007). Practical approach to diagnosis and treatment of ocular allergy: A 1-year systematic review. *Curr. Opin. Allergy and Clin. Immunol.*, *7*, 446–449.

Bousquet, J., et al., (2008). Allergic rhinitis and its impact on asthma (ARIA) 2008 update (in collaboration with the World Health Organization, GA2LEN, and Allergen). *Allergy*, *63*, 8–160.

Brzoska, M., et al., (2004). Incorporation of biodegradable nanoparticles into human airway epithelium cells-in vitro study of the suitability as a vehicle for drug or gene delivery in pulmonary diseases. *Biochem. Biophys. Res. Commun.*, *318*, 562–570.

Butler, J. C., et al., (1996). The continued emergence of drug-resistant *Streptococcus pneumoniae* in the United States: An update from the Centers for Disease Control and Prevention's pneumococcal sentinel surveillance system. *J. Infect. Dis.*, *174*, 986–993.

Cazzola, M., et al., (2013). Comorbidities of asthma: Current knowledge and future research needs. *Curr. Opin. Pulm. Med.*, *19*, 36–41.

Chace, K. V., Flux, M., & Sachdev, G. P., (1985). Comparison of physicochemical properties of purified mucus glycoproteins isolated from respiratory secretions of cystic fibrosis and asthmatic patients. *Biochem.*, *24*, 7334–7341.

Chen, J., et al., (2008). Galactose-poly(ethylene glycol)-polyethylenimine for improved lung gene transfer. *Biochem. Biophys. Res. Commun.*, *375*, 378–383.

Chougule, M. B., Padhi, B. K., & Misra, A., (2006). Nano-liposomal dry powder inhaler of amiloride hydrochloride. *J. Nanosci. Nanotechnol.*, *6*, 3001–3009.

Chung, K. F., Caramori, G., & Groneberg, D. A., (2004). Airway obstruction in chronic obstructive pulmonary disease. *N. Engl. J. Med.*, *351*, 1459–1461.

Clark, A., (2002). Formulation of proteins and peptides for inhalation. *Drug Deliv. Syst. Sci.*, *2*, 73–77.

Courrier, H. M., Butz, N., & Vandamme, T. F., (2002). Pulmonary drug delivery systems: Recent developments and prospects. *Crit. Rev. Ther. Drug Carr. Syst.*, *19*, 425–498.

Craparo, E. F., et al., (2016). Pegylated polyaspartamide-polylactide-based nanoparticles penetrating cystic fibrosis artificial mucus. *Biomacromol.*, *17*, 767–777.

Cruz, D. D., Tanoue, L. T. C. S., & Matthay, R. A., (2011). Lung cancer: Epidemiology, etiology, and prevention. *Clin. Chest Med.*, *32*, 605–644.

Cryan, S. A., (2005). Carrier-based strategies for targeting protein and peptide drugs to the lungs. *AAPS J.*, *7*, E20–E41.

Dailey, L. A., et al., (2003). Surfactant-free, biodegradable nanoparticles for aerosol therapy based on the branched polyesters, DEAPA-PVAL-g-PLGA. *Pharm. Res.*, *20*, 2011–2020.

Dang, J. M., & Leong, K. W., (2006). Natural polymers for gene delivery and tissue engineering. *Adv. Drug Deliv. Rev.*, *58*, 487–499.

Davies, J. R., et al., (1996). Mucins in airway secretions from healthy and chronic bronchitic subjects. *Biochem. J.*, *313*, 431–439.

Du Toit, L. C., Pillay, V., & Danckwerts, M. P., (2006). Tuberculosis chemotherapy: Current drug delivery approaches. *Respir. Res.*, *7*, 118.

Edwards, D. A., et al., (1997). Large porous particles for pulmonary drug delivery. *Science*, *276*, 1868–1871.

Ely, L., et al., (2007). Effervescent dry powder for respiratory drug delivery. *Eur. J. Pharm. Biopharm.*, *65*, 346–353.

Enarson, D. A., & Chretien, J., (1999). Epidemiology of respiratory infectious diseases. *Curr. Opin. Pulm. Med.*, *5*, 128–135.

Engelhardt, J. F., et al., (1993). Submucosal glands are the predominant site of CFTR expression in the human bronchus. *Nat. Genet.*, *2*, 240–248.

Evora, C., et al., (1998). Relating the phagocytosis of microparticles by alveolar macrophages to surface chemistry: The effect of 1, 2-dipalmitoylphosphatidylcholine. *J. Cont. Rel.*, *51*, 143–152.

Eyles, J. E., et al., (1998). Analysis of local and systemic immunological responses after intra-tracheal, intra-nasal and intra-muscular administration of microsphere co-encapsulated *Yersinia pestis* sub-unit vaccines. *Vaccine*, *16*, 2000–2009.

Fink, T. L., et al., (2006). Plasmid size up to 20 kbp does not limit effective in vivo lung gene transfer using compacted DNA nanoparticles. *Gene Ther.*, *13*, 1048–1051.

Funaro, A., et al., (2000). Monoclonal antibodies and therapy of human cancers. *Biotechnol. Adv.*, *18*, 385–401.

Garnett, M. C., (2001). Targeted drug conjugates: Principles and progress. *Adv. Drug Deliv. Rev.*, *53*, 171–216.

Gautam, A., Densmore, C. L., & Waldrep, J. C., (2000). Inhibition of experimental lung metastasis by aerosol delivery of PEI-p53 complexes. *Mol. Ther.*, *2*, 318–323.

Gautam, A., et al., (2002). Aerosol delivery of PEI-p53 complexes inhibits B16-F10 lung metastases through regulation of angiogenesis. *Cancer Gene Ther.*, *9*, 28–36.

Gill, S., et al., (2007). Nanoparticles: Characteristics, mechanisms of action and toxicity in pulmonary drug delivery-a review. *J. Biomed. Nanotechnol.*, *3*, 107–119.

Global Tuberculosis Report, (2016). WHO. http://www.who.int/tb/publications/global report/en/. (accessed Dec 12, 2016).

Gonda, I., et al., (1998). Inhalation delivery systems with compliance and disease management capabilities. *J. Cont. Rel.*, *53*, 269–274.

Gong, F., et al., (2005). Gene transfer of vascular endothelial growth factor reduces bleomycin-induced pulmonary hypertension in immature rabbits. *Pediatr. Int.*, *47*, 242–247.

Gopalan, B., et al., (2004). Nanoparticle-based systemic gene therapy for lung cancer: Molecular mechanisms and strategies to suppress nanoparticle-mediated inflammatory response. *Technol. Cancer Res. Treat.*, *3*, 647–657.

Green, M. R., et al., (2006). Abraxane®, a novel Cremophor®-free, albumin-bound particle form of paclitaxel for the treatment of advanced non-small cell lung cancer. *Ann. Oncol.*, *17*, 1263–1268.

Grenha, A., Seijo, B., & Remunan-Lopez, C., (2005). Microencapsulated chitosan nanoparticles for lung protein delivery. *Eur. J. Pharm. Sci.*, *25*, 427–437.

Groneberg, D. A., et al., (2002). Expression of respiratory mucins in fatal status asthmaticus and mild asthma. *Histopathology*, *40*, 367–373.

Groneberg, D. A., et al., (2003). Distribution of respiratory mucin proteins in human nasal mucosa. *Laryngoscope*, *113*, 520–524.

Gupta, R., & Jentoft, N., (1992). The structure of tracheobronchial mucins from cystic fibrosis and control patients. *J. Biol. Chem.*, *267*, 3160–3167.

Howard, K. A., et al., (2006). RNA interference in vitro and in vivo using a novel chitosan/siRNA nanoparticle system. *Mol Ther.*, *14*, 476–484.

Huang, Y. Y., & Wang, C. H., (2006). Pulmonary delivery of insulin by liposomal carriers. *J. Cont. Rel.*, *113*, 9–14.

Hunt, S. A., et al., (2001). ACC/AHA Guidelines for the evaluation and management of chronic heart failure in the adult: Executive summary a report of the American College of Cardiology/American Heart Association Taskforce on practice guidelines (Committee to revise the 1995 guidelines for the evaluation and management of heart failure): Developed in collaboration with the International Society for Heart and Lung Transplantation, endorsed by the Heart Failure Society of America. *Circulation*, *104*, 2996–3007.

Hwang, S. K., et al., (2007). Aerosol-delivered programmed cell death four enhanced apoptosis, controlled cell cycle and suppressed AP–1 activity in the lungs of AP–1 luciferase reporter mice. *Gene Ther.*, *14*, 1353–1361.

Hyduk, A., et al., (2005). Pulmonary hypertension surveillance-United States, 1980–2002. *MMWR Surveill Summ.*, *54*, 1–28.

Issa, M. M., et al., (2006). Targeted gene delivery with trisaccharide-substituted chitosan oligomers in vitro and after lung administration *in vivo. J. Cont. Rel.*, *115*, 103–112.

Jarvis, D., et al., (2012). Asthma in adults and its association with chronic rhinosinusitis: the GA2LEN survey in Europe. *Allergy*, *67*, 91–98.

Jiang, L., et al., (2014). Molecular characterization of redox mechanisms in allergic asthma. *Ann. Allergy Asthma Immunol.*, *113*, 137–142.

John, A. E., et al., (2003). Discovery of a potent nanoparticle P-selectin antagonist with anti-inflammatory effects in allergic airway disease. *FASEB J.*, *17*, 2296–2298.

Kaipel, M., et al., (2008). Increased biological half-life of aerosolized liposomal recombinant human Cu/Zn superoxide dismutase in pigs. *J. Aerosol Med. Pulm. Drug Deliv.*, *21*, 281–290.

Kariyawasam, H. H., & Rotiroti, G., (2013). Allergic rhinitis, chronic, and asthma: Unraveling a complex relationship. *Curr. Opin. Otolaryngology. Head and Neck Surg.*, *21*, 79–86.

Kaul, G., & Amiji, M., (2005). Tumor-targeted gene delivery using poly(ethylene glycol)-modified gelatin nanoparticles: *In vitro* and *in vivo* studies. *Pharm. Res.*, *22*, 951–961.

Kimura, S., et al., (2009). Nanoparticle-mediated delivery of nuclear factor {kappa}B decoy into lungs ameliorates monocrotaline-induced pulmonary arterial hypertension. *Hypertension*, *53*, 877–883.

Kohler, G., & Milstein, C., (1975). Continuous cultures of fused cells secreting antibody of predefined specificity. *Nature*, *256*, 495–497.

Kolek, V., Kasak, V., & Vasakova, M., (2014). *Pneumologie* (2nd edn.). Prague: Maxdorf.

Konduri, K. S., et al., (2003). Efficacy of liposomal budesonide in experimental asthma. *J. Allergy Clin. Immunol.*, *111*, 321–327.

Kong, X., et al., (2007). Respiratory syncytial virus infection in Fischer 344 rats is attenuated by short interfering RNA against the RSV-NS1 gene. *Genet Vaccines Ther.*, *5*, 4.

Koping-Hoggard, M., et al., (2004). Improved chitosan mediated gene delivery based on easily dissociated chitosan polyplexes of highly defined chitosan oligomers. *Gene Ther.*, *11*, 1441–1452.

Koshkina, N. V., et al., (2003). Biodistribution and pharmacokinetics of aerosol and intravenously administered DNA-polyethyleneimine complexes: Optimization of pulmonary delivery and retention. *Mol. Ther.*, *8*, 249–254.

Kumar, M., et al., (2002). Intranasal gene transfer by chitosan-DNA nanospheres protects BALB/c mice against acute respiratory syncytial virus infection. *Hum. Gene Ther.*, *13*, 1415–1425.

Kumar, M., et al., (2003). Chitosan IFN-gamma-pDNA nanoparticle (CIN) therapy for allergic asthma. Genet. *Vaccines Ther.*, *1*, 3.

Kumaresh, S., et al., (2000). Biomaterials in drug delivery and tissue engineering: One laboratory's experience. *Acc. Chem. Res.*, *33*, 94–101.

Kunin, C. M., (1993). Resistance to antimicrobial drugs-a worldwide calamity. *Ann. Intern. Med.*, *118*, 557–561.

Lagranderie, M., Ravisse, P., & Marchal, G., (1993). BCG-induced protection in guinea pigs vaccinated and challenged via the respiratory route. *Tuber. Lung Dis.*, *74*, 38–46.

Lambert, J. M., (2005). Drug-conjugated monoclonal antibodies for treatment of cancer. *Curr. Opin. Pharmacol.*, *5*, 543–549.

Li, D., et al., (1997). Localization and upregulation of mucin (MUC2) gene expression in human nasal biopsies of patients with cystic fibrosis. *J. Pathol.*, *181*, 305–310.

Li, S. D., & Huang, L., (2006). Targeted delivery of antisense oligodeoxynucleotide and small interference RNA into lung cancer cells. *Mol. Pharm.*, *3*, 579–588.

Lipsitch, M., & Samore, M. H., (2002). Antimicrobial use and antimicrobial resistance: A population perspective. *Emerg. Infect. Dis.*, *8*, 347–354.

Lynch, J., Behan, N., & Birkinshaw, C., (2007). Factors controlling particle size during nebulization of DNA-polycation complexes. *J. Aerosol Med.*, *20*, 257–268.

Mohapatra, S. S., et al., (2003). *Gene Expression Vaccine*. U.S. Patents 2003068333.

Mykhaylyk, O., Dudchenko, N., & Dudchenko, A., (2005). Doxorubicin magnetic conjugate targeting upon intravenous injection into mice: High gradient magnetic field inhibits the clearance of nanoparticles from the blood. *J. Magn. Mater.*, *293*, 473–482.

Nguyen, J., et al., (2008). Effects of cell-penetrating peptides and pegylation on transfection efficiency of polyethyleneimine in mouse lungs. *J. Gene Med.*, *10*, 1236–1246.

Niven, R. W., (1995). Delivery of biotherapeutics by inhalation aerosol. *Crit. Rev. Ther. Drug Carr. Syst.*, *12*, 151–231.

Pandey, R., & Khuller, G. K., (2005). Solid lipid particle based inhalable sustained drug delivery against experimental tuberculosis. *Tuberculosis*, *85*, 227–234.

Pandey, R., et al., (2003). Poly (DL-lactide-co-glycolide) nanoparticle-based inhalable sustained drug delivery system for experimental tuberculosis. *J. Antimicrob. Chemother.*, *52*, 981–986.

Pandey, R., Sharma, S., & Khuller, G. K., (2006). Chemotherapeutic efficacy of nanoparticle-encapsulated antitubercular drugs. *Drug Deliv.*, *13*, 287–294.

Park, J. H., et al., (2008). Polymeric nanomedicine for cancer therapy. *Prog. Polym. Sci.*, *33*, 113–137.

Parkin, D. M., et al., (2005). Global cancer statistics, 2002. *CA Cancer J. Clin.*, *55*, 74–108.

Patel, R. R., et al., (2015). Rationally developed core-shell polymeric-lipid hybrid nanoparticles as a delivery vehicle for cromolyn sodium: Implications of lipid envelop on *in vitro* and *in vivo* behavior of nanoparticles upon oral administration. *RSC Adv.*, *5*, 76491–76506.

Patel, R. R., et al., (2016). Cromolyn sodium encapsulated PLGA nanoparticles: An attempt to improve intestinal permeation. *Int. J. Bio. Macromol.*, *83*, 249–258.

Patel, R. R., (2016). Highly water-soluble mast cell stabilizer-encapsulated solid lipid nanoparticles with enhanced oral bioavailability. *J. Micriencap.*, 33, 209-220.

Patton, J. S., & Byron, P. R., (2007). Inhaling medicines: Delivering drugs to the body through the lungs. *Nat. Rev. Drug Discov.*, *6*, 67–74.

Patton, J. S., & Platz, R. M., (1992). Pulmonary delivery of peptides and proteins for systemic action. *Adv. Drug Deliv. Rev.*, *8*, 179–196.

Penaloza, D., & Arias-Stella, J., (2007). The heart and pulmonary circulation at high altitudes: Healthy highlanders and chronic mountain sickness. *Circulation*, *115*, 1132–1146.

Petrini, B., & Hoffner, S., (1999). Drug-resistant and multi-drug resistant tubercle bacilli. *Int. J. Antimicrob. Agents*, *13*, 93–97.

Plumley, C., et al., (2009). Nifedipine nanoparticle agglomeration as a dry powder aerosol formulation strategy. *Int. J. Pharm.*, *369*, 136–143.

Prabakaran, D., et al., (2004). Osmotically regulated asymmetric capsular systems for simultaneous sustained delivery of anti-tubercular drugs. *J. Cont. Rel.*, *95*, 239–248.

Preston, D. L., et al., (2007). Solid cancer incidence in atomic bomb survivors: 1958–1998. *Radiat. Res.*, *168*, 1–64.

Quattrin, T., (2004). Inhaled insulin: Recent advances in the therapy of Type 1 and 2 diabetes. *Exp. Opin. Pharmacother.*, *5*, 2597–2604.

Ramsey, B. W., (1996). Management of pulmonary disease in patients with cystic fibrosis. *N. Engl. J. Med.*, *335*, 179–188.

Respiratory Tract Infections, (2009). http://www.healthinsite.gov.au/topics/ "Respiratory Tract Diseases."

Roger, V. L., et al., (2004). Trends in heart failure incidence and survival in a community-based population. *JAMA*, *292*, 344–350.

Rom, W. N., & Markowitz, S., (2007). *Environmental and Occupational Medicine* (4th edn.), Lippincott Williams & Wilkins Philadelphia, PA.

Rosenstein, B. J., & Zeitlin, P. L., (1998). Cystic fibrosis. *Lancet*, *351*, 277–282.

Roussel, P., Lamblin, G., & Degand, P., (1975). Heterogeneity of the carbohydrate chains of sulfated bronchial glycoproteins isolated from a patient suffering from cystic fibrosis. *J. Biol. Chem.*, *250*, 2114–2122.

Ruiz, M., et al., (1999). Etiology of community-acquired pneumonia: Impact of age, co-morbidity, and severity. *Am. J. Resp. Crit. Care. Med.*, *160*, 397–405.

Schmoljohann, D., (2006). Thermo- and pH-responsive polymers for drug delivery. *Adv. Drug Deliv. Rev.*, *58*, 1655–1670.

Shahiwala, A., & Misra, A., (2005). A preliminary pharmacokinetic study of liposomal leuprolide dry powder inhaler: A technical note. *AAPS Pharm Sci. Tech.*, *6*, E482–E486.

Shahiwala, A., Vyas, T. K., & Amiji, M. M., (2007). Nanocarriers for systemic and mucosal vaccine delivery. *Recent Pat. Drug Deliv. Formul.*, *1*, 1–19.

Sham, J. O., et al., (2004). Formulation and characterization of spray-dried powders containing nanoparticles for aerosol delivery to the lung. *Int. J. Pharm.*, *269*, 457–467.

Sharma, P., et al., (1998). MUC5B and MUC7 are differentially expressed in mucous and serous cells of submucosal glands in human bronchial airways. *Am. J. Respir. Cell Mol. Biol.*, *19*, 30–37.

Shephard, M. J., et al., (2003). Immunogenicity of bovine parainfluenza type 3 virus proteins encapsulated in nanoparticle vaccines, following intranasal administration to mice. *Res. Vet. Sci.*, *74*, 187–190.

Silverstein, M. D., et al., (1998). Trends in the incidence of deep vein thrombosis and pulmonary embolism: A 25-year population-based study. *Arch. Intern. Med.*, *158*, 585–593.

Singer, M., et al., (2004). A MARCKS-related peptide blocks mucus hypersecretion in a mouse model of asthma. *Nat. Med.*, *10*, 193–196.

Smith, R. A., & Kirkpatrick, W., (1980). *Ribavirin: Structure and Antiviral Activity Relationships Ribavirin: Abroad Spectrum Antiviral Agent* (pp. 1–21). New York, NY: Academic Press.

Song, W., et al., (2015). Substantial contribution of extrinsic risk factors to cancer development. *Nature*, *529*, 43–47.

Stern, R. C., (1997). The diagnosis of cystic fibrosis. *N. Engl. J. Med.*, *336*, 487–491.

Sung. J. C., Pulliam, B. L., & Edwards, D. A., (2007). Nanoparticles for drug delivery to the lungs. *Trends Biotechnol.*, *25*, 563–570.

Taylor, G., & Kellaway, I., (2001). Pulmonary drug delivery. In: Hillery, A., Lloyd, A., & Swarbrick, J., (eds.), *Drug Delivery and Targeting* (p. 269). Taylor & Francis: New York.

Thornton, D. J., et al., (1990). Mucus glycoproteins from 'normal' human tracheobronchial secretion. *Biochem. J.*, *265*, 179–186.

Thornton, D. J., et al., (1991). Mucus glycoproteins from cystic fibrotic sputum. Macromolecular properties and structural 'architecture.' *Biochem. J.*, *276*, 667–675.

Todd, G. R., et al., (2002). Survey of adrenal crisis associated with inhaled corticosteroids in the United Kingdom. *Arch. Dis. Child.*, *87*, 457–461.

Tomoda, K., et al., (2009). Preparation and properties of inhalable nanocomposite particles for treatment of lung cancer. *Colloids Surf. B Biointerf.*, *71*, 177–182.

Torchilin, V., (2006). Multifunctional nanocarriers. *Adv. Drug Deliv. Rev.*, *58*, 1523–1555.

Trail, P. A., & Bianchi, A. B., (1999). Monoclonal antibody drug conjugates in the treatment of cancer. *Curr. Opin. Immunol.*, *11*, 584–588.

Trail, P. A., King, H. D., & Dubowchik, G. M., (2003). Monoclonal antibody drug immunoconjugates. *Cancer Immunol. Immunother.*, *52*, 328–337.

Truong-Le, V. L., et al., (1999). Gene transfer by DNA-gelatin nanospheres. *Arch. Biochem. Biophys.*, *361*, 47–56.

Tsapis, N., et al., (2002). Trojan particles: Large porous carriers of nanoparticles for drug delivery. *Proc. Natl. Acad. Sci.*, *99*, 12001–12005.

Videira, M. A., et al., (2002). Lymphatic uptake of pulmonary delivered radiolabeled solid lipid nanoparticles. *J. Drug Target.*, *10*, 607–613.

Vila, A., et al., (2004). PEG-PLA nanoparticles as carriers for nasal vaccine delivery. *J. Aerosol Med.*, *17*, 174–185.

Vogelstein, B., & Kinzler, K. W., (2004). Cancer genes and the pathways they control. *Nat. Med.*, *10*, 789–799.

Wernig, K., et al., (2008). Depot formulation of vasoactive intestinal peptide by protamine-based biodegradable nanoparticles. *J. Cont. Rel.*, *130*, 192–198.

Wickstrom, C., et al., (1998). MUC5B is a major gel-forming, oligomeric mucin from human salivary gland, respiratory tract and endocervix: Identification of glycoforms and C-terminal cleavage. *Biochem. J.*, *334*, 685–693.

World Health Organization, (2002). WHO strategy for prevention and control of chronic respiratory diseases. Geneva: World Health Organization.

Wright, A., (2006). Emergence of *Mycobacterium tuberculosis* with extensive resistance to second-line drugs worldwide. *JAMA*, *295*, 2349–2351.

Wu, T., et al., (2007). Effects of external magnetic field on biodistribution of nanoparticles: A histological study. *J. Magn. Magn. Mater.*, *311*, 372–375.

Xie, C., et al., (2007). Oral respiratory syncytial virus (RSV), DNA vaccine, expressing RSV F protein delivered by attenuated *Salmonella typhimurium*. *Hum. Gene Ther.*, *18*, 746–752.

Yang, W., Peters, J. I., & Williams, R. O., (2008). Inhaled nanoparticles–a current review. *Int. J. Pharm.*, *356*, 239–247.

Zhang, Q., Shen, Z., & Nagai, T., (2001). Prolonged hypoglycemic effect of insulin-loaded polybutylcyanoacrylate nanoparticles after pulmonary administration to normal rats. *Int. J. Pharm.*, *218*, 75–80.

Zhang, Y., et al., (1995). Genotypic analysis of respiratory mucus sulfation defects in cystic fibrosis. *J. Clin. Invest.*, *96*, 2997–3004.

Ziady, A. G., et al., (2003). Minimal toxicity of stabilized compacted DNA nanoparticles in the murine lung. *Mol. Ther.*, *8*, 948–956.

Ziady, A. G., et al., (2003). Transfection of airway epithelium by stable PEGylated poly-L-lysine DNA nanoparticles in vivo. *Mol. Ther.*, *8*, 936–947.

CHAPTER 4

Nanomedicine for the Treatment of Neurological Disorders

SHRINGIKA SONI and BIKASH MEDHI[*]

Department of Pharmacology, Postgraduate Institute of Medical Education and Research (PGIMER), Chandigarh-160012, India

[*]Corresponding author. Prof. Bikash Medhi

ABSTRACT

The brain is the most dedicated part of our body and the highly complex structure of BBB, control the molecular movement of the compounds or other bioproducts from blood to the brain. The BBB not only protect the brain via regulating its homeostasis, but it also precludes drug delivery to the brain and blocks the therapies in several neurological disorders. The new molecular entities always go through harmful physiochemical and biopharmaceutical properties during large-scale development that may cause poor pharmacokinetics and distribution in the human. Therefore nanotechnology-based drug delivery system can consider as an intriguing tool to enhance drug transport in the brain without altering its therapeutic efficacy. In this chapter, the recently born nanotechnology-based drug delivery system and their application in neuronal disorders are described. This chapter also includes strategies for drug transport across the BBB, nanocarriers, and their structure modification, and globally marketed nanomedicine for CNS treatment. The main concept of this chapter is to summarize the potential application of nanotherapeutics in the treatment of CNS disorder and promote the research in this area.

4.1 INTRODUCTION

What if we could identify the very first signal of neurological disorder? What if some tiny machines enter in our body to treat the disease? And what if, targeted drug delivery can be enhanced without exceeding other side effects and toxicity?

Needless to say, diagnostic, therapeutic, and regenerative possibilities of nanotechnology have boosted our lives to one upper level in the 21st century. After the idea of nanomedicine for the first time by Robert Freitas in 1999, currently more than 200 companies, approximately 100 products and more than 10 billion dollars per year in R&D are all over the global market of nanomedicine (Yadav et al., 2012). People often confused with the term 'nanomedicine' and 'nanodrug.' Nanomedicine includes a complete set of nanotechnology-based devices, diagnostic tools, drugs, and nanoparticles used in imaging techniques. On the other hand, nanodrug is an effective therapeutic intervention of nanometer size for targeted drug delivery in the brain.

Nowadays, brain disorders are the major burden to the society, and several drug delivery strategies have been opted, but their invasiveness and less target specificity cause failure of conventional drug delivery methods. These trial and error methods are impossible for successful transport of therapeutic agents without altering physical and functional ability of brain. Blood-brain barrier (BBB) creates major hurdle in drug transport to the brain; along with binding of drug to nontransporters, surface activity of the compound, control of ion channel and enzymatic activity to convert the drug into nontherapeutic intermediate molecule; conventional drug delivery methods was utter failure till '90s (Alam et al., 2010; Upadhyay, 2014). Therefore, biocompatibility of compound, high penetration, high absorption, structure-activity protection, and targeted drug delivery was very much needed in the treatment of neuronal disorders. It calls for 'smart' drug delivery system, and nanotechnology-based drug delivery system has opted as a perfect solution.

Despite of new interdisciplinary field, nanotechnology has broadened its endpoints to diagnostic medical sciences too. It has been using in medical devices and biosensors as key element for neuronal disorders diagnosis. Their 'ultra-selective' design also prompts their use in drug delivery and gene delivery in highly complex brain disorders.

Therefore, the main target of nanomedical science is to develop a novel and systemically administered diagnostic and therapeutic nanoparticles that could diagnose and treat Parkinson's disease (PD), Alzheimer's disease (AD), multiple sclerosis (MS), epilepsy, trauma, seizures, tumor, neuro-AIDS, neuropathy, and some other brain disorders (Illum, 2004; Sriramoju et al., 2014; Reynolds and Mahato, 2017).

In this chapter, we will focus on drug transport to the brain, nanocarrier properties, and their preparation, nanomedicine in diagnostics and therapeutics in neuronal disorder and currently available nanomedicine in the global market.

4.2 THE BLOOD BRAIN BARRIER (BBB)

BBB is the major barrier to prevent crossing of most circulatory cells and other molecules from blood to brain. Along with pericytes, perivascular astrocytes, neurons, and basal lamina, microvessel epithelial cells form structural component of BBB (Bendayan et al., 2002; Obermeier et al., 2013; Zhao et al., 2015; Freese et al., 2017). This structure and transportation of endogenous compounds are maintained by tight junction and proteins present within it: claudins and occludin, along with adherent junction proteins (made up of cadherins and catenins) (Weksler et al., 2005). Other than these structural units, few specific efflux/influx channels, receptors, and protein transporters also act as metabolism-driven barrier at BBB; which includes ATP-Binding cassette (ABC) associated transporters, multidrug resistance-associated proteins (MRPs), P-glycoprotein (P-gp), and breast cancer resistance protein (BCRP, ABCG2), organic anion transporters (OATs) and organic anion transporting polypeptide family (OATPs), where OATs and OATPs act as both influx and efflux system for brain (Kusuhara and Sugiyama, 2001; Loscher and Potschka, 2005; Bendayan et al., 2006; Dallas et al., 2006).

4.2.1 STRATEGIES OF DRUG DELIVERY IN BRAIN

As mentioned above, BBB tightly regulates transportation of endogenous compounds, hormones, xenobiotics, nutrients, and some pharmacological agents. In spite of having strict mechanism, several metabolic enzymes and membrane transporters can modulate the BBB permeability via permitting or restricting transcellular trafficking of the substrate. Transport mechanisms are following:

4.2.1.1 PASSIVE DIFFUSION

Passive diffusion is the transport of the molecule down to their concentration gradient without the requirement of metabolic energy; which depend upon weight, surface charge, flexibility, and lipophilicity of the molecules (Misra et al., 2003; Begley, 2004; Zlokovic, 2008; Mikitsh and Chacko, 2014).

Only lipid-soluble molecules with molecular weight of <400 Daltons can cross the barrier via passive diffusion (Pardridge, 2010).

4.2.1.2   CARRIER-MEDIATED TRANSPORT

Carrier-mediated transport includes facilitate diffusion, down to concentration gradient; and active transport, against electrochemical gradient via using energy. Glucose transporter–1 (Glut1), large neutral amino acid carrier–1 (LAT1), ABC transporter, equilibrative nucleoside transporter–1 (ENT1), and monocarboxylate transporter (MCT1) are well-studied transporter system of BBB (Begley, 2004; Zlokovic, 2008; Mikitsh and Chacko, 2014; Pardridge, 2015). But sometimes, continuously active efflux system and overexpression of the transporter system impede drug transport to the brain and pump back the drug to the blood (Domínguez et al., 2014).

Other than above mechanisms, molecule can transport into the brain via adsorptive and receptor-mediated endocytosis/transcytosis (Zlokovic, 2008; Mikitsh and Chacko, 2014).

To overcome the limitations of conventional drug therapy, invasive, and non-invasive methods of drug delivery were invented, and improvement of the therapeutic efficacy was reported to some extent. Invasive techniques of transcranial drug delivery are direct transportation of the molecule via intracerebroventricular (ICV) infusion into CSF, intracerebral implantation, convection-enhanced diffusion (CED) to intraparenchymal and BBB integrity disruption via solvent/adjuvant (Pardridge, 2005; Domínguez et al., 2014). Non-invasive techniques include drug delivery through transporters, transcytosis, oral, intranasal, and parenteral pathway (Begley, 2004; Pardridge, 2005; Domínguez et al., 2014).

Still, inconvenient invasive therapy with poor compliance and increased risk of infection and, transportation, and size limitation call for new therapeutic invention for targeted drug therapy in the brain. In answer to all of these, nanotechnology-based drug delivery and their application in disease diagnosis showed potential in treatment of CNS disorders.

4.3   PROPERTIES OF SELECTED NANOPARTICLE

From last two decades, global market is on boom with new and effective nanotechnology-based diagnostic and therapeutic interventions. Effective and targeted transportation of nanomedicine to the brain can be achieved by a molecule with specific size, shape, charge, and modified surface chemistry (Albanese et al., 2012). Surface modification of nanoparticles with hydrophilic/hydrophobic polymers and lipophilic polymer coated nanomedicine are direct methods to improve targeted drug delivery to the brain (Deng, 2010). The preparation methods of nanomedicine also based upon absorption, distribution, and elimination pattern of nanoscaled polymer or compound associated with drug (Reis et al., 2006). '*Drug ability*' of any compound can be measured by following properties:

4.3.1 SOLUBILITY

The targeted molecule should have high solubility in intrinsic human conditions. The solubility can be enhanced by using nonaqueous solution, surfactant, and reduced particle size with high bioavailability/high dissolution rate (IR, 2001).

4.3.2 PERMEABILITY

The permeability of the drug across the BBB occurs by several pathways including intranasal, invasive method, receptor-mediated endocytosis and lipophilic entrapped drug (Zhang et al., 2016). Thus, a custom-synthesized drug loaded nanocarrier can alter tight junction and enhance drug absorption in the brain (Devalapally et al., 2007).

4.3.3 LIPOPHILICITY

Lipophilic compounds show tissue specificity, protection of drug from undesired external conditions and high efficiency in binding with both hydrophilic and hydrophobic agents and permeable to BBB (Torchilin, 2005; Zhang et al., 2008). Therefore, lipophilic molecules are preferred over hydrophilic molecule to transport brain barrier via passive diffusion (Mikitsh and Chacko, 2014; Banks, 2009).

Other than above properties, selected nanosystem should have small diameter, high stability in blood, non-toxicity, non-inflammatory, long circulation time, biodegradability, and biocompatibility for effective and targeted drug delivery in the brain (Bhaskar et al., 2010).

On the basis of above properties, several attempts have been made to develop processes and techniques for nanomedicine synthesis. Both, nanoparticle encapsulated drug (therapeutic approach) and nanoparticle used in imaging techniques (diagnostic approach) have different preparation methods. Recently theranostic (thera(py) + (diag)nostics) has put a strong implication in healthcare, easy to synthesize, and cure neuronal disorder (Singh et al., 2012).

4.4 METHODOLOGY TO PREPARE NANOMEDICINE

Bottom-up and top-down approach produce nanostructured polymeric and non-polymeric compound via homogenization, emulsion techniques, phase inversion, ultrasonication, chemical vapor deposition (CVD), liquid phase techniques, laser pyrolysis, biocompound-polymer conjugates, hydrothermal synthesis, spray synthesis, plasma methods, molecular self assembly, emulsion, and interfacial polymerization (Geckeler and Nishide, 2009; Allouche, 2013; Kumar et al., 2013; Talegaonkar et al., 2013; Nikalje, 2015). Some the methods are mentioned below:

4.4.1 POLYMERIZATION

Polymerization of nanoparticle is fastest chemical methods which are classified into nanocapsules and nanospheres. Nanocapsules are cavity enclosed with polymer membrane, whether nanospheres are 'drug sponges' that on which drug is uniformly and physically dispersed (Soppimath et al., 2001; Mohanraj and Chen, 2006). The polymeric nanoparticles (PNP) show high efficiency and effectiveness with ability of higher stability, targeted delivery, biocompatibility, biodegradability, easy modification and easy association with targeted pathway (Nagavarma et al., 2012). Polymers in use in nanotechnology are: chitosan, gelatin, albumin, sodium alginate, polylactic-co-glycolic acid (PLGA), poly(ethylene glycol) (PEG), poly(vinyl alcohol), poly(acrylic acid), polyacrylamide, poly malic acid, polycaprolactone, polylactides (PLA), poly(methacrylic acid), poly(methyl methacrylate) (PMMA), poly(n-butyl cyanoacrylate) (PBCA), poly(ethyl cyanoacrylate) (PECA) (Rao and Geckele, 2011; Carné-Sánchez et al., 2013).

4.4.1.1 DISPERSION

Dispersion of the compounds/drugs can be performed by solvent evaporation, salting out, nanoprecipitation, emulsification/solvent diffusion, dialysis, and supercritical fluid technology (SCF) (Rao and Geckele, 2011; Nagavarma et al., 2012).

4.4.1.2 POLYMERIZATION

Preformed polymers are the raw material in dispersion methods. The desired properties for specific application can be achieved by polymerization of the monomers. Interfacial polymerization, controlled/living radical polymerization (C/LRP), mini-, and microemulsion polymerization are the major techniques currently in use (Nagavarma et al., 2012; Kumar et al., 2013).

4.4.2 VAPOR DEPOSITION METHODS

Vapor deposition of the specific metals at very high temperature onto compounds/drugs to be transported (Kumar et al., 2013). This method can also use as tracking material in imaging of the tissue (Kumar et al., 2013).

4.4.3 SPRAY SYNTHESIS

Spray pyrolysis can be exploited as rapid, scalable, size controllable, reproducible, and low-cost method for the preparation and self-assembly of nanometal oxides and mixed metal oxides (Kumar et al., 2013; Carné-Sánchez et al., 2013); which are normally used as contrast agents in brain imaging. Metal acetate, nitrate, and chloride are considered as best precursor solution in metal oxide synthesis (Kumar et al., 2013). The high drug payloads, controlled release, easily surface modification, and detectable properties of porous nanometal oxide enhance their use in theranostics (Horcajada et al., 2010).

Plasma and hydrothermal methods are high-temperature dependent methods to synthesize metal oxide nanoparticles in therapeutic application (Kumar et al., 2013).

4.5 NANOCARRIERS

Polymeric nanostructures like dendrimers, liposomes, micelles, solid-liquid nanoparticle, and polymeric nanoparticle, as well as non-polymeric nanostructures like carbon nanotubes (CNT), silica nanoparticles, quantum dots and metallic nanoparticles are used in both therapeutic and diagnostic intervention in several disorders (Beg et al., 2011; Nikalje, 2015). Surface modified nanocarriers with dextrans, gangliosides, and hydrophilic polymers, such as polyvinyl alcohol, poly-N-vinylpyrrolidones, and PEG (Brewer et al., 2007) can be prepared by direct conjugation (Torchilin, 2008; Hazra et al., 2011), surface functionalization (Marqués-Gallego and Kroon, 2014), enzymatic modification (Marqués-Gallego and Kroon, 2014) and chelating strategy (Marqués-Gallego and Kroon, 2014) to increase efficacy of the nanocarrier.

4.5.1 LIPOSOMES

Alec D. Bangham discovered the liposomes, in 1961, as nothing but self-assembled closed bilayer structure of phospholipids with capacity of site-specific targeting, amphiphilic nature and less toxicity to the neuronal cells (Bangham et al., 1964). It can be synthesized by dispersion, emulsification, sonication, and lipid drying techniques (Akbarzadeh et al., 2013).

4.5.2 SOLID LIQUID NANOPARTICLES (SLN)

This lipid-based nanocarrier has advantage over liposomes and lipid emulsion with higher stability and no drug leakage due to solid matrix (Ekambaram et al., 2012). It can be synthesized from homogenization, ultrasonication, evaporation, emulsification, and precipitation techniques (Uner and Yener, 2007; Almeida and Souto, 2007; Mukherjee et al., 2009; Puri et al., 2009; Ekambaram et al., 2012).

4.5.3 POLYMERIC NANOPARTICLES (PNP)

This 10–1000nm sized nanocarrier offer complete drug protection, high biodegradability and controlled drug release into brain. Collagen, chitosan, poly(lactic acid) and poly(lactic-*co*-glycolic acid) are the common polymers used in synthesis of PNPs (Castro and Kumar, 2013). Solvent evaporation, salting out, nanoprecipitation, emulsification/solvent diffusion, dialysis, and SCF as well as monomers. Interfacial polymerization, controlled/living radical polymerization (C/LRP), mini-, and microemulsion polymerization are the major and effective techniques to synthesized PNP (Nagavarma et al., 2012, Kumar et al., 2013).

4.5.4 POLYMERIC MICELLES

Polymeric micelles are supramolecular core-shell synthetic macromolecule of self-assembled amphiphilic copolymers of size <100 nm. It can be synthesized by micro-emulsification and self-assembly of polymers with amines or carboxylic acid group which help in drug introduction into micelles core (Gupta and Vyas, 2011).

4.5.5 POLYMERIC DENDRIMERS

Dendrimers are three-dimensional multivalent branched macromolecule of nanodomain network, discovered by Fritz Vogtle in 1978 (Baig et al., 2015). It can be synthesized by monodispersion, self-assembly, thiolyne reactions, Michael addition reaction, Diels-Alder reactions and azide-alkyne reactions of the amphiphilic polymers (Baig et al., 2015; Abbasi et al., 2014).

4.5.6 CNT AND FULLERENE

This nanosized buckyball structure of rolled graphite with a diameter of approximately $1/50,000^{th}$ of human hair was first to come into the limelight in 1952. They are typically synthesized by vapor deposition, electric-arc discharge, laser vaporization, catalytic pyrolysis of hydrocarbon, laser ablation and combustion process (Awasthi et al., 2005; Revathi et al., 2015). Its strength, hardness, thermal stability, and electrical property enhance their chances in biomedical application (Revathi et al., 2015).

Fullerene is spherical molecular structure with diameter of ~1nm with super-conductivity and antioxidant property. It can be synthesized by laser ablation, gas combustion method and laser vaporization (Smalley, 2012; Charitidis et al., 2014).

Functionalized CNT and fullerenes make themselves hydrophilic that is easy to bind with drug and contrast agents (He et al., 2013). Their use in vaccine, gene, DNA delivery and act as antimicrobial and antioxidant agent, have increased their value in medical sciences (Yamashita et al., 2012).

Nanocapsulated drugs can be characterized by electron microscopy, atomic force microscope, dynamic light scattering, powder X-ray diffraction, spectroscopy (X-ray photoelectron, Fourier transform infrared, matrix-assisted laser desorption, ionization time-of-flight mass spectroscopy, UV-visible), dual polarization interferometry and nuclear magnetic resonance (Kumar et al., 2013).

4.6 NANOCARRIER SURFACE MODIFICATION

Although all nanocarriers are designed to cross BBB, but their surface still needs to be modified to increase targeted delivery, binding affinity, drug payload, BBB permeability, biological, and chemical stability of the drug into the brain. Other than this, functionalizations of nanocarrier surface can also avoid non-specific interaction of nanocarrier with circulating proteins of the bloodstream. Surface modification generally focused on multivalency of the nanocarriers that requires prior knowledge of target organ and transport mechanism. There are following techniques to refine the surface of the nanocarrier:

4.6.1 SURFACTANTS

Nanomedicine and biological cells interaction directly depend upon nanocarrier surface functionality. Some hydrophilic surfactants- polyethylene glycol 1000 succinate, poloxamer 188, PEG-D-α-tocopheryl and polyvinyl alcohol, and polysorbate 80; coated nanoparticle inhibit the P-gp efflux transporter via binding with BBB surface receptor specific proteins and show high permeability (Sriramoju et al., 2014; Domínguez et al., 2014). Polysorbate 80 coated PBCA encapsulated doxorubicin (Steiniger et al., 2004) and gemcitabine (Wang et al., 2009a) also showed high penetration to the brain with high therapeutic efficiency.

4.6.2 ANTIBODY/PROTEIN/PEPTIDE CONJUGATION

Specific binding of the antibody makes it a very promising approach in drug delivery to the brain. Apolipoprotein, transferrin, trans-activating transcriptor (TAT) peptide and insulin receptor antibody are widely used antibodies to transport growth factors, neurotrophic factors and various drugs (Wong et al., 2012; Sriramoju et al., 2014). A brain-specific low-density lipoprotein-related protein–2 (LRP–2), megalin, also demonstrated improved drug delivery to the brain (Starr and Gabathuler, 2005). It was patented by Starr and group. Similarly, B6 peptide, LRP–1 mediated endocytosis and glutathione transporters also reported potential methods for customization of target drug delivery to brain (Demeule et al., 2008; Liu et al., 2013, Gaillard et al., 2012).

4.6.3 PEGYLATION

Covalent linkage and adsorption are appropriate techniques to attach PEG onto nanocarriers. PEG has following properties to enhance safe drug transport to the brain: flexibility, less toxicity, no degradation in systemic circulation, immunogenicity, higher solubility, improved pharmacokinetics and biodistribution (Juillerat-Jeannere, 2008; Park et al., 2008).

4.6.4 IONIC MODIFICATION

Cationic molecules have more capability to bind BBB with electrostatic force and improve cellular uptake in brain. Loperamide loaded cationic peptide derivatized poly (lactic-*co*-glycolic acid) nanoparticles showed high drug release in the brain (Michaelis et al., 2006). PEGylated cationic albumin also reported improved release of DNA plasmid to the neural cells (Xiang et al., 2003; Lu et al., 2006).

4.7 NANOMEDICINE IN NEURONAL DISORDER

Rapidly growing field of nanotechnology in medical research directly focused on developing nanotechnology-based compounds for therapeutic and diagnostic applications in neuronal disorders (Ventola, 2012). Although there are many barriers, but many nanomedicines for CNS disorders are already approved by FDA and on the global market, and many more are in clinical trials and in research use (Zhang et al., 2008; Ventola, 2012; DiNunzio and Williams, 2008; Marcato and Durán, 2008; Weissig et al., 2014; Beg et al., 2011).

Pharmaceutical agents can be transported via nanocarriers either encapsulated and conjugated or adsorbed onto the surface. Liposomes, lipid-based nanoparticles, polymeric nanoparticle, micelles, dendrimers, CNT, nanoemulsion, nanogel, and nanosuspension are widely used nanocarrier in neuronal disorder therapeutics.

4.7.1 ALZHEIMER'S DISEASE (AD)

Alzheimer's is disease of aging brain which can be identified by the accumulation of amyloid-β plaques and neurofibrillary tangles of hyper-phosphorylated tau protein in the brain. AD people also show low level of acetylcholine and less activity of P-glycoprotein in their brain (Hartz et al., 2016; Nalivaeva et al., 2016). Although, AD has been studied for over a century now, still approved anti-AD therapeutics is unable to inhibit neuro-degenerative process and, multifactorial pathogenesis of AD and limitation of drug transport due to brain barrier make it more difficult. Therefore, use of nanoparticle-based AD therapeutics has been increased from last few years.

Nanomedicine in AD use as β-Amyloid targeted drugs, cholinesterase inhibitors, and metal chelators. Curcumin is considered as potential thera-peutic to inhibit $A\beta_{1-42}$ polymerizations in aging brain and poor bioavail-ability of the compound can be avoided by nanotechnology-based delivery. Nanoliposomes encapsulated phytochemical curcumin improved its bioavailability while maintaining ability of alleviation of Aβ polymerization and cytotoxicity in-vitro (Taylor et al., 2011; Mourtas et al., 2011). Targeted delivery of the compound can also be improved by attaching peptides, proteins, antibodies, and specific ligands of brain cell receptor, onto the drug-loaded nanocarriers. In this direction, Mulik et al. (2010) performed in-vitro study of curcumin loaded poly(butyl) cyanoacrylate nanoparticles along with apolipoprotein E (ApoE) on the surface 2010), and high penetration to BBB was reported. Similarly, peptide Tet–1 targeted PLGA-Curcumin nanoparticles showed antioxidant properties and destroyed amyloid-β aggregates via retrograde transport (Mathew et al., 2012). PEGylated nano-micelles were reported to inhibit amyloid-β sheet formation and aggregation along with attenuated neurotoxicity in SHSY–5Y human neuroblastoma cell line (Pai et al., 2006). ICV administration of hydrated fullerene C60 showed anti-amyloidogenic ability and inhibition of amyloid–25–35 fibrillization which leads to improving cognitive task (Podolski et al., 2007). Polyami-doamine dendrimers and Copolymeric N-isopropylacrylamide: N-tert-butyl acrylamide is kind off nanotherapies that were reported to inhibit amyloid-β aggregates and temporarily reversion of amyloid-β fibrillization respectively (Klajnert et al., 2007; Cabaleiro-Lago et al., 2008).

Cholinergic nanocarrier has an advantage over conventional drug therapy to treat AD as external administration of AcH or cholinesterase inhibitor has a short half-life and cannot cross BBB. Rivastigmine loaded nanoparticles were used in treatment of dementia (Srikanth and Kessler, 2012). Cholinesterase inhibitor loaded, apolipoprotein functionalized PBCA nanopolymers bind to BBB endothelial cells and release drug via transcytosis (Kim et al., 2007). Similar study was performed to enhance the rivastigmine transport to the brain via polysorbate–80 functionalized PBCA in AD model of rats (Wilson et al., 2008). This nanocarrier system also reported in improved spatial learning and memory of scopolamine-lesioned AD mice (Joshi et al., 2008).

Nanoparticle-chelator conjugates are in use from some time against transition metal ions induced Aβ aggregate formation. D-penicillamine, a copper chelator, conjugated with 1,2-dioleoyl-syglycero–3-phosphoethanol-amine-N-[4-(*p*-maleimidophenyl)butyramide] and 1,2-dioleoyl-syglycero–3-phosphoethanolamine-N-[3-(2-pyridyldithio)-propionate] nanoparticles had the ability to dissolve copper-Aβ aggregates (Cui et al., 2005). Quinoline derivative clioquinol (CQ) functionalized PBCA nanopolymer was reported Cu^{2+}/Zn^{2+} chelation at Aβ aggregates *in-vitro* and easy to cross BBB in wild mice (Cherney et al., 2001).

4.7.2 PARKINSON'S DISEASE (PD)

PD is neurodegenerative disease which can be identified by loss of dopaminergic neurons in substantia nigra and Lewy bodies, aggregation of α-synuclein and ubiquitin, accumulation in brain stem of patient (Lees et al., 2009; Blesa et al., 2012). Immune-activated glial cells induced neurotoxicity causes BBB disruption and allows leukocytes entry that speeds up the neuroinflammatory cascade into the brain and alters cognition (Domínguez et al., 2014). L-dopa, dopamine agonist, and monoamine oxidase B inhibitors are common drug therapy for PD (Kincses and Vecsej, 2011). However, direct dopamine administration is complete failure to treat PD; therefore alternative approach of nanotechnology-based therapeutic delivery is currently being explored.

Currently, PD nanomedicine formulation focused on DNA or gene therapy because of their ability to reverse the PD progression (Denyerand and Douglas, 2012). Liposomes encapsulated tyrosine hydroxylase (TH), a synthetic dopamine enzyme, encoding plasmids, targeted with anti-transferrin receptor antibody (OX26) was reported to normalize the TH activity and restored rotational behavior in PD rodents (Zhang et al., 2003). Glial-derived neurotrophic factor (GDNF) gene loaded lactoferrin modified nanoparticles also demonstrated increased monoamine level, improved locomotor activity and less dopaminergic neuronal loss in PD rat brain (Huang et al., 2008).

Urocortin, a cryoprotectant, showed inhibition of the PD like feature but it was unable to cross the BBB; therefore urocortin-loaded lactoferrin modified PEG-PLGA functionalized nanoparticles was reported to enhance drug delivery in PD rats (Hu et al., 2011). Similar study showed significant reduction in dopaminergic cell loss and rotational behavior in intranasal administered urocortin loaded odorranalectin modified nanoparticles (Wen et al., 2011). SLN coated apomorphine, dopamine receptor agonist, also reported in PD treatment (Tsai et al., 2011).

Fullerene showed neuroprotective property in 6-OHDA and MPP+ cell lines of PD (Lotharius et al., 1999). Antioxidant carboxy fullerenes also prevented TH degradation in striatal neurons (Lin et al., 1999). Antioxidant and radical scavenging property of polyhydroxylated fullerene derivative was observed in 1-methyl-4-phenylpyridinium (MPP) induced PD model in human neuroblastoma cell line (Cai et al., 2008). Although ascorbic acid functionalized C60 fullerene showed protection against levodopa toxicity (Santos et al., 2008), but free radicals generated in the presence of visible light and damage the neuronal cells (Kolosnjaj et al., 2007). The nerve growth factor-loaded CNT promoted nerve outgrowth in PC12 cells *in-vitro* (Cellot et al., 2009) and electrical stimulation of CNT in deep brain (Mazzatenta et al., 2007) were considered as advanced method for PD treatment, but cytotoxicity of nanotubes was also reported (Jia et al., 2005; Manna et al., 2005).

Phosphatidylglycerol based phospholipid nanoparticles, VP025 (Vasogen Inc., ON, Canada), showed neuroprotective property against free radicals, apoptosis, and microglial inflammation in neurodegenerative PD (Fitzgerald et al., 2008; Cook and Petrucelli, 2009). This product has completed Phase-I clinical trial in 2005 and currently in Phase-II clinical development (Nowacek et al., 2009).

4.7.3 BRAIN TUMORS

Despite aggressive multidimensional therapy, including chemotherapy, radiation therapy and surgery of the brain tumor, the prognosis, and treatment malignant tumor has always been challenging. Inability of conventional drug therapy to cross BBB can be overcome by nanotechnology-based targeted drug therapy. Methotrexate and temozolomide loaded PBCA nanoparticles, and temozolomide, etoposide, and paclitaxel-loaded SLN nanoparticles were reported increased intracerebral drug concentration and decreased cytotoxicity of astrocytes (Huang et al., 2008; Gao et al., 2006; Lamprecht and Benoit, 2006; Garcion et al., 2006; Tian et al., 2011). Doxorubicin-loaded polysorbate-80 coated PBCA nanoparticles also demonstrated anti-tumor potential alongside diminishing p-glycoprotein efflux in the glioblastoma rat model (Steiniger et al., 2006).

PEGylated poly (trimethylene carbonate) and magnetic nanoparticles of paclitaxel were identified for improved survival rate of tumor-bearing mice (Steiniger et al., 2004; Jiang et al., 2011). Application of amino silane-coated supermagnetic iron oxide nanoparticles in radiotherapy and hyperthermic therapeutics in glioblastoma patients, are in clinical trial II (Landeghem et al., 2009). Similarly, CGKRK, a tumor homing peptide, and ($_D$[KLAKLAK]2), a pro-apoptotic peptide coated iron oxide nanoparticles showed theranostic potential in lentivirus mice model of glioblastoma (Agemy et al., 2011).

D-glucosamine loaded polyether-copolyester, PEPE, dendrimers transportation was increased via conjugation on glutamine transporter type 1 on *in-vitro* BBB model (Dhanikula et al., 2008). Doxorubicin-loaded PEGylated polyamidoamine dendrimers also showed wide therapeutic window against C6 glioma spheroid proliferation in *in-vitro* study (He et al., 2010). Epidermal growth factor receptor (EGFR) inhibitor, cetuximab, a monoclonal antibody, coated polyamidoamine dendrimers encapsulated with methotrexate showed targeted binding and cytotoxicity to EGFR-expressing rat glioma cell line F98 (EGFR) (Wu et al., 2006).

Similarly, sodium borocaptate loaded EGFR antibodies and transferrin functionalized liposomes improved the survival rate of mouse model of glioblastoma (Doi et al., 2008; Feng et al., 2009). EGFR antisense therapy was also reported in brain tumor therapeutics (Zhang et al., 2002). Similarly, laminin–411 antisense mRNA loaded poly (β-l-malic acid) nanoconjugates was delivered to glioma cells and, *in-vitro* apoptosis and *in-vivo* xenograft growth inhibition were reported (Ding et al., 2010).

Other investigations on pluronic–108 and doxorubicin encapsulated with Angiopep–2 peptide PEGylated nanotubes showed advanced therapeutic activity and survival rates in glioblastoma mouse model (VanHandel et al., 2009; Ren et al., 2012). Similar study of paclitaxel-loaded angiopep conjugated PEG-co-poly (ε-caprolactone) nanoparticles showed high therapeutic efficiency against glioma cells in *in-vitro* (Xin et al., 2011). A similar dual targeting approach of nucleolin targeted DNA aptamer, and paclitaxel-loaded PEG-PLGA nanoparticles has taken advantage against C6 glioma cells *in-vitro* (Guo et al., 2011).

Gene therapy with proapoptotic Apo2 ligand/tumor necrosis factor–related apoptosis-inducing ligand (Apo2L/TRAIL) loaded cationic albumin-conjugated PEGylated, CBSA, nanoparticles is one of the more tailored therapies in treatment of C6 intracranial xenografts of PEG-nude mice (Lu et al., 2006). Similarly, TRAIL gene loaded, modified candid candoxin, CDX, functionalized poly (ethylene glycol)-poly(d,l-lactide) (PEG–PLA) micelles effectively interacted with nAChR and showed anti-tumor effect on animal model of (Zhan et al., 2011). Angiopep–2 anchored liposomes also showed anti-tumor potential via co-delivery of pEGFP-hTRAIL and paclitaxel onto glioblastoma cells (Sun et al., 2012).

4.7.4 SPINAL CORD INJURY (SCI)

Damage of specific parts of spinal cord or extended nerve deterioration came from spinal cord that causes permanent alteration in sensory or motor abilities below the point of injury are visible in spinal cord injury (SCI) (Dixon and Budd, 2017). Blood flow disruption and hemorrhage are main results of primary injury that involves decrease of grey matter in spinal cord. Similarly, Ca^{2+} overload in neuronal cells, oxidative stress, malnutrition of the tissues, thrombosis, hypoxia, ischemia, and edema are cascade of deleterious reaction of secondary injury (Tyler et al., 2013). Limited efficacy of neuroprotective and regenerative therapy for SCI and adverse reactions of conventional therapy needs new techniques to tackle the problem.

Micelles have been used widely in SCI from last few decades. Methylprednisolone (MP) encapsulation in poly (ethylene oxide)-poly (propylene oxide)-poly (ethylene oxide), pluronic (PEO-PPO-PEO), polymeric micelles showed increased level of mRNA and Bcl-xl anti-apoptotic protein level, and improved bioavailability of MP in rabbit and mice model of SCI SCI (Chen et al., 2008). Similarly, monomethoxy PEG-poly(D,L-lactic acid) (mPEG-PDLLA) di-block copolymers micelles had the ability to seal the damaged cell membrane, high neuroprotection property and enhanced action potential in *ex-vivo* and *in-vivo* rat model of SCI (Shi et al., 2010).

Polymeric and silica nanoparticles are extensively used nanoparticles in SCI. MP loaded poly[lactic-co-glycolide] (PLGA) nanoparticles resultant in reduced Calpain, iNOS, Bax reactivity and enhanced Bcl2 activity, reduced immune response and functional recovery in rat model of SCI (Kim et al., 2009). Monosialotetrahexosylganglioside (GM–1) loaded amphiphilic Pluronic chains, coated on anionic superparamagnetic iron oxide nanoparticles were reported for high efficacy in drug release at very low temperature (Chen et al., 2007). Similarly, PEG decorated silica nanoparticles, SiNP, reported less reactive oxygen species, somatosensory evoked potential recovery and restored membrane integrity in guinea pig model of SCI (He et al., 2007). In another study mesoporous silica nanoparticles (MSNs) encapsulated hydralazine, functionalized with PEG, restored the membrane integrity, mitochondrial function and reduced free oxygen radicals in *in-vitro* acrolein challenged neuron cell model of spinal injury (Cho et al., 2008). Past studies on poly (butyl cyanoacrylate), PBCA, nanoparticles conjugated superoxide dismutase (SOD) and anti-glutamate N-methyl D-aspartate receptor 1 (NR1) antibody demonstrated neuroprotection from glutamatergic toxicity and oxidative stress in *in-vitro* cerebellar neuronal cells of spinal injury (Reukov et al., 2011). Another investigation on ceria nanoparticles neuroprotection in adult rat spinal cord culture also showed oxidative catalytic recovery (Das et al., 2007).

Instead of toxicity reports of carbon-based nanomaterial, fullerenes are in use in SCI treatment. Jin et al. demonstrated neuroprotective activity of C60 derivative fullerenols via lowering intracellular Ca^{2+} and blocking glutamate pathways in *in-vitro* neuronal cells (Jin et al., 2000). Another study also reported anti-oxidant property via using C60-based ebselen derivative to *in-vitro* neuronal cells model of SCI (Liu et al., 2007). PEG functionalized CNT was reported with less lesion volume and increased corticospinal tract fibers and neurofilament-positive fibers in animal model of SCI (Roman et al., 2011).

Neural cells regeneration is very difficult and complex process, as damaged oligodendrocytes, axonal loss, degraded myelin, and cyst and glial scar formation resulting physical and chemical barrier to regenerate neuronal cells (Fehlings and Tator, 1995; Buss et al., 2004; Yiu and He, 2006). Therefore, promoting neurotrophic factors by new therapeutic intervention of nanotechnology is potential techniques in SCI. Polyamide nanofiber scaffold decorated with neurite outgrowth promoting tenascin-C derived peptides (Ahmed et al., 2006) and poly(L-lactic acid) nanofiber scaffold incorporated with extracellular matrix (ECM) protein leminin (Kuo and Kuo, 2008) reported neurite outgrowth in *in-vitro* cell of spinal injury. Another study on biodegradable poly (ε-caprolactone), PCL, nanofiber scaffold differentiates embryonic cells into astrocytes and promoted neurite outgrowth in *in-vitro* study of SCI (Xie et al., 2009). Schwann cells or neural progenitor cells loaded self-assembly peptide nanofiber scaffolds (SAPNS) system, RADA–1 SAPNS and bone marrow homing motif (BMHP1) modified RADA–1 SAPNS transplantation in *in-vivo* model of rat SCI demonstrated neural regeneration and reduced inflammation (Guo et al., 2007; Cigognini et al., 2011).

4.7.5 STROKE

Stroke, "brain attack," is world's second leading cause of mortality over the globe. Blood supply interruption in the brain generally causes ischemia and hemorrhage, resulting into inflammatory reaction, free radical generation, and BBB dysfunction, and eventually lead to death. Although recanalization and neuroprotection therapy, their application are very limited by its inability to cross BBB, rapid renal clearance, acute renal toxicity and low tissue accessibility in stroke patient (Hishida, 2007; Kamouchi et al., 2013).

Currently, very limited use of nanomedicine was reported in stroke. Hemoglobin loaded liposomes (Urakami et al., 2009; Kawaguchi et al., 2010), echogenic liposomes filled with therapeutic gas, as xenon, (Britton et al., 2010), transferrin functionalized, vascular endothelial growth factor-loaded liposomes (Zhao et al., 2011) and citicoline loaded liposomes (Ramos-Cabrer et al., 2011) demonstrated diagnostic and therapeutic approach as neuroprotection and promote regeneration in stroke model. In another study puerarin loaded SLN nanoparticles showed protection from cerebral ischemic-reperfusion injury in an *in-vivo* gerbil model of ischemia (Zhu et al., 2010).

PAMAM dendrimer-N-acetyl cysteine also had a potential role in neuro-inflammation and anti-oxidant activity in clinical stroke patients (Gao et al., 2011; Reddy et al., 2010). In another study PAMAM dendrimer loaded with siRNA demonstrated high biodegradability, high BBB transportation and neuroprotection in postischemic tissue (Kim et al., 2010). Similarly, anti-fibrin monoclonal antibodies conjugated perfluorocarbon (PFC) dendrimer nanoparticles showed thrombolytic property in stroke model of dogs (Marsh et al., 2011).

Instead of showing toxicity effect, CNT were used in detection and moni-toring the stroke in animal models. Amine-modified SWNT encapsulated with stem cells reported neuroprotection via lowering of angiogenic, inflam-mation, and apoptotic markers in ischemic-injured rats (Lee et al., 2011). Similarly, SWNT modified glassy carbon electrode (Zhang et al., 2005) and SWNT based biosensor (Lin et al., 2009) were used to detect ascorbate deple-tion and, glucose, and lactate levels in global cerebral ischemia model of rats.

Anti-mouse transferrin receptor monoclonal antibody (TfRMAb) conjugated with PEG-coated chitosan nanospheres reported decreased neurological deficit, caspase–3 activity and infarct volume in mouse model of cerebral ischemia (Karatas et al., 2009). In other studies SOD loaded poly(D,L-lactide *co*-glycolide) nanoparticles, Reddy, and Labhasetwar, 2009), PX–18 loaded nanocrystals (Wang et al., 2009 (b)), indomethacin loaded nanocapsules (Bernardi et al., 2010) and nitroxyl radicals containing nanoparticles (Yoshitomi, and Nagasaki, 2011) demonstrated anti-inflam-matory properties, regained neurological function, glial cells activation and anti-oxidant activity in *in-vivo* and *in-vitro* stroke models.

Other than these, several nanomaterials were used in stroke progression and diagnosis. Superparamagnetic iron oxide nanoparticles (SPIO), feru-moxides-labeled human neural stem cells loaded SPIO, chitosan-SPIO, ultra small SPIO, iron oxide loaded polymeric micelles, and P-selectin targeted iron oxide nanoparticles were used in MRI or imaging techniques of cerebral ischemia (Nair et al., 2012).

4.7.6   NEURO-AIDS

Human immunodeficiency virus (HIV) is a global problem for humanity since its discovery and after its characterization in the 1980s. In 10% cases, patient complains for neurological complications due to viral meningitis (Almeida et al., 2006), and it 50% cases reported severe neuropathological symptoms (McArthur et al., 2005) as the disease progressed. Other than BBB impermeability of the HIV therapy other comorbid conditions, CNS lymphoma, inflammation, neuropathic pain, neurosyphilis, etc. has called for nanotechnology-based new multifunctional therapeutic interventions in HIV-AIDS (Ruiz et al., 2015).

Although there are so many techniques for drug therapy in NeuroAIDS currently, there are still so many areas need to be open. Poly (DL-lactide-co-glycolide), PLGA, encapsulated ritonavir, lopinvie, and efavirenz was reported in high concentration in brain tissue of BALB/C mice model of HIV (Destache et al., 2010). It was the first study to evaluate the *in-vivo* pharmacokinetics of triple- antiretroviral drugs loaded nanoparticles. In another study, poly (L-lactide), PLA, conjugated TAT peptide enhanced the ritonavir transport through BBB via bypassing P-glycoprotein efflux action in *in-vitro* (Rao et al., 2008). Al-Ghananeem et al. (2010) also reported higher drug concentration in CSF and brain by transporting didanosine loaded chitosan polymer in rat model of AIDS 2010). 2G-NN16 amino-terminated carbosilane dendrimer encapsulated siRNA also showed reduced via load in *in-vitro* human astrocytoma cells of HIV-associated dementia cell lines (Jiménez et al., 2010).

Silver nanoparticles showed drug-free vial inhibition method to treat HIV in *in-vitro*. 1–10nm sized silver nanoparticles attached with exposed sulfur bearing residues at HIV–1 gp120 glycoprotein and block viral replication (Elechiguerra et al., 2005). Similarly, fragmented HIV inhibitor TAK–779, SDC–1721, conjugated gold nanoparticles displayed allosteric inhibitor of CCR5 receptor and act as potent therapeutics for neuroAIDS (Bowman et al., 2008). Transferrins functionalized quantum rods loaded with SQV (Mahajan et al., 2010) and PEG coated-transferrin functionalized albumin-based nanoparticles, encapsulated with AZT (Mishra et al., 2006), enhanced BBB permeation via binding with TfR–1 receptor. Similarly, stavudine, saquinavir, indinavir, and darunavir loaded and bradykinin analog RMP–7 functionalized methylmethacrylate-sulfopropyl methacrylate nanoparticles targeted bradykinin type II receptor (B2-R) and increased drug loading into brain (Kuo and Lee, 2012).

Other than polymeric nanoparticle and dendrimers based drug delivery in neuroAIDS, micelles, and lipid-based nanocarriers are currently in use. Foscarnet, a selvage therapy based drug, loaded liposomes reported anti-HIV–1 activity via inhibiting immunopathogenesis in Sprague-Dawley female rats (Dusserre et al., 1995). In another study dioleoyl lecithin, dipalmitoyl phosphatidylglycerol, cholesterol, and triolein formulated liposome encapsulated zalcitabine showed increased half-life in *an in-vivo* rat model (Kim et al., 1990). Pluronic P85 loaded ritonavir also demonstrated higher drug transportation in *in-vitro* BBMEC and Caco–2 cell monolayers (Batrakova et al., 1999). In extension of the previous study P82 loaded with zidovudine, lamivudine, and nelfinavir also showed less MDM in *an in-vivo* model of neuroAIDS (Spitzenberger et al., 2007). AZT-myristate loaded liposomes also showed high encapsulation efficiency, targeted delivery and increased half-life in rats (Jin et al., 2005). Lipid E80 encapsulated indinavir NanoART loaded into bone marrow macrophages prompt BBB permeability and increased antiretroviral activity in a mouse model of HIV–1 encephalitis (Dou et al., 2009).

Magnetic nanoparticle also played a significant role in antiretroviral therapy of HIV. Azidothymidine, nucleotide analog reverse transcriptase inhibitors, loaded phosphatidylcholine (PC)-cholesterol-magnetite based liposomes demonstrated monocyte-mediated transport into BBB resulting in high permeability in *in-vitro* BBB model in the presence of external magnetic field (Saiyed et al., 2010). Similarly, tenofovir, anti-HIV therapy, and vorinostat, a latency breaking agents, coencapsulated into magnetic nanoparticles showed sustained drug release in *the in-vitro* model of HIV (Jayant et al., 2015).

Another lipid-based nanocarrier, SLN encapsulated with stavudine, delavirdine, saquinavir, and atazanavir demonstrated increased BBB permeability and higher dose in neural cells of AIDS model (Kuo and Su, 2007; Chattopadhyay et al., 2008). Saquinavir loaded SLN reported increased drug release under the influence of electromagnetic force and recommended for combinational therapy in clinic application (Kuo and Kuo, 2008). SQV loaded and polysorbate 80 stabilized cationic SLN also demonstrated higher entrapment efficiency and effective drug release in the BBB model (Kuo and Chen, 2009).

4.8   NANOMEDICINE IN MARKET

Just the word "nano" become very fashionable in pharmaceutical science to develop advanced therapeutics in neuronal disorder. National Institute of Health (NIH) has started Nanotechnology Initiative (NNI) in the year 2000 in order to promote nanopharmaceutical based research and development. Several nanomedicines are already FDA approved, commercially available in the market and many more in clinical trials. A succinct list of currently available nanomedicines in market has been provided in Table 4.1.

TABLE 4.1 List of Nanomedicines Currently Available in Market and in Various Phases of Clinical Trials

S.No.	Drug/ Compound name	Industry	Nanotechnology	Status	Indication
1	NCT00029523	Pacira	Liposomal cytosine arabinoside	Phase IV	Neoplastic meningitis
2	NCT02340156	Syner Gene	Temozolomide-loaded immunoliposomes functionalized with the human wild-type p53 DNA plasmidic sequence SGT–53	Phase II	Glioblastoma
3	NCT0734682	—	Topoisomerase I inhibitor CPT–11/irinotecan-loaded liposomes	Phase I	Gliomas
4	Copaxone®/ Glatopa	Teva Pharmaceutical Industries Ltd.	Random copolymer of L-glutamate, L-alanine, L-lysine, and L-tyrosine	Approved in 1996/2014	Multiple Sclerosis (MS)
5	Gliadel®/ Carmustine	Guilford Pharm	Wafer Polyanhydride polymeric nanoparticles	Approved in 1996	Glioblastoma multiform
6	DepoCyt®	Sigma-Tau	Cytarabine encapsulated in multivesicular liposomes	Approved in 1999/ 2007	Lymphomatous malignant meningitis
7	Avinza®	Pfizer	Morphine sulfate nanocrystal	Approved in 2002	Psychostimulant
8	Ritalin LA®	Novartis	Methylphenidate HCl nanocrystals	Approved in 2002	Psychostimulant
9	Risperdal Consta® (risperidone)	Johnson and Johnson	Albumin microspheres polymeric nanoparticles	Approved in 2002 (Germany)/ 2004	Schizophrenia
10	Rienso®/Ferumoxytol	AMAG/ Takeda	Iron polyglucose sorbitol carboxymethyl ether nanopolymer colloid	Approved in 2009	Neuroinflammation in epilepsy, brain metastases
11	Invega® Sustenna®	Janssen Pharms	Paliperidone Palmitate nanocrystals	Approved in 2009 and 2014	Schizophrenia Schizoaffective Disorder
12	Nanotherm®	MagForce	Iron oxide nanocrystals	Approved in 2010	Glioblastoma

TABLE 4.1 *(Continued)*

S.No.	Drug/ Compound name	Industry	Nanotechnology	Status	Indication
13	Opaxio®	CTI BioPharma	Solid polyglutamate nanoparticles linked paclitaxel	Approved in 2012	Glioblastoma
14	NanoTherm®	Cambridge Nanotherm Ltd	Aminosilane-coated superparamagnetic iron oxide nanoparticles	Approved in 2013 (Europe)	Local ablation in glioblastoma
15	Plegridy®	Biogen	Polymer-protein conjugate (PEGylated IFN beta–1a)	Approved in 2014	Multiple Sclerosis
16	Focalin XR®	Novartis	Dexamethyl-phenidate HCl nanocrystals	Approved in 2015	Psychostimulant

Currently, total nanomedicine of pre-clinical phase is higher in numbers than clinical phases or approved drugs. However its promising properties and their advanced role in mainstream neuromedicine, it still has some drawback, like toxicity, BBB permeability, higher selectivity, solubility, and optimization of nanosystem design. Other than these, complex drug approval rules of FDA, difficulty in finding strong investors and high expenses to bring the medicine in market itself a challenge. Therefore, future research needs to overcome the drawbacks of nanomedicine in neuronal disorders.

KEYWORDS

- **Alzheimer's disease**
- **blood-brain barrier**
- **brain tumor**
- **CNS disorder**
- **nanocarrier**
- **nanomedicine**
- **nanotechnology**
- **neuro-AIDS**
- **Parkinson's disease**
- **spinal cord injury**
- **stroke**

REFERENCES

Abbasi, E., et al., (2014). Dendrimers: Synthesis, applications, and properties. *Nanoscale Res. Lett.*, *9*(1), 247.

Agemy, L., et al., (2011). Targeted nanoparticle enhanced proapoptotic peptide as potential therapy for glioblastoma. *Proc. Natl. Acad. Sci. U.S.A*, *108*(42), 17450–17455.

Ahmed, I., et al., (2006). Three-dimensional nanofibrillar surfaces covalently modified with tenascin-C-derived peptides enhance neuronal growth in vitro. *J. Biomed. Mater. Res. A.*, *76*(4), 851–860.

Akbarzadeh, A., et al., (2013). Liposome: Classification, preparation, and applications. *Nanoscale Res. Lett.*, *8*(1), 102.

Alam, M. I., et al., (2010). Strategies for effective brain drug delivery. *Eur. J. Pharm. Sci.*, *40*, 385–403.

Albanese, A., Tang, P. S., & Chan, W. C., (2012). The effect of nanoparticle size, shape, and surface chemistry on biological systems. *Annu. Rev. Biomed. Eng., 14*, 1–6.

Al-Ghananeem, A. M., et al., (2010). Intranasal drug delivery of didanosine-loaded chitosan nanoparticles for brain targeting, an attractive route against infections caused by AIDS viruses. *J. Drug Target., 18*(5), 381–388.

Allouche, J., (2013). Synthesis of organic and bioorganic nanoparticles: an overview of the preparation methods. In: Roberta, B., Fernand, F., & Thibaud, C., (eds.), *Nanomaterials: A Danger or a Promise?* (pp. 27–74). Springer: London.

Almeida, A. J., & Souto, E., (2007). Solid lipid nanoparticles as a drug delivery system for peptides and proteins. *Adv. Drug Deliv. Rev., 59*(6), 478–490.

Almeida, S. M. D., Letendre, S., & Ellis, R., (2006). Human immunodeficiency virus and the central nervous system. *Braz. J. Infect. Dis., 10*(1), 41–50.

Awasthi, K., Srivastava, A., & Srivastava, O. N., (2005). Synthesis of carbon nanotubes. *J. Nanosci. Nanotechnol., 5*(10), 1616–1636.

Baig, T., et al., (2015). A review about dendrimers: Synthesis, types, characterization, and applications. *Int. J. Adv. Pharm. Biol. Chem., 4*, 44–59.

Bangham, A. D., & Horne, R. W., (1964). Negative staining of phospholipids and their structural modification by surface-active agents as observed in the electron microscope. *J. Mol. Biol., 8*(1), 660IN2–8IN10.

Banks, W. A., (2009). Characteristics of compounds that cross the blood-brain barrier. *BMC Neurol., 9*(1), S3.

Batrakova, E. V., et al., (1999). Pluronic P85 increases permeability of a broad spectrum of drugs in polarized BBMEC and Caco-2 cell monolayers. *Pharm. Res., 16*(9), 1366–1372.

Beg, S., et al., (2011). Advancement in carbon nanotubes: Basics, Biomedical applications, and toxicity. *J. Pharm Pharmacol., 63*, 141–163.

Beg, S., Samad, A., Alam, M. I., & Nazish, I., (2011). Dendrimers an effective and novel targeting agent for the neuropharmaceuticals to brain. *CNS Neurol. Dis. Drug Target., 10*(5), 576–588.

Begley, D. J., (2004). Delivery of therapeutic agents to the central nervous system: The problems and the possibilities. *Pharmacol. Ther., 104*(1), 29–45.

Bendayan, R., Lee, G., & Bendayan, M., (2002). Functional expression and localization of P-glycoprotein at the blood-brain barrier. *Microsc. Res. Tech., 57*(5), 365–380.

Bendayan, R., Ronaldson, P. T., Gingras, D., et al., (2006). In situ localization of p-glycoprotein (ABCB1) in human and rat brain. *J. Histochem. Cytochem., 54*(10), 1159–1167.

Bernardi, A., et al., (2010). Protective effects of indomethacin-loaded nanocapsules against oxygen-glucose deprivation in organotypic hippocampal slice cultures: involvement of neuroinflammation. *Neurochem. Int., 57*(6), 629–636.

Bhaskar, S., et al., (2010). Multifunctional nanocarriers for diagnostics, drug delivery and targeted treatment across blood-brain barrier: Perspectives on tracking and neuroimaging. *Part. Fiber Toxicol., 7*(1), 3.

Blesa, J., et al., (2012). Classic and new animal models of Parkinson's disease. *BioMed. Res. Int.*, 2012.

Bowman, M. C., et al., (2008). Inhibition of HIV fusion with multivalent gold nanoparticles. *J. Am. Chem. Soc., 130*(22), 6896–6897.

Brewer, M., et al., (2007). Future approaches of nanomedicine in clinical science. *Med. Clin. North Am., 91*(5), 963–1016.

Britton, G. L., et al., (2010). In vivo therapeutic gas delivery for neuroprotection with echogenic liposomes. *Circulation, 122*(16), 1578–1587.

Buss, A., et al., (2004). Gradual loss of myelin and formation of an astrocytic scar during Wallerian degeneration in the human spinal cord. *Brain, 127*(1), 34–44.

Cabaleiro-Lago, C., et al., (2008). Inhibition of amyloid β protein fibrillation by polymeric nanoparticles. *J. Am. Chem. Soc., 130*(46), 15437–15443.

Cai, X., et al., (2008). Polyhydroxylated fullerene derivative C60OH24 prevents mitochondrial dysfunction and oxidative damage in an MPP+-induced cellular model of Parkinson's disease. *J. Neurosci. Res., 86*(16), 3622–3634.

Carné-Sánchez, A., et al., (2013). A spray-drying strategy for synthesis of nanoscale metal-organic frameworks and their assembly into hollow superstructures. *Nat. Chem., 5*(3), 203–211.

Castro, E., & Kumar, A., (2013). Nanoparticles in drug delivery systems. In: Kumar, A., Mansour, H. M., Friedman, A., & Blough, E. R., (ed.), *Nanomedicine in Drug Delivery* (pp. 1–22). CRC Press.

Cellot, G., et al., (2009). Carbon nanotubes might improve neuronal performance by favoring electrical shortcuts. *Nat. Nanotechnol., 4*(2), 126–133.

Charitidis, C. A., et al., (2014). Manufacturing nanomaterials: From research to industry. *Manuf. Rev., 1*, 11.

Chen, C. L., et al., (2008). Bioavailability effect of methylprednisolone by polymeric micelles. *Pharm. Res., 25*(1), 39–47.

Chen, S., e al., (2007). Temperature-responsive magnetite/PEO– PPO– PEO block copolymer nanoparticles for controlled drug targeting delivery. *Langmuir, 23*(25), 12669–12676.

Cherny, R. A., et al., (2001). Treatment with a copper-zinc chelator markedly and rapidly inhibits β-amyloid accumulation in Alzheimer's disease transgenic mice. *Neuron, 30*(3), 665–676.

Cho, Y., et al., (2008). Functionalized mesoporous silica nanoparticle-based drug delivery system to rescue acrolein-mediated cell death. *Nanomed., 3*(4), 507–519.

Cigognini, D., et al., (2011). Evaluation of early and late effects into the acute spinal cord injury of an injectable functionalized self-assembling scaffold. *PloS One, 6*(5), e19782.

Cook, C., & Petrucelli, L., (2009). A critical evaluation of the ubiquitin-proteasome system in Parkinson's disease. *Biochim. Biophys. Acta, 1792*(7), 664–675.

Cui, Z., et al., (2005). Novel D-penicillamine carrying nanoparticles for metal chelation therapy in Alzheimer's and other CNS diseases. *Eur. J. Pharm. Biopharm., 59*(2), 263–272.

Dallas, S., Miller, D. S., & Bendayan, R., (2006). Multidrug resistance-associated proteins: Expression and function in the central nervous system. *Pharmacol. Rev., 58*(2), 40–161.

Das, M., et al., (2007). Auto-catalytic ceria nanoparticles offer neuroprotection to adult rat spinal cord neurons. *Biomaterials, 28*(10), 1918–1925.

Demeule, M., et al., (2008). Involvement of the low-density lipoprotein receptor-related protein in the transcytosis of the brain delivery vector angiopep-2. *J. Neurochem., 106*(4), 1534–1544.

Deng, C. X., (2010). Targeted drug delivery across the blood-brain barrier using ultrasound technique. *Ther. Deliv., 1*(6), 819–848.

Denyer, R., & Douglas, M. R., (2012). Gene therapy for Parkinson's disease. *Parkinson's Dis., 2012*, 757305.

Destache, C. J., et al., (2010). Antiretroviral release from poly (DL-lactide-co-glycolide) nanoparticles in mice. *J. Antimicrob. Chemother., 65*(10), 2183–2187.

Devalapally, H., Chakilam, A., & Amiji, M. M., (2007). Role of nanotechnology in pharmaceutical product development. *J. Pharm. Sci., 96*(10), 2547–2565.

Dhanikula, R. S., et al., (2008). Methotrexate loaded polyether-copolyester dendrimers for the treatment of gliomas: Enhanced efficacy and intratumoral transport capability. *Mol. Pharm.*, *5*(1), 105–116.

Ding, H., et al., (2010). Inhibition of brain tumor growth by intravenous poly (beta-l-malic acid) nanobioconjugate with pH-dependent drug release. *Proc. Natl Acad. Sci. USA.*, *107*(42), 18143–18148.

DiNunzio, J. C., & Williams, III, R. O., (2008). CNS disorders-current treatment options and the prospects for advanced therapies. *Drug Dev. Ind. Pharm.*, *34*(11), 1141–1167.

Dixon, T. M., & Budd, M. A., (2017). Spinal cord injury. In: Budd, M. A., Hough, S., Wegener, S. T., & Stiers, W., (eds.), *Practical Psychology in Medical Rehabilitation* (pp. 127–136). Springer International Publishing.

Doi, A., et al., (2008). Tumor-specific targeting of sodium borocaptate (BSH) to malignant glioma by transferrin-PEG liposomes: A modality for boron neutron capture therapy. *J. Neurooncol.*, *87*(3), 287–294.

Domínguez, A., Suárez-Merino, B., & Goñi-de-Cerio, F., (2014). Nanoparticles and blood-brain barrier: The key to central nervous system diseases. *J. Nanosci. Nanotechnol.*, *14*(1), 766–779.

Dou, H., et al., (2007). Laboratory investigations for the morphologic, pharmacokinetic, and anti-retroviral properties of indinavir nanoparticles in human monocyte-derived macrophages. *Virology*, *358*(1), 148–158.

Dusserre, N., et al., (1995). Encapsulation of foscarnet in liposomes modifies drug intracellular accumulation, in vitro anti-HIV–1 activity, tissue distribution, and pharmacokinetics. *AIDS*, *9*(8), 833–842.

Ekambaram, P., Sathali, A. A., & Priyanka, K., (2012). Solid lipid nanoparticles: A review. *Scientific Rev. Chem. Comm.*, *2*(1), 80–102.

Elechiguerra, J. L., et al., (2005). Interaction of silver nanoparticles with HIV–1. *J. Nanobiotechnol.*, *3*(1), 6.

Fehlings, M. G., & Tator, C. H., (1995). The relationships among the severity of spinal cord injury, residual neurological function, axon counts, and counts of retrogradely labeled neurons after experimental spinal cord injury. *Exp. Neurol.*, *132*(2), 220–228.

Feng, B., et al., (2009). Delivery of sodium borocaptate to glioma cells using immunoliposome conjugated with anti-EGFR antibodies by ZZ-His. *Biomaterials*, *30*(9), 1746–1755.

Fitzgerald, P., et al., (2008). Treatment with phosphatidylglycerol based nanoparticles prevents motor deficits induced by proteasome inhibition: Implications for Parkinson's disease. *Behav. Brain Res.*, *195*(2), 271–274.

Freese, C., et al., (2017). Identification of neuronal and angiogenic growth factors in an in vitro blood-brain barrier model system: Relevance in barrier integrity and tight junction formation and complexity. *Microvasc. Res.*, *111*, 1–11.

Gaillard, P. J., et al., (2012). Enhanced brain drug delivery: Safely crossing the blood-brain barrier. *Drug Discov. Today Technol.*, *9*(2), 155–160.

Gao, G. H., et al., (2011). pH-responsive polymeric micelle-based on PEG-poly(β-amino ester)/(amidoamine) as intelligent vehicle for magnetic resonance imaging in detection of cerebral ischemic area. *J. Control. Release*, *155*(1), 11–17.

Garcion, E., et al., (2006). A new generation of anticancer, drug-loaded, colloidal vectors reverses multidrug resistance in glioma and reduces tumor progression in rats. *Mol. Cancer Ther.*, *5*(7), 1710–1722.

Premkumar, T. & Geckeler, K.E., (2009). Gold nanoparticles and carbon nanotubes: Precursors for novel composite materials. In Geckeler, K.E. & Nishide, H. (eds.). *Advanced Nanomaterials* (p. 249-297). John Wiley & Sons, Inc., New York.

Guo, J., et al., (2007). Reknitting the injured spinal cord by self-assembling peptide nanofiber scaffold. *Nanomed. Nanotech. Biol. Med.*, *3*(4), 311–321.

Guo, J., et al., (2011). Aptamer-functionalized PEG–PLGA nanoparticles for enhanced anti-glioma drug delivery. *Biomaterials*, *32*(31), 8010–8020.

Gupta, M., & Vyas, S. P., (2011). Role of polymeric nanocarriers for cancer chemotherapy. In: Vyas, S. P., Murthy, R. S. R., & Narang, R. K. (eds.), *Nanocolloidal Carriers: Site Specific and Controlled Drug Delivery* (pp. 365–397). CBS publishers.

Hartz, A. M., Zhong, Y., Wolf, A., LeVine, H., Miller, D. S., & Bauer, B., (2016). Aβ40 reduces P-glycoprotein at the blood-brain barrier through the ubiquitin-proteasome pathway. *J. Neurosci.*, *36*, 1930–1941.

Hazra, M., Singh, S. K., & Ray, S., (2011). Surface modification of liposomal vaccines by peptide conjugation. *J. Pharm. Sci. Tech.*, *1*, 41–47.

He, G., et al., (2007). ABA and BAB-type triblock copolymers of PEG and PLA: A comparative study of drug release properties and "stealth" particle characteristics. *Int. J. Pharm.*, *334*(1), 48–55.

He, H., et al., (2010). PEGylated Poly(amidoamine) dendrimer-based dual-targeting carrier for treating brain tumors. *Biomaterials*, *32*(2), 478–487.

He, H., et al., (2013). Carbon nanotubes: Applications in pharmacy and medicine. *BioMed. Res. Int.*, *2013*.

Hishida, A., (2007). Clinical analysis of 207 patients who developed renal disorders during or after treatment with edaravone reported during post-marketing surveillance. *Clin. Exp. Nephrol.*, *11*(4), 292–296.

Horcajada, P., et al., (2010). Porous metal-organic-framework nanoscale carriers as a potential platform for drug delivery and imaging. *Nat. Mater.*, *9*(2), 172–178.

Hu, K., et al., (2011). Lactoferrin conjugated PEG-PLGA nanoparticles for brain delivery: Preparation, characterization, and efficacy in Parkinson's disease. *Int. J. Pharm.*, *415*(1), 273–283.

Huang, R., et al., (2008). The use of lactoferrin as a ligand for targeting the polyamidoamine-based gene delivery system to the brain. *Biomaterials*, *29*(2), 238–246.

Illum, L., (2004). Is nose-to-brain transport of drugs in man a reality? *J. Pharma. Pharmacol.*, *56*(1), 3–17.

IR, W., (2001). *In Research of Simple Solution for Complex Molecule* (pp. 9–11). Scrip Magazine.

Jayant, R. D., et al., (2015). Sustained-release nano-ART formulation for the treatment of neuro-AIDS. *Int. J. Nanomed.*, *10*, 1077–1093.

Jia, G., et al., (2005). Cytotoxicity of carbon nanomaterials: Single-wall nanotube, multi-wall nanotube, and fullerene. *Environ. Sci. Technol.*, *39*(5), 1378–1383.

Jiang, X., et al., (2011). PEGylated poly(trimethylene carbonate) nanoparticles loaded with paclitaxel for the treatment of advanced glioma: *In vitro* and *in vivo* evaluation. *Int. J. Pharm.*, *420*(2), 385–394.

Jiménez, J. L., et al., (2010). Carbosilane dendrimers to transfect human astrocytes with small interfering RNA targeting human immunodeficiency virus. *BioDrugs.*, *24*(5), 331–343.

Jin, H., et al., (2000). Polyhydroxylated C60, fullerenols, as glutamate receptor antagonists and neuroprotective agents. *J. Neurosci. Res.*, *62*(4), 600–607.

Jin, S. X., et al., (2005). Pharmacokinetics and tissue distribution of zidovudine in rats following intravenous administration of zidovudine myristate loaded liposomes. *Pharmazie.*, *60*(1), 840–843.

Joshi, S. A., Chavhan, S. S., & Sawant, K. K., (2010). Rivastigmine-loaded PLGA and PBCA nanoparticles: Preparation, optimization, characterization, in vitro and pharmacodynamic studies. *Eur. J. Pharm. Biopharm.*, *76*(2), 189–199.

Juillerat-Jeanneret, L., (2008). The targeted delivery of cancer drugs across the blood-brain barrier: Chemical modifications of drugs or drug-nanoparticles? *Drug Discov. Today.*, *13*(23), 1099–1106.

Kamouchi, M., et al., (2013). Acute kidney injury and edaravone in acute ischemic stroke: The Fukuoka stroke registry. *J. Stroke Cerebrovasc. Dis.*, *22*(8), 470–476.

Karatas, H., et al., (2009). A nanomedicine transports a peptide caspase–3 inhibitor across the blood-brain barrier and provides neuroprotection. *J. Neurosci.*, *29*(44), 13761–13769.

Kawaguchi, A. T., et al., (2010). Liposome-encapsulated hemoglobin ameliorates ischemic stroke in nonhuman primates: An acute study. *J. Pharmacol. Exp. Ther.*, *332*(2), 429–436.

Kim, H. R., et al., (2007). Translocation of poly(ethylene glycol-co-hexadecyl)cyanoacrylate nanoparticles into rat brain endothelial cells: Role of apolipoproteins in receptor-mediated endocytosis. *Biomacromolecules*, *8*(3), 793–799.

Kim, I. D., et al., (2010). Neuroprotection by biodegradable PAMAM ester (e-PAM-R)-mediated HMGB1 siRNA delivery in primary cortical cultures and in the postischemic brain. *J. Control. Release.*, *142*(3), 422–430.

Kim, S., et al., (1990). Direct cerebrospinal fluid delivery of an antiretroviral agent using multivesicular liposomes. *J. Infect. Dis.*, *162*(3), 750–752.

Kim, Y. T., Caldwell, J. M., & Bellamkonda, R. V., (2009). Nanoparticle-mediated local delivery of methylprednisolone after spinal cord injury. *Biomaterials*, *30*(13), 2582–2590.

Kincses, Z. T., & Vecsej, L., (2011). Pharmacological therapy in Parkinson disease: Focus on neuroprotection. *CNS Neurosci. Ther.*, *17*(5), 345–367.

Klajnert, B., et al., (2007). EPR study of the interactions between dendrimers and peptides involved in Alzheimer's and prion diseases. *Macromol. Biosci.*, *7*(8), 1065–1074.

Kolosnjaj, J., Szwarc, H., & Moussa, F., (2007). Toxicity studies of fullerenes and derivatives. *Adv. Exp. Med. Biol.*, (pp. 168–180). Springer, New York, NY.

Kumar, A., Saienni, A. E., & Dixit, N., (2013). Synthesis and characterization of nanocapsulated drugs. In: Kumar, A., Mansour, H. M., Friedman, A., & Blough, E. R., (eds.), *Nanomedicine in Drug Delivery* (pp. 23–42). CRC Press.

Kuo, Y. C., & Chen, H. H., (2009). Entrapment and release of saquinavir using novel cationic solid lipid nanoparticles. *Int. J. Pharm.*, *365*(1), 206–213.

Kuo, Y. C., & Kuo, C. Y., (2008). Electromagnetic interference in the permeability of saquinavir across the blood-brain barrier using nanoparticulate carriers. *Int. J. Pharm.*, *351*(1), 271–81.

Kuo, Y. C., & Lee, C. L., (2012). Methylmethacrylate–sulfopropyl methacrylate nanoparticles with surface RMP–7 for targeting delivery of antiretroviral drugs across the blood-brain barrier. *Colloids Surf. B: Biointerfaces*, *90*, 75–82.

Kuo, Y. C., & Su, F. L., (2007). Transport of stavudine, delavirdine, and saquinavir across the blood-brain barrier by polybutylcyanoacrylate, methylmethacrylate-sulfopropyl methacrylate, and solid lipid nanoparticles. *Int. J. Pharm.*, *340*(1), 143–152.

Kusuhara, H., & Sugiyama, Y., (2001). Efflux transport systems for drugs at the blood-brain barrier and blood-cerebrospinal fluid barrier (Part 2). *Drug Discov. Today.*, *6*(4), 206–212.

Lamprecht, A., & Benoit, J. P., (2006). Etoposide nanocarriers suppress glioma cell growth by intracellular drug delivery and simultaneous P-glycoprotein inhibition. *J. Control Release.*, *112*(2), 208–213.

Lee, H. J., et al., (2011). Amine-modified single-walled carbon nanotubes protect neurons from injury in a rat stroke model. *Nat. Nanotechnol.*, *6*(2), 121–125.

Lees, A. J., Hardy, J., & Revesz, T., (2009). Parkinson's disease. *Lancet*, *373*, 2055–2066.

Lin, A. M., et al., (1999). Carboxyfullerene prevents iron-induced oxidative stress in rat brain. *J. Neurochem.*, *72*(4), 1634–1640.

Lin, Y., et al., (2009). Physiologically relevant online electrochemical method for continuous and simultaneous monitoring of striatum glucose and lactate following global cerebral ischemia/reperfusion. *Anal. Chem.*, *81*(6), 2067–2074.

Liu, X. F., Guan, W. C., & Ke, W. S., (2007). Synthesis and enhanced neuroprotective activity of C60-based epsilon derivatives. *Can. J. Chem.*, *85*(3), 157–163.

Liu, Z., et al., (2013). B6 peptide-modified PEG-PLA nanoparticles for enhanced brain delivery of neuroprotective peptide. *Bioconjug. Chem.*, *24*(6), 997–1007.

Loscher, W., & Potschka, H., (2005). Drug resistance in brain diseases and the role of drug efflux transporters, *Nat. Rev. Neurosci.*, *6*(8), 591–602.

Lotharius, J., Dugan, L. L., & O'Malley, K. L., (1999). Distinct mechanisms underlie neurotoxin-mediated cell death in cultured dopaminergic neurons. *J. Neurosci.*, *19*(4), 1284–1293.

Lu, W., Sun, Q., Wan, J., et al., (2006). Cationic albumin-conjugated pegylated nanoparticles allow gene delivery into brain tumors via intravenous administration. *Cancer Res.*, *66*(24), 11878–11887.

Mahajan, S. D., et al., (2010). Enhancing the delivery of antiretroviral drug "Saquinavir" across the blood-brain barrier using nanoparticles. *Curr. HIV Res.*, *8*(5), 396–404.

Manna, S. K., et al., (2005). Single-walled carbon nanotube induces oxidative stress and activates nuclear transcription factor-κB in human keratinocytes. *Nano. Lett.*, *5*(9), 1676–1684.

Marcato, P. D., & Durán, N., (2008). New aspects of nanopharmaceutical delivery systems. *J. Nanosci. Nanotechnol.*, *8*(5), 2216–2229.

Marqués-Gallego, P., & de Kroon, A. I., (2014). Ligation strategies for targeting liposomal nanocarriers. *BioMed. Res. Int.*, 2014.

Marsh, J. N., et al., (2011). A fibrin-specific thrombolytic nanomedicine approach to acute ischemic stroke. *Nanomed.*, *6*(4), 605–615.

Mathew, A., et al., (2012). Curcumin-loaded-PLGA nanoparticles conjugated with Tet–1 peptide for potential use in Alzheimer's disease. *PLoS One*, *7*(3), e32616.

Mazzatenta, A., et al., (2007). Interfacing neurons with carbon nanotubes: Electrical signal transfer and synaptic stimulation in cultured brain circuits. *J. Neurosci.*, *27*(26), 6931–6936.

McArthur, J. C., Brew, B. J., & Nath, A., (2005). Neurological complications of HIV infection. *Lancet Neurol.*, *4*(9), 543–555.

Michaelis, K., et al., (2006). Covalent linkage of apolipoprotein E to albumin nanoparticles strongly enhances drug transport into the brain. *J. Pharmacol. Exp. Ther.*, *317*(3), 1246–1253.

Mikitsh, J. L., & Chacko, A. M., (2014). Pathways for small molecule delivery to the central nervous system across the blood-brain barrier. *Perspect. Medicin. Chem.*, *6*, 11–24.

Mishra, V., et al., (2006). Targeted brain delivery of AZT via transferrin anchored pegylated albumin nanoparticles. *J. Drug Target.*, *14*(1), 45–53.

Misra, A., et al., (2003). Drug delivery to the central nervous system: A review. *J. Pharm. Pharm Sci.*, *6*(2), 252–273.

Mohanraj, V. J., & Chen, Y., (2006). Nanoparticles-a review. *Trop. J. Pharm. Res.*, *5*(1), 561–573.

Mourtas, S., et al., (2011). Curcumin-decorated nanoliposomes with very high affinity for amyloid-β1–42 peptide. *Biomaterials*, *32*(6), 1635–1645.

Mukherjee, S., Ray, S., & Thakur, R. S., (2009). Solid lipid nanoparticles: A modern formulation approach in drug delivery system. *Ind. J. Pharm. Sci.*, *71*(4), 349.

Mulik, R. S., et al., (2010). ApoE3 mediated poly (butyl) cyanoacrylate nanoparticles containing curcumin: Study of enhanced activity of curcumin against beta-amyloid-induced cytotoxicity using in vitro cell culture model. *Mol. Pharm.*, *7*(3), 815–825.

Nagavarma, B. V., et al., (2012). Different techniques for preparation of polymeric nanoparticles-a review. *Asian J. Pharm. Clin. Res.*, *5*(3), 16–23.

Nair, S. B., Dileep, A., & Rajanikant, G. K., (2012). Nanotechnology-based diagnostic and therapeutic strategies for neuroscience with special emphasis on ischemic stroke. *Curr. Medicin. Chem.*, *19*(5), 744–756.

Nalivaeva, N. N., & Turner, A. J., (2016). AChE and the amyloid precursor protein (APP)–Cross-talk in Alzheimer's disease. *Chem. Bio. Interact.*, *259*, 301–306.

Nikalje, A. P., (2015). Nanotechnology and its applications in medicine. *Med. Chem.*, *5*(2), 81–89.

Nowacek, A., Kosloski, L. M., & Gendelman, H. E., (2009). Neurodegenerative disorders and nanoformulated drug development. *Nanomed.*, *4*(5), 541–555.

Obermeier, B., Daneman, R., & Ransohoff, R. M., (2013). Development, maintenance, and disruption of the blood-brain barrier. *Nat. Med.*, *19*(12), 1584–1596.

Pai, A. S., Rubinstein, I., & Önyüksel, H., (2006). PEGylated phospholipid nanomicelles interact with β-amyloid (1–42) and mitigate its β-sheet formation, aggregation, and neurotoxicity *in vitro*. *Peptides*, *27*(11), 2858–2866.

Pardridge, W. M., (2010). Biopharmaceutical drug targeting to the brain. *J. Drug Target*, *18*(3), 157–167.

Pardridge, W. M., (2015). Blood-brain barrier endogenous transporters as therapeutic targets: A new model for small molecule CNS drug discovery. *Expert Opin. Ther. Targets.*, *19*(8), 1059–1072.

Park, J. H., et al., (2008). Polymeric nanomedicine for cancer therapy. *Prog. Polym. Sci.*, *33*, 113–137.

Podolski, I. Y., et al., (2007). Effects of hydrated forms of C60 fullerene on amyloid-peptide fibrillization *in vitro* and performance of the cognitive task. *Nanosci. Nanotechnol.*, *7*(4–1), 1–7.

Puri, A., et al., (2009). Lipid-based nanoparticles as pharmaceutical drug carriers: From concepts to clinic. *Crit. Rev. Ther. Drug Carrier Syst.*, *26*(6), 523–580.

Ramos-Cabrer, P., et al., (2011). Serial MRI study of the enhanced therapeutic effects of liposome-encapsulated citicoline in cerebral ischemia. *Int. J. Pharm.*, *405*(1 & 2), 228–233.

Rao, J. P., & Geckeler, K. E., (2011). Polymer nanoparticles: Preparation techniques and size-control parameters. *Prog. Polym. Sci.*, *36*(7), 887–913.

Rao, K. S., et al., (2008). TAT-conjugated nanoparticles for the CNS delivery of anti-HIV drugs. *Biomaterials*, *29*, 4429–4438.

Reddy, A. M., et al., (2010). In vivo tracking of mesenchymal stem cells labeled with a novel chitosan-coated superparamagnetic iron oxide nanoparticles using 3.0T MRI. *J. Korean Med. Sci.*, *25*(2), 211–219.

Reddy, M. K., & Labhasetwar, V., (2009). Nanoparticle-mediated delivery of superoxide dismutase to the brain: An effective strategy to reduce ischemia-reperfusion injury. *FASEB J.*, *23*(5), 1384–1395.

Reis, C. P., et al., (2006). Nanoencapsulation I. Methods for preparation of drug-loaded polymeric nanoparticles. *Nanomed. Nanotech. Biol. Med.*, *2*(1), 8–21.

Ren, J., et al., (2012). The targeted delivery of anticancer drugs to brain glioma by PEGylated oxidized multi-walled carbon nanotubes modified with angiopep–2. *Biomaterials*, *33*(11), 3324–3333.

Reukov, V., Maximov, V., & Vertegel, A., (2011). Proteins conjugated to poly (butyl cyanoacrylate) nanoparticles as potential neuroprotective agents. *Biotechnol. Bioeng.*, *108*(2), 243–252.

Revathi, S., Vuyyuru, M., & Dhanaraju, M. D., (2015). Carbon nanotube: A flexible approach for nanomedicine and drug delivery. *Carbon*, *8*(1), 25–31.

Reynolds, J. L., & Mahato, R. I., (2017). Nanomedicines for the Treatment of CNS Diseases. *J. Neuroimmune. Pharmacol.*, *12*(1), 1–5.

Roman, J. A., et al., (2011). Single-walled carbon nanotubes chemically functionalized with polyethylene glycol promote tissue repair in a rat model of spinal cord injury. *J. Neurotrauma.*, *28*(11), 2349–2362.

Ruiz, A., Nair, M., & Kaushik, A., (2015). Recent update in Nano-Cure of neuro-AIDS. *Sci. Lett. J.*, *4*, 172.

Saiyed, Z. M., Gandhi, N. H., & Nair, M. P., (2010). Magnetic nanoformulation of azidothymidine 5'-triphosphate for targeted delivery across the blood-brain barrier. *Int. J. Nanomed.*, *5*, 157–166.

Santos, S. G., et al., (2008). Adsorption of ascorbic acid on the C60 fullerene. *J. Phys. Chem. B*, *112*(45), 14267–14272.

Shi, Y., et al., (2010). Effective repair of traumatically injured spinal cord by nanoscale block copolymer micelles. *Nat. Nanotechnol.*, *5*(1), 80–87.

Singh, A. V., et al., (2012). Theranostic implications of nanotechnology in multiple sclerosis: A future perspective. *Autoimmune Dis.*, 2012.

Smalley, R. E., (2012). Fundamentals of nanotechnology. In: Burgess, R., (ed.), *Understanding Nanomedicine: An Introductory Textbook* (pp. 1–44). Pan Stanford Publishing, Singapore.

Soppimath, K. S., et al., (2001). Biodegradable polymeric nanoparticles as drug delivery devices. *J. Control. Release.*, *70*(1), 1–20.

Spitzenberger, T. J., et al., (2007). Novel delivery system enhances efficacy of antiretroviral therapy in animal model for HIV–1 encephalitis. *J. Cereb. Blood Flow Metab.*, *27*(5), 1033–1042.

Srikanth, M., & Kessler, J. A., (2012). Nanotechnology-novel therapeutics. *Nat. Rev. Neurol.*, *8*(6), 307–318.

Sriramoju, B. K., Kanwar, R. R., & Kanwar, J., (2014). Nanomedicine based nanoparticles for neurological disorders. *Curr. Med. Chem.*, *21*(31), 4154–4168.

Zankel, T., Starr, C.M. & Gabathuler, R., Biomarin Pharmaceutical Inc., 2005. *Delivery of therapeutic compounds to the brain and other tissues.* Patent publication no. WO/2005/002515

Steiniger, S. C., et al., (2004). Chemotherapy of glioblastoma in rats using doxorubicin-loaded nanoparticles. *Int. J. Cancer, 109*(5), 759–767.

Sun, X., et al., (2012). Co-delivery of pEGFP-hTRAIL and paclitaxel to brain glioma mediated by an angiopep-conjugated liposome. *Biomaterials, 33*(3), 916–924.

Talegaonkar, S., Tariq, M., & Iqbal, Z., (2013). Formulations of nanoparticles in drug delivery. In: Kumar, A., Mansour, H. M., Friedman, A., & Blough, E. R., (eds.), *Nanomedicine in Drug Delivery* (pp. 239–285). CRC Press.

Taylor, M., et al., (2011). Effect of curcumin-associated and lipid ligand-functionalized nanoliposomes on aggregation of the Alzheimer's Aβ peptide. *Nanomed. Nanotechnol. Biol. Med., 7*(5), 541–550.

Tian, X. H., et al., (2011). Enhanced brain targeting of temozolomide in polysorbate-80 coated polybutylcyanoacrylate nanoparticles. *Int. J. Nanomed., 6*, 445–452.

Torchilin, V. P., (2005). Recent advances with liposomes as pharmaceutical carriers. *Nat. Rev. Drug Discov., 4*(2), 145–60.

Torchilin, V., (2008). Antibody-modified liposomes for cancer chemotherapy. *Expert Opin. Drug Deliv., 5*(9), 1003–1025.

Tsai, M. J., et al., (2011). Oral apomorphine delivery from solid lipid nanoparticles with different monostearate emulsifiers: Pharmacokinetic and behavioral evaluations. *J. Pharm. Sci., 100*(2), 547–557.

Tyler, J. Y., Xu, X. M., & Cheng, J. X., (2013). Nanomedicine for treating spinal cord injury. *Nanoscale, 5*(19), 8821–8836.

Uner, M., & Yener, G., (2007). Importance of solid lipid nanoparticles (SLN) in various administration routes and future perspectives. *Int. J. Nanomed., 2*(3), 289.

Upadhyay, R. K., (2014). Drug delivery systems, CNS protection, and the blood-brain barrier. *BioMed. Res. Int., 2014*, p. 37.

Urakami, T., et al., (2009). *In vivo* distribution of liposome-encapsulated hemoglobin determined by positron emission tomography. *Artif. Organs., 33*(2), 164–168.

Van Handel, M., et al., (2009). Selective uptake of multi-walled carbon nanotubes by tumor macrophages in a murine glioma model. *J. Neuroimmunol., 208*(1), 3–9.

Van Landeghem, F. K., et al., (2009). Post-mortem studies in glioblastoma patients treated with thermotherapy using magnetic nanoparticles. *Biomaterials, 30*(1), 52–57.

Ventola, C. L., (2012). The nanomedicine revolution: Part 2: Current and future clinical applications. *Pharm. Ther., 37*(10), 582.

Wang, C. X., et al., (2009a). Antitumor effects of polysorbate–80 coated gemcitabine polybutylcyanoacrylate nanoparticles in vitro and its pharmacodynamics in vivo on C6 glioma cells of a brain tumor model. *Brain Res., 1261*, 91–99.

Wang, Q., et al., (2009b). Neuroprotective effects of a nanocrystal formulation of sPLA(2) inhibitor PX–18 in cerebral ischemia/reperfusion in gerbils. *Brain Res., 1285*, 188–195.

Weissig, V., Pettinger, T. K., & Murdock, N., (2014). Nanopharmaceuticals (part 1): Products on the market. *Int. J. Nanomed., 9*, 4357.

Weksler, B. B., et al., (2005). Blood-brain barrier-specific properties of a human adult brain endothelial cell line. *FASEB J., 19*(13), 1872–1874.

Wen, Z., et al., (2011). Odorranalectin-conjugated nanoparticles: Preparation, brain delivery and pharmacodynamic study on Parkinson's disease following intranasal administration. *J. Control. Release, 151*(2), 131–138.

Wilson, B., et al., (2008). Poly(n-butyl cyanoacrylate) nanoparticles coated with polysorbate 80 for the targeted delivery of rivastigmine into the brain to treat Alzheimer's disease. *Brain Res., 1200*, 159–168.

Wong, H. L., Wu, X. Y., & Bendayan, R., (2012). Nanotechnological advances for the delivery of CNS therapeutics. *Adv. Drug Deliv. Rev.*, *64*(7), 686–700.

Wu, G., et al., (2006). Targeted delivery of methotrexate to epidermal growth factor receptor-positive brain tumors by means of cetuximab (IMC-C225) dendrimer bioconjugates. *Mol. Cancer Ther.*, *5*(1), 52–59.

Xiang, J. J., et al., (2003). IONP-PLL: A novel non-viral vector for efficient gene delivery. *J. Gene Med.*, *5*(9), 803–817.

Xie, J., et al., (2009). The differentiation of embryonic stem cells seeded on electrospun nanofibers into neural lineages. *Biomaterials*, *30*(3), 354–362.

Xin, H., et al., (2011). Angiopep-conjugated poly(ethylene glycol)-co-poly(ε-caprolactone) nanoparticles as dual-targeting drug delivery system for brain glioma. *Biomaterials*, *32*(18), 4293–4305.

Yadav, S., et al., (2012). Nanomedicine: Current status and future implications. *J. Nanomed. Nanotechnol.*, *6*, 11–14.

Yamashita, T., et al., (2012). Carbon nanomaterials: Efficacy and safety for nanomedicine. *Materials*, *5*(2), 350–363.

Yiu, G., & He, Z., (2006). Glial inhibition of CNS axon regeneration. *Nat. Rev. Neurosci.*, *7*(8), 617.

Yoshitomi, T., & Nagasaki, Y., (2011). Nitroxyl radical-containing nanoparticles for novel nanomedicine against oxidative stress injury. *Nanomed (Lond).*, *6*(3), 509–518.

Zhan, C., et al., (2011). Micelle-based brain-targeted drug delivery enabled by a nicotine acetylcholine receptor ligand. *Angew. Chem. Int. Ed.*, *50*(24), 5482–5485.

Zhang, L., et al., (2008). Nanoparticles in medicine: Therapeutic applications and developments. *Clin. Pharm. Ther.*, *83*(5), 761–769.

Zhang, M., et al., (2005). Continuous on-line monitoring of extracellular ascorbate depletion in the rat striatum induced by global ischemia with carbon nanotube modified glassy carbon electrode integrated into a thin-layer radial flow cell. *Anal. Chem.*, *77*(19), 6234–6242.

Zhang, T. T., et al., (2016). Strategies for transporting nanoparticles across the blood-brain barrier. *Biomater. Sci.*, *4*(2), 219–229.

Zhang, Y., et al., (2002). Receptor-mediated delivery of an antisense gene to human brain cancer cells. *J. Gene Med.*, *4*(2), 183–194.

Zhang, Y., et al., (2003). Intravenous nonviral gene therapy causes normalization of striatal tyrosine hydroxylase and reversal of motor impairment in experimental parkinsonism. *Hum. Gene Ther.*, *14*(1), 1–12.

Zhao, H., et al., (2011). Postacute ischemia vascular endothelial growth factor transfer by transferrin-targeted liposomes attenuates ischemic brain injury after experimental stroke in rats. *Hum. Gene Ther.*, *22*(2), 207–215.

Zhao, Z., et al., (2015). Establishment and dysfunction of the blood-brain barrier. *Cell.*, *163*(5), 1064–1078.

Zhu, L., et al., (2010). Molecular mechanism of protective effect of puerarin solid lipid nanoparticle on cerebral ischemia-reperfusion injury in gerbils. *Zhong Yao Cai.*, *33*(12), 1900–1904.

Zlokovic, B. V., (2008). The blood-brain barrier in health and chronic neurodegenerative disorders. *Neuron.*, *57*(2), 178–201.

CHAPTER 5

Nanomedicinal Genetic Manipulation: Promising Strategy to Treat Some Genetic Diseases

BISWAJIT MUKHERJEE*, IMAN EHSAN, DEBASMITA DUTTA, MOUMITA DHARA, LOPAMUDRA DUTTA, and SOMA SENGUPTA

Department of Pharmaceutical Technology, Jadavpur University, Kolkata–700032, India, Tel. & Fax: +91-33-24146677, E-mail: biswajit55@yahoo.com

Corresponding author. E-mail: biswajit55@yahoo.com

ABSTRACT

Genetic disease is characterized as an ailment due to a defect in an individual's genome, chromosomal abnormalities, and gene mutations. A successful therapeutic and diagnostic intervention on genetic diseases has been unrevealed predominately, and it demands technological development. Nanomedicine is a boon in various fields of medical applications from the development of diagnostic devices, contrast agents, *in vivo* imaging, analytical tools to drug delivery vehicles. Versatilities of nanomedicines owing to the dimension, biocompatibility, and manipulating capability to achieve successful delivery at the targeted site of action and their lower toxicity have made this field a technological revolution to combat different unmet needs in health. This is a possible alternative for combating chronic genetic disorders by targeting a particular gene(s) or tissue or cell, which creates a pathological state. Thus, nanomedicine could be a potential therapeutic tool to treat different genetic pathological conditions, such as diabetes, heart diseases, lysosomal storage disorders, Alzheimer's disease, cystic fibrosis, etc. In this chapter, we will discuss several strategies to manipulate or knock down the expression of the lethal gene(s) specifically responsible for some of such diseases. Here we mainly focus on diseases such as cancer and its fate while manipulating genetic alteration by the nanomedicinal strike.

5.1   INTRODUCTION

Genetic diseases are upcoming well-known health hazards in the entire world, and improved technological diagnostics and therapeutics are required to combat with these problems efficiently. Genetic diseases are defined as any diseases or disorders caused by an imperfection in individual's genome, deformities in chromosome and mutation in pre-existing one or more genes. Nanomedicine offers innumerable prospects in the health care system, and it is an emerging field in genetic diseases through nanodiagnostics as well as nanotherapeutics and by applying nanotechnology in recombinant DNA technology and specially in molecular biology. Nanomedicine may also play a dynamic role in target specific gene delivery (Liu et al., 2007). Nanostructures have some exclusive physicochemical characteristics such as increased surface area, higher surface area to volume ratio, a resistor of quantum effects related with small sizes that prepare them as a potential candidate in genetic diseases (Alex et al., 2013). Various types of nanoformulations such as nanoliposomes, nanoparticles, nanospheres, dendrimers, nanocapsules, quantum dots (QD), nanorobots, etc. expand a new era to deliver active agents or genes to get proper therapeutic outcome avoiding higher toxicity along with enhanced biocompatibility.

In this chapter, we have focused on three genetic diseases (namely, cancer, diabetes, and hypertension), diagnostics of these diseases by various tests, advances of nano drug carriers with their mode of administration in concern of genetic diseases, and several targeting approach.

5.2   CATEGORIZATION OF GENETIC DISEASES OR DISORDERS

Genetic diseases may be categorized as single gene Mendelian inheritance, single gene non-Mendelian inheritance, chromosomal abnormalities and multifactorial inheritance (Stoppler, 2017; Robinson et al., 2017; National Human Genome Research Institute, 2017)

   a.   Single gene Mendelian inheritance: It is also known as monogenetic inheritance. This is due to alterations and mutations in DNA sequence of a single gene only. This inheritance is again divided into three groups such as autosomal dominant, autosomal recessive and sex-linked. In every 200 births, one birth affects by single gene Mendelian inheritance causing genetic diseases. Some examples of such diseases are sickle cell anemia, cystic fibrosis, Marfan syndrome, hemochromatosis, Huntington's disease, etc.

b. Single gene non-Mendelian inheritance: It occurs because of triple recurrence expansions within or near specific genes, mutations within imprinted genes or mitochondrial DNA, etc. Fragile-x syndrome, Prader-Willi syndrome, Angelman syndrome, etc. belong to this category.

c. Chromosomal abnormalities: Chromosomes, which are made by DNA and protein, are the transporters of genetic substances. Numerical and structural abnormalities in autosome or abnormalities in sex chromosomes are the probable reason for this disorder. Examples of such genetic diseases or disorders are Down syndrome (due to copies of chromosome 21 thrice), Klinefelter syndrome (47, XXY), Turner syndrome (45, X0), etc.

d. Multifactorial inheritance: It is related to the mutations in multiple genes as well as blend of environmental influences. This is also acknowledged as polygenic inheritance. Diabetes, cancer, obesity, high blood pressure, arthritis, Alzheimer's disease, etc. are the examples of this category. As mentioned earlier, in this chapter, we have selected and discussed on genetic manipulation of three diseases by nanomaterials of this category (Table 5.1).

TABLE 5.1 Sign and Symptoms of Some Genetic Diseases or Disorders (Robinson et al., 2017; Mayo Clinic, 2017; Boslaugh, 2017; Encyclopedia.com, 2017)

Disease or disorder	Cause	Sign and symptoms
Sickle cell anemia	Single gene Mendelian inheritance	Breathing problem, tiredness, hindered growth, abdominal, and muscle pain
Cystic fibrosis		Chronic intestinal and lung indications
Marfan syndrome		Cardiovascular and eye complications, thin as well as long extremities and fingers
Hemochromatosis		Liver damage, excess quantity of iron accretion
Huntington's disease		Dementia, emotional disturbance, involuntary movement
Fragile-x syndrome	Single gene non-Mendelian inheritance	Typical facial expression, mental retardation
Prader-Willi syndrome		Muscular hypotonia, strabismus, distinctive facial features
Angelman syndrome		Intellectual disability, frequent smiling, and laughter, movement problem

5.3   DIAGNOSIS OF GENETIC DISEASE

A broad clinical investigation comprising of three important components (Genetic Alliance, 2010) is essential to diagnosis of genetic diseases. These three vital components are:

   i.   physical investigation;
   ii.   family history in depth at medical aspect; and
   iii.   laboratory and clinical testings as per accessibility.

A conclusive diagnosis of a genetic disease is not always possible by primary health care providers at all times. In those cases, the activities of primary health care providers are very crucial for accumulation of family history in details, contemplation of genetic disease occurring probability through variable diagnosis, permission of specified testing or investigation as per accessibility and at last, referring patients properly to the genetic specialists.

Various factors are pointed out through variable diagnosis which indicates the probability of a genetic disease. During the gathering of family history, existence of similar condition in family members (more than one family member and mainly first-degree family relation) is a crucial aspect, indicating a genetic disease such as several miscarriages, deaths in childhood, stillbirths, etc. If heart disease, dementia, cancer, etc. are observed in family history (two or more close family relatives) in adult phase, it may also direct a genetic susceptibility. Congenital abnormalities, mental retardation or developmental delay, may also contemplate as a genetic disease after investigating clinical symptoms. Certain physical characteristics such as droopy or wide-set eyes, short fingers, flat face, and tall stature may seem as a considerably diverse manner which may not be recommended as a genetic disease by the primary health care provider, but it may be easy to decide whether it is a genetic disease or not, after investigating through the genetics specialist. Some genetic conditions seem in childhood but disappear in adults, on the other hand occasionally genetic diseases may be unnoticed for quite a few years, and pregnancy or puberty activates the disease as genetic (Genetic Alliance, 2010).

5.3.1   USING GENETIC TESTING TO DETERMINE GENETIC DISEASE (GENETIC ALLIANCE, 2010)

   a.   *Carrier testing:* It helps to acquire the information that parents convey the threat of genetic disease which permits to their child as a genetic disease such as Tay-Sachs disease, sickle cell anemia, cystic fibrosis, etc.

b. *New-born screening:* It is a frequently used genetic testing because initial detection of this disease helps to reduce rigorousness of disease or inhibit the starting of symptoms.

c. *Prenatal diagnosis:* It identifies the chromosomal or gene alterations in fetus.

d. *Predictive/predispositional:* Mutations which may be the cause of genetic disease such as cancer can be recognized by predictive tests.

e. *Diagnostic/prognostic tests:* Such tests may be utilized to check a diagnosis or monitor prognosis of a genetic disease.

5.3.2 CATEGORIES OF GENETIC TESTING

There are three types of existing genetic testing to determine genetic diseases. They are mentioned below (Genetic Alliance, 2010).

a. *Cytogenetic testing:* This is nothing but an investigation of entire chromosome responsible for causing anomalies.

b. *Biochemical testing:* Various kinds of proteins such as transporters, enzymes, regulatory proteins, structural proteins, hormones, and receptors necessitate for monotonous biochemical reactions in cells. Mutation of any such protein causing disaster of function of protein may be responsible for genetic disease. Those proteins may be tested for screening genetic diseases or disorders.

c. *Molecular testing:* When specific protein function is unknown and biochemical tests cannot be established, straight DNA testing may be applied for recognizing small DNA mutations.

5.4 LIMITATIONS OF CONVENTIONAL THERAPY

i. Rapid clearance of therapeutic agents from body;
ii. Poor bioavailability and pharmacokinetic profile of therapeutic agent;
iii. Increasing toxicity;
iv. Lower biocompatibility; and
v. Unable to reach in site-specific area of action.

5.5 NANOFORMULATIONS

A successful therapeutic and diagnostic intervention on genetic diseases has been predominately understood, and it demands technological development. Nanomedicine is a boon in various fields of medical applications from development of diagnostic devices, contrast agents, *in vivo* imaging, and analytical tools to drug delivery vehicles (Hsu and Muro, 2011). Versatilities of nanomedicines owing to the dimension, biocompatibility, and manipulating capability to achieve successful delivery at the targeted site of action and their lower toxicity made this field a technological revolution to combat different unmet needs in health. Encapsulation of active ingredients and altering the pharmacokinetics to enhance the bioavailability and stability of the drug is the most promising area in nanomedicine (Ahmad et al., 2017). Site-specific delivery of active molecules results in less toxicity, reduced dose and better therapeutic index of the drug over the conventional medications. Better patient compliance is also achieved by using nanoformulations since nanomaterials can modulate the rate of drug release by acting as a depot for sustained action (Bhatia, 2016). This is a possible alternative for combating chronic genetic disorders by targeting particular gene(s) or tissue or cell which creates pathological state. Some predominant polymeric nanomedicinal formulations are tabulated in Table 5.2 (Bhatia, 2016).

TABLE 5.2 Some Polymeric Nanoformulations

Types of nanocarrier	Size (nm)	Applications
Liposome	50–100	Long-circulating targeted delivery for proteins, genes, etc.
Polymer micelles	10–100	Poorly water-soluble pharmaceutical active ingredients can be delivered
Dendrimers	<10	Long-circulating controlled releases for liver targeting
Polymer nanoparticles	10–1000	Controlled and sustained delivery of bioactive for targeted delivery

5.5.1 *FEATURES OF THE NANOFORMULATIONS*

The ability to modify the surface properties and the size range of nanoformulations has opened a new era in medical sciences. The different features to be considered for designing the nanoformulations are discussed in the following sections.

Nanoformulations can be targeted to general as well as specific tissue/ organ/cellular level depending upon the situation by attaching suitable ligand molecules. They can be designed to alter the cellular membrane permeability so as to target a variety of intracellular components such as lysosome, nuclei, etc. (Langer, 1998; Sahay et al., 2010).

After administration into the body, active ingredients are quickly cleared from the biological systems being treated as foreign particles. Clearance mainly happened by the immune system, reticuloendothelial system and renal filtration (Moghimi and Szebeni, 2003). Rapid clearance hinders the drug to achieve its minimum effective concentration in the targeted site of action. Nanoformulations can be designed to extend the half-life of the drug to improve its effectiveness. This can be done in various ways like PEGylation, i.e., attaching polyethylene glycol (PEG) to the nanoformulation. PEG reduces the interaction of the formulation(s) with phagocytes, opsonins, and lymphocytes and alters the hydrophobicity of the drug molecule (Campbell and Hope, 2005). As a result, circulation time increases and enhanced therapeutic properties by reduced dose and reduced frequency of administration are achieved. On the other hand, CD47 is a transmembrane protein, and its binding to SIRPα prevents phagocytosis. Addition of CD47 on nanocarriers prolongs circulation time by reducing engulfment by macrophages (Stachelek et al., 2011).

5.5.2 DIFFERENT TARGETING APPROACHES

Targeting is required to maximize the drug concentration in the required area of the biological system for enhanced effectiveness and less toxicity (Hsu and Muro, 2011). There are various genetic disorders related to both central and peripheral nervous systems where general distribution in the body is preferred. Some strategies can be applied for specific targeting. Such as positively charged hydrophilic polymers show attraction towards negatively charged plasma cell membranes (El-Sayed, 2005). On the other hand, targeting is required is required for cell/tissue/organ-specific diseases to minimize side-effects and therapeutic efficiency (Pardridge, 2010). For example, liver is considered as a target organ for diseases related to many metabolic organ disorders. Different drug targeting strategies are depicted in Figure 5.1.

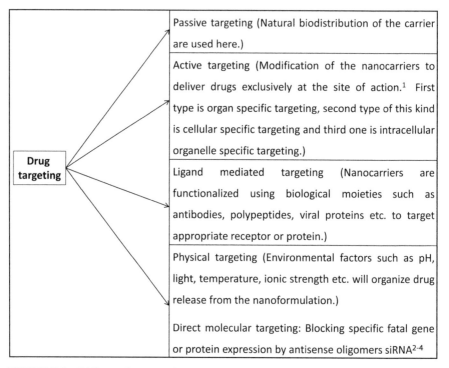

FIGURE 5.1 Different drug targeting strategies for gene therapy. (1. Satapathy et al., 2016, 2. Ghosh et al., 2014, 3. Mukherjee et al., 2005, 4. Mukherjee et al., 2007a)

5.5.3   *VARIOUS NANOFORMULATIONS FOR GENETIC DISORDERS*

Application of nanomedicines is not extensively explored in the field of genetic disorders. However, it is a very much promising and fast-emerging field with high possibilities. Nanoparticle-based probes are used to detect genetic diseases (Jia et al., 2012). The unusual sensitivity of different biomarkers reflects the disease conditions. By detecting those biomarkers in different biological fluids such as serum can indicate the occurrence of the pathological state. The levels of expression of some biomarkers are also used to decide the effectiveness of the therapy.

Exogenous enzymes can be administered to substitute defective and impaired production of endogenous enzymes due to diseased condition. For example, in case of lysosomal storage disorders due to genetic defects (Burrow et al., 2007) or prolidase deficiency (Colonna et al., 2008) can be treated by this option. Delivery of enzymes by nanocarriers increases the effectiveness by increasing stability, decreasing immunogenicity and specific targeting.

Encapsulation of small active ingredients prevents speedy clearance and assists increased therapeutic index and stability of the molecules. For example, intravenous administration of liposome containing coagulation factors for the treatment of hemophilia increases therapeutic efficacy (Yatuv et al., 2010). Another example is iron-chelating dendrimers for thalassaemias (Zhou et al., 2006). Oral delivery of solid lipid nanoparticles of anti-diabetic agents reduced glucose level in the blood (Nnamani et al., 2010). Chitosan is used for enhanced bioadhesion and pH-sensitive release resulting increased efficacy of anti-diabetic drugs (Wong, 2010). Nanocarriers are proved to be effective to cross the blood-brain barrier for different small molecules (Liu et al., 2009). For examples, chelator nanoparticles and supplement metal were explored for Alzheimer disease (Liu et al., 2009).

Healthy cells from healthy volunteers can be implanted after suitable modifications to produce a deficient molecule *in vivo* in a sustained manner in a patient with genetic disorder. For example, alginate microparticles with recombinant fibroblasts for sustained release of α-glucuronidase were tested for the treatment of mucopolysaccharidosis VII (Ross et al., 2010). Cell encapsulated in the nanocarriers was found to act as protective barriers to the immune system and unnecessary cell escape and neoplastic growth (Ross et al., 2010). Porous structural design of the nanocarriers is suitable for sufficient oxygen and nutrient exchange for cell viability (Desai et al., 2004).

Genetic disorders can be tackled using gene therapy. Delivery of cDNA for the affected gene, or delivery of oligonucleotides for amendment of mRNA transcripts by mRNA insertion/deletion, etc. are the examples of gene therapy (Hsu and Muro, 2011). Gene delivery requires viral vectors (Campbell and Hope, 2005) and suffers the safety concerns due to immunogenicity and random inclusion in the host genome. These limitations can be completed with nanomedicine therapy. For example, PEGylation of DNA nanoparticles exhibited effectual therapy for retinitis pigmentosa with simultaneous avoidance of immune appreciation (Cai et al., 2008).

5.6 THREE GENETIC DISEASES AND THEIR NANOTHERAPEUTIC STRATEGIES

5.6.1 CANCER

Cancer remains one of mankind's most dreaded diseases with more than 10 million new cases every year (Jemal et al., 2011). It is one of the leading causes of death worldwide with the second common cause of mortality in the

US and the reason of 70% of deaths in developing and underdeveloped countries according to a recent survey (AACR Cancer Progress Report, 2016). Cancer is considered a genetic disease as its way of progression is written in the coding sequence of some faulty genes containing lethal mutation(s). Mutations happen within genes may be beneficial, lethal or may not make any phenotypic changes. These mutations may sometimes act for either growth-promoting or growth inhibiting over time. Depending on their functional outcome, genes, where these mutations occur, can be broadly categorized into two main groups: tumor suppressor genes and oncogenes. Tumor suppressor genes are genes that, in their wild-type, protect cells from cancer development. Inactivation of tumor suppressor genes due to characteristic mutations causes progression of cancer. Oncogenes are growth promoting and responsible for cancer development. Oncogenes are formed by acquiring mutations from their wild-type form, known as proto-oncogenes. Mutations can also be classified in two types. When the genetic mutations are inherited from parents, they are called germline mutations and found in all the cells of offspring. A person can also acquire mutations during one's lifetime, promoted by exposure to environmental agents (some chemicals such as carcinogens or mutagens, radiation, ultraviolet ray, tobacco smoking, etc.) that cause DNA damage (Parsa, 2012). These types of mutations are called somatic mutations, and they are cancer tissue-specific. Cancer-causing mutations can be identified by comparing DNA sequences isolated from cancer tissue with the DNA sequence of normal blood cells or saliva. Hereditary genetic mutations that are found both in tumor cells, as well as normal cells of a cancer patient, play a pivotal role in about 5 to 10% of all cancer. Researchers have been engaged to report the genetic alterations related to more than fifty different types of cancer (Jurel et al., 2014). The cancer-specific genetic mutation data bank may pave the way to genetic testing for a person who has family history for a particular cancer type. The result determines cancer development risk of the person that helps in advance medication or genetic therapy to prevent occurrence of cancer (Jurel et al., 2014).

5.6.1.1 THERAPEUTIC STRATEGIES USING GENETIC MANIPULATION FOR CANCER TREATMENT

Despite the continuous improvement of cancer-combating strategies throughout the world, conventional anticancer therapies such as surgery chemotherapy or radiotherapy, and their combinations have a number of limitations. Prolonged treatment protocol, alarming increase of drug resistance property, severe adverse side-effects towards normal healthy cells and in majority of cases the death of patients ultimately owing to multiple organ failure, make such therapies difficult as successful treatment option. In last few decades, one of the major developments in modern biomedical research for health care is genetic therapy. Gene therapy has drawn a great attention in experimental therapy for cancers. Gene therapy approaches are targeted at replacing cancer-causing mutated gene with wild-type or delivering antisense oligonucleotides (AONS) that bind at the DNA level to silence the gene or to specific messenger RNAs to degrade them. Oncolytic viruses that infect and kill malignant cells, as well as the recently approved chimeric antigen receptor T cells that selectively find and kill the cancer cells, are worth mentioning (Cartellieri et al., 2010). These therapies mainly accomplished by selectively blocking the expression of a disease-causing gene. The lethal gene is transcribed to its corresponding mRNA which can be blocked by delivering its complementary sequence to pause the synthesis of damaging protein. However, there are lots of challenges and limitations in the treatment procedure such as severe side-effects due to transgene specific toxicities, inefficient transfection, rapid clearance by host immune responses, inadequate transgene expression and poor efficacy at the desired site of action (Zhang et al., 2017). Apart from that, naked genetic materials are often highly unstable in blood and unable to reach at target site due to large size. Negative charge makes it difficult to achieve successful gene delivery through penetrating cell membrane. A carrier or vector is required for successful delivery of genetic material into the cells (Gore, 2003). Viruses are mostly used as efficient gene therapeutic vectors as they can recognize the specific host cells and can be genetically engineered by replacing its disease-causing gene with the gene of interest to interrupt the cancer progression. However, potential risk factors such as serious immune response due to incorporation of the virus may cause extreme inflammation in body even organ failure whereas wrong targeting of virus initiates damage to healthy cells. The virus may mutate inside the body to create a pathological version, and uncontrolled insertion site inside cellular genome with the potential for iatrogenic tumor formation (Peng et al., 2016). Here, we elaborate some approaches using nanotechnology that are being used to counter these obstacles.

5.6.1.2 NANOFORMULATED GENE THERAPY FOR CANCER TREATMENT

Nanomedicine has become an emerging field of research for advancement in healthcare specially for anticancer drug and gene delivery (Yasuzaki et al., 2015; Sriraman et al., 2016). Nanodrug delivery systems are formulated with biocompatible polymers, have enhanced retention time in solid tumor due to EPR (Enhanced Permeability and Retention) effect and have controlled drug release action at the target site making it potential therapeutic tool for cancer treatment. Further modification of nanocarriers are possible by conjugating a homing ligand (such as antibodies, antibody fragments, aptamers etc.) against receptors, antigens, and other molecular marker specifically present on the surface of tumor cells (Jhaveri et al., 2016; Mukherjee et al., 2013) but not in normal tissue to reduce cytotoxic outcome. The nucleic acid-mediated nanomedicine drug delivery system accommodates AONS, microRNA (miRNA), short hairpin RNA (shRNA), and small interfering RNA (siRNA) to stop the expression of disease-causing mRNA or degrade it in small pieces. In case of AONS, a single-stranded oligonucleotide directly binds to its complementary mRNA sequence whereas miRNA, shRNA, and siRNA mediated gene silencing are propagated through the formation of Dicer and RNA-induced silencing (RISC) complex (Dash et al., 2015; Frank, 2009). The composition of the AONS, miRNA, rRNA, Pre-RNA is determined by the sequence of targeted mRNA of interest. The nucleotide sequence of the targeted oligo is complementary to the "sense" sequence of the DNA/mRNA (Sattler, 2010; Mukherjee et al., 2001). If the expression of a dysfunctional gene product is known to cause the disease, AONS can be simply designed from the "sense" sequence of the gene. There are numerous possible oligonucleotide modifications that enhance their blocking efficiency (Das et al., 2010). The most advanced oligonucleotide modification for antisense drug is phosphorothioate modified oligodeoxynucleotide that is commercially available, cost-effective, easily synthesized, RNase H activity supportive, exhibited acceptable pharmacokinetics for systemic as well as local delivery and has not exhibited toxicities that limit their use in humans. The modifications of oligonucleotides on the basis of base sequence, sugar, and backbone such as heterocyclic modifications, modifications to pyrimidine bases with a methyl group, 5-propynyl modification, increase the binding affinity to RNA (Mukherjee et al., 2005). Earlier, we reported that phosphorothioate AONS against three coding exons (exon–1/exon–2/exon–3) of IGF-II (Insulin-like growth factor-II) overexpressed in hepatocarcinoma, significantly controls the development of Hepatocellular carcinoma by formation of mRNA-antisense oligomer duplex. Nanoformulation mediated gene therapy has been investigated as a promising therapeutic agent for the treatment of a variety of disease including cancer, viral infections, various inflammatory diseases and autosomal dominant genetic diseases (Dean and

Bennett, 2003). In case of these diseases, a single base point mutation in the coding region may generate undesirable phenotype. So, nanomedicinal gene therapy is necessary to deliver the intact oligonucleotide which specifically inhibits the expression of mutant genes without affecting the expression of the normal gene in a sustained manner. The following nanoformulated gene delivery strategies are discussed for cancer treatment:

5.6.1.3 *POLYMER BASE NANOFORMULATION FOR GENE THERAPY*

Polymers available for nanoformulations provide a wide range of advantages over viral carriers for gene therapy and provide an opportunity for convenient manipulation of physiochemical characteristics to make a potent targeted delivery system. Polylactic-co-glycolic acid (PLGA), a FDA approved polymer of great interest to use as a device for drug/gene delivery, is biocompatible and biodegradable polymer with a variety of surface modifications possibility (Lundin et al., 2015). PLGA encapsulated nanoformulation having a property of sustaining drug release in a controlled manner at the targeted site of interest, is highly appreciated in clinical application. PEG is another broadly used hydrophilic polymer for the surface modification of nanoliposomes (Ghosh et al., 2014). Nanoliposomes has an important advantage of delivering more than one active ingredients such as drugs in combination with genetic material and stimuli-sensitive substances to synergistically manage cancer progression.

5.6.1.4 *NANOMATERIALS FOR NUCLEIC ACID DELIVERY*

DNA can form complexes with metal like gold to form gold nanoparticles (Au-NP) with adequate size, shape, charge density and colloidal stability to target cancer cells. Moreover, Au-NP can be coated with desired peptide sequence for targeted delivery or to enhance the cell penetrating capability. Upon entering into the cells, a conformational change in the nucleic acid structure causing endosomal escape followed by releasing the siRNA, at the same time promotes aggregation of the Au-NP. Interestingly, Au-NP aggregation, a useful material to induce photothermal ablation using a laser to generate heat locally at the desired site shows a synergistic effect to destroy cancer cells (Freier and Altmann, 1997). Iron oxide-based magnetic nanoparticles such as superparamagnetic iron oxide nanoparticles (SPIONs) are largely used for siRNA mediated stimuli-sensitive gene delivery to a specific location monitored by a strong external magnet (McLendon et al., 2015). Nonmetals like carbon nanotube, graphene similar to QD are also considered as attractive gene delivery vehicles.

5.6.1.5   *PROTEINS AND PEPTIDE NANOMATERIALS FOR GENE DELIVERY*

Proteins like albumin, gelatin, elastin are biocompatible as well as biodegradable material to be used as a gene delivery material. The amphiphilic nature of proteins provides a great advantage to interact it with hydrophobic solvent as well as encapsulating hydrophilic DNA or RNA based oligonucleotides. Among the different types of available proteins, gelatin is most advantageous used for gene delivery. It is one of the easy available cost-effective protein synthesized by collagen hydrolysis. Gelatin is mainly categorized into Type-A and Type-B depending upon the differences in their isoelectric point. So, the value is significant for surface modification (e.g., Thiolated gelatin nanoparticles can crosslink with siRNA) (Sinha et al., 2013). Albumin found in blood plasma can also be modified due to presence of reactive groups on its surface. Apart from that its tendency to accumulate in tumor sites makes it an attractive carrier for gene delivery. Elastin present in connective tissue, Zein, a storage protein present in maize has been reported as a potential gene delivery vehicle. Another major class of biomolecules, lipids is associated with delivery of genetic material. The amphiphilic nature of lipid pledges its ability to form vesicles like liposome. Cationic lipids are mostly preferred to deliver DNA into the cells due to their penetrating ability across the cellular membrane.

5.6.1.6   *COMBINED CHEMO AND GENE THERAPY FOR CANCER TREATMENT*

A nanoformulated combined chemotherapy and genotherapy may be an effective treatment protocol to combat against cancer. Nowadays, a major problem in cancer treatment is that the success of chemotherapy is confined up to the primary stage of tumor cells. After certain time period, the tumor cells acquire resistance to chemotherapy (Mukherjee et al., 2007; Son et al., 2014). The main reason to develop drug resistance property to cancer cells is the activation of cellular antiapoptotic proteins. Therefore, nanoformulation of siRNA responsible for knockdown of genes producing antiapoptotic proteins like Bcl2, surviving in combination with apoptosis-inducing chemo-therapeutic drug may be a potential therapy to prevent development of drug resistance in cancer as well as progression of the disease toward metastasis.

5.6.2 DIABETES

It is approximated that more than 285 million people are suffering from chronic and pandemic diabetes which often demotes to the abnormality in the pancreas to control glycaemic conditions (Mukherjee et al., 2005). Diabetes is a complex metabolic disease diagnosed by a persistently high blood glucose level resulting from progressive impairment of insulin secretion which may lead to insulin resistance also (Mukherjee et al., 2001). This constant elevation of glucose leads to the generation of reactive oxygen species (ROS), resulting in increased oxidative stress in β-cells of pancreas (Mukherjee et al., 2013). In addition to macromolecular damage to concerned DNA, proteins, and lipids, ROS can activate a number of cellular stress-sensitive pathways and cytokines that have been linked to insulin resistance and decreased insulin secretion (Mukherjee et al., 2013). Compared to other tissues, β-cells have a lower abundance of antioxidant and some enzymes such as superoxide dismutase (SOD), catalase, and glutathione peroxidase (GPx). Residual *β*-cell function declines faster following diagnosis in patients carrying increased genetic load of islet-expressed and cytokine (IL–1β + IFNγ + TNFα) regulated candidate genes (Mukherjee et al., 2013). Although environmental changes play an essential part, β-cell dysfunction has a clear genetic component to both type 1 diabetes and type 2 diabetes (National Human Genome Research Institute, 2017).

5.6.2.1 GENETIC MANIPULATION IN DIABETES

Subsequently, secondary forms of the disease may arise due to defects/mutations in genome of the organism (Table 5.3).

Mutations occurring in mitochondrial genome referred to as mitochondrial diabetes MELAS syndrome (mitochondrial myopathy, stroke-like syndrome encephalopathy, and lactic acidosis).

However, the disease together with its multiple complications put forward the immediate requirement to act with a well-defined strategy. The underlying platform is aimed at achievement of complete glycaemic regulation.

TABLE 5.3   Gene Mutations in Diabetes (Newton-Cheh et al., 2009)

Gene mutations	Chromosome effected
MODY1-hepatocyte nuclear factor–4-alpha (HNF4α)	Chromosome 20q12-q13.1
MODY2-glucokinase (GCK) gene mutation	Chromosome 7p15-p13
MODY3-hepatocyte nuclear factor–1-alpha (HNF1A) gene	Chromosome 12q24.2
MODY4-insulin promoter factor–1 (IPF1) gene	Chromosome 13q12.1
MODY5-hepatic transcription factor–2 (TCF2) gene	Chromosome 17cen-q21.3
MODY6- neurogenic differentiation 1 (NEUROD1) gene	Chromosome 2q32
MODY7-Kruppel-like factor 11 (KLF11) gene mutation	Chromosome 2p25
MODY8 (or diabetes pancreatic exocrine dysfunction syndrome)-carboxyl ester lipase (CEL) gene	Chromosome 9q34
MODY9- paired box gene 4 (PAX4) mutation	Chromosome 7q32

5.6.2.2   NANOTECHNOLOGY AND DIABETES

The edge of nanotechnology in the treatment of diabetes has introduced novel strategies for glucose measurement and insulin delivery. Researchers have demonstrated the advantages of glucose sensors and closed-loop insulin delivery, gene delivery approaches through nanosize delivery system in facilitating the diabetes treatment to make it (Newton-Cheh et al., 2009) beneficial in both type 1 and types 2 diabetes.

5.6.2.2.1   Development of Insulin Delivery Using Nanotechnology

(a) Non-invasive delivery: Glycaemic control is recommended through oral treatment for diabetes treatment along with physical activity and diet. The past few decades it frequently fails to mimic glucose homeostasis observed in healthy individuals (Niaz et al., 2016). In this pathway, insulin is delivered to the peripheral circulation, not the portal circulation (Niu et al., 2008). Therefore, injecting insulin numerous times a day is referring to poor patient compliance with subcutaneous route of treatment. Researchers have been working for better and safer route for insulin administration.

Oral, inhalable, and transdermal delivery can offer painless and simple methods relative to traditional insulin injections (Niu et al., 2011). However, poor, and unpredictable bioavailability has limited the success of insulin delivery via these alternative routes of administration (Niu et al., 2011). In this case, the application of nanotechnology could be a solution to overcome this problem for the delivery of insulin in a proper pathway. All of these systems may improve stability of insulin against low pH and digestive enzymes of the stomach and intestine where intestinal epithelium is a major barrier for absorption of insulin; in addition to those long-term redistributions and release of insulin that facilitates persistent lowering of blood glucose (Niu et al., 2009).

Use of nanotechnology for oral delivery of insulin includes Pre-drugs (conjugated insulin-polymer), micelles, and liposomes, solid lipid nanoparticles and biodegradable polymer nanoparticles. The formulation of insulin is often tried with PEG (PEGylation) to increase the solubility, stability, and permeability. Insulin–PEG pre-drug formulations have shown many advantages of oral insulin delivery (Nnamani et al., 2010). The liposomal formulation containing glycol recently has been developed for oral delivery of insulin, and it has enhanced ion permeate which showed excellent insulin ionic shield that protects the insulin from enzymatic degraded by pepsin, trypsin, and chymotrypsin (Nunes et al., 2010). Biodegradable polymers such as (polylactic glycolic acid, PLGA) and polycaprolactone were studied for the oral delivery of insulin as nanoformulations. However, delivery of hydrophobic insulin through the oral route is still a large challenge. Intestinal penetrating nanocarrier such as chitosan has been used to aid the absorption of insulin (Orive et al., 2003). Dextran nanoparticle conjugated with vitamin B12 was examined to overcome the intestinal digestion of insulin (Ortega et al., 2014). In addition, gold nanoparticles have also been applied for delivery of insulin. Gold nanoparticles synthesized with chitosan as a regenerative have been studied in order to bring insulin to its target (Pardridge, 2010). However, none of them was found to be commercially successful. Multiple investigations are required.

(b) Insulin gene therapy: By gene therapy insulin gene could be introduced, perhaps under the control of a tissue-specific promoter, or in a gene encoding for a factor that activates the insulin gene. In insulin gene therapy, one of the key issues is the development of efficient delivery systems (Parsa, 2012).

Chitosan-coated nanoparticles, in addition to their good biocompatibility and non-toxicity, are economically available (Paulis and Unger, 2010). The transfection efficiency of chitosan can be regulated by changing its molecular weight, plasmid concentration, and chitosan/plasmid ratio. After the plasmid is embedded in chitosan, it can resist the degradation of nucleases.

The human insulin gene thus can be transfected successfully by chitosan nanoparticles *in vitro* and *in vivo*, which was found to be expressed efficiently in NIH3T3 cells and the gastrointestinal tracts of diabetic rats, indicating that chitosan is a promising safe and effective non-viral vector for gene on clinics, and it may be willingly accepted by patients (Paulis et al., 2015).

5.6.2.2.2 Treatment of Diabetes with Nanotechnology Enabled Glucose Sensing

Accurate and frequent glucose measurements are the basis of contemporary diabetes management. However, it is commonly acknowledged that contemporary clinical glucose measurement systems are a nuisance to the patient as a result of frequent and painful needle sticks, and the current standard of intermittent testing can miss dangerous fluctuations in blood glucose concentration (Niu et al., 2008; Peng et al., 2016). Therefore, one of the most significant challenges in diabetes research is the development of glucose sensors which achieve accurate glucose measurements painlessly and frequently, with the goal of continuous glucose measurement. A wide variety of glucose sensing modalities have been reported in the last two decades, and commonly used Concanavalin A, phenylboronic acid (PBA), or most commonly, employed glucose oxidase as a sensor for detecting glucose in solution, are worthy to mention.

5.6.2.2.3 Treatment of Diabetes Using Antioxidant Property of Nanoparticles

In diabetes oxidative stress is a major pathophysiological complication which causes delay in wound healing also. The problem could be addressed by the use of some nanoparticles (aluminum oxide, cerium oxide, gold, vanadium, and zinc) that act as an ROS absorbent.

Vanadium element in streptozotocin-induced diabetic rats has shown the anti-diabetic effects (Peng et al., 2016; Pridgen et al., 2014). This could be due to the physical and biochemical changes of nanoparticles that may have an effect on the properties of bioavailable, insulin-like insulin tropic ammonium vanadate (Pridgen et al., 2014). Cerium oxide (CeO_2) plays a major role in the inhibition of free radicals due to its strong potential (Prina et al., 2014) as antioxidant. A study in the streptozotocin-diabetic rats showed that oral zinc nanoparticles improve glucose tolerance, increasing serum insulin and decrease of blood glucose (Revanasiddappa and Bhadauria, 2013). The oral use of gold nanoparticles exerts their role by the inhibition of formation of free radicals (Robinson et al., 2017). These kinds of nanoparticles provide for a stringent control on the anti-oxidant enzymes such as GSH, SOD, GPX, and catalase, responsible for the inhibition of lipid peroxidation and also inhibit the production of free radicals produced due to hypergly-cemia. Therefore, they could be used as a successful tool in treating diabetes (Robinson et al., 2002).

5.6.2.2.4 Nanotechnology in Combination with sRNA or Gene Transfer Approach to Treat Diabetes

Generate a "glucose sensor" in skeletal muscle through co-expression of glucokinase (Gck) and insulin (Ins), was found to increase glucose uptake and correcting hyperglycemia in diabetic mice. A one-time intramuscular administration of adenoviral vectors of serotype 1 (AAV1) encoding for Gck and Insulin in diabetic dogs resulted in normalization of fasting glycemia, accelerated disposal of glucose after oral challenge, and no episodes of hypoglycemia during exercise for >4 years after gene transfer. This was associated with recovery of body weight, reduced glycosylated plasma proteins levels, and long-term survival without secondary complications. Conversely, exogenous insulin or gene transfer for insulin or Gck alone failed to achieve complete correction of diabetes, indicating that a syner-gistic action of insulin and Gck are needed for full therapeutic effect. This study provides the first proof-of-concept in a large animal model for a gene transfer approach to treat diabetes (Ross et al., 2010).

Microcapsules containing replacement islets of Langerhans cells, mostly derived from pigs, could be implanted beneath the skin of diabetes patients. This could temporarily restore the body's delicate glucose control feedback loop without the need for powerful immunosuppressants that can leave the patient at serious benefits risk for infection (Newton-Cheh et al., 2009).

Hyperglucagonemia is implicated in the etiology of uncontrolled hepatic glucose production. The CS-coated PSi nanoparticles showed the highest glucagon-like peptide–1 (GLP–1) permeation across the intestinal *in vitro* models (Sahay et al., 2010). Also, to block glucagon action, glucagon receptor (GCGR) in rodent was targeted with 2′- methoxyethyl–modified phosphorothioate-antisense oligonucleotide (ASO) inhibitors (Satapathy et al., 2016). Thus, miRNAs may use as excellent pharmacological targets in diabetes by regulating glucose and lipid metabolism within the liver as well as an excellent, promising controller that regulates insulin biosynthesis and secretion in pancreatic beta cells and in skeletal muscle or adipose tissues (Sattler et al., 2010).

There is a close relationship between variations in circulating miRNAs and T2D and their potential consequence in insulin sensitivity; however, it subjects to age and sex-dependent (Shah et al., 2014).

Nanoparticle-mediated drug delivery approach may benefit the diabetic patients substantially as it improves bioavailability of drug. However, one of the biggest technological challenges is the scalability of a nanoparticle and effect on the patients upon long-term treatment.

5.6.3　HYPERTENSION

Hypertension commonly known as elevated blood pressure is a common chronic disease. The flow of blood through arteries with required force is termed as blood pressure. In normal conditions, the blood pressure measured at rest is within the range of 100–140 mmHg systolic (top reading) and 60–90 mmHg diastolic (bottom reading). High Blood Pressure is when the pressure of the blood is persistently at or above 140/90 mmHg. In the Hypertensive crisis, blood pressure elevates to 180/110 mmHg or above.

5.6.3.1　HYPERTENSION ASSOCIATED SIGNS AND SYMPTOMS

Hypertension involves important risk factor such as serious cardiovascular disorders, including myocardial infarction, heart failure, stroke, and peripheral artery disease. Some of them are mentioned below.

- **Aneurysms:** It arises when an abnormal bulge is developed in the wall of an artery. It is noticed as it grows and presses against a body part either it gets ruptured or blocks the flow of blood.
- **Chronic kidney disease:** Narrowing of the artery might result in kidney failure.
- **Cognitive changes:** A prolonged period of hypertension may result in cognitive changes such as symptoms like improper speech, Amnesia.
- **Visual impairment:** Hypertension can often result in rupturing vessels in the eye resulting in vision changes or blindness
- **Heart attack:** The deficit of oxygen in the heart muscle due to blockage of the flow of oxygenated blood leads to ischemia of the cardiac muscle.
- **Heart failure:** The heart fails to pump sufficient blood, resulting in the deficient supply of blood to the organs. Heart failure is preceded

by shortness of breath, muscle fatigue, edema of ankle, feet, and veins in the neck.

- **Peripheral artery disease:** Plaque deposits are seen in arteries as a result of hypertension which circulates blood to the head, limbs, and other organs followed by pain, cramps, numbness, and heaviness in the legs, feet on exertion.
- **Stroke:** The symptoms are ischemia in the brain with sudden onset of weakness, difficulty in speech, paralysis, difficulty in recognizing and blurred vision.

Consequences of elevated blood pressure are that the heart has to work harder than normal for proper flow of blood through blood vessels.

5.6.3.2 CLASSIFICATION OF HYPERTENSION

Hypertension can be classified as either primary hypertension or secondary hypertension.

- *Primary Hypertension/Essential hypertension:* 90% of essential hypertension is caused due to factors that include weight, age, sex, ethnicity, physical activity, diet, cigarette smoke, stress, hormones, other medical conditions such as diabetes, by the consumption of high salt intake, alcohol, fat products. For adult primary hypertension, early life events like low birth weight, maternal smoking, lack of breastfeeding, etc. can be implicated as risk factors. Genetics is one of the important factors. Studies suggest that essential hypertension has a strong genetic component (Dickson & Sigmund, 2006).
- *Secondary hypertension:* Some risk factors for secondary hypertension are endocrine conditions, sleep apnea, obesity, excessive liquories, illegal drugs, herbal medicines, etc. Hypertension may result from a complex interaction of genes and environmental factors.

5.6.3.3 THE GENETIC BASIS OF HYPERTENSION

The family history of hypertension and heritability of blood pressure plays a vital role. Various studies suggest that hypertension is a familial disease. Identification of the genes that contribute to hypertension has been very challenging for the researchers, as it is complicated involving dynamic interplay among heterogeneous genetic backgrounds. Identification of a gene or genetic region has a large effect on phenotype. Researchers realized that all diseases do not follow the Mendelian laws of genetics. According to them these genes individually do not express as autosomal dominant or recessive. But there is a set of genes with complex interaction that causes the disease, and this is referred to as quantitative trait loci.

Genome-wide association study (GWAS) is used widely to identify genetic factors that influence common and complex diseases. It is the method to study candidate genes in any disease. GWAS mainly focuses on single nucleotide polymorphisms (SNPs) and major diseases. With the help of GWAS, eight loci associated with hypertension are identified. The loci identified were CYP17A1, CYP1A2, FGF5, SH2B3, MTHFR, c10orf107, ZNF652, and PLCD3 (Newton-Cheh et al., 2009).

Many evidences support association of angiotensinogen (AGT) gene with hypertension. The participation of the renin-angiotensin-aldosterone system (RAAS) is well established in the etiology of hypertension, which will be discussed further. The first gene associated essential hypertension is the AGT gene and protein. It is the most scrutinized gene linked to hypertension. AGT is expressed in liver, adipose tissue, heart, vessel wall, brain, and kidney. It is equally diverse in its cell specificity. Particular attention will be given to the role of AGT gene polymorphisms in the etiology of essential hypertension.

Some individual genes are studied further in association with hypertension is given in Table 5.4.

5.6.3.4 DIAGNOSIS OF HYPERTENSION

Persistent high blood pressure requires several tests such as the system tests, renal microscopic urinalysis, proteinuria, a complete screening of thyroid, metabolic fasting blood glucose, creatinine, endocrine, serum sodium, calcium, potassium, high density lipoprotein and high density lipoprotein, total cholesterol, triglycerides, radiograph of the chest, and electrocardiogram.

5.6.3.5 ANTIHYPERTENSIVE DRUG THERAPY

Pentaquine was the first drug which was developed for treating hypertension in 1946 along with hypertensive activity. But it showed several side-effects. It was followed by ganglionic blocking agent Hexamethonium,

Veratrum, Hydralazine, and Reserpine. Later, diuretics and b-blockers were introduced and widely prescribed. As the first-line therapy either alone or in combination calcium channel blockers, angiotensin-converting enzyme inhibitors, and angiotensin blockers are widely utilized owing to therapeutic efficacy.

TABLE 5.4 Individual Genes and Their Association with Hypertension (Revanasiddappa & Bhadauria, 2013)

Genes	Association with hypertension
Angiotensin-converting enzyme gene	Polymorphism of angiotensin-converting enzyme ACE genes are supposed to be important in hypertension, but a study showed that there was no as such relationships between ACE gene polymorphism and antihypertensive treatment response.
Angiotensinogen	Variants of angiotensinogen and hypertension were found to be associated with each other. Review showed that Mitochondrially Encoded TRNA Threonine variants of angiotensinogen are linked with hypertension.
SGK1	Serum- and glucocorticoid-inducible kinase 1 (SGK1) has a pivotal role which has been associated with BP variation, salt sensitivity of HTN and response of HTN to therapy.
Alpha adducing	A study revealed that a point mutation, G460W in human alpha Adducing gene is related to HTN.
ADAMTS16	A study confirmed that ADAMTS16 and regulation of BP were linked with each other.
ATP2B1	ATP2B1 gene was found to be associated with HTN in Japanese and Caucasians. Reduced expression of ATP2B1 showed risk of hypertension.
Apolipoprotein E genotype	Associated with hypertension in elderly.
KLHL3	Mutation of apolipoprotein E genotype can lead to hypertension.
Na$^+$-ATPase	Regulation of Na$^+$-ATPase was found to be angiotensin. Studies confirmed the link between Angiotensin 1 and Na$^+$-ATPase regulation.
Myosin heavy chain 9 genes (MYH9)	Myosin heavy chain 9 gene (MYH9) has a significant association with HTN.
WNK kinases	No direct relationship was found between WNK kinase and hypertension. One SNPrs1468326, located near to WNK1 promoter, was found to be nominally associated with severity of hypertension.
Endothelial NO synthase Gen (eNOS)	Studies explored the role of eNOS in HTN (Niu et al., 2009).

Thorough understanding of renin angiotensinogen aldosterone system (RAAS) has led to the development of several antihypertensives. There is dramatic progress in the development of novel therapeutics, the target of which is also related to RAAS (Paulis & Unger, 2010). Figure 5.2 shows the novel targets, which have opened the new possibility for the successful development of drug for treating hypertension.

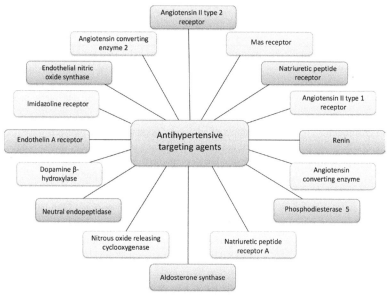

FIGURE 5.2   Novel molecular targets for antihypertensive therapy (Adapted from Alam et al., 2017).

5.6.3.5.1   Renin Inhibitor

The kidney releases rennin, which is the first step in the RAAS cascade, thus, making it as a viable target for antihypertensive therapy. The first-marketed renin inhibitor is 'aliskiren' introduced into the market in 2007. A new molecule VTP27999 is under Phase II clinical trial (Paulis et al., 2015).

5.6.3.5.2   Angiotensin II Type 2 Receptor Agonist

AT2R has action opposite to that of AT1R. It opposes the AT1R-mediated vasoconstrictor action of angiotensin II. AT2R shows the vasodilatory action, which is formed of bradykinin, nitric oxide, and cGMP. AT2R also mediates natriuresis (Carey & Padia, 2008). Some AT2R agonist, which are under clinical trial for its antihypertensive action. It acts on the sodium/hydrogen exchanger 3 (NHE 3) and the Na+/K+ATPase in the proximal tubules, thus showing natriuresis (Paulis et al., 2015).

5.6.3.5.3   Phosphodiesterase 5 (PDE–5) Inhibitor

The cyclic GMP is degraded by PDE–5, which is the intermediate step in vasodilatory action. cGMP degradation is inhibited by PDE–5 thereby causing vasodilatation (Nunes et al., 2010). Example is Tadalafil which shows vasodilatory effect; KD027 is under phase II clinical trial of study (Paulis et al., 2015).

5.6.3.5.4   Natriuretic Peptide Receptor an (NPRA) Agonist

NPRA agonist causes increase in cGMP level which leads to decrease in the blood pressure and induces natriuresis. PL3994 is under phase II trial of study (Paulis et al., 2015).

5.6.3.5.5   Mas Receptor Modulator

Mas receptor-like AT2 receptor causes the release of nitrous oxide. Blockage of either Mas receptor or AT2 causes the blockage of another receptor due to their hetero-dimerization. Natural ligand for Mas receptor is Ang (1–7), which possess low bioavailability. The bioavailability can be enhanced by complexing it with hydroxyl-propyl beta-cyclodextrin. (Paulis et al., 2015).

5.6.3.5.6   Angiotensin Converting Enzyme 2 (ACE2) Modulator

ACE2 is responsible for the metabolism of angiotensin I and angiotensin II, key peptides for RAAS. APN01 (rhACE2) an (ACE2) modulator to reduce hypertension is under Phase II clinical trial (Paulis et al., 2015).

5.6.3.5.7   Endothelin A Receptor (ETA) Antagonist

Macitentan and Ambrisentan are approved Endothelin A receptor (ETA) antagonist utilized for pulmonary hypertension, which prevents the binding of ET1 to both ETA and ETB (Paulis et al., 2015; Iglarz et al., 2008).

5.6.3.5.8   Combined AT1R Blocker and NEP Inhibitor

The bioavailability of Natriuretic peptides is increased by NEP inhibitors contributing to lowering of blood pressure. Daglutril and LCZ696 (phase II and III trial) are NEP inhibitor with AT1R blocking activity (Paulis et al., 2015; McMurray, 2015).

5.6.3.5.9   Imidazoline-Receptor Blocker

Imidazoline receptors are of three types: First two types are present on the rostral ventrolateral medulla oblongata (RVLM), which are stimulated by a first-generation imidazolines clonidine. It mediates the sympathoinhibitory actions, thus lowering the blood pressure. Second-generation imidazolines are centrally acting antihypertensives like monoxidine and are less toxic (Ernsberger et al., 1994; Head & Mayorov, 2006).

5.6.3.5.10   Endothelial Nitric Oxide Synthase (eNOS) Coupler

Cicletanine is a diuretic such as a thiazide which also acts as eNOS coupler and makes eNOS active thereby, increasing the production of NO and decreasing ROS formation (Alam et al., 2017).

5.6.3.5.11   NO-Releasing Cyclooxygenase (COX) Inhibitor

Naproxcinod is NSAID which acts as cyclooxygenase (COX) inhibiting nitric oxide donor (CINOD). The produced NO has the vasodilatory effect, making it an eligible entity for lowering blood pressure (Townsend et al., 2011).

5.6.3.5.12   Aldosterone Synthase Inhibitor

Non-genomic effect of aldosterone is not effectively reduced by mineralo-corticoid receptor antagonists. Aldosterone synthase helps in mediating the blood pressure. Aldosterone synthase inhibitor may result in disruption of RAAS, thus control the rise of blood pressure. ASI LCI699 is in phase II trial (Niaz et al., 2016).

5.6.3.5.13   *Dopamine Beta-Hydroxylase (DbH) Inhibitor*

Etamicastat causes vasodilation along natriuresis, and dieresis (Nunes et al., 2010).

5.6.3.6   *RATIONALE FOR USING NANOCARRIERS*

Oral route is the most preferred and convenient route for the administration of the drugs. Drugs with low aqueous solubility and low permeability (BCS class II or IV) or high first-pass metabolism or drugs undergoing enzymatic degradation cannot be administered via oral route. Oral administration of anti-hypertensive drugs such as candesartan, cilexetil leads to chemical degradation of these drugs at acidic pH. As pH of the gastrointestinal tract varies from acidic (pH 1–2) in the stomach to basic (pH 8) (Koziolek et al., 2015). The Pharmacological activity of drug is severely hampered with the wide difference in the pH by oxidation, deamidation, or hydrolysis of protein drugs.

Many enzymes such as liver esterase and cytochrome P450 present in the route degrades anti-hypertensives. Barriers such as intestinal mucosa which consists of an extrinsic barrier and intrinsic barrier, hinders drug permeation. A various mechanism like transcellular, paracellular, and transcytosis are utilized by a molecule to cross this barrier. The charge and its large size of the particles restrict its permeation through transcytosis. Once the molecules cross lamina propria, the molecule can get entry into the bloodstream (Turner, 2009; Pridgen et al., 2014).

Mucoadhesive formulations can be utilized to overcome the intestinal barrier by increasing the contact time of the formulation with mucus. Thereby, they increase drug concentration at the site of absorption.

Transport through M cells enhances GI permeability. The amount of protease enzyme in the M cells is low; hence it lacks mucus secretion. M cell transport is improved via lipophilic molecules (Pridgen et al., 2014). Lipid nanoparticles like SLN and NLC are transferred through the intestinal barrier by clathrin-mediated transport. SLN is also transcytosed by caveolae-mediated endocytosis.

Nanoformulations or nanocarriers are preferred over oral administration of antihypertensives due to several limitations of antihypertensives administration via the oral route. Nanoformulations such as liposomes, polymeric nanoparticles, nanocrystals, nanostructured lipid carrier and much more are widely used to improve the bioavailability of antihypertensives as drug delivery systems (Table 5.5).

TABLE 5.5 Some of the Nanoformulations Containing Antihypertensives

Types of nano-delivery system	Therapeutic system	Excipients used	Comments
Polymeric nanoparticle	Lercanidipine	HPMC, TPGS	2.47 fold increase in oral bioavailability than raw drug without TPGS (Ha et al., 2015)
	Ramipril	Lecithin/Chitosan	A 1.6-fold decrease in systolic blood pressure (Chadha et al., 2013)
	Nifedipine	PCL, PLAGA Eudragit RL/RS	Initially a decrease in systolic blood pressure was rapid for PEG solution followed by PCL NP and PLAGA NP. Blood pressure was within normal range after 10 h of dosing using all three NPs while PEG solution failed to achieve such sustained effect (Kim et al., 1997).
	Felodipine,	PLGA, Pluronic F–68	Systolic blood pressure became normal, and elevated ST segment of ECG became normal up to a period of 3 days as compared to drug suspension (Shah et al., 2014).
	Aliskiren	Magnetite, poly (D, L-lactide), Pluronic F–68	A significant drop in mean systolic blood pressure by aliskiren nanoparticle as compared to aliskiren suspension and placebo
Solid lipid nanoparticle	Candesartan	GMS, soya lecithin, Tween 80	12 times increase in oral bioavailability
	Nisoldipine	Trimyristin (TM; Dynasan–114; glyceryl trimyristate), egg lecithin, Poloxamer–188	2.17 times increase in oral bioavailability, a significant reduction in systolic blood pressure for a period of 36 h
	Isradipine Cilexetil	Trimyristin or GMS, poloxamer 188	A significant decrease in the systolic blood pressure with SLN formulation using two different lipids
Nanostructured lipid Carrier	Lercanidipine	Labrafil 2130M, GMS, linseed oil and Tween 80	24 h control on the blood pressure by NLC as compared to plain drug Suspension
	Lacidipine	GMS, Linoleic acid and poloxamer 407	3.9 times enhancement in the relative bioavailability

TABLE 5.5 *(Continued)*

Types of nano-delivery system	Therapeutic system	Excipients used	Comments
Nanoemulsion	Amlodipine	DE (Labrafilm 1944 CS and Dextrin)	In vitro release studied showed the higher release of amlodipine from DE than the powdered drug. 2.6 to 2.9 times increase in Cmax and AUC (0–24h) from DE than powder. Marked reduction in the photo-degradation of drug in DE than powdered drug (5.6% versus 66.9%)
	Ramipril	Sefsol 218, Tween 80, carbitol	About 229.62% increase in relative bioavailability of ramipril nanoemulsion as compared to ramipril marketed capsule and 539.49% increase in bioavailability of formulation as compared to drug suspension.
	Olmesartan, Me-doxomil (Beg et al., 2015)	SNEDDS (SNEOF and CSNEOF	After 0.5 h of dosing, significant reduction in arterial blood pressure (180 to 189mm Hg) was seen with SNEOF (141±1.36 189mm Hg), CSNEOF (136±1.45 189mm Hg), and Marketed formulation (138±1.98 189mm Hg). After 48 h of study, rats were found normotensive (BP5130mm Hg) with SNEOF and CSNEOF.
	Valsartan	S-SNEDDS (Capmul MCM, Labrasol, Tween 20)	3–3.5-time increase in the rate of dissolution, significant reduction in the mean systolic blood pressure after 0.5 h and 2 h of dosing of S-SNEDDS as compared to valsartan suspension showing faster onset of action of SSNEDDS thus showing it to have the potential of the bioavailability enhancement of Valsartan
	Carvedilol	S-SNEDDS (Capmul MCM, Nikkol HCO 50) L-SNEDDS (Cremophor EL, Transcutol HP)	2.34 and 1.85 times increase in Cmax and AUC, respectively of SSNEDDS, thus showing enhanced bioavailability
	Lacidipine	S-SNEDDS (Labrafil and capmul as oil, Cremophor, and Tween 80 as surfactant and transcutol as co-surfactant)	significant increase in rate of dissolution

TABLE 5.5 *(Continued)*

Types of nano-delivery system	Therapeutic system	Excipients used	Comments
Lipotome	Lacidipine	Cetyl alcohol and Tween 80	540.11% increase in relative bioavailability of enteric-coated capsule of lipotome
Dendrimers	Candesartan cilexetil Nifedipine	polyamidoamine, poly(propylene imine)	Enhanced water solubility
Nanocrystals	Nitrendipine Nifedipine	Pluronic F127	Improvement in physical stability, *in vitro* drug release, and bioavailability, enhanced dissolution profile

5.6.3.7 NANOPARTICLE-MEDIATED GENE THERAPY OF HYPERTENSION

It is the most recent technology that uses small interfering RNA to silence those receptors which are implicated in increasing the blood pressure. The delivery of siRNA is administered via intravenous route in treating hypertension. Incorporation of siRNA in the nanoformulation delivery system is required to prevent their degradation by exonuclease activity present in the blood (McLendon et al., 2015).

The target mRNA is made nonfunctional by its cleavage. RNA (siRNA) is produced by RNAse III, and the endonuclease is also called Dicer. The duplex siRNA is incorporated into RISC complex. RNA helicase unwinds siRNA duplex with the help of RNA helicase yielding in antisense strand which remains with RISC (called activated RISC) while exonuclease degrades the sense strand. The target mRNA is bound by the activated form of RISC. Hence, RNAase activity is initiated by antisense strand of the activated RISC. The inactive fragments (cleaved from mRNA) become nonfunctional for protein synthesis. Hence, gene silencing occurs, and receptor protein is not synthesized. Further, mRNA is destroyed once the activated RISC becomes free mRNA (Koenig et al., 2013).

The basic hindrance with siRNA is the rapid degradation upon administration. Therefore, to prevent the degradation of siRNA by endo and exonuclease present in blood and cells, nanosystems is utilized, making i.v., route aptly for its delivery. Lipoplex, a cationic liposome made of DOTAP (N-[1-(2,3dioleoyloxy)]-N-N-N trimethylammonium propane), reduces the expression of b1-adrenoreceptor and controls the blood pressure for 12 days when given through intravenous route (Koenig et al., 2013).

5.7 CONCLUSION

Improved understanding of the diversity of the human genome now enables us to understand the possible cause of various rare genetic diseases like cancer, diabetes, hypertension, obesity, and anemia. Suitable nanocarriers are required to overcome challenges such as cellular uptake and pharmacokinetics. Nanotechnology is one of the most promising fields in medical science to meet this type of expected progress and benefits. A range of therapeutic advantages of Nanomedicine over conventional therapy certainly is a revolution in medical science. Gene silencing technology plays the major role in the treatment of various genetic diseases. siRNA technology is being utilized and developing very fast from preclinical to clinical trial level. It might be the most successful tool for early-stage diagnosis to effective treatment for genetic diseases. In conclusion, nano-medicine is a premise for future research on the benefits and applicability of nanotechnology in medicine to treat chronic genetic diseases.

KEYWORDS

- antisense oligonucleotides
- cyclooxygenase
- polylactic-co-glycolic acid
- quantum dots
- reactive oxygen species
- RNA-induced silencing

REFERENCES

Ahmad, N., Khan, Z., & Hasan, N., (2017). A review on nanomedicine as a modern form of drug delivery. *Glob. J. Nanomed., 2*(1), 1–2.

Alam, T., et al., (2017). Nanocarriers as treatment modalities for hypertension. *Drug. Deliv., 24*, 358–369.

Alex, S. M., & Sharma, C. P., (2013). Nanomedicine for gene therapy. *Drug Deliv. Transl. Res., 3*(5), 437–445.

American Association for Cancer Research, (2016). AACR cancer progress report. *Clin. Cancer Res., 22*, SI–S137.

Barathmanikanth, S., et al., (2010). Anti-oxidant effect of gold nanoparticles restrains hyperglycemic conditions in diabetic mice. *J. Nanobiotechnology., 8*(16), 1–15.

Bhatia, S. *Natural Polymer Drug Delivery Systems: Nanoparticles, Plants, and Algae;* Springer International Publishing: Switzerland, (e-book), pp33-93, 2016. DOI 10.1007/978-3-319-41129-3

Bhumkar, D. R., et al., (2007). Chitosan reduced gold nanoparticles as novel carriers for transmucosal delivery of insulin. *Pharm. Res., 24*(8), 1415–1426.

Boslaugh, S. E., (2017). *Encyclopedia Britannica.* Prader-Willi syndrome (PWS). https://www.britannica.com/science/Prader-Willi-syndrome (accessed Aug. 17, 2017).

Brorsson, C. A., et al., (2016). Hvidoere study group on childhood diabetes. Genetic risk score modeling for disease progression in new-onset type 1 diabetes patients: Increased genetic load of islet-expressed and cytokine-regulated candidate genes predicts poorer glycemic control. *J. Diabetes. Res., 2016*(9570424), 1–8.

Burrow, T. A., et al., (2007). Enzyme reconstitution/replacement therapy for lysosomal storage diseases. *Curr. Opin. Pediatr.*, *19*(6), 628–635.

Cai, X., Conley, S., & Naash, M., (2008). Nanoparticle applications in ocular gene therapy. *Vision. Res.*, *48*(3), 319–324.

Calceti, P., Salmaso, S., Walker, G., & Bernkop-Schnürch, A., (2004). Development and *in vivo* evaluation of an oral insulin–PEG delivery system. *Eur. J. Pharm. Sci.*, *22*(4), 315–323.

Callejas, D., et al., (2013). Treatment of diabetes and long-term survival after insulin and glucokinase gene therapy. *Diabetes*, *62*(5), 1718–1729.

Cam, M. C., Rodrigues, B., & McNeill, J. H., (1999). Distinct glucose lowering and beta-cell protective effects of vanadium and food restriction in streptozotocin-diabetes. *Eur. J. Endocrinol.*, *141*(5), 546–554.

Campbell, E. M., & Hope, T. J., (2005). Gene therapy progress and prospects: Viral trafficking during infection. *Gene. Ther.*, *12*(18), 1353–1359.

Carey, R. M., & Padia, S. H., (2008). Angiotensin AT 2 receptors: Control of renal sodium excretion and blood pressure. *Trends. Endocrinol. Metab.*, *19*, 84–87.

Cartellieri, M., et al., (2010). Chimeric antigen receptor-engineered T cells for immunotherapy of cancer. *Hindawi Publishing Corporation Journal of Biomedicine and Biotechnology*, *10*, 1–9.

Chadha, R., et al., (2013). Exploring lecithin/chitosan nanoparticles of ramipril for improved antihypertensive efficacy. *J. Nanopharm. Drug. Deliv.*, *1*, 173–181.

Chae, S. Y., et al., (2008). Preparation, characterization, and application of biotinylated and biotin–PEGylated glucagon-like peptide–1 analog for enhanced oral delivery. *Bioconjugate. Chem.*, *19*(1), 334–341.

Colonna, C., et al., (2008). Site-directed PEGylation as a successful approach to improving the enzyme replacement in the case of prolidase. *Int. J. Pharm.*, *358*(1 & 2), 230–237.

Couvreur, P., & Vauthier, C., (2006). Nanotechnology: Intelligent design to treat complex disease. *Pharm. Res.*, *23*(7), 1417–1450.

Das, T., et al., (2010). Effect of antisense oligomer in controlling c-RAF. Overexpression during diethylnitrosamine-induced hepatocarcinogenesis in rat. *Cancer Chemother. Pharmacol.*, *65*, 309–318.

Dash, B. C., et al., (2015). An injectable elastin-based gene delivery platform for dose-dependent modulation of angiogenesis and inflammation for critical limb ischemia. *Biomaterials*, *65*, 126–139.

Dean, N. M., & Bennett, C. F., (2003). Antisense oligonucleotide-based therapeutics for cancer. *Oncogene.*, *22*, 9087–9096.

Desai, T. A., et al., (2004). Nanoporous microsystems for islet cell replacement. *Adv. Drug. Deliv. Rev.*, *56*(11), 1661–1673.

Dickson, M. E., & Sigmund, C. D., (2006). Genetic basis of hypertension. *Hypertension, [Online]*, *48*, 14–20. http://hyper.ahajournals.org/content/48/1/14 (accessed Aug 18, 2017).

El-Sayed, M. E., Hoffman, A. S., & Stayton, P. S., (2005). Smart polymeric carriers for enhanced intracellular delivery of therapeutic macromolecules. *Expert. Opin. Biol. Ther.*, *5*(1), 23–32.

Encyclopedia.com, (2017). *Genetic Diseases*. http://www.encyclopedia.com/medicine/diseases-and-conditions/pathology/genetic-diseases (accessed Aug 17, 2017).

Ernsberger, P., et al., (1994). A novel mechanism of action for hypertension control: Moxonidine as a selective I1-imidazoline agonist. *Cardiovasc. Drugs. Ther.*, *8*(1), 27–41.

Frank, A. S., (2009). Somatic evolutionary genomics: Mutations during development cause highly variable genetic mosaicism with risk of cancer and neurodegeneration, *PNAS, 107,* 1725–1730.

Freier, S. M., & Altmann, K. H., (1997). The ups and downs of nucleic acid duplex stability: Structure-stability studies on chemically-modified DNA: RNA duplexes. *Oxford Univ. Press Nucleic Acids Res., 25*(22), 4429–4443.

Genetic Alliance, A District of Columbia Department of Health, *Understanding Genetics: A District of Columbia Guide for Patients and Health Professionals.* Washington (DC): Genetic Alliance; 2010 Feb. Genetic Alliance Monographs and Guides. https://www. ncbi. nlm.nih.gov/pubmed/23586106 (accessed Aug 17, 2017).

Ghosh, M. K., et al., (2014). Antisense oligonucleotides directed against insulin-like growth factor-II messenger ribonucleic acids delay the progress of rat hepatocarcinogenesis. *J. Carcinog., 13,* 2.

Ghosh, M. K., et al., (2014b). Antisense oligonucleotides directed against insulin-like growth factor-II messenger ribonucleic acids delay the progress of rat hepatocarcinogenesis. *J. Carcinog., 13,* 2.

Gore, M. E., (2003). Gene, therapy can cause leukemia: No shock, mild horror but a probe, *Gene Therapy, 10,* 1–4.

Guo, D., et al., (2013). Zinc oxide nanoparticles decrease the expression and activity of plasma membrane calcium ATPase, disrupt the intracellular calcium homeostasis in rat retinal ganglion cells. *Int. J. Biochem. Cell Biol., 45*(8), 1849–1859.

Ha, E. S., et al., (2015). Dissolution and bioavailability of lercanidipine–hydroxypropylmethylcellulose nanoparticles with surfactant. *Int. J. Biol. Macromolec., 72,* 218–222.

Han, J., et al., (2011). Remission of diabetes by insulin gene therapy using a hepatocyte-specific and glucose-responsive synthetic promoter. *Mol. Ther., 19*(3), 470–478.

Head, G. A., & Mayorov, D. N., (2006). Imidazoline receptors, novel agents and therapeutic potential. *Cardiovasc. Hematol. Agents. Med. Chem., 4,* 17–32.

Hsu, J., & Muro, S., (2011). Nanomedicine and drug delivery strategies for treatment of genetic diseases. In: Plaseska-Karanfilska, D., (ed.), *Human Genetic Diseases* (pp. 241–266). InTech: Europe.

Iglarz, M., et al., (2008). Pharmacology of macitentan, an orally active tissue-targeting dual endothelin receptor antagonist. *J. Pharmacol. Exp. Ther., 327,* 736–745.

Jemal, A., et al., (2011). Global cancer statistics. *CA: A Cancer J. Clinicians., 61*(2), 69–90.

Jhaveri, A., & Torchilin, V., (2016). Intracellular delivery of nanocarriers and targeting to subcellular organelles. *Expert Opin. Drug Deliv., 13,* 49–70.

Jia, C., Jin, Q., & Zhao, J., (2012). Nanoparticle-based immunoassays and their applications in nervous system biomarker detection. In: Martin, C. R., Preedy, V. R., & Hunter, R. J., (eds.), *Nanomedicine and the Nervous System* (pp. 17–37). Science Publisher: USA.

Jurel, S. K., et al., (2014). Genes and oral cancer. *Indian J. Hum. Genet., 20*(1), 4–9.

Kahn, S. E., Cooper, M. E., & Prato, S. D., (2014). Pathophysiology and treatment of type 2 diabetes: Perspectives on the past, present, and future. *Lancet, 383*(9922), 1068–1083.

Karunakaran, U., & Park, K., (2013). A systematic review of oxidative stress and safety of antioxidants in diabetes: Focus on islets and their defense. *Diabetes. Metab. J., 37*(2), 106–112.

Kayser, O., Lemke, A., & Hernandez-Trejo, N., (2005). The impact of nanobiotechnology on the development of new drug delivery systems. *Curr. Pharm. Biotechnol., 6*(1), 3–5.

Keyshams, N., et al., (2013). Effect of ammonium van date nanoparticles on experimental diabetes and biochemical factors in male Spargue-Dawly rats. *Zahedan. J. Res. Med. Sci.*, *15*(10), 59–64.

Kim, Y. I., et al., (1997). The antihypertensive effect of orally administered nifedipine-loaded nanoparticles in spontaneously hypertensive rats. *Br. J. Pharmacol.*, *120*, 399–404.

Koenig, O., et al., (2013). New aspects of gene-silencing for the treatment of cardiovascular diseases. *Pharmaceuticals*, *6*, 881–914.

Kolfschoten, I. G., et al., (2009). Role and therapeutic potential of microRNAs in diabetes. *Diabetes Obes. Metab.*, *11*(4), 118–129.

Koziolek, M., et al., (2015). Investigation of pH and temperature profiles in the GI tract of fasted human subjects using the Intellicap® system. *J. Pharm. Sci.*, *104*, 2855–2863.

Langer, R., (1998). Drug delivery and targeting. *Nature*, *392*, 5–10.

Liu, G., et al., (2009). Metal chelators coupled with nanoparticles as potential therapeutic agents for Alzheimer's disease. *J. Nanoneurosci.*, *1*(1), 42–55.

Liu, Y., Miyoshi, H., & Nakamura, M., (2007). Nanomedicine for drug delivery and imaging: A promising avenue for cancer therapy and diagnosis using targeted functional nanoparticles. *Int. J. Cancer.*, *120*(12), 2527–2537.

Lundin, K. E., et al., (2015). Oligonucleotide therapies: The past and the present. *Hum. Gene Ther.*, *26*(8), 475–485.

Mayo Clinic, (2017). *Angelman Syndrome.* http://www.mayoclinic.org/diseases-conditions/ angelman-syndrome/symptoms-causes/dxc–20307383 (accessed Aug 17, 2017).

McLendon, J. M., et al., (2015). Lipid nanoparticle delivery of a microRNA–145 inhibitor improves experimental pulmonary hypertension. *J. Control. Release.*, *210*, 67–75.

McMurray, J. J., (2015). Neprilysin inhibition to treat heart failure: A tale of science, serendipity, and second chances. *Eur. J. Heart. Fail.*, *17*, 242–247.

Moghimi, S. M., & Szebeni, J., (2003). Stealth liposomes and long circulating nanoparticles: Critical issues in pharmacokinetics, opsonization, and protein-binding properties. *Prog. Lipid. Res.*, *42*(6), 463–478.

Mukherjee, B., et al., (2001). Effect of selenomethionine on N-methylnitronitrosoguanidine-induced colonic aberrant crypt foci in rats. *Eur. J. Cancer Prev.*, *10*, 347–355.

Mukherjee, B., et al., (2005). Characterization of insulin-like growth factor II (IGF II) mRNA positive hepatic altered foci and IGF II expression in hepatocellular carcinoma during diethylnitrosamine-induced hepatocarcinogenesis in rats. *J. Carcinog.*, *4*(12), 1–14.

Mukherjee, B., et al., (2007a). Changes in the antioxidant defense and hepatic drug metabolizing enzyme and isoenzyme levels, 8-hydroxydeoxyguanosine formation and expressions of c-raf.1 and insulin-like growth factor II genes during the stages of development of hepatocellular carcinoma in rats. *Eur. J. Cancer. Prev.*, *16*(4), 363–371.

Mukherjee, B., et al., (2007b). Sustained release of acyclovir from nanoliposomes and nanoliposomes: An in vitro study. *Int. J. Nanomed.*, *2*, 213–225.

Mukherjee, B., et al., (2013a). Obesity and insulin resistance: An abridged molecular correlation. *Lipid. Insights.*, *6*, 1–11.

Mukherjee, B., et al., (2013b). Potentials and challenges of active targeting at the tumor cells by engineered polymeric nanoparticles. *Curr. Pharm. Biotechnol.*, *14*, 1250–1263.

National Human Genome Research Institute, (2017). *Frequently Asked Questions About Genetic Disorders.* https://www.genome.gov/19016930/faq-about-genetic-disorders/ (accessed Aug. 17, 2017).

Newton-Cheh, C., et al., (2009). Genome-wide association study identifies eight loci associated with blood pressure. *Nat. Genet.*, *41*, 666–676.

Niaz, T., et al., (2016). Antihypertensive nanoceuticales based on chitosan biopolymer: Physico-chemical evaluation and release kinetics. *Carbohydr. Polym.*, *142*, 268–274.

Niu, L., et al., (2008). Gene therapy for type 1diabetes mellitus in rats by gastrointestinal administration of chitosan nanoparticles containing human insulin gene. *World J. Gastroenterol.*, *14*(26), 4209–4215.

Niu, M., et al., (2011). Liposomes containing glycocholate as potential oral insulin delivery systems: Preparation, *in vitro* characterization, and improved protection against enzymatic degradation. *Int. J. Nanomedicine.*, *6*, 1155–1166.

Niu, W. G., et al., (2009). Endothelial nitric oxide synthase genetic variation and essential hypertension risk in Han Chinese: The Fangshan study. *J. Hum. Hypertens.*, *23*, 136–140.

Nnamani, P. O., et al., (2010). SRMS142-based solid lipid microparticles: Application in oral delivery of glibenclamide to diabetic rats. *Eur. J. Pharm. Biopharm.*, *76*(1), 68–74.

Nunes, T., et al., (2010). Safety, tolerability, and pharmacokinetics of Etamicastat, a novel dopamine-β-hydroxylase inhibitor, in a rising multiple-dose study in young healthy subjects. *Drugs. R. D.*, *10*, 225–242.

Orive, G., et al., (2003). Drug delivery in biotechnology: Present and future. *Curr. Opin. Biotechnol.*, *14*(6), 659–664.

Ortega, F. J., et al., (2014). Profiling of circulating microRNAs reveals common microRNAs linked to type 2 diabetes that change with insulin sensitization. *Diabetes. Care*, *37*(5), 1375–1383.

Pardridge, W. M., (2010). Biopharmaceutical drug targeting to the brain. *J. Drug. Target.*, *18*(3), 157–167.

Parsa, N., (2012). Environmental factors inducing human cancers. *Iranian J. Publ. Health.*, *41*(11), 1–9.

Paulis, L., & Unger, T., (2010). Novel therapeutic targets for hypertension. *Nat. Rev. Cardiol.*, *7*, 431–441.

Paulis, L., Rajkovicova, R., & Simko, F., (2015). New developments in the pharmacological treatment of hypertension: Dead-end or a glimmer at the horizon? *Curr. Hypertens. Rep.*, *17*, 557.

Peng, L. et al., (2016). Integration of antimicrobial peptides with gold nanoparticles as unique non-viral vectors for gene delivery to mesenchymal stem cells with antibacterial activity: *Biomaterials*, *103*, 137–149.

Pridgen, E. M., Alexis, F., & Farokhzad, O. C., (2014). Polymeric nanoparticle technologies for oral drug delivery. *Clin. Gastroenterol. Hepatol.*, *12*, 1605–1610.

Prina, M. R., et al., (2014). Diabetes gene therapy by means of nanoparticles. *Boletim. Informativo. Geum.*, *5*(2), 100–112.

Revanasiddappa, M., & Bhadauria, D., (2013). The role of genetics in hypertension. *Clinical Queries: Nephrology*, *2*, 120–125.

Robinson, A., Fridovich-Keil, J. L., & Fridovich, I., (2017). *Encyclopedia Britannica.* Human genetic disease. https://www.britannica.com/science/human-genetic-disease (accessed Aug 17, 2017).

Robinson, R. D., et al., (2002). Visible thermal emission from sub-band-gap laser excited cerium dioxide particles. *J. Appl. Phys.*, *92*, 1936–1941.

Ross, C. J., et al., (2010). Treatment of a lysosomal storage disease, mucopolysaccharidosis VII, with microencapsulated recombinant cells. *Hum. Gene. Ther.*, *11*(15), 2117–2127.

Sahay, G., Alakhova, D. Y., & Kabanov, A. V., (2010). Endocytosis of nanomedicines. *J. Control. Release.*, *145*(3), 182–195.

Satapathy, B. S., et al., (2016). Lipid nanocarrier-based transport of docetaxel across the blood-brain barrier. *RSC. Adv.*, *6*(88), 85261–85274.

Sattler, K. D., (2010). *Handbook of Nanophysics: Nanomedicine and Nanorobotics* (p. 887). CRC Press: Boca Raton, FL, USA.

Shah, U., Joshi, G., & Sawant, K., (2014). Improvement in antihypertensive and antianginal effects of felodipine by enhanced absorption from PLGA nanoparticles optimized by factorial design. *Mater. Sci. Eng. C.*, *35*, 153–163.

Shaw, J. E., Sicree, R. A., & Zimmet, P. Z., (2010). Global estimates of the prevalence of diabetes for 2010 and 2030. *Diabetes. Res. Clin. Pract.*, *87*(1), 4–14.

Sinha, B., et al., (2013). Poly-lactide-co-glycolide nanoparticle containing voriconazole for pulmonary delivery: *In vitro* and *in vivo* study. *Nanomed. Nanotech. Biol. Med.*, *9*(1), 94–104.

Sloop, K. W., et al., (2004). Hepatic and glucagon-like peptide–1-mediated reversal of diabetes by glucagon receptor antisense oligonucleotide inhibitors. *J. Clin. Invest.*, *113*(11), 571–581.

Son, S., et al., (2014). i-Motif-driven Au nanomachines in programmed siRNA delivery for gene-silencing and photothermal ablation. *ACS Nano.*, *8*, 5574–5584.

Sonaje, K., et al., (2010). Biodistribution, pharmacodynamics, and pharmacokinetics of insulin analogs in a rat model: Oral delivery using pH-responsive nanoparticles vs. subcutaneous injection. *Biomaterials*, *31*(26), 6849–6858.

Sriraman, K., et al., (2016). Anti-cancer activity of doxorubicin-loaded liposomes co-modified with transferrin and folic acid. *Eur. J. Pharm. Biopharm.*, *105*, 40–49.

Stachelek, S. J., et al., (2011). The effect of CD47 modified polymer surfaces on inflammatory cell attachment and activation. *Biomaterials*, *32*(19), 4317–4326.

Stoppler, M. C., (2017). MedicineNet.com. *Genetic Diseases Overview*. http://www.medicinenet.com/genetic_disease/article.htm (accessed Aug 17, 2017).

Tao, S. L., & Desai, T. A., (2003). Microfabricated drug delivery systems: From particles to pores. *Adv. Drug Deliv. Rev.*, *55*(3), 315–328.

Tiwari, P., (2015). Recent trends in therapeutic approaches for diabetes. Management: A comprehensive update. *J. Diabetes. Res.*, *2015*(340838), 1–11.

Townsend, R., et al., (2011). Blood pressure effects of naproxcinod in hypertensive patients. *J. Clin. Hypertens.*, *13*, 376–384.

Turner, J. R., (2009). Intestinal mucosal barrier function in health and disease. *Nat. Rev. Immunol.*, *9*, 799–809.

Veiseh, O., et al., (2015). Managing diabetes with nanomedicine: Challenges and opportunities. *Nat. Rev. Drug Discov.*, *14*(1), 45–57.

Ward, P. D., Tippin, T. K., & Thakker, D. R., (2000). Enhancing paracellular permeability by modulating epithelial tight junctions. *Pharm. Sci. Technol. Today.*, *3*(10), 346–358.

Wong, T. W., (2010). Design of oral insulin delivery systems. *J. Drug. Target.*, *18*(2), 79–92.

Yasuzaki, Y., et al., (2015). Validation of mitochondrial gene delivery in liver and skeletal muscle via hydrodynamic injection using an artificial mitochondrial reporter DNA vector. *Mol. Pharm.*, *12*(12), 4311–4320.

Yatuv, R., et al., (2010). The use of PEGylated liposomes in the development of drug delivery applications for the treatment of hemophilia. *Int. J. Nanomedicine*, *5*, 581–591.

Zhang, K., et al., (2017). Germline mutations of PALB2 gene in a sequential series of Chinese patients with breast cancer. *Breast Cancer Res Treat.*, *17*, 1–9.

Zhou, T., et al., (2006). Iron binding dendrimers: A novel approach for the treatment of hemochromatosis. *J. Med. Chem.*, *49*(14), 4171–4182.

CHAPTER 6

Nanomedicine: Could It Be a Boon for Pulmonary Fungal Infections?

BISWAJIT MUKHERJEE*, ASHIQUE AL HOQUE,
SHREYASI CHAKRABORTY, LEENA KUMARI, SOMDATTA ROY, and
PARAMITA PAUL

*Department of Pharmaceutical Technology, Jadavpur University,
Kolkata–700032, India, Tel.: +91-33-2457-2588/2414-6677/2457-2274,
E-mail: biswajit55@yahoo.com*

*Corresponding author. E-mail: biswajit55@yahoo.com

ABSTRACT

Pulmonary fungal infections cover a broad spectrum related to a variety of fungal sources. They can particularly affect immune-compromised individuals. Fungi may cause lung infection when fungal material or fungal spores are inhaled. The incidence of pulmonary fungal infection has been increasing globally in the last few decades. There are a variety of fungal infections such as histoplasmosis, sporotrichosis, blastomycosis, coccidioidomycosis, para-coccidioidomycosis, cryptococcosis, aspergillosis, candidiasis, pneumonia, etc. that can cause severe pulmonary injury. The advancement of nanotechnology has revolutionized the field of pharmacotherapy because of its therapeutic potential specifically in the treatment of pulmonary fungal infections. Nanotechnology offers a plethora of advantages over conventional therapy including smaller size, larger surface area, capability of surface modification, site-specific targeting to increase local drug concentration, thereby reducing dose-related side effects, potential to entrap both hydrophilic and hydrophobic drugs, and improved pharmacokinetic profile such as extended retention time, increased half-life of drugs, etc. In this chapter, we will mainly focus on the current treatment regimen, approaches of various nanoformulations for the treatment of pulmonary fungal infections via different routes of administration, factors to be considered for designing nanoformulations and their nanotoxicological aspects.

6.1 INTRODUCTION

Pulmonary fungal infection is one of the leading causes of deaths in developing countries. They mainly affect immunocompromised patients (i.e., hematological, solid organ transplant or intensive care unit patients) and people residing in certain particular geographic areas. The infections of the persons due to their geographic location are referred to as endemic mycoses, whereas the persons having immune deficiency are found to develop opportunistic infections (Smith and Kauffman, 2012). Although the rate of incidence of invasive pulmonary fungal infections are less common as compared to bacterial and viral infections, but still they are responsible for significant morbidity and mortality in infected patients (Erjavec et al., 2009). The most common type of fungal pneumonia is invasive pulmonary aspergillosis. Some other types of angioinvasive molds, including Fusarium and Zygomycetes species, also infect the susceptible patients (Wahba et al., 2008).

In spite of several progresses made in this field, any vaccine for fungal infections is not yet available clinically. Several antifungal agents are currently available in the market for the treatment of pulmonary fungal infections. Among them, amphotericin B had been the drug of choice for most cases of fungal infections. However, its usage is limited due to serious adverse effects, the most significant one being nephrotoxicity. Liposomal formulation of amphotericin B is also developed, and it greatly reduces nephrotoxicity, but still, it encounters numerous drawbacks due to its non-targeted delivery via parenteral route. In addition, it gets rapidly cleared from the systemic circulation due to uptake by reticuloendothelial system, which cannot be overlooked (Bowden et al., 2002; Moen et al., 2009). Other chemotherapeutic agents found to be therapeutically effective for the treatment of pulmonary fungal infections include the newer second-generation triazole antifungals (voriconazole, posaconazole) and flucytosine. However, these drugs also suffer from several dose-dependent adverse effects such as hepatotoxicity, gastric disturbances, bone marrow suppression-characterized as leucopenia, thrombocytopenia, and/or pancytopenia, etc. (Cook & Confer, 2011).

The present scenario demands the development of formulations which can directly deliver the desired dose of drug specifically to the site of action, thereby minimizing the dose-related adverse effects of drug in other parts of the body. Nanotechnology in the field of drug delivery is emerging rapidly and holds great promise in the significant improvements in human health. Nanoparticle-based system for pulmonary delivery may help to retain the particles at the desired site (lungs) for prolonged period of time, thus becoming more efficacious and less toxic as compared to conventional formulations (Mansour et al., 2009; Watts & Williams, 2011).

This chapter provides insights into the various nanoscale size drug delivery systems as an effective diagnostic and therapeutic tool for the treatment of pulmonary fungal infections, factors to be considered for designing nanoformulations and their nanotoxicological aspects. The chapter also summarizes the recent findings and applications of nanoparticulate systems via different routes of administration in the field of pulmonary delivery.

6.2 ANATOMY OF HUMAN LUNGS

Lung is a vital organ to sustain life. Lungs are responsible for taking in oxygen and expelling carbon dioxide and thus optimize gas exchange between our blood and the air (Figure 6.1). The lungs are a pair of spongy organs located in the thoracic cavity. Both lungs have a central recession called the hilum at the root of the lung, where the blood vessels and bronchus enter into the organ (Standring & Borley, 2008). The elastic fibers of the lungs aid it to expand and contract during breathing. Healthy lungs have a smooth, shiny surface because they are encased in a series membrane called the visceral pleura (Dorland, 2011). There is also a pleural covering called the parietal pleura that lines the inner surface of the chest cavity. Both the pleurae are normally slippery and slide easily against each other. The lung is connected to the heart and trachea by the root. It consists of bronchus, pulmonary artery, pulmonary veins, bronchial arteries and veins, pulmonary plexuses of nerves, lymphatic vessels, bronchial lymph glands, and areolar tissue.

Each lung is conical in shape with an apex, a base, three borders, and two surfaces (Drake et al., 2014). Apex is round and extends into the root of the neck, reaching from 2.5 to 4 cm above the level of the sternal end of the first rib. It is covered by the cervical pleura and supra pleural membrane and grooved by subclavian artery and vein. Base-is a large, bottom part of the lung that rests on the diaphragm (Drake et al., 2014). Borders are of different types. Anterior border is thin and sharp. The left lung has a notch in its border called the cardiac notch of the left lung to make space for the heart (Standring & Borley, 2008; Dorland, 2011). Inferior border is thin and sharp and separates base from coastal surface. It extends into the phrenicocostal sinus. Posterior border is thick and ill-defined. It fits into deep paravertebral

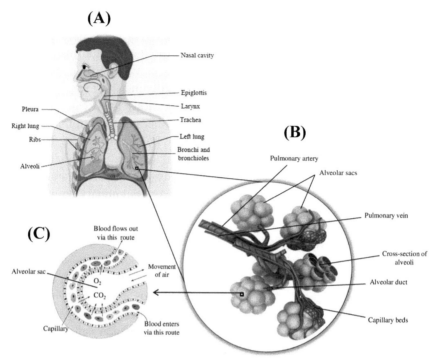

FIGURE 6.1   (See color insert.) Human respiratory organ (A) Structure of the human respiratory system; (B) An enlarged view of terminal alveolar region and dense capillary network; (C) Close up view of gaseous exchange of carbon dioxide and oxygen between the capillaries and alveoli.

gutter. Surfaces of the lung have two varieties: (a) costal surface are in contact with costal pleura and overlying thoracic wall; (b) medial surface consists of posterior and anterior part. Lobes and fissures of the right and left lung are separated by mediastinum. Unlike some other paired organs in the body, the two lungs are not identical. The right lung is normally larger than the left. It is divided into three sections or lobes, the upper (also called superior), middle, and lower (also called inferior) by two fissures, one oblique and other horizontal (Drake et al., 2014). The left lung is divided by the oblique fissure into two lobes, the upper and the lower one. The left lung has a homologous feature, the lingual. Each lung lobe Gets divided into broncho-pulmonary segments that are aerated by segmental/tertiary bronchus (Arakawa et al., 2000). Each segment is a separate distinct respiratory unit, having its own separate artery. The bronchial airway system is called a 'tree' for a very practical reason. The airways branch approximately 20 times in the lungs. At each branch point, the airways become smaller and more in number much like the branches of trees. The main bronchi branch into lobar bronchi then into segmental bronchi that again branch into smaller airways called bronchioles. The final branches of the bronchial tree are called atria. The atria end in alveoli (Drake et al., 2014). Alveoli resemble clusters of grapes under the microscope. There are about 300 million alveoli in each lung. The alveolar walls are extremely thin and fragile thus making them susceptible to damage and this damage is irreversible. Alveoli are surrounded by capillaries. There are about one billion capillaries in the lungs, more than three for each air sac. The blood in the capillaries is separated from the air in the alveoli only by the extremely thin alveolar and capillary walls. Each alveolus is Made up of type I and type II alveolar cells, alveolar macrophages (Koeppen, 2008). The inner layer of the alveolar epithelium is lined by the surfactant, synthesized by type II alveolar cells (Koeppen, 2008).

6.3 PULMONARY FUNGAL INFECTIONS

The virulence of pulmonary fungal infections can be seen when fungi or their spores are inhaled. Invasive fungal infections (termed mycoses) occur in both immunocompetent and immunocompromised patients (Limper et al., 2011). Mycoses can be divided into two broad categories: the opportunistic and the endemic mycoses. Opportunistic fungal infections primarily infects the immunocompromised patients through a congenital or acquired disease process. Whereas, endemic fungal infections can affect both healthy and immunocompromised patients; and are seen in specific geographical locations around the world (Table 6.1).

6.3.1 DIAGNOSIS AND CURRENT TREATMENT OF PULMONARY FUNGAL INFECTIONS

In the 21st century, there have been important approaches in the diagnosis and treatment of both the endemic and the opportunistic mycoses, in depiction of the epidemiology of invasive pulmonary mold infections in immunocompromised patients and in the taxonomy of both yeasts and molds that cause pulmonary fungal infection. The extent of invasive fungal infection has been continuously increasing as an outcome of factors such as aggressive chemotherapy for cancer, bone marrow and organ transplantation, AIDS, and advanced critical care (Perfect and Schell, 1996). Here we will briefly discuss about various effective tools for the diagnosis of pulmonary fungal infections (Table 6.2).

TABLE 6.1 Various Types of Lung Fungal Infection with Causative Organism and Their Signs, Symptoms

S No.	Pulmonary Fungal Infections	Causative Organisms	Signs and Symptoms	References
A.	**Opportunistic Fungal Infections**			
A.1.	Aspergillosis	*Aspergillus fumigatus, Aspergillus flavus,* and *Aspergillus niger*	Fever unresponsive to antibiotics, tracheobronchitis associated with ulcerations and plaque formation, pleuritic chest pain (due to vascular invasion leading to small pulmonary infarcts), hemoptysis, chills, shock, delirium, seizures, and blood clots	(Limper et al., 2011; Zmeili & Soubani, 2007)
A.2.	Cryptococcosis	*Cryptococcus neoformans*	Fever, malaise, pleuritic chest pain, the cough usually nonproductive, hemoptysis, headache, nausea vomiting	(Limper et al., 2011; Perfect et al., 2010)
A.3.	Candidiasis	*Candida albicans*	Hypersensitivity reactions (allergic asthma, bronchopulmonary aspergillosis, extrinsic allergic alveoli tides), dry cough, fever	(Limper et al., 2011; Pappas et al., 2009)
A.4.	Mucormycosis	*Mucor* spp	Fever, cough, chest pain, shortness of breathing	(Limper et al., 2011; Goldstein et al., 2009)
A.5.	Pneumocystis Pneumonia (PCP)	*Pneumocystis jirovecii*	Fever, cough, difficult breathing, chest pain, chills, fatigue	(Limper et al., 2011; Harris et al., 2010)
B.	**Endemic fungal infections**			
B.1.	Histoplasmosis	*Histoplasma capsulatum*	Fever, chills, headache, muscles aches, dry cough, chest discomfort	(Limper et al., 2011; Wheat et al., 2007; Kauffman et al., 2006)
B.2.	Blastomycosis	*Blastomyces dermatitidis*	Cough, fever, night sweats, dyspnea, aches, chest pain, hemoptysis	(Limper et al., 2011; Kauffman et al., 2006)
B.3.	Sporotrichosis	*Sporothrix schenckii*	Cough, shortness of breath, chest pain, fever	(Kauffman et al., 2006)
B.4.	Coccidioidomycosis	*Coccidioides immitis, Coccidioide sposadasii*	Wheezing, headache, muscle aches, blood tinged sputum, loss appetite, shortness of breath	(Galgiani et al., 2005)

TABLE 6.2 Diagnostic Tools for Various Pulmonary Fungal Infections (James et al., 2006; Meshram and Mishra, 2011; Nguyen et al., 2013)

Fungal Infections	Diagnostic Technique
Aspergillosis	Culture, histopathology, computed tomography finding of halo sign, galactomannan enzyme immunoassay, in serum and bronchoalveolar lavage
Histoplasmosis	Serology, culture, Histopathology, antigen enzyme immunoassay in urine, serum bronchoalveolar lavage
Mucormycosis	RT-PCR in blood and BAL (likely to play an important role in future), histopathology, culture, CT finding of reverse halo sign,
Fusariosis	Histopathology, culture, characteristic skin lesions.
Cryptococcosis	Culture, histopathology, Antigen latex agglutination assay in serum, cerebrospinal fluid, body fluids
Scedosporiosis	Culture, histopathology, PCR, and in situ hybridization (probably both will play an important role in future)
Blastomycosis	Antigen enzyme immunoassay in urine, serum, and perhaps body fluids, culture, histopathology.
Coccidioidomycosis	Serology, antigen enzyme immunoassay in urine, serum culture, histopathology.

To overcome the above mentioned fungal infections, the first-line treatment, step-down therapy, and second-line therapy are summarized in Table 6.3.

6.4 NANODOSAGE FORMS UNDER INVESTIGATION, INTENDED FOR THE PULMONARY LUNG FUNGAL INFECTIONS ALONG WITH THEIR ADVANTAGES AND DISADVANTAGES

Influenced by the progress in nanomedicine field, many researchers have shown innovation in formulation development with improved biopharmaceutical properties. The drugs which could not be administered for lung fungal infections directly can be delivered by a pulmonary delivery system with a formulation of nanoscale size. Besides, being the alternative to conventional therapy, those formulations also have promising activity. Those inventions have opened a door in the treatment of pulmonary fungal infections. Here those different types of formulations with their limitations are emphasized.

TABLE 6.3 Recommended Treatment Regimens for Pulmonary Fungal Infections (Smith & Kauffman, 2012)

Fungal infection	Favored treatment	Step-down therapy	Second line therapy
Aspergillosis	Voriconazole, 4 mg/kg twice daily (intravenous) until stable.	Voriconazole, 200 mg, twice daily (oral), until lesions resolved.	Lipid amphotericin B; 5 mg/kg/day (intravenous), until stable, or posaconazole, 400 mg, twice daily (oral).
Histoplasmosis	Drastic or immunocompromised: lipid Amphotericin B, 3–5 mg/kg/day (intravenous), until stable.	Itraconazole, 200 mg, twice daily (oral), A loading dose of 200 mg three times a day (oral) for the first 3 days should be given for 6–12 months.	Amphotericin B-deoxycholate, 0.7–1 mg/kg/day, Intravenous, until stable, then itraconazole, 200 mg twice daily (oral), A loading dose of 200 mg total for the first 3 days should be given for 6–12 months.
	Mild to moderate infection: itraconazole, 200 mg twice daily (oral), A loading dose of 200 mg three times a day (oral) for the first 3 days should be given for 6–12 months.	—	Voriconazole, 200 mg twice daily (oral), A loading dose of 400 mg twice daily (oral) for the first day should be given or fluconazole, 800 mg every day (oral), for 6–12 months.
Mucormycosis	Lipid amphotericin B, 5 mg/kg/day (intravenous), or higher doses until lesions resolved.	Posaconazole 400 mg, twice daily (oral)	Amphotericin B-deoxycholate, 1 mg/kg/day (intravenous), until lesions resolved (possible addition of echinocandin to AmB formulation).
Fusariosis	Lipid amphotericin B, 5 mg/kg/day (intravenous), or voriconazole, 4 mg/kg bid IV, A loading dose of 6 mg/kg twice daily IV for the first day should be given or combination of lipid amphotericin B, 5 mg/kg/day (intravenous), plus voriconazole, 4 mg/kg twice daily (intravenous)	Voriconazole, 200 mg twice daily (oral), until lesions resolved	Posaconazole, 400 mg bid oral, Posaconazole can also be dosed as 200 mg 4 times a day (oral) as long as the patient is able to follow with frequent dosing.

TABLE 6.3 *(Continued)*

Fungal infection	Favored treatment	Step-down therapy	Second line therapy
Cryptococcosis	Drastic or with dissemination: lipid Amphotericin B, 3–5 mg/kg/day (intravenous) with flucytosine, 25 mg/kg, 4 times daily (oral), for about 15 to 30 days	Fluconazole, 400 mg every day (oral), for 8–10 weeks, then fluconazole, 200 mg for 6–12 months	Amphotericin B-deoxycholate, 0.7–1 mg/kg/day (intravenous) with flucytosine, 25 mg/kg 4 times a day (oral), for 2–4 weeks, followed by same step-down therapy.
Scedosporiosis	Voriconazole, 4 mg/kg twice daily (intravenous). A loading dose of 6 mg/kg twice daily (intravenous) for the first day should be given until stable.	Voriconazole, 200 mg twice daily (oral), until lesions resolved	Posaconazole, 400 mg twice daily (oral), Posaconazole can also be dosed as 200 mg 4 times daily (oral) as long as the patient is able to comply with frequent dosing.
Blastomycosis	Drastic or immunocompromised: lipid Amphotericin B, 3–5 mg/kg/day (intravenous). Mild to moderate infection: itraconazole, 200 mg bid oral, A loading dose of 200 mg 3 times a day (oral) for the first 3 days should be given for 6–12 months.	Itraconazole, 200 mg bid oral, A loading dose of 200 mg 3 times a day (oral) for the first 3 days should be given for 6–12 months.	Amphotericin B-deoxycholate 0.7–1 mg/kg/day, until stable, then itraconazole, 200 mg twice daily (oral), A loading dose of 200 mg 3 times a day (oral) for the first 3 days should be given for 6–12 months. Voriconazole, 200 mg bid oral, A loading dose of 400 mg twice daily (oral) for the first day should be given or fluconazole, 800 mg every day (oral), for 6–12 months.
Coccidioidomy-cosis	Drastic or immunocompromised: lipid Amphotericin B, 3–5 mg/kg/day (intravenous), until stable.	Itraconazole, 200 mg bid oral, A loading dose of 200 mg 3 times a day (oral) for the first 3 days should be given or fluconazole, 400 mg every day (oral), for 12 months	Amphotericin B-deoxycholate 0.7–1 mg/kg/day (intravenous), until stable, then itraconazole, 200 mg bid oral, A loading dose of 200 mg 3 times a day (oral) for the first 3 days should be

TABLE 6.3 (*Continued*)

Fungal infection	Favored treatment	Step-down therapy	Second line therapy
	Mild to moderate infection: itraconazole, 200 mg twice daily (oral), A loading dose of 200 mg three times a day (oral) for the first 3 days should be given or fluconazole, 400 mg every day (oral), for 6 months		given or fluconazole, 400 mg every day (oral), for 12 months. Voriconazole, 200 mg bid oral, A loading dose of 400 mg twice daily (oral) for the first day should be given or posaconazole, 400 mg twice daily (oral), Posaconazole can also be dosed as 200 mg 4 times a day (oral) as long as the patient is able to comply with frequent dosing.

6.4.1 POLYMERIC NANOPARTICLES

A major part of nanotherapeutics includes the use of different types of polymers via different routes for pulmonary fungal infections. In many studies, polyvinyl alcohol and poly (lactide-co-glycolide) (PLGA) are used as sustained release agents because of their biocompatibility and minimal toxicity in both *in vitro* and *in vivo* (Salama et al., 2009; Roberts et al., 2013; Dailey et al., 2006). Das et al. (2015) showed that voriconazole formulation with PLGA has a sustained effect on pulmonary fungal infections (2015). However, these undissolved particles in the lungs may trigger pulmonary inflammation and fibrosis which is a serious concern. As these are foreign particles, they could be removed either by macrophage present in the lungs or mucociliary clearance (Tomashefski et al., 1988). Therefore, *in vivo* safety and efficacy are still needed to be explored.

6.4.2 LIPID-BASED NANOPARTICLES

Lipid-based nanoparticles are widely used for pulmonary fungal infections. Many lipids are amphiphilic in nature, thus can form various structures and easily penetrate the biological membranes or cells. The bioavailability of many hydrophobic drugs can be improved by encapsulating them in lipid formulations. The biocompatibility of lipids limits the toxicity, which is also a desirable property for therapeutic applications. Lipid particles usually are well taken by the lungs providing for a prolonged retention of carriers and drugs.

6.4.2.1 LIPOSOMES

Many drugs are readily water-soluble, well absorbed in the systemic circulation and cleared subsequently. These drugs fail to maintain the concentration above minimum inhibitory concentration in respiratory tract. The liposomal formulations of those drugs are effective because of sustained release of drug from the formulation, thus prolonging the action at the site of infection. This reduces the frequent dosing and patient compliance. The above-mentioned mechanism can be considered for targeted delivery (Vyas & Khatri, 2007).

A major drawback of liposomal formulations is to retain some molecules into the liposome core (Gregoriadis, 1973; Juliano & Stamp, 1978; Mayhew et al., 1978). The drug release from liposomes was shown to be affected by serum proteins (Scherphof et al., 1978; Allen & Cleland, 1980; Senior & Gregoriadis, 1982) Leakage of drug from liposomes is also a problem which can be reduced by incorporation of cholesterol that tightens fluid bilayer (Allen & Cleland, 1980; Cullis, 1976; McIntosh, 1978). Another problem of liposomes is fast clearance by mononuclear phagocyte system, especially in the liver and lungs (Kimelberg et al., 1976; Gregoriadis & Neerunjan, 1974; Juliano & Stamp, 1975).

6.4.2.2   SOLID LIPID NANOPARTICLES (SLN)

The disadvantages of polymeric nanoparticles have driven the researchers towards the development of solid lipid nanoparticles (SLN) since the beginning of the nineties. Lipids are well tolerated in the body, and it is also cost-effective in large-scale production. Advantages of SLN in lungs thus include prolonging drug release, faster degradation of particles than polymeric nanoparticles. Disadvantages of SLN include low drug loading capacity and drug expulsion during storage.

6.4.2.3   NANOSTRUCTURED LIPID CARRIER (NLC)

To overcome the potential difficulties of SLN, NLC was introduced (Muller et al., 2002; Radtke & Muller, 2000; Uner, 2006). The object of NLC preparations was to prevent drug leakage and increase drug loading. NLCs are prepared by melt emulsification technique where mixture of solid lipid and spatially incompatible liquid lipids are used. Hydrophobic drugs can be well entrapped into the core. The surface of NLC can be modified with target moieties. In a 12 day study of fungal infection treatment using itraconazole: polysorbate 80: poloxamer 407 nanostructured aggregates administered by inhalation showed higher lung concentration and lower toxicity in infected mice. (McConville et al., 2006; Vaughn et al., 2006; Alvarez et al., 2007; Hoeben et al., 2006). No histological changes in lungs were detected (Vaughn et al., 2007).

6.4.3   DENDRIMERS

Dendrimers are highly branched structures with diameters range from 4–20 nm. This size of dendrimers provides efficient internalization by different cells, but rapid penetration into systemic circulation minimized retention time in lungs (Kuzmov & Minko, 2015).

6.5   ADVANTAGES OF NOVEL DRUG DELIVERY SYSTEM OVER CONVENTIONAL THERAPY

Nanoformulations are gaining increasing importance for pulmonary drug delivery as compared to conventional therapy owing to their site-specific delivery, bioadhesive properties, sustained drug release and minimized dosing frequency for improved patient compliance. Due to their smaller size, nanoparticles escape from getting cleared by the alveolar macrophage clearance. The deposition of drugs into the lungs enhances the possibility of reduced side-effects and systemic toxicity, and also helps in achieving higher concentration of drugs at the site of action (Sung et al., 2007; Rytting et al., 2008).

6.6   MODE OF DRUG ADMINISTRATION FOR THE TREATMENT OF PULMONARY FUNGAL INFECTIONS

The effective treatment of lung fungal infection desires a suitable drug along with drug delivery mode or device. Oral and intravenous routes are still popular choices for the treatment of lung fungal infection. However, pulmonary or nasal routes are preferred nowadays because of its direct delivery to the diseased site without affecting other vital organs. For example, products of amphotericin B are available as parenteral formulations; whereas, lipid-based formulations of amphotericin B are increasingly being used in clinical practice because of its reduced side effects (http://www.cps.ca/documents/position/anti fungal-agents-fungal-infections).

6.6.1   ORAL ROUTE

This route is generally the most convenient and cost-effective. Majority of the antifungal agents are delivered in the patients using oral route of drug administration. Azole antifungal agents such as fluconazole, itraconazole, voriconazole, posaconazole, ravuconazole, etc. and flucytosine are available as oral formulations such as tablets, capsules, solution or suspensions.

6.6.2   INTRAVENOUS ROUTE

Intravenous route is used for easy access to rapid administration of solutions containing drug or other chemicals which allows rapid and more predictable delivery. For some drugs, i.v. route allows higher doses than would be tolerated orally. Furthermore, this route helps to bypass the first-pass metabolism. It is a route of choice for a wide range of drugs like amphotericin B, fluconazole, voriconazole, itraconazole, ravuconazole, caspofungin, etc. Liposomal amphotericin B or lipid complex of amphotericin B are also available and are effective than the amphotericin B deoxycholate, but are highly expensive (Allen, 2010).

6.6.3   PULMONARY ROUTE

Drug may be delivered through pulmonary route either by nebulization, metered dose inhalers or by dry powder inhalers. The pulmonary route, owing to a noninvasive method of drug administration, for both local and systemic delivery of a drug forms an ideal environment for drugs acting on pulmonary disorders. Additionally, this route offers many advantages, such as a high surface area with rapid absorption due to high vascularization and circumvention of the first pass effect. Aerosolization or inhalation of colloidal systems is currently being extensively studied and has huge potential for targeted drug delivery in the treatment of various diseases particularly for lung fungal infections. Furthermore, the surfactant-associated proteins present at the interface enhance the effect of these formulations by decreasing the surface tension and allowing the maximum effect. The most challenging part of developing a colloidal system for nebulization is to maintain the critical physicochemical parameters for successful inhalation (Paranjpe & Muller-Goymann, 2014). Voriconazole loaded nanoparticles prepared with biodegradable and biocompatible polymers like PLGA has been successfully delivered in mice lung by nebulization and dry powder inhalation (DPI) method (Sinha et al., 2013, Das et al., 2015).

6.7   MECHANISM OF DEPOSITION OF THE PARTICLES IN THE LUNG

The deposition of particles in the different regions of the lungs depends on the particle size of the formulation (Figure 6.2). Based on the particle size, three different mechanisms of drug deposition are defined, namely impaction, sedimentation, and diffusion/ Brownian motion.

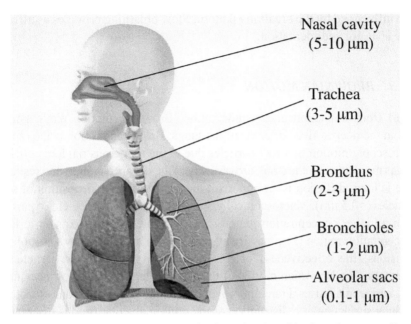

Nasal cavity
(5-10 μm)

Trachea
(3-5 μm)

Bronchus
(2-3 μm)

Bronchioles
(1-2 μm)

Alveolar sacs
(0.1-1 μm)

FIGURE 6.2 **(See color insert.)** Particle-size dependent deposition in various parts of lungs.

6.7.1 IMPACTION

In impaction, the aerosol particles pass through the oropharynx and upper respiratory passages at a high velocity. Due to the centrifugal force, the particles collide with the respiratory wall and are deposited in the oropharynx region. This mechanism is generally observed for DPI, and metered dose inhalators (MDI), with particles, sizes greater than 5 μm. In case of the DPI, the inspiratory effort of the patient plays an important role in the deposition. If the force of inhalation is insufficient, the dry powder deposits in the upper airways, owing to the mass of the particles and the inertial forces. For the MDI, despite of high speed of the generated aerosol, large particle sizes also lead to the deposition of the particles mostly in the upper respiratory tract region.

6.7.2 SEDIMENTATION

Gravitational forces are predominantly responsible for the sedimentation of particles. Particles with sufficient mass and sizes between 1 to 5 μm are deposited in the smaller airways and bronchioles, where they are deposited slowly, provided a sufficiently long time span. Therefore, sedimentation is also influenced by the breathing pattern. Slow breathing provides a sufficient time span for sedimentation.

6.7.3   BROWNIAN MOTION

Apart from impaction and sedimentation, Brownian motion plays a major role in the deeper alveolar areas of the lungs. Brownian motion is the random microscopic motion of small particles due to the numerous random collisions by gas molecules. In the small airways where the distance is short and residence time is long, diffusion is an important mechanism for the deposition of small particles (<0.5 μm). Macroscopically, we see the overall movement of particles from a higher concentration region (i.e., the center of air stream) to a lower concentration region (i.e., the airway wall). Since it is caused by gas molecule collisions, the effectiveness of this mechanism increases as particle size decreases. The Brownian motion of the surrounding molecules of the aqueous lung surfactant causes a random movement of the particles. Upon contact with the lung surfactant, the dissolution of the drug in alveolar fluid is essential for diffusion. Additionally, the concentration gradient also influences the diffusion process. Particles smaller than 1 to 0.5 μm are deposited in the alveolar region, while most of the particles, owing to smaller sizes, are exhaled.

Apart from the mechanisms, parameters such as the particle size, particle morphology, and geometry of the aerosol, along with surface properties play an important role in deposition phenomena. Furthermore, breathing frequency and the holding of breaths, humidity, air velocity, and tidal volume also are vital factors influencing the deposition (http://aerosol.ees.ufl.edu/respiratory/section04.html; Patil & Sarasija, 2012).

6.8   FEATURES TO BE CONSIDERED IN DESIGNING NANOFORMULATIONS

To escape the clearance mechanisms in lung and to be effective for delivering therapeutic agents, there are some factors needed to be considered for designing nanoformulations.

6.8.1   PARTICLE SIZE

The diameter of particles is one of the primary determinants in designing nanoformulation for pulmonary delivery. Particles having diameter larger than 5–10 μm, are normally exhaled, but smaller particle can be delivered by pulmonary route. Ultrafine particles (1–2 μm) are usually deposited in the bronchioles, and particles at the nanoscale (<1 μm) can be delivered to the lower respiratory tract including the alveoli. Ultra small-sized nanoparticles, such as dendrimers (<20 nm) showed efficient delivery to the alveoli. The size-dependent deposition of particles in lung is represented in Figure 6.2 (Kumar et al., 2013).

6.8.2   PARTICLE SHAPE AND ORIENTATION

Besides the diameter of the particles, particle shape is also an important factor that influences the alveolar macrophage clearance. Champion & Mitragotri (2006). Showed in their study that spherical particles were rapidly engulfed by the macrophages whereas, macrophages attached to the minor axis or flat surface of elliptical disks cannot engulf the particles even after 2 h (2006). Another study carried by the same research group reported that worm-like particles exhibit negligible phagocytosis compared to spherical particles due to low curvature region of worm-like particles (Champion & Mitragotri, 2009).

6.8.3   STEALTH CHARACTERISTICS

In order to decrease alveolar clearance and increase half-life of drugs, stealth characteristics were developed by coating the drug with stealth materials. For example, hyaluronic acid (HA), a mucoadhesive polysaccharide present in lungs, is combined with inhaled drugs. HA-conjugated drugs suppress the phagocytosis and provide a prolonged drug effect in the lung (Suzuki & Yamaguchi, 1993).

6.9   TOXICOLOGICAL CONCERN

Nanoparticulate based drug delivery system offers promising results in the treatment of pulmonary fungal infections. However, *in vivo* biofate of nano-formulations for pulmonary delivery and their translocation from pulmonary sites to systemic circulation must be taken into consideration. The nanocarriers might be taken up into by cells and cross epithelial and endothelial region by the process of transcytosis into the systemic circulation, thereby leading to their distribution in various organs of the body such as liver, bone marrow, spleen, lymph nodes, heart, and also to central nervous system (Sung et al., 2007). The *in vivo* performance of nanocarriers and their toxicological concern mainly depends of their surface charge (Geys et al., 2008).

The amount of the nanocarriers at the specific site in the lungs determines its potential toxicity. The probability of the deposition of nanoformulations in the lungs increases greatly with the decrease in particle size. In case the particles are not soluble and biodegradable in the pulmonary region, then there may be chances of their local accumulation upon sustained release. The defense mechanisms of the lungs such as mucociliary clearance (upper airways) and macrophage clearance (lower airways, alveoli) extensively remove the deposited nanocarriers. However, in few special cases such as long nanofibers are not engulfed by alveolar macrophages, and hence suffers from longer residence time in the lungs than short nanofibers, leading to potential toxicity (Borm & Kreyling, 2004).

Therapeutic nanocarriers composing of biomaterials such as poly-lactide-co-glycolide (PLGA), poly (lactic acid), chitosan, gelatin, polycaprolactone, and poly-alkyl-cyanoacrylates which will eventually degrade into biocompatible components might pose reduced risk of toxicity as compared to non-biodegradable materials. According to recent studies, PLGA nanoparticles exhibits lower inflammatory response as compared to non-biodegradable polystyrene nanoparticles (Kumari et al., 2010; Dailey et al., 2006).

The present knowledge regarding mechanism of interaction of biodegradable nanocarriers with biological systems is scarce; therefore there is an urgent need for thorough investigation of health impact upon pulmonary delivery of nanomedicine.

6.10 CONCLUSION AND FUTURE PERSPECTIVES

It is evident that nanoformulations have proved to the promising carriers with novel applications for their utilization in the diagnosis, detection, imaging, and treatment of pulmonary fungal infections. These novel delivery systems are quite efficacious to overcome several vital issues related to conventional therapies such as poor bioavailability, nonspecific targeting, unpredictable side-effects, and drug resistance. Further, prolonged, and sustained release of drug in lungs from nanoparticles directly administered in lungs is a great hope for effective controlling of fungal growth and proliferation of fungal spores. Short treatment of systemic antifungal agents (7–12 days) at a stretch is insufficient to kill fungal spores for rapid removal of drug from lungs due to its high and fast blood turnover. However, the major challenge for nanomedicine for pulmonary delivery is to attain clear understanding of the fate of nanocarriers. Moreover, investigation of the role of physicochemical properties of the nanocarriers, i.e., particle size, surface area, mucoadhesion, etc. in the lungs and their impact on the fate of controlling/inhibiting fungal infection are quite essential for the design and development of these nanocarriers. In the current scenario, insufficient knowledge on the basic mechanism of action of pathogenic fungi affecting lungs prevents the scientists from attaining great success on antifungal therapy in lungs is at the clinical level. Therefore, researchers must work toward identifying new targets to interrupt the life cycle of pathogenic fungi and find possible solutions to strengthen the immune system of the host. Furthermore, nanotoxicological aspects of nanoformulations for pulmonary fungal infections must be taken into consideration to foresee effective targeting to pulmonary fungal infections.

KEYWORDS

- **poly (lactide-co-glycolide)**
- **solid lipid nanoparticles**
- **nanostructured lipid carrier**
- **dry powder inhalation**
- **metered dose inhalators**
- **hyaluronic acid**

REFERENCES

Allen, T. M., & Cleland, L. G., (1980). Serum-induced leakage of liposome contents. *Biochim. Biophys. Acta.*, *597*(2), 418–426.

Allen, U. D., (2010). Antifungal agents for the treatment of systemic fungal infections in children. *Paediatrics Child Health*, *15*(9), 603–608.

Alvarez, C. A., Wiederhold, N. P., McConville, J. T., Peters, J. I., Najvar, L. K., & Graybill, J. R., (2007). Aerosolized nanostructured itraconazole as prophylaxis against invasive pulmonary aspergillosis. *J. Infect.*, *55*(1), 68–74.

Arakawa, H., Niimi, H., Kurihara, Y., Nakajima, Y., & Webb, W. R., (2000). Expiratory high-resolution CT: Diagnostic value in diffuse lung diseases. *Amer. J. Roentgenology*, *175*(6), 1537–1543.

Borm, P. J., & Kreyling, W., (2004). Toxicological hazards of inhaled nanoparticles-potential implications for drug delivery. *J. Nanosci. Nanotech.*, *4*(5), 521–531.

Bowden, R., Chandrasekar, P., White, M. H., Li, X., Pietrelli, L., Gurwith, M., et al., (2002). A double-blind, randomized, controlled trial of amphotericin B colloidal dispersion versus amphotericin B for treatment of invasive aspergillosis in immunocompromised patients. *Clin. Infect. Diseases*, *35*(4), 359–366.

Champion, J. A., & Mitragotri, S., (2006). Role of target geometry in phagocytosis. *Proc. Nat. Acad. Sci. USA.*, *103*(13), 4930–4934.

Champion, J. A., & Mitragotri, S., (2009). Shape-induced inhibition of phagocytosis of polymer particles. *Pharm. Res.*, *26*(1), 244–249.

Cook, S., & Confer, J., (2011). Assessment and treatment of fungal lung infections. *US Pharm*, *36*(7), HS17–HS24.

Cullis, P. R., (1976). Lateral diffusion rates of phosphatidylcholine in vesicle membranes: Effects of cholesterol and hydrocarbon phase transitions. *FEBS Lett.*, *70*(1 & 2), 223–228.

Dailey, L. A., Jekel, N., Fink, L., Gessler, T., Schmehl, T., Wittmar, M., Kissel, T., & Seeger, W., (2006). Investigation of the proinflammatory potential of biodegradable nanoparticle drug delivery systems in the lung. *Toxicol. Appl. Pharmacol.*, *215*(1), 100–108.

Das, P. J., Paul, P., Mukherjee, B., Mazumder, B., Mondal, L., Baishya, R., Debnath, M. C., & Dey, K. S., (2015). Pulmonary delivery of voriconazole loaded nanoparticles providing a prolonged drug level in lungs: A promise for treating fungal infection. *Mol. Pharm.*, *12*(8), 2651–2664.

Dorland, W. A. N., (2011). *Dorland's Illustrated Medical Dictionary 32: Dorland's Illustrated Medical Dictionary* (p. 1077). Elsevier Health Sciences.

Drake, R. L., Vogl, W., & Mitchell, A. W. M., (2014). In: *Gray's Anatomy for Students* (3rd edn., pp. 167–174). Edinburgh: Churchill Livingstone/Elsevier. ISBN 978–0–7020–5131–9.

Erjavec, Z., Kluin, N. H., & Verweij, P. E., (2009). Trends in invasive fungal infections, with emphasis on invasive aspergillosis. *Clin. Microbiol. Infect.*, *15*(7), 625–633.

Galgiani, J. N., Ampel, N. M., Blair, J. E., Catanzaro, A., Johnson, R. H., Stevens, D. A., & Williams, P. L., (2005). Coccidioidomycosis. *Clin. Infect. Diseases*, *41*(9), 1217–1223.

Geys, J., Nemmar, A., Verbeken, E., Smolders, E., Ratoi, M., Hoylaerts, M. F., et al., (2008). Acute toxicity and prothrombotic effects of quantum dots: Impact of surface charge. *Environ. Health Perspec.*, *116*(12), 1607.

Goldstein, E. J., Spellberg, B., Walsh, T. J., Kontoyiannis, D. P., Edwards, Jr, J., & Ibrahim, A. S., (2009). Recent advances in the management of mucormycosis: From bench to bedside. *Clin. Infect. Diseases*, *48*(12), 1743–1751.

Gregoriadis, G., & Neerunjun, D. E., (1974). Control of the rate of hepatic uptake and catabolism of liposome-entrapped proteins injected into rats. Possible therapeutic applications. *The FEBS J.*, *47*(1), 179–185.

Gregoriadis, G., (1973). Drug entrapment in liposomes. *FEBS Lett.*, *36*(3), 292–296.

Harris, J. R., Balajee, S. A., & Park, B. J., (2010). Pneumocystis jirovecii pneumonia: Current knowledge and outstanding public health issues. *Curr. Fungal Infect. Reports*, *4*(4), 229–237.

Hoeben, B. J., Burgess, D. S., McConville, J. T., Najvar, L. K., Talbert, R. L., Peters, J. I., et al., (2006). *In vivo* efficacy of aerosolized nanostructured itraconazole formulations for prevention of invasive pulmonary aspergillosis. *Antimicrob. Agents Chemother.*, *50*(4), 1552–1554.

http://aerosol.ees.ufl.edu/respiratory/section04.html (accessed on 22 January 2019).

http://www.cps.ca/documents/position/antifungal-agents-fungal-infections) (accessed on 22 January 2019).

https://radiopaedia.org/articles/lung-fissures (accessed on 22 January 2019).

James, W. D., Berger, T. G., Elston, D. M., & Odom, R. B., (2006). *Andrews' Diseases of the Skin: Clinical Dermatology*. Saunders Elsevier. ISBN 0–7216–2921–0.

Juliano, R. L., & Stamp, D., (1975). The effect of particle size and charge on the clearance rates of liposomes and liposome encapsulated drugs. *Biochem. Biophys. Res. Commun., 63*(3), 651–658.

Juliano, R. L., & Stamp, D., (1978). Pharmacokinetics of liposome-encapsulated anti-tumor drugs: Studies with vinblastine, actinomycin D, cytosine arabinoside, and daunomycin. *Biochem. Pharmacol., 27*(1), 21–27.

Kauffman, C. A., (2006). Endemic mycoses: Blastomycosis, histoplasmosis, and sporotrichosis. *Infect. Disease Clin., 20*(3), 645–662.

Kimelberg, H. K., Tracy, T. F., Biddlecome, S. M., & Bo.urke, R. S., (1976). The effect of entrapment in liposomes on the in vivo distribution of [3H] methotrexate in a primate. *Cancer Res., 36*(8), 2949–2957.

Koeppen, B. M., & Stanton, B. A., (2008). *Berne and Levy Physiology, 6,* 338–340.

Kumar, A., Chen, F., Mozhi, A., Zhang, X., Zhao, Y., Xue, X., Wang, P. C., & Liang, X. J., (2013). Innovative pharmaceutical development based on unique properties of nanoscale delivery formulation. *Nanoscale, 5*(18), 8307–8325.

Kumari, A., Yadav, S. K., & Yadav, S. C., (2010). Biodegradable polymeric nanoparticles based drug delivery systems. *Colloids Surf. B. Biointerfaces, 75*(1), 1–18.

Kuzmov, A., & Minko, T., (2015). Nanotechnology approaches for inhalation treatment of lung diseases. *J. Control. Rel., 219,* 500–518.

Limper, A. H., Knox, K. S., Sarosi, G. A., Ampel, N. M., Bennett, J. E., Catanzaro, A., et al., (2011). An official American thoracic society statement: Treatment of fungal infections in adult pulmonary and critical care patients. *Am. J. Respir. Crit. Care Med., 183*(1), 96–128.

Mansour, H. M., Rhee, Y. S., & Wu, X., (2009). Nanomedicine in pulmonary delivery. *Int. J. Nanomed., 4,* 299.

Mayhew, E., Papahadjopoulos, D., Rustum, Y. M., & Dave, C., (1978). Use of liposomes for the enhancement of the cytotoxic effects of cytosine arabinoside. *Ann. New York Acad. Sci., 308*(1), 371–386.

McConville, J. T., Overhoff, K. A., Sinswat, P., Vaughn, J. M., Frei, B. L., Burgess, D. S., et al., (2006). Targeted high lung concentrations of itraconazole using nebulized dispersions in a murine model. *Pharm. Res., 23*(5), 901–911.

McIntosh, T. J., (1978). The effect of cholesterol on the structure of phosphatidylcholine bilayers. *Biochim. Biophys. Acta, 513*(1), 43–58.

Meshram, S. H., & Mishra, G. P., (2011). Pulmonary scedosporiosis: a rare entity. *Asian Pacific J. Trop. Disease, 1*(4), 330–332.

Moen, M. D., Lyseng-Williamson, K. A., & Scott, L. J., (2009). Liposomal amphotericin B. *Drugs, 69*(3), 361–392.

Müller, R. H., Radtke, M., & Wissing, S. A., (2002). Nanostructured lipid matrices for improved microencapsulation of drugs. *Int. J. Pharm., 242*(1), 121–128.

Nguyen, C., Barker, B. M., Hoover, S., Nix, D. E., Ampel, N. M., Frelinger, J. A., Orbach, M. J., & Galgiani, J. N., (2013). Recent advances in our understanding of the environmental, epidemiological, immunological, and clinical dimensions of coccidioidomycosis. *Clin. Microbiol. Rev., 26*(3), 505–525.

Pappas, P. G., Kauffman, C. A., Andes, D., Benjamin, D. K., Calandra, T. F., Edwards, J. E., et al., (2009). Clinical practice guidelines for the management candidiasis: 2009 update by the infectious diseases. Society of America. *Clin. Infect. Diseases, 48*(5), 503–535.

Paranjpe, M., & Müller-Goymann, C. C., (2014). Nanoparticle-mediated pulmonary drug delivery: A review. *Int. J. Mol. Sci., 15*(4), 5852–5873.

Patil, J. S., & Sarasija, S., (2012). Pulmonary drug delivery strategies: A concise, systematic review. *Lung India: Official Organ of Indian Chest Society, 29*(1), 44.

Perfect, J. R., & Schell, W. A., (1996). The new fungal opportunists are coming. *Clin. Infect. Diseases, 22*(2), S112–S118.

Perfect, J. R., Dismukes, W. E., Dromer, F., Goldman, D. L., Graybill, J. R., Hamill, R. J., et al., (2010). Clinical practice guidelines for the management of cryptococcal disease: 2010 update by the infectious diseases. Society of America. *Clin. Infect. Diseases, 50*(3), 291–322.

Radtke, M., & Muller, R. H., (2000). Comparison of structural properties of solid lipid nanoparticles (SLN) versus other lipid particles. In: *Proc. Int. Symp. Control Rel. Bioact. Mater.*, (27, 309–310).

Roberts, R. A., Shen, T., Allen, I. C., Hasan, W., DeSimone, J. M., & Ting, J. P., (2013). Analysis of the murine immune response to pulmonary delivery of precisely fabricated nano-and microscale particles. *PLOS One, 8*(4), e62115.

Rytting, E., Nguyen, J., Wang, X., & Kissel, T., (2008). Biodegradable polymeric nanocarriers for pulmonary drug delivery. *Expert Opin. Drug Deliv., 5*(6), 629–639.

Salama, R. O., Traini, D., Chan, H. K., Sung, A., Ammit, A. J., & Young, P. M., (2009). Preparation and evaluation of controlled release microparticles for respiratory protein therapy. *J. Pharm. Sci., 98*(8), 2709–2717.

Scherphof, G., Roerdink, F., & Waite, M., & Parks, J., (1978). Disintegration of phosphatidylcholine liposomes in plasma as a result of interaction with high-density lipoproteins. *Biochim. Biophys. Acta, 542*(2), 296–307.

Senior, J., & Gregoriadis, G., (1982). Is half-life of circulating liposomes determined by changes in their permeability? *FEBS Lett., 145*(1), 109–114.

Sinha, B., Mukherjee, B., & Pattnaik, G., (2013). Poly-lactide-co-glycolide nanoparticles containing voriconazole for pulmonary delivery: *In vitro* and *in vivo* study. *Nanomed. Nanotech. Biol. Med., 9*(1), 94–104.

Smith, J. A., & Kauffman, C. A., (2012). Pulmonary fungal infections. *Respirology, 17*(6), 913–926.

Standring, S., & Borley, N. R., (2008). In: *Gray's Anatomy: The Anatomical Basis of Clinical Practice* (40th edn., pp. 992–1000). Edinburgh: Churchill Livingstone/Elsevier. ISBN 978-0-443-06684-9.

Standring, S., Ellis, H., Healy, J., Johnson, D., Williams, A., Collins, P., & Wigley, C., (2005). Gray's anatomy: The anatomical basis of clinical practice. *Am. J. Neuroradiol., 26*(10), 2703.

Sung, J. C., Pulliam, B. L., & Edwards, D. A., (2007). Nanoparticles for drug delivery to the lungs. *Trends Biotechnol., 25*(12), 563–570.

Suzuki, Y., & Yamaguchi, T., (1993). Effects of hyaluronic acid on macrophage phagocytosis and active oxygen release. *Inflam. Res., 38*, 32–37.

Tomashefski, J. F., Cohen, A. M., & Doershuk, C. F., (1988). Long-term histopathologic follow-up of bronchial arteries after therapeutic embolization with polyvinyl alcohol (Ivalon) in patients with cystic fibrosis. *Human Pathol., 19*(5), 555–561.

Üner, M., (2006). Preparation, characterization and physicochemical properties of solid lipid nanoparticles (SLN) and nanostructured lipid carriers (NLC): Their benefits as colloidal drug carrier systems. *Die Pharmazie, 61*(5), 375–386.

Vaughn, J. M., McConville, J. T., Burgess, D., Peters, J. I., Johnston, K. P., Talbert, R. L., & Williams, R. O., (2006). Single dose and multiple dose studies of itraconazole nanoparticles. *Eur. J. Pharm. Biopharm., 63*(2), 95–102.

Vaughn, J. M., Wiederhold, N. P., McConville, J. T., Coalson, J. J., Talbert, R. L., Burgess, D. S., et al., (2007). Murine airway histology and intracellular uptake of inhaled amorphous itraconazole. *Int. J. Pharm.*, *338*(1), 219–224.

Vyas, S. P., & Khatri, K., (2007). Liposome-based drug delivery to alveolar macrophages. *Expert Opin. Drug Deliv.*, *4*(2), 95–99.

Wahba, H., Truong, M. T., Lei, X., Kontoyiannis, D. P., & Marom, E. M., (2008). Reversed halo sign in invasive pulmonary fungal infections. *Clin. Infect. Diseases*, *46*(11), 1733–1737.

Watts, A. B., & Williams, III, R. O., (2011). Nanoparticles for pulmonary delivery. In: *Controlled Pulmonary Drug Delivery* (pp. 335–366). Springer New York.

Wheat, L. J., Freifeld, A. G., Kleiman, M. B., Baddley, J. W., McKinsey, D. S., Loyd, J. E., & Kauffman, C. A., (2007). Clinical practice guidelines for the management of patients with histoplasmosis: 2007 update by the infectious diseases. Society of America. *Clin. Infect. Diseases*, *45*(7), 807–825.

Zmeili, O. S., & Soubani, A. O., (2007). Pulmonary aspergillosis: A clinical update. *J. Assoc. Physicians*, *100*(6), 317–334.

FURTHER READING

Mukherjee, B., Paul, P., Dutta, L., Chakraborty, S., Dhara, M., Mondal, L., & Sengupta, S., (2017). Pulmonary administration of biodegradable drug nanocarriers for more efficacious treatment for fungal infections in lungs: Insights based on recent findings. In: Grumezescu, A. M., (ed.), *Multifunctional System for Combined Delivery, Biosensing and Diagnostic* (pp. 261–280). Elsevier: Netherlands.

CHAPTER 7

Nanotechnological Approaches for Management of NeuroAIDS

DIVYA SUARES[1], MOHD. FAROOQ SHAIKH[2], and CLARA FERNANDES[1*]

[1]*Shobhaben Pratapbhai Patel School of Pharmacy & Technology Management, SVKM'S NMIMS, V.L. Mehta Road, Vile Parle (W), Mumbai–400056, India, E-mail: clara_fern@yahoo.co.in*

[2]*Neuropharmacology Research Laboratory, Jeffrey Cheah School of Medicine and Health Sciences, Monash University Malaysia, Selangor Darul Ehsan, Malaysia*

Corresponding author. E-mail: clara_fern@yahoo.co.in

ABSTRACT

It is a well-known fact that the human immunodeficiency virus (HIV) is a neurotropic virus that enters the central nervous system (CNS) early in the course of infection. Despite the advancement brought about in therapy by the use of highly active antiretroviral therapy (HAART), controlling HIV infections still remains a major challenging task. The primary challenge of the treatment is the poor penetrability of most antiretroviral drugs (ARVs) across the blood-brain barrier (BBB) making it difficult to eradicate the viral brain reservoirs. Thus, there is an urgent demand for the development of a target-specific, efficacious drug-delivery approach to curb the AIDS menace. In this context, nanoformulations majorly comprising of polymeric, inorganic, and lipidic origin materials of constructs and recently explored, cellular domain confined nanoformulations are emerging as the promissory approach. This chapter will highlight the barriers to the delivery and advancements in nanomedicine to design and develop size-specific therapeutic cargo for targeting HIV brain reservoirs. This chapter would provide insight about the nano-enabled multidisciplinary research to formulate efficient nanomedicine for the management of neuroAIDS.

7.1 INTRODUCTION

NeuroAIDS describes an invasion of human immunodeficiency virus (HIV) in the intra-central nervous system (CNS) region leading to HIV encephalitis (HIVE), which is diagnosed as microglial nodular inflammation associated with multinucleated giant cells. The clinical neurobehavioral abnormalities of HIVE include HIV–1 associated cognitive/motor complex or HIV–1 associated progressive encephalopathy (Black, 1996). Another most distressing complication arising due to neuroAIDS is dementia (Al-Ghananeem et al., 2013). In accordance to World Health Organization (WHO) AIDS Epidemic Update of 2016, it was reported that at the end of 2016, 36.7 million (30.8–42.9 million) HIV infected people were estimated to be living worldwide. Among this, 1.8 million (1.6–2.1 million) people were assumed to be infected in the year 2016 itself (www.unaids.org, UNAIDS, 2017). These alarming statistics warrant the need for the effective management of the HIV and HIV related infections for better management of the disease. Earlier it was presumed that the neuroAIDS occur only in the later stages of disease progression. However, incriminating evidence shows the presence of HIV-related components and the anti-HIV antibodies in the brain even during the early stages of infection. There have been assumptions made that the virus would be trespassing the BBB by the passage of infected monocytes, macrophages, and/or T-cells across the tight junction of BBB (Sagar et al., 2014). Widely known therapy includes use of ARVs for complete eradication of viral cargoes. However, to date, the most challenging aspect of treating neuroAIDS is to achieve effective concentrations of ARVs in the infected regions of the brain. Although there have been many approaches, conventional formulations lack the desired characteristics to enable the delivery of highly active ARVs across the BBB (Varghese et al., 2017). Thus, in this review, the authors have focused on the novel delivery approaches using nanotechnology for enhanced delivery of ARVs to the CNS.

7.2 NEUROPATHOLOGY OF AIDS

A HIV infection typically begins without the development of any symptoms. This asymptomatic stage usually persists for several years before ultimately progressing to symptomatic end stages. These terminal end stages are known collectively as AIDS and progression to these stages was almost entirely inevitable before advent of highly active antiretroviral therapy (HAART). Some of the symptoms of late-stage HIV in a neurological context are opportunistic infections coupled with a baffling assortment of neurological signs and symptoms. This implies that the CNS is involved in the pathology of AIDS (Anthony and Bell, 2008). It is thus noteworthy that the CNS trails behind only the lungs in terms of being the most affected by AIDS. This is because opportunistic CNS infections tend to develop in AIDS patients due to the suppression of their immune system by HIV. Among the more common CNS, opportunistic agents are *Toxoplasma gondii*, *Cryptococcus neoformans*, and cytomegaloviruses, although factors such as the environment as well as geographic and social economic factors may play a role in determining the dominant infectious agents. The later stages of HIV CNS infection are marked by the appearance of HIV-specific lesions in the CNS together with giant multinucleated cells. The lesions are most commonly the result of protozoan, fungal, bacterial or viral infections, although non-infectious lesions in the form of neoplasms or vascular lesions also occur. After the development of HAART, the number of AIDS patients with opportunistic infections and hence neurological complications, have decreased. However, this trend is mostly true only in developed countries. The remaining 90% of AIDS patients in other less developed countries often succumb before the appearance of neurological complications (Silva et al., 2012). Besides the neurological problems typically associated with a HIV-infection of the CNS, neurological symptoms such as motor and cognitive dysfunction also occur regularly. These neurological symptoms are termed HIV-Associated Neurocognitive Disorders (HAND) and occur in humans infected with HIV–1, regardless of their age. HAND is defined by three categories of increasingly severe disorders, beginning with asymptomatic neurocognitive impairment, followed by mild cognitive disorder and HIV-associated dementia (HAD) (Sanchez and Kaul, 2017). As the macrophages and microglia of the brain are the main cell types which are infected by HIV–1 in the CNS, they likely facilitate the neurodegeneration seen in HAD patients; although the exact pathogenesis of the neurodegeneration remains elusive (Gonzalez-Scarano and Martin-Garcia, 2005).

After infecting the periphery, HIV–1 is believed to quickly reach the brain and become localized primarily to perivascular macrophages and microglia. Infection of macrophages and lymphocytes by HIV–1 in the macrophages and lymphocytes of the periphery and microglia in the brain occurs via the binding of the viral envelop protein gp120 to CD4 in combination with one of several possible chemokine receptors depending on the strain of HIV–1. The pathological features of a HIV brain infection are termed HIVE and include activated resident microglia, microglial nodules, multinucleated giant cells, infiltration predominantly by monocytoid cells, including blood-derived macrophages, and decreased synaptic and dendritic density, combined with selective neuronal loss, widespread reactive astrocytosis and myelin pallor. In living HIV patients, increased microglia numbers, decreased synaptic and dendritic density, selective neuronal loss, elevated tumor necrosis factor (TNF-α), mRNA in microglia and astrocytes and evidence of excitatory neurotoxins in CSF and serum, are the pathological features which provide the best correlation with measures of cognitive dysfunction. Distinct brain regions such as the frontal cortex, substantia nigra, cerebellum, and putamen also show neuronal damage and loss attributable to a HIV infection. Neuronal apoptosis has also been found in the brains of HAD patients, particularly in areas with signs of structural damage and microglial activation. While the use of HAART has reduced the incidence of opportunistic infections and HAD, it has increased the prevalence of HIVE. In addition, macrophage/microglia infiltration and activation in the hippocampus and basal ganglia have been found to be greater in HAART treated HIV patients (Sanchez and Kaul, 2017). In addition to the increased prevalence of HIVE in the post HAART era, minor cognitive motor disorder (MCMD) has become more prevalent as compared to HAD. Far from being the lesser of two evils, the presence of MCMD is usually correlated with a worse overall prognosis for HIV patients (Gonzalez-Scarano and Martin-Garcia, 2005).

One possible reason for occurrence of MCMD is that even when HAART falls just short of being wildly successful, a low level of HIV replication stills occurs and progressively causes neurodegeneration over an extended time frame. While still conjecture, this theory is consistent with the greatly increased lifespan of treated patients and that certain antiretroviral drugs (ARVs) are unable to reach the brain in satisfactory amounts due to their poor penetration of the BBB. The BBB is a defensive mechanism used by the CNS to protect against infection by physically separating the CNS from the rest of the body. In addition, the cerebrospinal fluid (CSF) is separated from the rest of the periphery by the blood-CSF barrier found in the epithelium of the choroid plexus. However, even with the barriers and the diverse cell types in the CNS which increases the complexity of the virus-cell interactions in the brain, the CNS is still vulnerable to infection by various species of retroviruses and members of the lentivirus family in particular. HIV is one such virus which the CNS is vulnerable to, as it can enter the CNS. While immune cells do penetrate the BBB as part of the body's immune surveillance system, this process is usually tightly regulated. This raises the question of how HIV manages to enter the CNS, with several models being proposed as a result of studies involving animal models and tissue-culture preparations designed to mimic the BBB. The most intuitive theory is well supported with compelling evidence and states that HIV and other lentiviruses penetrate the CNS by acting as passengers in the immune cells which penetrate the BBB (Gonzalez-Scarano and Martin-Garcia, 2005).

7.3 BARRIERS TO CNS DELIVERY

For ARVs to be effective, they must first accumulate in the CNS in a sufficiently high dose. While ARVs have little difficulty in accumulating to therapeutic levels in blood plasma, the BBB acts as a barrier to CNS delivery. This is because the accumulation of drugs is the result of an interplay between their influx from the blood to the CNS and the rate of their efflux. While the rate of influx is determined by lipid solubility and the presence of transport systems at the BBB which the drug can utilize, a variety of efflux transporters prevent most drugs from accumulating in the CNS, including HIV anti-viral drugs (Banks, 2000). The inability of HIV anti-viral drugs to accumulate to therapeutic levels in the CNS turns the CNS into a viral reservoir in which HIV can replicate and create latently infected cells with little interference from ARVs. If HAART is interrupted for any reason, HIV can re-enter the bloodstream and re-establish a high level of viremia. In addition, the low level of ARVs in the CNS may result in resistant mutants, rendering the drugs useless. Patients treated with ARVs demonstrating a low degree of CNS penetration have been found to be associated with a higher prevalence of neurocognitive disorders and poorer performance on neurophysiological tests, despite being treated with ARVs regularly and having undetectable levels of HIV RNA in their blood (Bertrand et al., 2016).

7.3.1 BBB AND BLOOD-CSF

The BBB acts as a barrier to CNS drug delivery because it is a selectively permeable continuous cellular layer consisting of brain microvascular endothelial cells linked to each other via tight junctions and functions to regulate the movement of cells and substances from the bloodstream into the CNS (Gonzalez-Scarano and Martin-Garcia, 2005). While the development of the BBB was the result of a need to protect the brain from toxic substances, the same BBB also hampers the delivery of diagnostic and therapeutic agents to the brain. Many drugs have poor permeability of the BBB despite the presence of billions of capillaries which together have a very large total surface area. The structure of the BBB itself contributes to its poor permeability as the capillaries within the CNS do not possess intercellular clefts and fenestrae, unlike those outside the CNS. While certain lipophilic and low molecular weight compounds can enter the BBB via passive diffusion, lipophilic compounds are quickly removed from the cytoplasm of the BBB cells via efflux systems. The BBB also possess several active and passive efflux systems such as P-glycoprotein which is a multidrug transporter specific to many antiretroviral compounds (Nowacek and Gendelman, 2009). While P-glycoprotein is responsible for reducing the accumulation of the protease inhibitor class of HIV anti-virals, drugs such as the nucleoside analog reverse-transcriptase inhibitor Zidovudine are removed from the brain via a probenecid-sensitive efflux system (Banks, 2000). In addition, enzymes which participate in metabolism and activation of endogenous compounds such as g-glutamyl transpeptidase, alkaline phosphatase, and aromatic acid decarboxylase have elevated activity and expression levels (Nowacek and Gendelman, 2009).

The active barrier between the blood and the CSF is found at the epithelium of the choroid plexuses (CPs). The CPs consequently establishes a direct route for drug delivery into the CSF, subarachnoid spaces and perivascular spaces which are primary locations for infected macrophages, more precisely in the early phase of the infection. The CPs is the sites, which are rich in CSF-to-blood efflux pumps and play a crucial role in regulating the cerebral biodisposition of various drugs (Everall and Levy, 2002).

7.3.2   PATHWAYS FOR UPTAKE OF NANOCARRIERS

With the success of HAART being limited by poor drug penetration of the CNS and the need for an uninterrupted and regular dosing schedule, there is a growing need for novel treatments. Nanotechnology may hold the key to the next breakthrough in antiretroviral therapy by utilizing the exceptional physiochemical parameters of materials at the nanometer scale. Nanodrugs are derived from nanotechnology and involves drugs and any target molecules being enclosed or absorbed on nanoparticles which act as carriers. Among the advantages of these nanodrugs is the remarkable increase in bioavailability due to reduced first pass hepatic metabolism and the ability of the nanosize particles to penetrate various physiological barriers including the BBB. Nanodrugs also do not require adherence to a strict dosing regimen as the higher surface area of drug nanocarriers also increases drug loading ability. When large amounts of a drug are loaded into a nanocarrier, there is an initial burst of drug release followed by a constant release over an extended period of time; which minimizes the dose frequency (Sagar et al., 2014).

Many different nanocarriers exist; among them are magnetic nanoformulations which allow the movement and speed of the carriers to be controlled with an external magnetic force. Certain dendrimer-based nanocarriers have been found to enhance the phagocytotic activity of HIV infected macrophages, in particular, promoting the uptake of the loaded drug. Recent studies have also found that certain dendrimer themselves possess antiretroviral activity, although erratic drug release kinetics and long-term safety issues need to be resolved before dendrimer nanocarriers can gain widespread use (Nair et al., 2016).

Nanocarriers typically enter a cell via one of several main endocytotic pathways such as phagocytosis, pinocytosis, macropinocytosis, clathrin-mediated endocytosis, and caveolin-mediated endocytosis. Other receptor-dependent trafficking pathways also exist, such as the transferrin receptor and low-density lipoprotein receptor (LDLR) trafficking which bind iron ions and LDL respectively before undergoing clathrin-mediated endocytosis. Epidermal growth factor receptor (EGFR) trafficking binds EGF and undergoes endocytosis as well, although multiple endocytotic pathways may be used, possibly depending on the extracellular ligand concentration. Examples of non-receptor endocytotic pathways include circular dorsal ruffles, Clathrin-independent carriers (CLIC) and GPI-anchored protein-enriched early endosomal compartment (GEEC) pathways (Kettiger et al., 2013; Xu et al., 2013).

While there are many potential pathways available to nanocarriers to gain entry into a cell, the characteristics of the nanocarriers themselves may influence the pathway that is ultimately used. One such characteristic is the size of the nanocarriers as nanocarriers with a diameter of 50 nm are more efficiently internalized by cells than smaller (about 15–30 nm) or larger (about 70–240 nm) particles. This is because nanocarriers with a diameter of 30–50 nm are easily taken up by receptor-mediated endocytosis as they efficiently recruit and interact with membrane receptors. The shape of the nanocarrier also plays a role in the uptake pathway as spherical nanocarriers are more easily taken up than their road-shaped counterparts, possibly due to the increased time needed to wrap the membrane around a longer rod-shaped nanocarrier. In addition, the properties of the nanocarrier surface such as surface charge and the presence of functional groups also determine the route of uptake. Positively charged nanocarriers interact well with the negatively charged plasma membrane and are taken up via adsorptive mediated pinocytosis, whereas negatively charged nanocarriers use alternative uptake routes, and non-ionic nanocarriers do not interact with the cell membrane at all (Kettiger et al., 2013).

7.4   ADVANTAGES OF NANOMEDICINE IN NEUROAIDS

Conventional HAART therapy is fraught with issues such as poor patient compliance, poor targetability, development of resistance to therapy and toxicity such as increased incidences of coronary disorders, hepatic disorders, metabolic disorders like diabetes, cancer, and accelerated aging (Richman et al., 2009). This primarily arises due to decrease *in vivo* residence time of ARVs which results in lower drug concentrations in viral load reservoir sites or sanctuary sites such as lymphatic system, macrophages, lymphocytes, CNS, and lungs (Varatharajan and Thomas, 2009). For neuroAIDS, things are further complicated as the virus is located in the microglial cells of the CNS. This is an impediment to most of the ARVs to enter the brain in particular. However, there are few exceptions, zidovudine, lamivudine, and protease inhibitors. Further, the involvement of CNS in HIV infection is associated with many other complications such as dementia and other neurological disorders. As a consequence, it becomes difficult to treat the CNS associated infections with the drugs that are otherwise effective at suppressing the viral replication in the rest of the body. Thus, the need for targeting the brain for delivering ARVs has been realized worldwide by various scientists mainly by developing nanocarrier-based systems. The nanocarrier systems overcome the limitations of conventional drug therapy such as inaccessibility to challenging HIV reservoir sites like CNS, CSF, lymphatic system and macrophages, as well as non-availability of adequate drug payload at the infected site for desired period, multi-drug resistance and so on (Vyas et al., 2006). As reiterated earlier, studies have established that the influx of ARVs across the BBB is very minimal as compared to its efflux, owing to the presence of efflux transporters such as ATP binding cassette (ABC) viz. P-gp and multidrug resistance-associated proteins (MRP) (Ene et al., 2011; Saksena et al., 2010). The other factors responsible for preventing the entry of ARVs through BBB could be attributed to the presence of metabolic enzymes and tight endothelial cell junctions in the brain capillaries (Varatharajan and Thomas, 2009; Ronaldson et al., 2008). In addition to this, the development of nanocarrier-based systems can be engineered to increase the concentration of ARVs in the CNS, thereby reducing dose and length of conventional therapy. This can directly minimize the risks of peripheral side effects (Ojewole et al., 2008). Further, these flexible and versatile nanocarrier-based systems eliminate the necessity to modify the ARVs, thus avoiding the changes in their pharmacological activity (Wong et al., 2010).

Prominently, numerous benefits proffered by nanocarrier due to their submicron size and resourcefulness in compositions researched by various scientists can be summarized as follows;

i.   Tunable particle size aids in superior concentration of nanoparticles in target tissue (Parboosing et al., 2012).
ii. By virtue of their large surface-to-volume ratio, they provide useful modality for targeting hydrophilic as well as hydrophobic drugs to the target size (Mahajan et al., 2012).
iii. Prominent reason for failure of drug delivery across the BBB is presence of P-glycoprotein (P-gp) efflux transporters. However, the literature cites nanoparticulate systems can effectively evade these efflux transporters, thus, ensuring high drug payload as well as prolonged drug release in the CNS (das Neves et al., 2010).
iv. Ease of surface biofunctionalization and/or multifunctionality by using ligands such as thiamine, transferrin, cell-penetrating peptides, such as HIV–1 transactivating transcriptional activator peptide and so on, thereby enabling targeting to CNS barriers (Mallipeddi and Rohan, 2010).
v.   Generally, they are constructed using biocompatible and biodegradable materials making this relatively less toxic approach and providing stability against enzymatic degradation (Parboosing et al., 2012).

Broadly, it can be stated that using nanoparticles, the drug can be concentrated in the HIV reservoirs in the brain by one of the following mechanisms, such as: i) passive targeting by promotion of higher concentration gradient of ARVs across the BBB, ii) receptor-mediated targeting by promotion of receptor-mediated endocytosis and iii) finally by evading drug efflux across the BBB (Wong et al., 2010).

7.5   NANOCARRIERS FOR THE TREATMENT OF NEUROAIDS

This section offers a comprehensive discussion of various nanocarriers developed for targeting ARVs for the treatment of neuroAIDS. As cited in the literature, the types of nanocarriers may be classified as polymer-based systems such as polymeric nanoparticles, polymeric micelles or dendrimers. Lipid-based systems which may be further classified as Class

I viz. microemulsions, nanoemulsions, solid lipid nanoparticles (SLNs), nanostructured lipid carriers, Class II viz. liposomes and Class III viz. lipid core micelles, targeted lipid nanocarriers (Fernandes et al., 2018), inorganic nanocarriers and cell-based nanoformulations (Figure 7.1). Herein, we have broadly categorized the lipid nanocarriers based on the nature of lipid as soft and hard lipid nanocarriers.

7.5.1   POLYMERIC NANOPARTICLES

Polymeric nanoparticles (less than 1μm) have been extensively researched due to their ability to target specific sites owing to their nanosize, their potential for sustained release and their ability to evade efflux pumps by undergoing endocytosis. They are composed of biologically safe and compatible polymers enclosing actives. Based on their constructs, they can be classified as nanocapsules; wherein the active is enclosed within the polymer membrane, nanospheres; wherein the active is embedded within the polymer matrix and nanoconjugates; wherein the actives are confined to the polymer domain either by surface/ conjugation (Govender et al., 2008). Literature cites, various safe and efficient polymers that have been explored for delivering ARVs to the brain include poly(lactic) acid (PLA) (Mainardes et al., 2009; Rao et al., 2008), poly(lactide-co-glycolides) (PLGA) (Destache et al., 2010), polybutylcyanoacrylate (PBCA) (Kuo, 2005; Kuo and Chung, 2012), methylmethacrylate-sulfopropyl methacrylate (MMASPM) (Kuo and Chen, 2006), poly(ethylene oxide)-modified poly(epsilon-caprolactone) (Shah and Amiji, 2006), poly(hexylcyanoacrylate) (Löbenberg et al., 1997), Pluronic 85 (Shaik et al., 2008), including some natural polymers such as chitosan (Gu et al., 2017; Al-Ghananeem et al., 2010), albumin (Mishra et al., 2006), gelatin (Kaur et al., 2008a) to name a few.

Löbenberg et al. (1997) developed azidothymidine nanoparticles (size 230±20 nm and zeta potential of –51.6 ± 3 mV) by emulsion polymerization of the n-hexylcyanoacrylate monomer. The radiolabeled drug-loaded nanoparticles and radiolabeled drug solution were administered orally to Wistar rats. The radioactivity study revealed that in comparison to plain drug, the nanoparticles enhanced the brain uptake drug by 33% within 1 h of administration and found to be sustained up to a period of 8 h. Furthermore, the researchers hypothesized the role of surfactants such as Tween® 80, polyethylene glycol (PEG) in influencing the brain uptake of nanoparticles either via endocytic uptake mechanism or modulation of tight junctions or inhibition of efflux transporters (Kreuter, 2004). Mainardes et al., (2009)

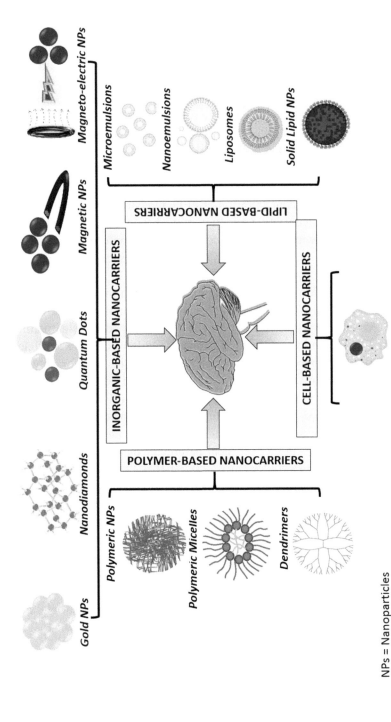

FIGURE 7.1 Nanocarriers for the treatment of neuroAIDS.

NPs = Nanoparticles

investigated the effect of surface modification to evade phagocytosis and promote brain uptake of zidovudine loaded PLA nanoparticles and PLA/ PEG blend nanoparticles in an *in vitro* cell model isolated from rat peritoneal exudate. The primary objective of this work was to correlate diminished phagocytosis with prolonged period of blood circulation time for achieving higher brain uptake. The study underlined the significance of surface charge and surface steric hindrance for cellular uptake by macrophages. It was observed that charged nanoparticles were opsonized more readily as compared to neutral ones. Further, the steric hindrance achieved by water-loving polymer chains of PEG hindered the electrostatic and hydrophobic interactions between the nanoparticles and opsonins thus preventing phagocytosis. Further, it was reported that concentration of PEG influenced steric barrier, at very low concentration of PEG, the barrier was incomplete, while low and high concentration resulted in mushroom-like and brush-like configuration, respectively. Both, these conformations conferred steric effect leading to increased circulation time in blood and consequently improved brain uptake.

Similar to PEGylation of polymeric nanoparticles, coating of nanocarriers with protein or antibody (Gu et al., 2017) or specific ligands can be advanta-geous for crossing the BBB. Rao et al. (2008) developed trans-activating transcriptor (TAT) peptide conjugated ritonavir-loaded PLA nanoparticles and observed a prolonged delivery of ARV in the brain, i.e., for about 2 weeks after single intravenous administration, without compromising the integrity of the BBB. On similar lines, Kuo, and Lee (2012) successfully fabricated ARVs loaded-MMASPM nanoparticles coupled RMP–7 (Cere-port, a synthetic pseudopeptide of bradykinin analog agonist) to deliver ARVs across the BBB. Another physical parameter that may influence the permeability of the nanoparticles across the BBB is the particle size. Kuo and Chen (2006) developed zidovudine and lamivudine nanoparticles using PBCA and MMASPM polymers, separately. The MMASPM based nanopar-ticles of diameter less than 70 nm demonstrated 15-fold increase in perme-ability particles across the BBB. Simultaneously, the investigators reported that the presence of alcohol (0.5%) increased permeability of ARV-loaded polymeric nanoparticles across the BBB by 4–12%, which was attributed to momentary unfolding of tight junctions. Furthermore, Kuo and Kuo (2008) studied the effect of an external physically-activated trigger such as electro-magnetic field on the permeability (Pandey et al., 2016) of saquinavir-loaded polymeric nanoparticle across the BBB.

Selection of route of delivery of ARV-loaded nanoparticles can also influence the uptake of nanoparticles across the BBB. Literature states the ability of ARVs to bypass BBB via nose-to-brain delivery involving olfactory or trigeminal nerve systems (Mittal et al., 2014). Al-Ghananeem et al. (2010) reported enhancement of uptake of didanosine-loaded mucoadhesive chitosan nanoparticles in the brain through intranasal delivery possibly due to modulation of tight junctions thereby promoting paracellular drug transport. Another research group, Belgamwar et al. (2017) also reported superior brain uptake of efavirenz-loaded chitosan-g-hydroxypropyl beta-cyclodextrin nanoparticles intranasally as evident from the pharmacokinetic study and gamma scintigraphy.

7.5.2 POLYMERIC MICELLES

Polymeric micelles are nanosized core/shell constructs comprising of amphiphilic block copolymers (Croy and Kwon, 2006). Pluronic block copolymer as micelles have been explored for CNS delivery of ARVs due to its capability of inhibiting P-gp efflux transporters (Spitzenberger et al., 2007). The study was carried out in a severely combined immunodeficiency mouse model of viral encephalitis. The antiretroviral activity of the micellar system, in a 2-week treatment, was found to be appreciable as compared to the control group. This inhibition of HIV replication by Pluronic 85 could be attributed to its interaction with P-gp or the glycolipid membrane of the virus, leading to virus destruction.

As stated earlier, intranasal being one of the well-researched routes for delivery of actives directly to the CNS, Chiapetta et al. (2013) developed efavirenz-loaded poly(ethylene oxide)–poly(propylene oxide) polymeric micelles for intranasal delivery and compared its pharmacokinetic profile post intravenous administration.

7.5.3 DENDRIMERS

Dendrimers are branched nanostructures of 1–10 nm in diameter, composed of repeated monomer units, arranged in a radially symmetric molecular manner characterized by tree-like arms or branches (Svenson, 2009; Abbasi et al., 2014). Several researchers are exploring the potential of dendrimers in crossing the BBB due to their capability of undergoing endocytosis-mediated cellular internalization, thus facilitating CNS delivery of ARVs at the niche of viral reservoir. This behavior may be due to their proficiency in modulating occludin and actin, tight junction proteins. However, the amount of dendrimers and their surface charge may reverse the modulation of these proteins (Xu et al., 2013). Few dendrimers that have been studied for anti-HIV therapy include poly(amidoamine) (PAMAM), polylysine (PPI), carbosilane, polypropylenimine tetrahexaconta-amine, polypropylenimine Gallic acid-triethylene glycol (PPI-GATG). The dendrimers may be further functionalized to inhibit binding of virus to the host cell and minimize its replication in host cell (Peng et al., 2013).

Dutta and Jain (2007) have successfully developed lamivudine-loaded mannosylated fifth generation poly(propylene imine) dendrimeric nanocarriers to target mannose receptors expressed on the surface of monocyte-macrophages, alveolar macrophages, hepatocytes as well as the astrocytes in brain. These nanocarriers showed tremendous increase in the cellular uptake of ARVs and equivalent reduction in the viral load in *in-vitro* MT2 cell culture. Moreover, Dutta et al. (2007) developed efavirenz-loaded fifth generation poly(propylene imine) dendrimeric nanocarriers which were found to be cytotoxic due to the presence of free primary amino group. Thus, the authors conjugated the poly(propylene imine) with t-Boc-glycine and mannose, separately to develop dendritic nanocontainers which exhibited negligible cytotoxicity. Further, there was significant increase in the uptake of efavirenz by monocyte-macrophages. With an intention to develop biologically safe and less toxic dendritic nanocontainers, Kumar et al. (2013) and Pyreddy et al. (2014) successfully developed PEGylated-PAMAM dendritic nanocarrier for delivery of ARVs. However, the brain uptake of these in an *in vivo* scenario has not been explored by the authors. Perisé-Barrios et al. (2014) reported effective transfection of 2nd generation carbosilane dendrimers in CD4 T-lymphocytes, wherein these non-viral vectors were explored for HIV infection. They formed dendriplexes with siRNA and did not exhibit cytotoxicity when investigated using MTT and 7AAD labeling assay. The developed dendrimers reduced macrophage phagocytosis, minimized the viral load in them and enhanced the efficacy of HAART. With the purpose of delivering the ARVs to the brain, non-viral vectors such as dendrimers have been employed to deliver and transfect siRNA to the target site (Jiménez et al., 2010). From their studies, they have established the capability of 2G-NN16 amino-terminated carbosilane dendrimers in transfecting non-CD4+ human astrocytoma cells with a time-controlled release of siRNA cargo. Additionally, authors have developed siP24/2G-NN16 and siNEF/2G-NN16 dendriplexes as HIV replication inhibitors and have compared with the activity of 2G-NN16 dendrimers alone wherein siP24/2G-NN16 showed 85% inhibition, while siNEF/2G-NN16 showed 50%.

Dendritic nanogels are found to be beneficial in the treatment of neuroAIDS as discussed by Vashisht et al. (2016). Vinogradov et al. (2010) encapsulated 5′-triphosphates of NRTIs within cationic nanogels and were successful in suppressing HIV–1 activity while minimizing mitochondrial toxicity (Vinogradov et al., 2005). Similarly, Gerson et al. (2014) conjugated the dendritic nanogels with peptide binding brain-specific apolipoprotein E molecules for delivery of ARVs, resulting in reduced apoptosis and reactive oxygen species levels, thus eliminating neuron death.

7.5.4 SOFT LIPID NANOCARRIERS

Soft lipid nanocarriers are characterized by flexible low melting lipid or oil-based core. Typically, they may be categorized into microemulsions or nanoemulsions or liposomes. Microemulsions are thermodynamically stable systems comprising of oil droplets of submicron size (less than 300 nm) in aqueous phase, which are stabilized using surfactants and co-surfactants (Jain et al., 2010). On the contrary, nanoemulsions are relatively thermodynamically unstable but kinetically stable systems and are generally fabricated by either low/high-energy based techniques (Fernandes et al., 2010; Fernandes et al., 2018; Pidaparthi and Suares, 2017). On the other hand, liposomes are miniature-sized spherical vesicles composed of phospholipids and cholesterol enclosing aqueous compartment to transport hydrophilic drugs, while carrying the lipophilic/amphiphilic drugs in the lipid bilayer, to the target tissue. Due to their small size and nature, they are taken up by mononuclear phagocytic cells and undergo rapid clearance (Vieira and Gamarra, 2016).

Vyas et al. (2008) prepared saquinavir-loaded nanoemulsions using two different polyunsaturated fatty acid rich oils viz. flax-seed and safflower oil with egg phosphatidylcholine as primary emulsifier and deoxycholic acid as co-surfactant using ultrasonication method to obtain globules of diameter < 200 nm. The pharmacokinetic studies in male albino mice revealed 86.69% and 63.6% availability of ARVs in the brain, from flax-seed and safflower oil-based nanoemulsion, respectively post oral administration, as compared to the drug suspension (39.5%). Whereas, the average relative bioavailability in the brain post intravenous administration was found to be 364.52% and 141.97% for flax-seed and safflower oil -based nanoemulsion. The higher bioavailability of flax-seed oil-based nanoemulsion was attributed to the higher content of omega–3-fatty acids, which possesses selective brain uptake when compared to omega–6-fatty acids. Prabhakar et al. (2011) studied the brain uptake of transferrin coupled indinavir-loaded lipid emulsions composed of stearylamine, which is capable of holding transferrin at the surface of the globule due to long C–18 lipid chain. This hypothesis is ascribed to transferrin transporting iron to the brain via receptor-mediated endocytosis and transcytosis.

As restated earlier, nose-to-brain delivery has been extensively explored for delivering actives directly to the CNS. Mahajan et al. (2014) successfully formulated intranasal nanoemulsion of saquinavir mesylate by spontaneous emulsification technique providing globules of diameter 176.3±4.21 nm. The *in vivo* distribution of ARVs from intranasal nanoemulsion was compared with intravenously administered drug suspension. The pharmacokinetic study demonstrated quick absorption of drug from the nanoemulsion in the brain and blood followed by sustained release for a period of approximately 3 hours. This behavior was attributed to the globule size of the nanoemulsion facilitating transcellular transport via olfactory neurons. Another contributing factor for enhanced bioavailability was credited to the presence of excipients having P-gp inhibitory activity such as Tween 80 and PEG 400. Furthermore, the uptake of radiolabeled complex of ARV loaded-nanoemulsion via intranasal delivery in rats was envisaged using gamma scintigraphy, where the scintigrams demonstrated a clear accumulation of material in the brain. Development of liposomal formulations of ARVs such as zidovudine-loaded surface engineered liposome (Kaur et al., 2008b), stavudine-loaded mannosylated liposome (Garg et al., 2006) and indinavir-loaded immunoliposomes (Gagne et al., 2002) have been extensively studied for targeting the HIV reservoirs. However, these systems have not been explored for their capability of crossing the BBB. Nonetheless, Jin et al. (2005) developed liposomes of zidovudine, wherein they observed low entrapment efficiency and quick leakage of the drug from the liposome. Hence the authors synthesized a prodrug of zidovudine (i.e., zidovudine myristate) and when converted to liposomes, they were successful in improving the entrapment efficiency to 98% and obtained an average particle size of 90 nm of reconstituted lyophilized liposomes. The tissue distribution study in rats demonstrated an improved uptake of the prodrug-loaded liposomes in the brain and RES organs.

Several scientists have done modifications to the liposomal surface by using polymers such as PEG and thus minimized the loss of actives and enhance the systemic circulation time. On similar lines, HIV latency activators may be incorporated into liposomes for selective targeting to primary CD4+ cells thus minimizing bystander (unrelated) cell activation. Thus, Kovochich and co-workers (2011) developed liposomes (Lip-Bry-Nel) containing HIV latency activator (viz. bryostatin–2) and protease inhibitor (viz. nelfinavir) by thin film evaporation method providing nanoparticles of average size 219 nm. This hypothesis was demonstrated by the researchers using peripheral blood mononuclear cell (PBMC) culture wherein the response of Lip-Bry-Nel was compared with isotype control antibody-coated liposomes, confirming selective targeting.

7.5.5 HARD LIPID NANOCARRIERS

SLNs are submicron-sized nanocarriers consisting of biocompatible lipid with aqueous surfactant solution (Fernandes et al., 2013; Suares and Prabhakar, 2016). These SLNs may be physicochemically altered with a liquid oil to obtain nanostructured lipid carriers (Salunkhe et al., 2015) or with a biocompatible polymer to obtain lipid polymeric nanoparticles (Suares and Prabhakar, 2017). Oliveira et al. (2005) studied the P-gp efflux inhibitory activity of Solutol® HS15 by formulating indinavir-loaded lipid nanocapsules using phase inversion process. The study showed that there was 1.9 fold-increase in the uptake of the nanocapsules in the brain and testes when compared to plain indinavir solution owing to the alteration in cell membrane fluidity and/or inhibition of indinavir excretion by inhibiting mrp2. Chattopadhyay et al. (2008) studied the brain uptake of Atazanavir-loaded SLNs (~167nm in size) using Human Brain Endothelial cell-line. The ARV-loaded SLNs showed higher accumulation in the CNS which was ascribed to bypassing of P-gp efflux pump or prevention of P-gp recognition through encapsulation of ARV by the SLN. Vyas et al. (2015) prepared efavirenz-loaded SLN (size 150 nm) attached with phenylalanine, an amino acid micro-nutrient as ligand, causing carrier-mediated transport via l-amino acid transporter for brain uptake. The developed SLNs showed a controlled release of ARV for 24 h and displayed a 2–3-fold and 7–8-fold ARV accumulation in the brain from ligated-SLN when compared to plain-SLN and ARV alone, respectively. Jindal et al. (2017) showed the brain targeting ability of *in situ* hybrid nano drug delivery system of nevirapine (70–1100 nm, size) on intravenous administration. The *in vivo* studies in rate showed simultaneous targeting of drug in liver, spleen, and brain of about 6.1, 5.8 and 3.7 fold, respectively.

For nose-to-brain delivery, Dandagi et al. (2014) formulated SLNs of stavudine using double emulsion solvent evaporation method producing particles in the size range of 175–393 nm. In spite of higher oral bioavailability of stavudine, the adverse effects such as a dose-dependent peripheral neuropathy, anemia, hypersensitivity, lactic acidosis, hepatitis, and/

or liver failure on prolonged drug use have motivated the investigators to formulate an intranasal delivery system. Delivery of ARVs directly to the brain may cause alterations in the osmotic pressure, leading to opening of tight junctions, thus compromising the safety and causing infiltration of other substances into the brain. In contrast to that, colloidal drug carrier-mediated SLNs undergo passive uptake through macrophages into mono-nuclear phagocyte system (MPS) and do not interfere with the structure of tight junctions. This hypothesis was demonstrated by performing *in vivo* organ distribution study of stavudine-loaded SLN and drug alone in the rat model. The study results revealed 11-times higher drug concentrations from the SLN as compared to drug alone, indicating accessibility of SLN into brain bypassing the BBB. Gupat et al. (2017) also investigated the brain uptake of efavirenz-loaded SLNs when administered via the intranasal route wherein they were able to achieve 150-times more brain uptake of the ARV and 70-times higher absorption when compared to the commercial oral product of the ARV.

7.5.6   INORGANIC NANOPARTICLES

Widely known inorganic nanoparticles have been explored for imaging and therapeutic applications. It is reported that by modulating the various attributes of nanoparticles viz., composition, size, shape, and structure, it is possible to manipulate the optical, magnetic, and electronic properties. Thus, inorganic nanoparticles provide an attractive alternative to enhance the accumulation of the drug in HIV reservoirs (Nam et al., 2013).

7.5.6.1   GOLD NANOPARTICLES

Gold nanoparticles are widely explored inorganic nanoparticles owing to their ease of production of nanoparticles with diverse sizes, chemical stability, and size-dependent unique optical properties. It is reported that gold nanoparticle surface exhibits strong affinity for thiol, disulfide, and amine groups; hence they are easily conjugated to targeting peptides, proteins, antibodies, and so on (Shilo et al., 2015). Besides this, it is also reported that gold nanoparticles have inherent antiviral activity. Vijaykumar and Ganesan (2012) demonstrated that gold nanoparticles inhibited the viral fusion to HeLa-CD4-LTR-B-gal cell line by binding with gp120. This could be ascribed to the core nanoparticles as well as the ligands bound to core.

Bowman et al. (2008) showed that SDC–1721, a fragment of the potent HIV inhibitor TAK–779, conjugated to gold nanoparticles had superior HIV infection inhibitory activity compared to free SDC–1721 and comparable to TAK–779. Garrido et al. (2015) have reported the significance of density of conjugates on gold nanoparticles for antiviral activity. The group had conjugated thiolated raltegravir to gold nanoparticles and studied its antiviral activity and cellular uptake across macrophages and brain microendothelial cells. They found inverse correlation between number of molecules of thiolated raltegravir with the antiviral activity. The plausible reasoning for this altered effect was assumed to be decreased van der Waals interactions (at higher density of thiolated raltegravir) between the gold nanoparticles and the target. Another interesting study was carried out by Di Gianvincenzo et al. (2010), the study entailed use of capping of gold nanoparticles with polysulfates, potent inhibitors of HIV envelope glycoprotein gp120. The study stated that 50% density of sulfated ligands on ~2 nm gold nanoparticles was comparable to 100% density in attaining *in vitro* antiviral activities in T-cells at nanomolar concentrations. Similar observations were made by Martinez et al. (2009), the study reported truncated (oligo) mannosides of the high-mannose undecasaccharide $Man_9GlcNAc_2$ conjugated gold nanoparticles displayed 100% inhibition at 115 nM as compared to the unconjugated monomeric disaccharide (100% inhibition at 2.2 mM), i.e., more than 20,000-fold enhancement of activity. Further, the increase in density from 50 to 100% did not show a corresponding increase in the level of inhibition. Nevertheless, the study revealed the synergistic anti-gp120 activity of gold nanoparticles in conjunction of disaccharides. Another group showcased the ability of hybrid glyceryl monostearate-nevirapine-gold nanoparticles for rapid, high, and sustained distribution in the HIV reservoir organs such as liver, spleen, lymph nodes, thymus, brain, ovary, and bone marrow (Dalvi et al., 2016).

7.5.6.2 NANODIAMONDS

As term suggests, nanodiamonds are carbonaceous nanomaterials characterized by three-dimensional cubic lattice comprising of tetrahedrally bonded carbon atoms. They provide advantage of ease of functionalization either by covalent or non-covalent techniques. Further, it is stated that they have superior biocompatibility over single walled and multi-walled carbon nanotubes as well as other carbon nanomaterials (Ansari et al., 2016). Although, there are limited reports on the application of nanodiamonds in treatment of neuroAIDS, research has been undertaken in the area of cancer for delivery of therapeutics such as doxorubicin, purvalanol A, 4-hydroxytamoxifen, and dexamethasone for human colon cancer, liver cancer, breast cancer and blood cancer, respectively (Moosa et al., 2014). It is reported that the drug loading is achieved by physical adsorption of monolayer of drug onto the hydroxyl and carboxyl groups on the carbon surfaces of nanodiamonds by electrostatic interaction (Moore et al., 2012; Mochalin et al., 2013). Hence, it is believed that this adsorption of monolayer onto nanodiamonds with particle size < 10 nm, results in maximal surface area of exposure to an aqueous environment rendering water-insoluble drug, soluble in water in comparison to its nanocrystalline form (Mochalin et al., 2013). Roy et al. (2018) investigated the effect of efavirenz (a non-nucleoside reverse transcriptase inhibitor) loaded nanodiamond for brain delivery. The researchers formulated efavirenz loaded unmodified nanodiamond, -COOH surface-modified nanodiamond, and -NH$_2$ surface-modified nanodiamond, respectively. Among these delivery systems, -COOH surface-modified nanodiamond was found to be relatively toxic. The *in vitro* study revealed that unmodified nanodiamond, and -NH$_2$ surface-modified nanodiamond showed similar drug loading. The developed nanodiamond was found to control viral replication up to seven days suggesting the suitability of nanodiamond loaded drug for long-term anti-HIV therapy.

7.5.6.3   QUANTUM DOTS

Quantum dots are fluorescent semiconductor nanoparticles in the size range <10 nm. They are endowed with tunable optical properties having emission spectra across both UV-visible and near-infrared (NIR) regions. Combining their narrow size range and amenability to surface modification, they are widely explored for real-time monitoring of nanocarrier- cellular interactions in the biological milieu (Probst et al., 2013). Mahajan et al. (2010) investigated the brain uptake and antiviral activity of transferring-quantum rod-Saquinavir bioconjugates. The study revealed due to the receptor-mediated uptake of the bioconjugates there was a significant reduction in p24 antigen production (62%) and expression of LTR/RU5 gene (91%) in the HIV–1 infected peripheral blood leukocytes in comparison to free drug. The group also reported the similar observations for transferrin-quantum dot-Amprenavir nanoplex to transverse the BBB and halt the HIV replication (Mahajan et al., 2012).

7.5.6.4   MAGNETIC NANOPARTICLES

Magnetic nanoparticles, i.e., magnetite (Fe_3O_4) and maghemite (γ-Fe_2O_3) have been widely used for delivery of drugs since it offers numerous advantages of site-specific drug delivery, sustained release and better tissue distribution. However, for brain delivery of these nanoparticles has been rather limited because of the requirement of 30–50 cm working distance for its use in humans and further, restriction of magnetic field strength by FDA of 8 teslas for adults and 4 teslas for children (Shapiro et al., 2010). Widely known strategies to increase the permeability of these magnetic nanoparticles across the BBB include (Shetty and Upadhya, 2018; Busquets et al., 2015);

i.   Surface functionalization of the nanoparticle with ligands that can be used to target specific receptors in brain cells.

ii. Application of an external magnetic field to govern the movement of nanoparticle into the brain.

iii. Perturbation of BBB by application of radio frequency to nanoparticles to generate localized heat.

Sometimes to confer stability in the presence of metabolizing enzyme (present in blood) drugs bound to magnetic nanoparticles are often encapsulated within liposomes (magnetized liposomes). Sayed et al. (2010) report the combined use of magnetic field (0.3 Tesla) and monocyte-mediated transport for transmigration of magnetic azidothymidine 5′-triphosphate (AZTTP) liposomal nanoformulation across an *in vitro* BBB model. Further, magnetic AZTTP was found to have relative sustained antiviral effect as compared to free drug in *in-vitro*. Sagar et al. (2015) have demonstrated the usefulness of 25 to 40 nm sized magnetic nanoparticles for transmigration of highly selective and potent morphine antagonist, CTOP (D-Pen-Cys-Tyr-DTrp-Orn-Thr-Pen-Thr-NH_2), which is impermeable across BBB for the mitigation of opioid-induced neuropathogenesis and neuroAIDS. Jayant et al. (2015) investigated layer-by-layer assembled an anti-HIV drug (tenofovir), and a latency-breaking agent (vorinostat) loaded ultrasmall (10±3 nm) Fe_3O_4 nanoparticles for the treatment of neuroAIDS. The developed nanoparticles were found to simultaneously provide sustained release of drugs over a period of 5 days with induction of expression of latent HIV in infected primary human astrocyte. Besides this, they were found to exhibit blood-brain barrier transmigration ability of 37.95% ± 1.5% and in vitro antiviral efficacy evident by ~33% reduction of the p24 level. Recently, Jayant et al. (2017) explored this approach for sustained and concurrent release of sigma σ1 antagonist Rimcazole and anti-HIV drug Nelfinavir for 8 days to overcome the damaging effects of drugs of abuse, methamphetamine or cocaine in HIV infection. Further, in comparison to cocaine and methamphetamine-treated group (~50%)., the nanoformulations demonstrated a dose-dependent reduction (~55%) in p24 levels in in-vitro HIV–1-infected primary astrocyte culture and transmigration ability of $38.8 \pm 6.5\%$ across in vitro BBB model on the application of an external magnetic field (0.8 tesla).

Recently, Tomitaka et al. (2018) reported the fabrication of multifunctional magneto-plasmonic liposomes (MPLs), a hybrid system which combines salient features of liposomes, magnetic, and gold nanoparticles for triple-modality image-guided drug delivery of Tenofovir disoproxil fumarate, an antiretroviral drug. In the presence of magnetic field (150 mTesla), there was significance increase in transmigration permeability (15%) of drug loaded MPLs across the in vitro BBB model.

7.5.6.5   MAGNETO-ELECTRIC NANOPARTICLES

Although useful, magnetic nanoparticles are plagued with issues such as possibility of phase change of nanoparticles occurring due to unfavorable interaction in biologic milieu; resulting in drug loss. Further, in the presence of biological changes induced by disease (pH, temperature, etc.) or interaction with body defenses, the nanoparticles may exhibit erratic drug release pattern thereby defeating the objective of site-specific drug release systems (Kaushik et al., 2014). To address these shortcomings, magneto-electric nanoparticles have been devised. This class of nanoparticles exhibits both ferromagnetic and ferroelectric effect in a single phase. Owing to this attribute, they are considered as heat dissipation-free, highly efficient and are enabled to on-demand targeted drug release even at low remote magnetic electric field (Guduru et al., 2013). Nair et al. (2013) explored this next generation nanoparticles to deliver azidothymidine 5'-triphosphate, antiviral drug, from barium titanate-cobalt ferrite (20–30 nm in size) magneto-electric nanoparticles in the presence of a low alternating current (44Oe at 1,000 Hz)-magnetic field (0.8 Tesla). Unlike magnetic nanocarrier, the release of the drug from the nanoparticles is controlled by exploiting its non-zero electric property (alternating current trigger) to break the symmetry of ionic bonding between drugs molecules and nanoparticles. The authors have reported that remote low-energy DC magnetic field releases ~40% of the bound antiviral drug, prompting ~3 times higher translocation of the drug across the BBB in comparison to free drug. Further, low alternating current field triggered the release of ~100% of bound drugs from the nanoparticles. Additionally, Kaushik et al. (2016) demonstrated the safety and efficacy of barium titanate-cobalt ferrite magneto-electric nanoparticles in mice model and suggested its applications for targeted, responsive, sustained delivery of across the BBB into the brain.

7.5.7 CELL-BASED NANOFORMULATION

This approach exploits the cellular domains; monocytes and monocyte-derived macrophages (MDM) to mediate the extended delivery of nanoret-rovirals (Guo et al., 2014; Edagwa et al., 2014). The advantages proffered by this unique approach include:

i.   the ability of nanocarriers to deliver retrovirals to infected tissues irrespective of the tenacity of the biological barriers,
ii. improved drug stability and half-life of ARVs by mitigating the risk of drug degradation by metabolizing enzymes (Araínga et al., 2015).
iii. target specificity to infected tissues owing to delivery of nanocar-riers in response to cytokine signaling (Kadiu, Nowacek, McMillan & Gendelman, 2011).
iv. sustained release of drug as they are confined with the cellular domain for extended times (Guo et al., 2014).

It is assumed that nanoretroviral drugs are introduced into endosomal organelles via clathrin-endosome pathways, after which the drug particles are sorted via recycling pathway avoiding macrophage degradation. As reported, this particulate sorting is distinct from HIV endocytic sorting thereby curbing viral replication (Kadiu, Nowacek, McMillan & Gendelman, 2011). Although, this concept is fascinating, yet much has to be known about how these sequestered drug nanoparticles remain unchanged in macrophages without losing its antiviral efficacy for extended periods. Kanmogne et al. (2012) established that nanoretrovirals (up to 52%) were transferred from cellular carriers (MDM) into human brain microvascular endothelial cells via cell-to-cell contacts. They hypothesized that this intimate interaction between the endothelial-macrophage resulted in efficient uptake of nanoretroviral released from MDM by the endothelial cells of the BBB, thereby potentiating the transfer of the drugs from the periphery into the viral brain reservoirs. Further, the researchers demonstrated the significant increase (up to 77%) in the intercellular transfer of folate coated nanoparticles from cellular carriers into human brain microvascular endothelial cells underlining the role of folate receptors. Nowacek et al. (2009) demonstrated that larger nanoparticles were preferentially taken up by monocyte-derived macrophage and were found to exhibit sustained released over 14 days. Further, it had good BBB permeability and provided dose-dependent antiviral efficacy as evident from decreased *in vitro* production of progeny virion and HIV–1 p24 antigen. Dou et al. (2009) reported the superior efficacy of indinavir nanoparticles loaded murine bone marrow macrophages in distribution of drug in subregions of the brain in HIV–1 encephalitis rodent model. The developed formulation provided sustained release for 14 days and reduction in HIV–1 replication in brain.

7.6   CONCLUDING REMARKS

To date, complete eradication of HIV from the different body reservoirs of the virus especially the brain, still remains the major roadblock for the treatment of this pandemic. Although, nanomedicines have made great strides to improve the overall therapeutic efficiency of the existing ARVs, the targeted delivery of ARVs into these formidable viral sanctuaries is still a challenge. Through this chapter, the authors have made an attempt to review the various barriers presented by the brain and the complications arising due to the HIV. Further, the chapter discusses diverse nanoapproaches present in the literature for efficient brain uptake of the ARVs. The nanoparticles explored for brain delivery broadly constitutes of polymeric, inorganic, and lipidic origin and also recently explored cellular domain confined nanoformulations. The results are promissory mostly in *in-vitro* cell lines based models and to an extent in a preclinical setting. However, there is a dearth of research which demands more conclusive and clinically authenticated research to expound the true potential of nanocarriers for the eradication of HIV load from the brain reservoirs.

KEYWORDS

- **cell-based nanoparticles**
- **inorganic nanocarriers**
- **lipid nanocarriers**
- **neuroAIDS**
- **polymeric nanocarriers**

REFERENCES

Abbasi, E., Aval, S. F., Akbarzadeh, A., Milani, M., Nasrabadi, H. T., Joo, S. W., Hanifehpour, Y., Nejati-Koshki, K., & Pashaei-Asl, R., (2014). Dendrimers: Synthesis, applications, and properties. *Nanoscale Res. Lett., 9*(1), 247.

Al-Ghananeem, A. M., Saeed, H., Florence, R., Yokel, R. A., & Malkawi, A. H., (2010). Intranasal drug delivery of didanosine-loaded chitosan nanoparticles for brain targeting, an attractive route against infections caused by AIDS viruses. *J. Drug Target., 18*(5), 381–388.

Al-Ghananeem, A. M., Smith, M., Coronel, M. L., & Tran, H., (2013). Advances in brain targeting and drug delivery of anti-HIV therapeutic agents. *Expert Opin. Drug Deliv., 10*(7), 973–985.

Ansari, S. A., Satar, R., Jafri, M. A., Rasool, M., Ahmad, W., & Zaidi, S. K., (2016). Role of nanodiamonds in drug delivery and stem cell therapy. *Iran J. Biotechnol., 14*(3), 130–141.

Anthony, I. C., & Bell, J. E., (2008). The neuropathology of HIV/AIDS. *Int. Rev. Psychiatry., 20*(1), 15–24.

Araínga, M., Guo, D., Wiederin, J., Ciborowski, P., McMillan, J., & Gendelman, H. E., (2015). Opposing regulation of endolysosomal pathways by long-acting nanoformulated antiretroviral therapy and HIV–1 in human macrophages. *Retrovirology, 12*(1), 5.

Banks, W. A., (2000). Protector, prey, or perpetrator: The pathophysiology of the blood-brain barrier in neuroAIDS. *Science Online: NeuroAIDS, 3*(3), 1–6.

Batrakova, E. V., Li, S., Alakhov, V. Y., Miller, D. W., & Kabanov, A. V., (2003). Optimal structure requirements for pluronic block copolymers in modifying P-glycoprotein drug efflux transporter activity in bovine brain microvessel endothelial cells. *J. Pharmacol. Exp. Ther., 304*(2), 845–854.

Belgamwar, A., Khan, S., & Yeole, P., (2017). Intranasal chitosan-g-HPβCD nanoparticles of efavirenz for the CNS targeting. *Artif. Cells, Nanomedicine, Biotechnol., 19*, 1–13.

Bertrand, L., Nair, M., & Toborek, M., (2016). Solving the blood-brain barrier challenge for the effective treatment of HIV replication in the central nervous system. *Curr. Pharm. Des., 22*(35), 5477–5486.

Bowman, M. C., Ballard, T. E., Ackerson, C. J., Feldheim, D. L., Margolis, D. M., & Melander, C., (2008). Inhibition of HIV fusion with multivalent gold nanoparticles. *J. Am. Chem. Soc., 130*(22), 6896–6897.

Busquets, M. A., Espargaró, A., Sabaté, R., & Estelrich, J., (2015). Magnetic nanoparticles cross the blood-brain barrier: When physics rises to a challenge. *Nanomaterials, 5*(4), 2231–2248.

Chattopadhyay, N., Zastre, J., Wong, H. L., Wu, X. Y., & Bendayan, R., (2008). Solid lipid nanoparticles enhance the delivery of the HIV protease inhibitor, atazanavir, by a human brain endothelial cell line. *Pharm. Res., 25*(10), 2262–2271.

Croy, S. R., & Kwon, G. S., (2006). Polymeric micelles for drug delivery. *Curr. Pharm. Des., 12*(36), 4669–4684.

Dalvi, B. R., Siddiqui, E. A., Syed, A. S., Velhal, S. M., Ahmad, A., Bandivdekar, A. B., & Devarajan, P. V., (2016). Nevirapine loaded core shell gold nanoparticles by double emulsion solvent evaporation: *In vitro* and *in vivo* evaluation. *Curr. Drug Deliv., 13*(7), 1071–1083.

Das Neves, J., Amiji, M. M., Bahia, M. F., & Sarmento, B., (2010). Nanotechnology-based systems for the treatment and prevention of HIV/AIDS. *Adv. Drug Deliv. Rev., 62*(4), 458–477.

De Oliveira, M. P., Garcion, E., Venisse, N., Benoît, J. P., Couet, W., & Olivier, J. C., (2005). Tissue distribution of indinavir administered as solid lipid nanocapsule formulation in mdr1a (+/+) and mdr1a (−/−) CF–1 mice. *Pharm. Res., 22*(11), 1898–1905.

Destache, C. J., Belgum, T., Goede, M., Shibata, A., & Belshan, M. A., (2010). Antiretroviral release from poly(DL-lactide-coglycolide) nanoparticles in mice. *J. Antimicrob. Chemother., 65*(10), 2183–2187.

Di Gianvincenzo, P., Marradi, M., Martínez-Ávila, O. M., Bedoya, L. M., Alcamí, J., & Penadés, S., (2010). Gold nanoparticles capped with sulfate-ended ligands as anti-HIV agents. *Bioorganic. Med. Chem. Lett., 20*(9), 2718–2721.

Dou, H., Grotepas, C. B., McMillan, J. M., Destache, C. J., Chaubal, M., Werling, J., et al., (2009). Macrophage delivery of nanoformulated antiretroviral drug to the brain in a murine model of neuroAIDS. *J. Immunol., 183*(1), 661–669.

Dutta, T., & Jain, N. K., (2007). Targeting potential and anti-HIV activity of lamivudine loaded mannosylated poly (propylene imine) dendrimer. *Biochim. Biophys. Acta., 1770*(4), 681–686.

Dutta, T., Agashe, H. B., Garg, M., Balasubramanium, P., Kabra, M., & Jain, N. K., (2007). Poly (propylene imine) dendrimer-based nanocontainers for targeting of efavirenz to human monocytes/macrophages *in vitro. J. Drug Target., 15*(1), 89–98.

Edagwa, B., Zhou, T., McMillan, J., Liu, X. M., & Gendelman, H., (2014). Development of HIV reservoir targeted long-acting nano-formulated antiretroviral therapies. *Curr. Med. Chem., 21*(36), 4186–4198.

Ene, L., Duiculescu, D., & Ruta, S. M., (2011). How much do antiretroviral drugs penetrate into the central nervous system? *Journal of Medicine and Life, 4*(4), 432.

Everall, I., & Levy, R., (2002). Clinical aspects of neuroAIDS. *Journal of Neurovirology, 8*(3), 87–95.

Fernandes, C. B., Mandawgade, S., & Patravale, V. B., (2013). Solid lipid nanoparticles of etoposide using solvent emulsification-diffusion technique for parenteral administration. *International Journal of Pharma Bioscience and Technology, 1*(1), 27–33.

Fernandes, C. B., Suares, D., & Dhawan, V. (2018). Lipid Nanocarriers for Intracellular Delivery. In *Multifunctional Nanocarriers for Contemporary Healthcare Applications* (pp. 129-156). IGI Global.

Fernandes, C., Soni, U., & Patravale, V., (2010). Nano-interventions for neurodegenerative disorders. *Pharmacol. Res., 62*(2), 166–178.

Gagné, J. F., Désormeaux, A., Perron, S., Tremblay, M. J., & Bergeron, M. G., (2002). Targeted delivery of indinavir to HIV–1 primary reservoirs with immunoliposomes. *Biochim. Biophys. Acta., 1558*(2), 198–210.

Garg, M., Asthana, A., Agashe, H. B., Agrawal, G. P., & Jain, N. K., (2006). Stavudine-loaded mannosylated liposomes: *In-vitro* anti-HIV-I activity, tissue distribution, and pharmacokinetics. *J. Pharm. Pharmacol., 58*(5), 605–616.

Garrido, C., Simpson, C. A., Dahl, N. P., Bresee, J., Whitehead, D. C., Lindsey, E. A., et al., (2015). Gold nanoparticles to improve HIV drug delivery. *Future Med. Chem., 7*(9), 1097–1107.

Gerson, T., Makarov, E., Senanayake, T. H., Gorantla, S., Poluektova, L. Y., & Vinogradov, S. V., (2014). Nano-NRTIs demonstrate low neurotoxicity and high antiviral activity against HIV infection in the brain. *Nanomedicine: Nanotechnology, Biology and Medicine, 10*(1), 177–185.

Gonzalez-Scarano, F., & Martin-Garcia, J., (2005). The neuropathogenesis of AIDS. *Nat. Rev. Immunol., 5*(1), 69–81.

Govender, T., Ojewole, E., Naidoo, P., & Mackraj, I., (2008). Polymeric nanoparticles for enhancing antiretroviral drug therapy. *Drug Delivery, 15*(8), 493–501.

Gu, J., Al-Bayati, K., & Ho, E. A., (2017). Development of antibody-modified chitosan nanoparticles for the targeted delivery of siRNA across the blood-brain barrier as a strategy for inhibiting HIV replication in astrocytes. *Drug Deliv. Transl. Res.*, *7*(4), 497–506.

Guduru, R., Liang, P., Runowicz, C., Nair, M., Atluri, V., & Khizroev, S., (2013). Magneto-electric nanoparticles to enable field-controlled high-specificity drug delivery to eradicate ovarian cancer cells. *Sci. Rep.*, *3*, 2953.

Guo, D., Zhang, G., Wysocki, T. A., Wysocki, B. J., Gelbard, H. A., Liu, X. M., et al., (2014). Endosomal trafficking of nanoformulated antiretroviral therapy facilitates drug particle carriage and HIV clearance. *J. Virol.*, *88*(17), 9504–9513.

Gupta, S., Kesarla, R., Chotai, N., Misra, A., & Omri, A., (2017). Systematic approach for the formulation and optimization of solid lipid nanoparticles of Efavirenz by high-pressure homogenization using design of experiments for brain targeting and enhanced bioavailability. *Biomed Res Int.* 2017:5984014. doi: 10.1155/2017/5984014.

Jain, J., Fernandes, C., & Patravale, V., (2010). Formulation development of parenteral phospholipid-based microemulsion of etoposide. *AAPS Pharm. Sci. Tech.*, *11*(2), 826–831.

Jayant, R. D., Atluri, V. S., Agudelo, M., Sagar, V., Kaushik, A., & Nair, M., (2015). Sustained-release nanoART formulation for the treatment of neuroAIDS. *Int. J. Nanomed.*, *10*, 1077–1093.

Jayant, R. D., Atluri, V. S., Tiwari, S., Pilakka-Kanthikeel, S., Kaushik, A., Yndart, A., & Nair, M., (2017). Novel nanoformulation to mitigate co-effects of drugs of abuse and HIV–1 infection: Towards the treatment of neuroAIDS. *J. Neurovirol.*, *23*(4), 603–614.

Jiménez, J. L., Clemente, M. I., Weber, N. D., Sanchez, J., Ortega, P., De la Mata, F. J., et al., (2010). Carbosilane dendrimers to transfect human astrocytes with small interfering RNA targeting human immunodeficiency virus. *BioDrugs.*, *24*(5), 331–343.

Jin, S. X., Bi, D. Z., Wang, J., Wang, Y. Z., Hu, H. G., & Deng, Y. H., (2005). Pharmacokinetics and tissue distribution of zidovudine in rats following intravenous administration of zidovudine myristate loaded liposomes. *Die Pharmazie*, *60*(11), 840–843.

Jindal, A. B., Bachhav, S. S., & Devarajan, P. V., (2017). In situ hybrid nanodrug delivery system (IHN-DDS) of antiretroviral drug for simultaneous targeting to multiple viral reservoirs: An *in vivo* proof of concept. *Int. J. Pharm.*, *521*(1 & 2), 196–203.

Kadiu, I., Nowacek, A., McMillan, J., & Gendelman, H. E., (2011). Macrophage endocytic trafficking of antiretroviral nanoparticles. *Nanomedicine*, *6*(6), 975–994.

Kandadi, P., Syed, M. A., Goparaboina, S., & Veerabrahma, K., (2011). Brain-specific delivery of pegylated indinavir submicron lipid emulsions. *Eur. J. Pharm. Sci.*, *42*(4), 423–432.

Kanmogne, G. D., Singh, S., Roy, U., Liu, X., McMillan, J., Gorantla, S., et al., (2012). Mononuclear phagocyte intercellular crosstalk facilitates transmission of cell-targeted nano-formulated antiretroviral drugs to human brain endothelial cells. *Int. J. Nanomed.*, *7*, 2373–2388.

Kaur, A., Jain, S., & Tiwary, A. K., (2008a). Mannan-coated gelatin nanoparticles for sustained and targeted delivery of didanosine: *In vitro* and *in vivo* evaluation. *Acta Pharm.*, *58*(1), 61–74.

Kaur, C. D., Nahar, M., & Jain, N. K., (2008b). Lymphatic targeting of zidovudine using surface-engineered liposomes. *J. Drug Target.*, *16*(10), 798–805.

Kaushik, A., Jayant, R. D., Nikkhah-Moshaie, R., Bhardwaj, V., Roy, U., Huang, Z., et al., (2016). Magnetically guided central nervous system delivery and toxicity evaluation of magneto-electric nanocarriers. *Sci. Rep.*, *6*, 25309.

Kaushik, A., Jayant, R. D., Sagar, V., & Nair, M., (2014). The potential of magneto-electric nanoparticles for drug delivery. *Expert Opin. Drug Deliv.*, *11*(10), 1635–1646.

Kettiger, H., Schipanski, A., Wick, P., & Huwyler, J., (2013). Engineered nanomaterial uptake and tissue distribution: From cell to organism. *Int. J. Nanomedicine.*, *8*, 3255–3269.

Kovochich, M., Marsden, M. D., & Zack, J. A., (2011). Activation of latent HIV using drug-loaded nanoparticles. *PLoS One.*, *6*(4), 18270.

Kreuter, J., (2004). Influence of the surface properties on nanoparticle-mediated transport of drugs to the brain. *J. Nanosci. Nanotechnol.*, *4*(5), 484–488.

Kumar, P. D., Kumar, P. V., Selvam, T. P., & Rao, K. S., (2013). PEG-conjugated PAMAM dendrimers with anti-HIV drug stavudine for prolong release. *Research in Biotechnology*, *4*(2), 10–18.

Kuo, Y. C., & Chen, H. H., (2006). Effect of nanoparticulate polybutylcyanoacrylate and methylmethacrylate–sulfopropyl methacrylate on the permeability of zidovudine and lamivudine across the in vitro blood–brain barrier. *Int. J. Pharm.*, *327*(1 & 2), 160–169.

Kuo, Y. C., & Chung, C. Y., (2012). Transcytosis of CRM197-grafted polybutylcyanoacrylate nanoparticles for delivering zidovudine across human brain microvascular endothelial cells. *Colloids Surf. B Biointerfaces.*, *91*, 242–249.

Kuo, Y. C., & Kuo, C. Y., (2008). Electromagnetic interference in the permeability of saquinavir across the blood-brain barrier using nanoparticulate carriers. *Int. J. Pharm.*, *351*(1 & 2), 271–281.

Kuo, Y. C., & Lee, C. L., (2012). Methylmethacrylate-sulfopropyl methacrylate nanoparticles with surface RMP-7 for targeting delivery of antiretroviral drugs across the blood-brain barrier. *Colloids Surf. B Biointerfaces.*, *90*, 75–82.

Kuo, Y. C., & Su, F. L., (2007). Transport of stavudine, delavirdine, and saquinavir across the blood-brain barrier by polybutylcyanoacrylate, methylmethacrylate-sulfopropyl methacrylate, and solid lipid nanoparticles. *Int. J. Pharm.*, *340*(1 & 2), 143–152.

Kuo, Y. C., (2005). Loading efficiency of stavudine on polybutylcyanoacrylate and methylmethacrylate-sulfopropyl methacrylate copolymer nanoparticles. *Int. J. Pharm.*, *290*(1 & 2), 161–172.

Löbenberg, R., Araujo, L., & Kreuter, J., (1997). Body distribution of azidothymidine bound to nanoparticles after oral administration. *Eur. J. Pharm. Biopharm.*, *44*(2), 127–132.

Mahajan, H. S., Mahajan, M. S., Nerkar, P. P., & Agrawal, A., (2014). Nanoemulsion-based intranasal drug delivery system of saquinavir mesylate for brain targeting. *Drug Delivery*, *21*(2), 148–154.

Mahajan, S. D., Aalinkeel, R., Law, W. C., Reynolds, J. L., Nair, B. B., Sykes, D. E., et al., (2012a). Anti-HIV–1 nanotherapeutics: Promises and challenges for the future. *Int. J. Nanomedicine.*, *7*, 5301–5314.

Mahajan, S. D., Law, W. C., Aalinkeel, R., Reynolds, J., Nair, B. B., Yong, K. T., et al., (2012b). Nanoparticle-mediated targeted delivery of antiretrovirals to the brain. In: *Methods in Enzymology* (Vol. 509, pp. 41–60). Academic Press.

Mahajan, S. D., Roy, I., Xu, G., Yong, K. T., Ding, H., Aalinkeel, R., et al., (2010). Enhancing the delivery of antiretroviral drug "Saquinavir" across the blood-brain barrier using nanoparticles. *Curr. HIV. Res.*, *8*(5), 396–404.

Mainardes, R. M., Gremião, M. P. D., Brunetti, I. L., Da Fonseca, L. M., & Khalil, N. M., (2009). Zidovudine-loaded PLA and PLA–PEG blend nanoparticles: Influence of polymer type on phagocytic uptake by polymorphonuclear cells. *J. Pharm. Sci.*, *98*(1), 257–267.

Mallipeddi, R., & Rohan, L. C., (2010). Progress in antiretroviral drug delivery using nanotechnology. *Int. J. Nanomedicine.*, *5*, 533–547.

Martínez-Ávila, O., Hijazi, K., Marradi, M., Clavel, C., Campion, C., Kelly, C., & Penadés, S., (2009). Gold manno-glyconanoparticles: Multivalent systems to block HIV-1 gp120 binding to the lectin DC-SIGN. *Chem. Eur. J.*, *15*(38), 9874–9888.

Mishra, V., Mahor, S., Rawat, A., Gupta, P. N., Dubey, P., Khatri, K., & Vyas, S. P., (2006). Targeted brain delivery of AZT via transferrin anchored pegylated albumin nanoparticles. *J. Drug Target*, *14*(1), 45–53.

Mittal, D., Ali, A., Md, S., Baboota, S., Sahni, J. K., & Ali, J., (2014). Insights into direct nose to brain delivery: Current status and future perspective. *Drug Delivery*, *21*(2), 75–86.

Mochalin, V. N., Pentecost, A., Li, X. M., Neitzel, I., Nelson, M., Wei, C., He, T., Guo, F., & Gogotsi, Y., (2013). Adsorption of drugs on nanodiamond: Toward development of a drug delivery platform. *Mol. Pharm.*, *10*(10), 3728–3735.

Moore, L. K., Gatica, M., Chow, E. K., & Ho, D., (2012). Diamond-based nanomedicine: Enhanced drug delivery and imaging. *Disrupt. Sci. Technol.*, *1*(1), 54–61.

Moosa, B., Fhayli, K., Li, S., Julfakyan, K., Ezzeddine, A., & Khashab, N. M., (2014). Applications of nanodiamonds in drug delivery and catalysis. *J. Nanosci. Nanotechnol.*, *14*(1), 332–343.

Nair, M., Guduru, R., Liang, P., Hong, J., Sagar, V., & Khizroev, S., (2013). Externally controlled on-demand release of anti-HIV drug using magneto-electric nanoparticles as carriers. *Nat. Commun.*, *4*, 1707.

Nair, M., Jayant, R. D., Kaushik, A., & Sagar, V., (2016). Getting into the brain: Potential of nanotechnology in the management of neuroAIDS. *Adv. Drug Deliv. Rev.*, *103*, 202–217.

Nam, J., Won, N., Bang, J., Jin, H., Park, J., Jung, S., Jung, S., Park, Y., & Kim, S., (2013). Surface engineering of inorganic nanoparticles for imaging and therapy. *Adv. Drug Deliv. Rev.*, *65*(5), 622–648.

Nowacek, A. S., Miller, R. L., McMillan, J., Kanmogne, G., Kanmogne, M., Mosley, R. L., et al., (2009). Nano-ART synthesis, characterization, uptake, release, and toxicology for human monocyte-macrophage drug delivery. *Nanomedicine*, *4*(8), 903–917.

Nowacek, A., & Gendelman, H. E., (2009). Nano-ART, neuroAIDS, and CNS drug delivery. *Nanomedicine*, *4*(5), 557–574.

Ojewole, E., Mackraj, I., Naidoo, P., & Govender, T., (2008). Exploring the use of novel drug delivery systems for antiretroviral drugs. *Eur. J. Pharm. Biopharm.*, *70*(3), 697–710.

Pandey, P. K., Sharma, A. K., & Gupta, U., (2016). Blood-brain barrier: An overview on strategies in drug delivery, realistic *in vitro* modeling and *in vivo* live tracking. *Tissue Barriers*, *4*(1), 1129476.

Parboosing, R., Maguire, G. E., Govender, P., & Kruger, H. G., (2012). Nanotechnology and the treatment of HIV infection. *Viruses*, *4*(4), 488–520.

Peng, J., Wu, Z., Qi, X., Chen, Y., & Li, X., (2013). Dendrimers as potential therapeutic tools in HIV inhibition. *Molecules*, *18*(7), 7912–7929.

Perisé-Barrios, A. J., Jiménez, J. L., Dominguez-Soto, A., De la Mata, F. J., Corbí, A. L., Gomez, R., & Muñoz-Fernandez, M. Á., (2014). Carbosilane dendrimers as gene delivery agents for the treatment of HIV infection. *J. Control. Release*, *184*, 51–57.

Pidaparthi, K., & Suares, D., (2017). Comparison of nanoemulsion and aqueous micelle systems of paliperidone for intranasal delivery. *AAPS Pharm. Sci. Tech.*, *18*(5), 1710–1719.

Prabhakar, K., Afzal, S. M., Kumar, P. U., Rajanna, A., & Kishan, V., (2011). Brain delivery of transferrin coupled indinavir submicron lipid emulsions-Pharmacokinetics and tissue distribution. *Colloids Surf. B Biointerfaces.*, *86*(2), 305–313.

Probst, C. E., Zrazhevskiy, P., Bagalkot, V., & Gao, X., (2013). Quantum dots as a platform for nanoparticle drug delivery vehicle design. *Adv. Drug Deliv. Rev.*, *65*(5), 703–718.

Pyreddy, S., Kumar, P. D., & Kumar, P. V., (2014). Polyethylene glycosylated PAMAM dendrimers-Efavirenz conjugates. *Int. J. Pharm. Investig.*, *4*(1), 15.

Rao, K. S., Reddy, M. K., Horning, J. L., & Labhasetwar, V., (2008). TAT-conjugated nanoparticles for the CNS delivery of anti-HIV drugs. *Biomaterials*, *29*(33), 4429–4438.

Richman, D. D., Margolis, D. M., Delaney, M., Greene, W. C., Hazuda, D., & Pomerantz, R. J., (2009). The challenge of finding a cure for HIV infection. *Science*, *323*(5919), 1304–1307.

Ronaldson, P. T., Persidsky, Y., & Bendayan, R., (2008). Regulation of ABC membrane transporters in glial cells: Relevance to the pharmacotherapy of brain HIV-1 infection. *Glia.*, *56*(16), 1711–1735.

Roy, U., Drozd, V., Durygin, A., Rodriguez, J., Barber, P., Atluri, V., et al., (2018). Characterization of nanodiamond-based anti-HIV drug delivery to the brain. *Sci. Rep.*, *8*, 1603.

Sagar, V., Pilakka-Kanthikeel, S., Atluri, V. S., Ding, H., Arias, A. Y., Jayant, R. D., Kaushik, A., & Nair, M., (2015). Therapeutical neuron targeting via magnetic nanocarrier: Implications to opiate-induced neuropathogenesis and neuroAIDS. *[J. Biomed. Nanotechnol.*, *11*(10), 1722–1733.

Sagar, V., Pilakka-Kanthikeel, S., Pottathil, R., Saxena, S. K., & Nair, M., (2014). Towards nanomedicines for neuroAIDS. *Rev. Med. Virol.*, *24*(2), 103–124.

Saiyed, Z. M., Gandhi, N. H., & Nair, M. P., (2010). Magnetic nanoformulation of azidothymidine 5'-triphosphate for targeted delivery across the blood-brain barrier. *[Int. J. Nanomed.*, *5*, 157–166.

Saksena, N. K., Wang, B., Zhou, L., Soedjono, M., Ho, Y. S., & Conceicao, V., (2010). HIV reservoirs *in vivo* and new strategies for possible eradication of HIV from the reservoir sites. *HIV/AIDS (Auckland, NZ)*, *2*, 103.

Salunkhe, S. S., Bhatia, N. M., Kawade, V. S., & Bhatia, M. S., (2015). Development of lipid-based nanoparticulate drug delivery systems and drug carrier complexes for delivery to brain. *J. Appl. Pharm. Sci.*, *5*(5), 110–129.

Sanchez, A. B., & Kaul, M., (2017). Neuronal stress and injury caused by HIV–1, cART and drug abuse: Converging contributions to HAND. *Brain Sci.*, *7*(3), 25.

Shah, L. K., & Amiji, M. M., (2006). Intracellular delivery of saquinavir in biodegradable polymeric nanoparticles for HIV/AIDS. *Pharm. Res.*, *23*(11), pp. 2638–2645.

Shaik, N., Pan, G., & Elmquist, W. F., (2008). Interactions of Pluronic block copolymers on P-gp efflux activity: Experience with HIV–1 protease inhibitors. *J. Pharm. Sci.*, *97*(12), 5421–5433.

Shapiro, B., Dormer, K., & Rutel, I. B., (2010). A two-magnet system to push therapeutic nanoparticles. In: *AIP Conference Proceedings* (Vol. 1311, No. 1, pp. 77–88).

Shetty, S. R., & Upadhya, A. (2018). Magnetic Nano-Systems in Drug Delivery and Biomedical Applications. In *Multifunctional Nanocarriers for Contemporary Healthcare Applications* (pp. 157-191). IGI Global.

Shilo, M., Sharon, A., Baranes, K., Motiei, M., Lellouche, J. P. M., & Popovtzer, R., (2015). The effect of nanoparticle size on the probability to cross the blood-brain barrier: An *in-vitro* endothelial cell model. *J. Nanobiotechnology.*, *13*(1), 19.

Silva, A. C., Rodrigues, B. S., Micheletti, A. M., Tostes, S., Meneses, A. C., Silva-Vergara, M. L., & Adad, S. J., (2012). Neuropathology of AIDS: An autopsy review of 284 cases from Brazil comparing the findings pre- and post-HAART (highly active antiretroviral therapy) and pre- and postmortem correlation. *AIDS Res. Treat.*, 186850.

Spitzenberger, T. J., Heilman, D., Diekmann, C., Batrakova, E. V., Kabanov, A. V., Gendelman, H. E., Elmquist, W. F., & Persidsky, Y., (2007). Novel delivery system enhances efficacy of antiretroviral therapy in animal model for HIV–1 encephalitis. *J. Cereb. Blood Flow Metab.*, *27*(5), 1033–1042.

Suares, D., & Prabhakar, B., (2016). Oral delivery of rosuvastatin lipid nanocarriers: Investigation of *in vitro* and *in vivo* profile. *International Journal of Pharmaceutical Sciences and Research*, *7*(12), 4856.

Suares, D., & Prabhakar, B., (2017). Cuboidal lipid polymer nanoparticles of rosuvastatin for oral delivery. *Drug Dev. Ind. Pharm. 43*(2), 213–224.

Svenson, S., (2009). Dendrimers as versatile platform in drug delivery applications. *Eur. J. Pharm. Biopharm. 71*(3), 445–462.

Tomitaka, A., Arami, H., Huang, Z., Raymond, A., Rodriguez, E., Cai, Y., Febo, M., Takemura, Y., & Nair, M., (2018). Hybrid magneto-plasmonic liposomes for multimodal image-guided and brain-targeted HIV treatment. *Nanoscale. 10*(1), 184–194.

UNAIDS Joint United Nations Programme on HIV/AIDS. World Aids Day Report; Geneva, Swizerland, 2017. URL: www.unaids.org.

Varatharajan, L., & Thomas, S. A., (2009). The transport of anti-HIV drugs across blood–CNS interfaces: summary of current knowledge and recommendations for further research. [*Antivir. Res. 82*(2), A99–A109.

Varghese, N. M., Senthil, V., & Saxena, S. K., (2017). Nanocarriers for brain specific delivery of anti-retro viral drugs: challenges and achievements. *J. Drug Target.* 1–13. DOI: 10.1080/1061186X.2017.1374389.

Vashist, A., Kaushik, A., Vashist, A., Jayant, R. D., Tomitaka, A., Ahmad, S., Gupta, Y. K., & Nair, M., (2016). Recent trends on hydrogel based drug delivery systems for infectious diseases. *Biomaterials Science. 4*(11), 1535–1553.

Vieira, D. B., & Gamarra, L. F., (2016). Getting into the brain: liposome-based strategies for effective drug delivery across the blood–brain barrier. *Int. J. Nanomedicine. 11*, 5381.

Vijayakumar, S., & Ganesan, S., (2012). Gold nanoparticles as an HIV entry inhibitor. *Curr. HIV. Res. 10*(8), 643–646.

Vinogradov, S. V., Kohli, E., & Zeman, A. D., (2005). Cross-Linked Polymeric Nanogel Formulations of 5'-Triphosphates of Nucleoside Analogues: Role of the Cellular Membrane in Drug Release. *Mol. Pharm. 2*(6), 449–461.

Vinogradov, S. V., Poluektova, L. Y., Makarov, E., Gerson, T., & Senanayake, M. T., (2010). Nano-NRTIs: efficient inhibitors of HIV type-1 in macrophages with a reduced mitochondrial toxicity. *Antivir. Chem. Chemother. 21*(1), 1-14.

Vyas, A., Jain, A., Hurkat, P., Jain, A., & Jain, S. K., (2015). Targeting of AIDS related encephalopathy using phenylalanine anchored lipidic nanocarrier. *Colloids Surf. B Biointerfaces. 131*, 155-161.

Vyas, T. K., Shah, L., & Amiji, M. M., (2006). Nanoparticulate drug carriers for delivery of HIV/AIDS therapy to viral reservoir sites. *Expert Opin. Drug Deliv. 3*(5), 613–628.

Vyas, T. K., Shahiwala, A., & Amiji, M. M., (2008). Improved oral bioavailability and brain transport of Saquinavir upon administration in novel nanoemulsion formulations. *Int. J. Pharm. 347*(1), 93–101.

Wong, H. L., Chattopadhyay, N., Wu, X. Y., & Bendayan, R., (2010). Nanotechnology applications for improved delivery of antiretroviral drugs to the brain. *Adv. Drug Deliv. Rev. 62*(4), 503–517.

Xu, S., Olenyuk, B. Z., Okamoto, C. T., & Hamm-Alvarez, S. F., (2013). Targeting receptor-mediated endocytotic pathways with nanoparticles: Rationale and advances. *Adv. Drug Deliv. Rev. 65*(1), 121–138.

CHAPTER 8

Nanomedicines for the Treatment of Cardiovascular Disorders

SHIVANI VERMA[1,2], PUNEET UTREJA[2,3], MAHFOOZUR RAHMAN[4] and and LALIT KUMAR[2,3,*]

[1]*Department of Pharmaceutics, Rayat-Bahra College of Pharmacy, Hoshiarpur, Punjab 146001, India.*

[2]*I. K. Gujral Punjab Technical University, Jalandhar-Punjab 144601, India.*

[3]*Faculty of Pharmaceutical Sciences, Department of Pharmaceutics, PCTE Group of Institutes, Ludhiana, Punjab 142021, India.*

[4]*Department of Pharmaceutical Science, Faculty of Health Science, Shalom Institute of Health and Allied Sciences SHUATS-State University (Formerly Allahabad Agriculture Institute) Naini, Allahabad-211007, India.*

*Corresponding author. Tel: +91-7837-344360; Fax: +91-161-2888505; E-mail: lkpharma27gmail.com

ABSTRACT

There are a large number of diseases that are imaged and treated in the current scenario by using a novel approach based on nanotechnology, also called as "nanomedicines." Nanomedicines are basically various nanocarrier systems combined with either a therapeutic agent or an imaging agent implemented for treatment and diagnosis purpose respectively. Currently, nanomedicines have been explored for various abnormalities of the body like cancer, infections, and cardiovascular disorders. There are various pharmacotherapeutic agents that have been employed successfully for the elimination of cardiovascular disorders like thrombosis, atherosclerosis, and myocardial infarction (MI). However, these agents may show deleterious effects in the body of the patient after administration at a dose required to produce desired pharmacological effects. Therefore, various nanomedicines like liposomes, nanoparticles (NPs), dendrimers, and carbon nanotubes (CNTs) have been investigated by scientists to minimize these drawbacks and high mortality rates due to cardiovascular disorders. The present chapter will provide information about recent developments in the field of nano-medicine envisaged to effectively eradicate cardiovascular disorders like thrombosis, atherosclerosis, and MI.

8.1   INTRODUCTION

Cardiovascular disorders are one of the major causes of disability and death of individuals around the globe (Palella and Phair, 2011). Therefore, treatment and prevention of various cardiovascular disorders is still a challenging task in the medical field (Chowdhury et al., 2013). Nanomedicine has emerged as a highly advanced technology for the treatment of various disorders over the past decade (Cormode et al., 2009). Scientific literature reveals various useful aspects of nanomedicines like higher penetration to inflamed tissues and prolonged circulation times irrespective of their limited clinical use (Chung et al., 2015). Nanomedicines are categorized under biological nanomaterials class having a size below 100 nm; however, the higher vascular permeability is even observed up to the size of 400 nm (Tinkle et al., 2014). Nanomedicines have the capability to entrap high payloads of various therapeutic and diagnostic molecules, and their easy surface modification makes them excellent alternatives for treatment of various properties (Lammers et al., 2011). Surface modifiers used in nanomedicine research are peptides, aptamers, and antibodies (Kunjachan et al., 2012). These surface modifiers significantly affect the biodistribution behavior of the nanomedicines and enhance their localization in the target tissues or organs, improving therapeutic efficacy against various human disorders (Rizzo et al., 2013). Therefore, in this chapter, our major aim was to explore the utility of nanomedicine-based techniques for the treatment of cardiovascular disorders like thrombosis, atherosclerosis, and myocardial infarction (MI).

8.2   NANOMEDICINES AND THROMBOSIS

Thrombosis forms the basis of various cardiovascular defects like pulmonary embolism and stroke (Undas, 2017). Vascular thrombosis represents a major health problem in the Western region of the world. Vascular thrombosis may be responsible for approximately 50% of death cases caused due to cardiovascular diseases there (Kunamneni et al., 2007). Early detection and highly efficient thrombolytic therapy against vascular thrombosis may improve patient survival and reduce morbidity rate due to this problem (Vyas and Vaidya, 2009). Main therapeutic agents used against thrombosis are urokinase, streptokinase, and tissue plasminogen activator. These agents have the capability to convert plasminogen into the activated plasmin followed by initiation of fibrinolytic cascade due to the vascular circulation of activated plasmin that leads to thrombus dissolvation (Collen et al., 1988). Thrombolytic agents discussed earlier show many side effects in the body, reducing patient compliance (Bennett, 2001). Currently, targeting strategies like drug-eluting stent and trans-catheter method for thrombosis elimination reveal disadvantages like uncontrolled drug concentration and early washout of the drug (Coutre and Leung, 1995). Therefore, there is a requirement of delivery system which shows early localization of bioactive molecules in thrombus area and limited effect in normal tissues (De Meyer et al., 2008). Surface modified nanocarriers systems loaded with thrombolytic drugs may be an alternative to achieve these needs efficient therapy (Figure 8.1) (Wadajkar et al., 2013). Table 8.1 enlists various glycoprotein receptors, which have a vital role in adhesion and activation of platelets for a thrombus generation (Vyas and Vaidya, 2009). Various nanocarriers explored for the treatment of thrombosis are liposomes, polymeric nanoparticles (NPs), lipidic nanoparticles, and dendrimers.

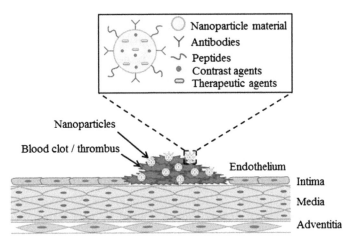

FIGURE 8.1 **(See color insert.)** Multifunctional nanoparticle system with target specific, diagnostic, and therapeutic capabilities for deep vein thrombosis (DVT) management (From Wadajkar et al., 2013, With permission © 2013 Elsevier).

TABLE 8.1   List of Various Glycoprotein Receptors Involved in Platelet Adhesion and Activation

Sr. No.	Receptor		Corresponding Ligand
1.	Integrins	GP IIb/IIIa (αIIbβ3)	Fibrinogen, vitronectin, fibronectin
		GP Ia/IIa (α2β1)	Collagen
		GP Ic/IIa (α5β1)	Fibronectin
		VnR (αvβ3)	Vitronectin, fibrinogen
2.	Leucin-rich protein	GP Ib/IX	vWF, Thrombin

8.2.1   LIPOSOMES AS EFFECTIVE NANOMEDICINES FOR TREATMENT OF THROMBOSIS

Liposomes are vesicular structures having a spherical shape formed due to self-assembling of lipid materials, and they have the capability to encapsulate various thrombolytic drugs (Koudelka et al., 2016). Liposomes can be manufactured by implementing simple techniques, and ease of surface modifications make them suitable for various therapeutic applications (Ding et al., 2006). Furthermore, efficacy of liposomes can be enhanced through steric stabilization using poly(ethylene)-glycol (PEG) coating which can modify their pharmacokinetics behavior (Brochu et al., 2004). PEG-coated liposomes cannot be cleared by the mononuclear phagocyte system (MPS), and hence, there is a prolonged blood circulation time of these carriers (Bakker-Woudenberg, 2002). Various targets in thrombosis like blood cells, clot, or endothelial cells can be easily targeted by using intravenous administration of the liposomes (Tan et al., 2010). Ligand conjugated liposomes represent an active targeting approach for thrombus targeting as these modified liposomal carriers have the capability to recognize various target receptors present on thrombus (Absar et al., 2013). Anti-platelet antibodies, anti-fibrin antibodies, and Arg-Gly-Asp (RGD) motifs are examples of ligand used for thrombus targeting (Crommelin et al., 1995). So, there are mainly three liposomal classes namely- simple liposomes, sterically stabilized liposomes (long-circulating liposomes), and ligand appended (targeted liposomes) liposomes used for thrombosis treatment (Koudelka et al., 2016). Nguyen et al., (1989) evaluated *in-vitro* thrombolytic efficacy of streptokinase loaded in conventional liposomes made up of palmitoyl-oleoyl phosphatidylcholine. Liposomal formulation caused a reduction in clot dissolving time up to 16 minutes, which was very low compared to free streptokinase (24 minutes) at the 37°C in the presence of poor platelet plasma. This indicated better thrombolytic efficacy of the streptokinase loaded liposomal formulation compared to free streptokinase. *In-vivo* efficacy of streptokinase loaded stealth (PEG conjugated) liposomes made up of distearoylphosphatidylcholine was evaluated by Kim et al., (1988) in rats. PEGylated liposomes loaded with streptokinase showed 16.3 fold increased the half-life and 6.1 fold increased area under curve infinity of streptokinase compared to free form in rats a dose of 15,000 IU/kg. Furthermore, Vaidya et al., (2011) investigated dioleoyl phosphatidyl ethanolamine and dipalmityol-c[RGD (Arg-Gly-Asp)fK] containing liposomes loaded with streptokinase for targeted drug delivery and thrombolytic activity *in-vitro*. These target oriented liposomes showed high *in-vitro* activity by dissolving 50% clot in a period of 30 minutes compared to free streptokinase which dissolved the same clot percentage in a 55 minute period. Fluorescence microscopic analysis revealed an efficient binding of c[RGD (Arg-Gly-Asp) fK] containing liposomes to the activated platelets compared to other liposomal formulation (Figure 8.2).

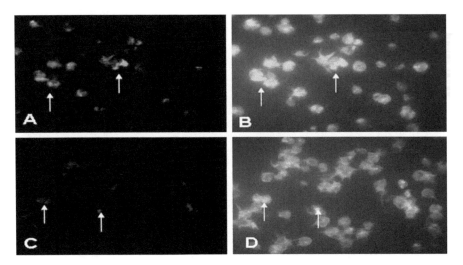

FIGURE 8.2 Microscopic observation of the platelets incubated with fluorescein–5-Isothiocyanate (FITC) labeled RGD- (A) and RAD-liposomes (C) under a fluorescence microscope. The same field was also observed under phase contrast mode (B) and (D), respectively. Figures show that RGD liposomes bind more efficiently to the activated platelets as compared to RAD liposomes (From Vaidya © 2011 Elsevier. et al., 2011, With permission).

Table 8.2 describes liposomes as effective nanomedicines for the treatment of thrombosis.

TABLE 8.2 Role of Liposomes in Treatment of Thrombosis

Composition of Liposomes	Drug	Entrapment/size/ zeta potential	Animal model/ Route of Administration	Key Findings	References
1, 2-distearoyl-*sn*-glycero-3-phosphocholine, cholesterol, RGD (Arg-Gly-Asp) motifs	Eptifibatide (integrilin®)	$37 \pm 5\%$ / 90 ± 10 nm / —	Sprague-Dawley rats/intravenous	RGD-coupled eptifibatide loaded nano-liposomes showed no cytotoxic effect on normal cells *in-vitro* and higher antiplate-let activity (81.63%) *in-vivo* compared to simple liposomes (66.67%) and free liposomes (46.17%)	Bardania et al., 2017
Dioleyl phosphatidylethanol-amine, c(RGD) peptide [CNPRGDY(OEt)RC]	Streptokinase	18% / 100–120 nm /—	Wistar rats/ intravenous	Intravital microscopy revealed higher accumulation of target-oriented liposomes in thrombus and their clot dissolving capacity was very high ($28.27 \pm 1.56\%$) compared to free streptokinase ($17.18 \pm 1.23\%$) *in-vivo*	Vaidya et al., 2016
Dipalmitoylphosphatidyl-choline, dioleoylphospha-tidylcholine, dipalmitoyl phosphatidylglycerol, cholesterol	Alteplase	—	New Zealand white rabbits/ intra-arterial	Developed echogenic liposomes showed efficient clot dissolving efficacy on exposure to color Doppler US (6MHz) for approximately half an hour compared to conventional liposomal system	Laing et al., 2012
Distearoyl-sn-glycero–3-phospho ethanolamine, poly(ethylene) glycol–2000, cholesterol	Low molecular weight heparin (LMWH)	$34 \pm 13\%$ to $47 \pm 4\%$ / 116–133 nm / —	Sprague –Dawley rats/ intratracheal or subcutaneous	Long-circulating liposomes loaded with LMWH administered through intratra-cheal route at a dose of 100 U/kg showed a same thrombolytic effect in-vivo as shown by 50 U/kg LMWH administered subcutaneously	Bai and Ahsan, 2010

TABLE 8.2 *(Continued)*

Composition of Liposomes	Drug	Entrapment/size/ zeta potential	Animal model/ Route of Administration	Key Findings	References
Dioleyl phosphatidylethanol-amine, Arg-Gly-Asp-Ser (RGDS), cholesterol	Subtilisin FS33	— / 111 ± 7 nm / —	Wistar rats/ intravenous	RGDS coupled liposomes showed very high antithrombotic activity at a dose of 4000 U/kg after 120 minutes of infusing *in-vivo* compared to the free Subtilisin FS33 at the same dose	Wang et al., 2010
Distearoyl-sn-glycero—3-phospho ethanolamine, cholesterol	Streptokinase	— / 108 ± 5 nm / —	Rabbits/ intravenous	Liposomes showed significantly higher thrombolytic efficacy (33.8 ± 1.5%) compared to free streptokinase (29.3 ± 2.1%) *in-vivo* when administered at a dose of 40,000 U/kg	Perkins et al., 1997

8.2.2   ROLE OF POLYMERIC NANOPARTICLES (NPS) IN TREATMENT OF THROMBOSIS

Polymeric nanoparticles (NPs) has emerged as potential candidates for early detection and treatment of various cardiovascular disorders (Agyare and Kandimalla, 2014). Polymeric nanoparticles are categorized under colloidal carriers having a particle size in 1–500 nm range used for therapeutic delivery of various bioactive molecules (Zhao et al., 2016). Polymeric nanoparticles have a very high surface area which makes them suitable for coupling with any targeting probe or ligand and their sufficient internal volume make them appropriate for entrapping various bioactive and imaging molecules (Park et al., 2005). Polymeric nanoparticles (NPs) coupled with target ligands have the capability to bind with endothelial cell receptors, which promotes their intracellular localization following specific endocytic pathways (Ding et al., 2006). Chitosan (CS), poly(vinyl alcohol), polylactic-co-glycolic acid (PLGA), and poly(l-lactic acid) (PLLA) are the examples of polymers utilized for nanoparticles preparation to deliver thrombolytic drugs (Chacko et al., 2011). Jogala et al., (2016) evaluated PEG-PLGA nanoparticles loaded with low molecular weight heparin (LMWH) for *in-vivo* thrombolytic effect in a rat model. The developed nanoparticulate system showed sustained release of LMWH for two days *in-vitro*, higher *in-vivo* thrombolytic effect, and reduced cellular toxicity compared to free LMWH. Li et al., (2009) performed platelet compatibility studies of CS, PLGA, and CS-PLGA complex nanoparticles. All developed nanoparticles showed high compatibility towards platelets at a concentration below 10 μg/ml. This result revealed them as efficient and nontoxic carriers for loading of various cardiovascular agents. Furthermore, Chen et al., (2011) prepared Actilyse® loaded CS coated magnetic nanoparticles and evaluated them in a rat embolic model. The developed nanoparticulate system loaded with Actilyse® showed a 58% reduction in blood clot lysis time *in-vivo* under the magnetic influence which was very higher compared to free Actilyse® administered at the same dose. CS magnetic nanoparticles loaded with Actilyse® showed increased hind limb perfusion up to 77 ± 15% under the magnetic effect for the period of two hours, which was very high compared to vehicle (39 ± 7%) (Figure 8.3). A literature review regarding use of polymeric nanoparticles for treatment of thrombosis is given in Table 8.3.

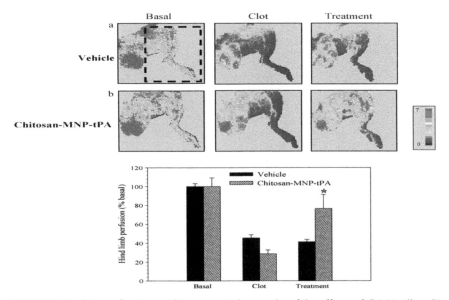

FIGURE 8.3 **(See color insert.)** Representative results of the effects of tPA (Actilyse®) immobilized to the chitosan-coated magnetic nanoparticle (chitosan-MNP–tPA) on tissue perfusion in a rat iliac embolic model. Tissue perfusion of hind limb was measured with a laser Doppler perfusion imager. Five minutes after clot lodging, vehicle (a) or tPA covalently bound to chitosan-coated magnetic particle C250 (b) was administered from the right iliac artery under magnet guidance and treated for 120 min. Laser Doppler signals in designated areas, as illustrated with the square in (a), were acquired for quantitative analysis of hindlimb perfusion. *p < 0.05 compared with vehicle. (From Chen et al., 2011, With permission. © 2011 Elsevier.).

8.2.3 *LIPIDIC NANOPARTICLES AS EFFECTIVE NANOMEDICINES FOR TREATMENT OF THROMBOSIS*

Lipidic nanoparticles used for the treatment of various diseases include solid lipid nanoparticles (SLNs) and nanostructured lipid carriers (NLCs) (Naseri et al., 2015). SLNs and NLCs were introduced during 1990 as substitutes for various carrier systems like liposomes and polymeric nanoparticles (Patidar et al., 2010). SLNs have a spherical shape with a size range 40–1000 nm which can be confirmed with advanced techniques like transmission electron microscopy (TEM) and scanning electron microscopy (SEM) (Thatipamula et al., 2011). Dispersion of solid fat in a concentration range 0.1–30% (w/w) in a medium aqueous result in the formation of the SLNs (Blasi et al., 2007). Addition of surfactants in a concentration range

TABLE 8.3 Polymeric Nanoparticles as Effective Nanomedicines for Treatment of Thrombosis

Composition of nanoparticles	Drug	Entrapment/size / zeta potential	Animal model/Route of Administration	Key Findings	References
Chitosan, PLGA, GRGD peptide (Gly-Arg-Gly-Asp)	Actilyse (Recombinant human tissue-type Plasminogen activator)	67.33 ± 1.21%/ 320.1 ± 5.7 nm/ 5.18 ± 1.42 mV	—	Chitosan-PLGA nanoparticles showed lowest clot lyses time of 20.7± 0.7 min and chitosan-PLGA- GRGD conjugated nanoparticles showed the maximum percentage of weight of digested clots (25.7± 1.3 weight%) *in-vitro* which indicated their better efficacy compared to free actilyse	Chung et al., 2008
PLGA	Streptokinase	86%/ 37 ± 12 nm / −43.6 ± 2.3 mV	—	Developed PLGA nanoparticles showed sustained release of streptokinase up to three days and higher stability with less enzyme leakage for up to 90 days	Yaghoobi et al., 2017
Poly (ε-caprolactone) (PCL)	Enoxaparin	76.98 ± 3.45%/ 112 ± 5 nm /—	Wistar rats/ subcutaneous	PCL nanoparticles loaded with enoxaparin showed significant thrombolytic activity and sustained release of the drug up to 12 hours with C_{max} 0.62 IU/ml *in-vivo* when administered at a dose of 1000 IU/kg	Pazzini et al., 2015
PLGA, GRGD peptide (Gly-Arg-Gly-Asp), chitosan	Gadolinium	49.5 ± 11.5%/ 893 ± 63.7 nm / 7.2 ± 3.8 mV	—	Developed PLGA-chitosan/ GRGD complex nanoparticles loaded with gadolinium showed *in-vitro* accumulation in a higher amount in thrombus as revealed through clinical magnetic resonance scanner	Zhang et al., 2013
Chitosan, sodium alginate	Enoxaparin sodium	70.6 ± 4.2% / 213 ± 3.8 nm / +39.3 mV	Sprague Dawley rats/oral	Alginate coated chitosan core-shell nanoparticles loaded with enoxaparin sodium showed three-fold increased oral bioavailability, and 60% reduced formation of thrombus in experimental animal model compared to free enoxaparin sodium	Bagre et al., 2013

0.5–5% may enhance their stability (Khatak and Dureja, 2015). Various lipids used in the manufacturing of SLNs are monoglycerides, triglycerides, waxes, and fatty acids. Encapsulation of both hydrophilic and lipophilic drugs can be carried out in SLNs (Pardeshi et al., 2012). Low production cost, ease of manufacturing, biodegradable nature, and high physical stability are some properties of SLNs which make them good carriers for drug delivery (Feng and Mumper, 2013). SLNs may show disadvantages like growth of lipid particles, gelation tendency, and low loading capacity (Yoon et al., 2013). Furthermore, NLCs are categorized under modified SLNs containing both solid and liquid lipids as parts of lipidic phase (Jaiswal et al., 2016). Implementation of solid or liquid lipids both in the NLCs generates a formless matrix which is having the very high loading capacity and stability (Iqbal et al., 2012). NLCs show distinct advantages like high loading capacity, less drug leakage, and less water content in the dispersion over the SLNs (Shidhaye et al., 2008). Paliwal et al., (2011) evaluated LMWH-lipid conjugate loaded SLNs for enhancement of oral bioavailability of LMWH to treat embolism and deep vein thrombosis. The Developed SLNs system was found non-toxic to tissues of intestinal epithelium as revealed in histological evaluation of the gastrointestinal tract (GIT). The results of anti-FXa chromogenic assay showed significantly higher plasma concentration of LMWH after oral administration of LMWH-lipid conjugate loaded SLNs compared to simple SLNs and free LMWH (Figure 8.4). Later on, Jain et al., (2013) developed NLCs loaded with enoxaparin for the transdermal delivery to treat thrombosis. Developed lipidic formulation showed very high skin penetration and reduced skin irritation compared to free enoxaparin. NLCs loaded with enoxaparin were found stable for up to two months at room temperature as they showed minimal drug leakage during this time period.

8.2.4 ROLE OF DENDRIMERS IN TREATMENT OF THROMBOSIS

Dendrimers are novel carrier systems which consist of a spherical polymer structure having multiple branches. The term 'dendrimer' is derived from Greek word 'Dendron' which means 'trees,' and these dendrimers have some unique properties which make them efficient drug vehicles (Yu et al., 2015). Dendrimers have three distinct parts in their structure, namely- focal core, building blocks, and peripheral functional groups. Focal core is composed of a single atom or an atomic group while several repeating units originating from the core form building blocks (Yang and Kao, 2006). Various properties of dendrimers like well-defined molecular weight reduced polydispersity, and functional variation makes them excellent choice for drug delivery in medical field (Bai et al., 2006).

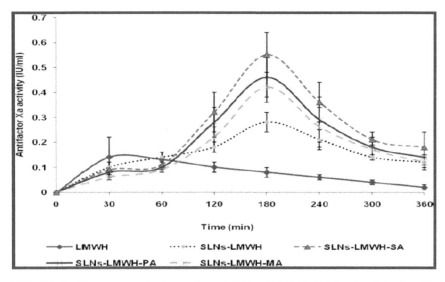

FIGURE 8.4   Plasma concentration profiles of different LMWH- lipid conjugate loaded SLN formulations after oral administration. Data is presented as mean ± SD (n = 4). LMWH-PA (PA = PALMITIC ACID), LMWH-SA (SA = stearic acid), LMWH-MA (MA = myristic acid) (From Paliwal et al., 2011, With permission. © 2011 American Chemical Society.).

Dendrimers are capable to encapsulate various drugs in the void spaces present in their structure or drugs can also be conjugated to various functional groups present in the outer region (Cheng et al., 2008). Dendrimers can enhance the oral absorption, solubility, and bioavailability of various therapeutic agents along with reduction of their side effects (Jain and Asthana, 2007). Bai et al. (2007) evaluated enoxaparin conjugated polyamidoamine (PAMAM) positively charged dendrimers for the treatment of deep vein thrombosis through pulmonary route in rats. Developed positively charged dendrimers showed 40% enhancement in relative bioavailability of enoxaparin which was very high compared to the neutral dendrimeric system. Positively charged dendrimer showed effective reduction in deep vein thrombosis *in-vivo* with least side effects on the lungs. Later on, Bai and Ahsan (2009) developed PEGylated PAMAM dendrimers conjugated LMWH and evaluated them in rats after pulmonary administration. Developed PEGylated dendrimers showed 40% LMWH entrapment, and 60.6%

increased relative bioavailability of LMWH compared to subcutaneously administered LMWH. The thrombolytic activity of PEGylated dendrimers conjugated to LMWH and administered at 48-hour intervals was equivalent to LMWH administered subcutaneously at 24-hour intervals (Figure 8.5).

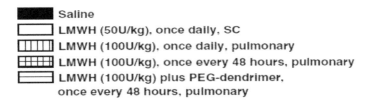

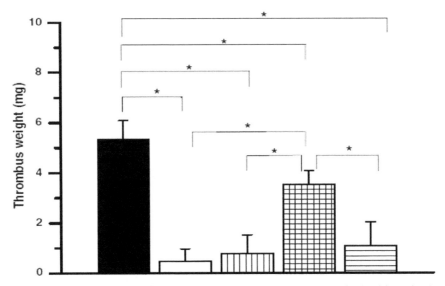

FIGURE 8.5 Efficacy of LMWH plus mPEG [poly(ethylene)-glycol]–dendrimer in the treatment of jugular vein thrombosis. Data represent mean ± SD, $n = 3$–5. *Results are significantly different, $p < 0.05$. (From Bai and Ahsan, 2009, With permission. © 2008 Springer.).

Furthermore, Durán-Lara et al., (2013) performed derivatization of PAMAM G4 dendrimers and evaluated them for hemotoxicity and cytotoxicity. Derivatives like PAMAM G4-Arginine-Tos and G4-Lysine-Cbz were developed, and both of them showed significant inhibition of platelet aggregation. However, PAMAM G4-Arginine-Tos derivative showed better hemotoxicity and cytotoxicity *in-vitro* compared to PAMAM G4-Lysine-Cbz derivative.

8.3   NANOMEDICINES AND ATHEROSCLEROSIS

Atherosclerosis is one of the most common cardiovascular disorder which originates from improper lipid metabolism leading to inflammation of the arterial wall. The maladaptive inflammatory response is also considered as a cause of generation of atherosclerosis (Schiener et al., 2014). Atherosclerosis complications are associated with levels of low-density lipoprotein (LDL) which has a great influence on various cardiovascular events. LDL promotes arterial leukocyte infiltration by the enhancement of endothelial cell adhesion expression (Weber and Noels, 2011). Atherosclerosis stages are recognized through the leukocyte accumulation in the arterial wall. Inflammatory cell adhesion on the arterial wall occurs through various adhesion molecules like vascular cell adhesion molecule (VCAM1), intercellular adhesion molecule (ICAM1), and P-selectin (Zernecke and Weber, 2005). Furthermore, neutrophils and monocytes accumulation on arterial wall involves the integral role of receptors like CC-chemokine receptor 1 and CXC-chemokine receptor 2 (Zernecke et al., 2008). At the end stage of atherosclerosis, excessive proliferation of plaque resident macrophages occurs which promote leukocyte accumulation in higher content on the wall of arteries leading to the formation of atherosclerotic lesions (Andrés et al., 2012). Furthermore, macrophages present in lesions engulf the modified lipoproteins through the activation of scavenger receptors, producing a new cellular structure called foam cells (de Vries and Quax, 2016). Figure 8.6 describes the various phases of development of atherosclerosis (Psarros et al., 2012). Being a multifactorial disease different approaches have been implemented for its treatment and prevention. Therefore, nanomedicines based strategies have been explored to target specific cells or tissues to treat atherosclerosis (Antoniades et al., 2010). Various targeting strategies for atherosclerosis treatment using nanomedicines is explained in Figure 8.7 (Lobatto et al., 2011).

8.3.1   ROLE OF LIPOSOMES AS NANOMEDICINES IN TREATING ATHEROSCLEROSIS

Prednisolone loaded liposomes were evaluated by Vander Valk et al., (2015) for their accumulation in atherosclerotic macrophages following intravenous administration in humans. Liposomal formulation, improved pharmacokinetic behavior of drug through increased half-life up to 63 hours when administered at a dose of 1.5 mg/kg. Macrophages separated from iliofemoral plaques of patients showed presence of 75% prednisolone of the administered dose without producing arterial wall inflammation. Later on, Miao et al., (2015) evaluated liposomal system containing rapamycin *in-vitro* for effective targeting of atherosclerotic plaques. The study revealed formation of the optimized formulation at phospholipid and cholesterol content at the ratio 8:1 with very high encapsulation efficiency of 82.11 ± 2.13%. This formulation sustained the drug release up to 30 hours following first order kinetics. PEGylated liposomes loaded with dexamethasone were investigated by Bartneck et al., (2014) to modify migration characteristics of human macrophages *in-vitro* to treat athero-sclerosis. Developed liposomal system targeted macrophages *in-vitro* in a dose-dependent manner with a significant reduction in toxicity against them. PEGylated liposomes also reduced migration of macrophages and monocytes *in-vitro*, which could be beneficial to block atherosclerotic plaque formation (Figure 8.8). Table 8.4 gives overview of liposomes for the effective treatment of atherosclerosis.

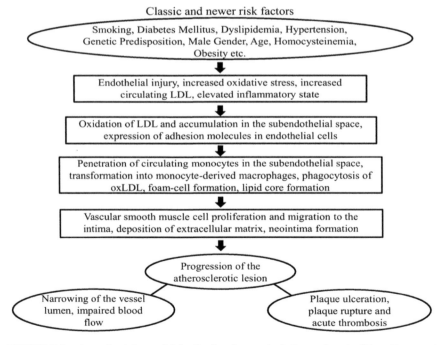

FIGURE 8.6 A mechanistic model for the development of atherosclerosis. (From Psarros et al., 2012, With permission. © 2012 Elsevier.).

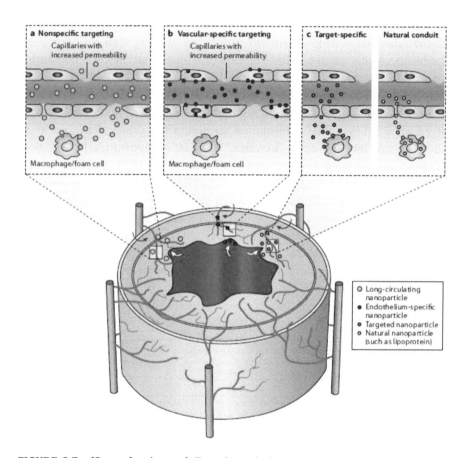

FIGURE 8.7 **(See color insert.)** Targeting principles in atherosclerotic plaques. The vessel walls of larger arteries are supplied with nutrients by the lumen and the vasa vasorum—a network of small microvessels. In the lesioned vessel wall the vasa vasorum undergoes angiogenic expansion, with neovessels reaching into the base of the plaque, which is accompanied by the upregulation of cell-surface receptors and increased permeability of the endothelium. The upregulation of receptors and the increased permeability also affect the endothelium on the luminal side of the plaque. The main targeting principles can be classified into nonspecific targeting of the plaque (part a), specific targeting of the vasculature (part b) and specific targeting of components (part c) of the plaque (for example, the extracellular matrix or macrophages) with either synthetic nanoparticles or via interaction through a natural conduit. The targeting of the plaque occurs via both the vasa vasorum and the main lumen at lesioned sites and is exemplified on the figure with corresponding arrows. Depending on the targeting principle applied, the cellular distribution of nanoparticles in the plaque will vary considerably. (From Lobatto et al., 2011, With permission. © 2011 Nature Publishing Group.).

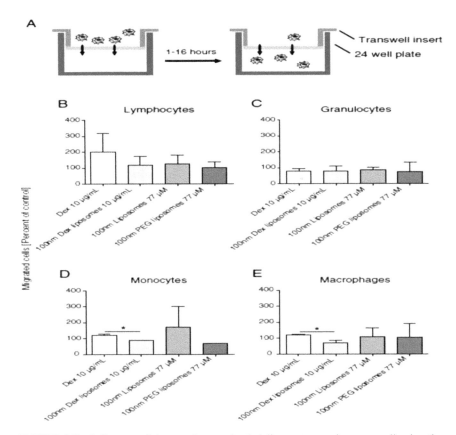

FIGURE 8.8 Influence of dexamethasone loaded liposomes on immune cell migration. Schematic depiction of the migration assay principle before (A) and after the experiment (B). Human primary blood leukocytes represented by lymphocytes (C), granulocytes (D), and monocytes (E) were isolated from heparinized blood of healthy volunteers and purified using dextran sedimentation. Human primary macrophages were generated by seven days of monocyte culture in Roswell Park Memorial Institute (RPMI1640) medium with 5% of autologous serum (F). Prior to the migration experiment, cells were incubated for one hour with the formulations in RPMI1640 medium with 5% fetal bovine serum (FCS). Blood cells were left to migrate for one hour, whereas macrophages were left to migrate for 16 h in RPMI1640 medium with 1% FCS. Mean data of n = 4. *P b 0.05, **P b 0.005, ***P b 0.001 (one way ANOVA). (From Bartneck et al., 2014, With permission. © 2014 Elsevier.).

8.3.2 POLYMERIC NANOPARTICLES AS EFFECTIVE NANOMEDICINES FOR TREATMENT OF ATHEROSCLEROSIS

D-Threo–1-phenyl–2-decanoylamino–3-morpholino–1-propanol loaded poly (ethyleneglycol)sebacicacidcopolymernanoparticleswereevaluatedbyMishra

TABLE 8.4   Liposomes as Nanomedicines in Treatment of Atherosclerosis

Composition of Liposomes	Drug	Entrapment/size / zeta potential	Animal model/ Route of Administration	Key Findings	References
N-(carbonyl-methoxy-PEG2000)–1,2-distearo-yl-sn-glycero-3-phospho-ethanolamine, cholesterol, methoxy poly(ethylene glycol)-b-(N-(2-benzo-yloxy propyl) methacry-lamide (for making poly-meric micelles)	Simvastatin	Liposomes: 71 ± 3% / 94 ± 10 nm / −15.1 ± 0.7 mV Polymeric micelles: 65 ± 8% / 80 ± 7 nm / −5.1 ± 0.9 mV	Apoe−/− mice/ intravenous	Polymeric micelles loaded with simvastatin showed higher uptake in plaque macrophages and longer circulation time *in-vivo* compared liposomes as revealed in pharmacokinetics and biodistribution using radioisotopes	Alaarg et al., 2017
1,2-Dipalmitoyl-sn-glyce-ro-3-phosphocholine, N-(carbonyl-methoxy-PEG2,000)–1,2-distearo-yl-snglycero-3-phospho ethanolamine, cholesterol	Docosahexaenoic acid	81.35 ± 3.24% / 99 ± 16 nm / −15.7 ± 2.5 mV	—	Developed liposomes showed reduction in generation of cytokines like TNFα along with reduced nitrogen and oxygen release from murine macrophages *in-vitro* indicating their effectiveness to treat inflammatory disease like atherosclerosis	Alaarg et al., 2016
Phosphatidylserine, cholesterol	—	— / 110 nm / —	ApoE-KO mice/ intraperitoneal	Prepared liposomes reduced level of inflammatory cytokines *in-vivo* by promoting immediate activation of B1a cells followed by enhancement of phagocytosis which could be beneficial to treat atherosclerosis	Hosseini et al., 2015

TABLE 8.4 *(Continued)*

Composition of Liposomes	Drug	Entrapment/size / zeta potential	Animal model/ Route of Administration	Key Findings	References
L-α-phosphatidyl choline, 1,2-dipalmitoyl-sn-glycero–3-phospho choline, 1,2-dipalmitoyl-sn-glycero–3 [phosphor-rac–1-glycerol], 1,2-dipalmitoyl-sn-glycero–3-phospho ethanol-amine, cholesterol	Nitric oxide	—	Yucatan Miniswine/ intra-arterial	Echogenic immunoliposomes loaded with nitric oxide promoted the delivery of anti-intercellular adhesion molecule–1 complexed echogenic immunoliposomes to arterial wall under the influence of ultrasound to improve the imaging effect	Kim et al., 2013
1-palmitoyl–2-oleo-yl-sn-glycero–3-Phosphocholine, 1,2-distearoyl-sn-glycero–3-phosphoethanolamine-N-[methoxy (polyethylene glycol)–2000], cholesterol	—	— / 183.9 ± 0.8 nm / −10.2 ± 0.2 mV	Balb/c Mice/ intravenous	Long-circulating liposomes coupled with interleukin–10 showed very high fluorescence intensity in the atherosclerotic plaque area in comparison to free interleukin–10 and simple liposomes as revealed in confocal laser-scanning microscopic analysis	Almer et al., 2013
Dipalmitoylphosphatidylcholine, dipalmitoyl phosphatidyl glycerol, 4-(p-maleimidolphenyl) butyrate phosphatidyleth-anolamine, cholesterol	Calcein	15% / 98 ± 4 nm / —	Yucatan miniswine/ intra-arterial	Echogenic immunoliposomes liposomes loaded with calcein showed more than 300% increase in the arterial uptake of calcein on the application of ultrasound of 1 MHz *in-vivo* compared to conventional liposomal system indicating their enhanced targeting effect	Laing et al., 2010

et al., (2015) in apoE-/- mice for the treatment of atherosclerosis and cardiac hypertrophy. Developed nanoparticulate system enhanced *in-vivo* residence time of D-Threo–1-phenyl–2-decanoylamino–3-morpholino–1-propanol up to 48 hours along with improved gastrointestinal absorption and efficient destruction of atherosclerosis plaque compared to free D-threo–1-phenyl–2-decanoylamino–3-morpholino–1-propanol. Furthermore, hyaluronic acid nanoparticles were investigated for active targeting of atherosclerotic plaque by Lee et al., (2015) and compared with hydrophobically modified glycol CS nanoparticles. Cells showing overexpression of stabilin–2 or CD44 showed high *in-vitro* uptake of hyaluronic acid nanoparticles compared to hydrophobically modified glycol CS nanoparticles. Results of *in-vivo* fluorescence imaging analysis revealed better accumulation of hyaluronic acid nanoparticles in atherosclerotic lesion compared to hydrophobically modified glycol CS nanoparticles (Figure 8.9). Later on, Sanchez-Gaytan et al. (2015) evaluated high-density lipoprotein–mimetic PLGA nanoparticles for the treatment of atherosclerosis through plaque macrophage targeting. The developed nanoparticulate system showed very high uptake in macrophages, excellent cholesterol efflux capacity, and prolonged accumulation in atherosclerotic plaque *in-vivo* in ApoE knockout mouse model. Table 8.5 describes the utility of polymeric nanoparticles to treat atherosclerosis.

8.3.3   ROLE OF OTHER NANOMEDICINES IN TREATMENT OF ATHEROSCLEROSIS

Other nanocarriers which may be explored for the treatment of atherosclerosis are carbon nanotubes (CNTs), dendrimers, SLNs, and gold nanoparticles (GNPs). CNTs are novel drug carrier systems which consist of thin plates/sheets of benzene ring carbons piled up into a smooth tubular structure (Ji et al., 2010). They were firstly described by Iijima in the year 1991 (Iijima, 1991). They come under the class fullerenes, which are recognized as a third allotropic form of carbon Sheikhpour et al., 2017). Three techniques, namely chemical vapor deposition (CVD) through thermal effect, electric discharge method, and laser ablation technique are implemented for the production of CNTs (Wong et al., 2013). If CNTs contain a single sheet of graphene in their structure, then they are called single-walled carbon nanotubes (SWCNTs), and if they contain a large number of concentric graphene sheets, then the name given is multi-walled carbon nanotubes (MWCNTs) (Gong et al., 2013). Various properties of CNTs like good mechanical strength, lightweight, excellent aspect ratio, good electrical and thermal conductivity make the effective nanocarriers for delivery of various bioactive molecules (Liang and Chen, 2010). Xu et al., (2012) evaluated antiatherosclerotic effect of MWCNTs in Sprague-Dawley (SD) rats. MWCNTs administered at a dose of 200 µg/kg showed increased aorta calcification in animal model indicating their efficacy to reduce the formation of atherosclerotic plaque. Furthermore, Lu et al., (2011) investigated *in-vitro* adsorption capacity of sulphonated porous CNTs -activated carbon composite (S-CNTs/AC) beads to adsorb LDL which plays an important role in atherosclerosis generation. The results predicted electrostatic interaction between LDL and S-CNTs/AC beads. S-CNTs/AC beads containing CNTs at concentration of 45 wt% showed maximum *in-vitro* adsorption of LDL (Figure 8.10).

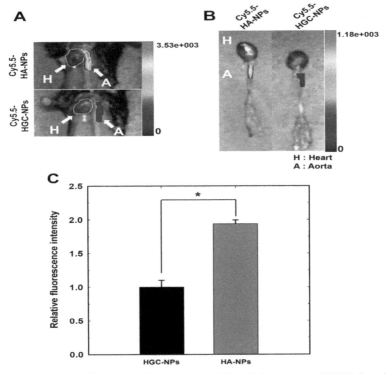

FIGURE 8.9 (See color insert.) *In-vivo* near-infrared fluorescence (NIRF) imaging of atherosclerotic lesion using cyanine (Cy) 5.5 - labeled hyaluronic acid nanoparticles (HA-NPs) and Cy5.5 - labeled hydrophobically modified glycol chitosan nanoparticles (HGC-NPs). (A) Fluorescence images of atherosclerotic lesion in ApoE KO mice. (B) Fluorescence images of isolated aorta from ApoE KO mice. (C) Relative fluorescence intensity of atherosclerotic lesion in (B). Asterisk (*) indicates difference at the $p < 0.05$ significance level. (From Lee et al., 2015, With permission. © 2015 Elsevier.).

TABLE 8.5 Role of Polymeric Nanoparticles in Effective Elimination of Atherosclerosis

Composition of nanoparticles	Drug	Entrapment/size / zeta potential	Animal model/ Route of Administration	Key Findings	References
Poly (lactic-co-glycolic acid) (PLGA), hyaluronan (HA), Reconstituted high-density lipoprotein (rHDL)	Simvastatin	90.64 ± 0.43% / 138.2 ± 2.3 nm / −28.38 ± 0.52 mV	New Zealand White rabbits/ intravenous	PLGA-HA-rHDL complex nanoparticles loaded with simvastatin 2.43 fold higher efflux capacity for cholesterol *in-vitro* and better accumulation in atherosclerotic plaque *in-vivo* compared to the conventional reconstituted high-density lipoprotein	Zhang et al., 2017
Polyaspartic acid (PAA), chitosan (CS),	(-)-Epigallocatechin gallate	25.0 ± 2.1% / 102.4 ± 5.6 nm / 33.3 ± 0.2 mV	New Zealand white rabbits/oral	Epigallocatechin gallate encapsulated PAA-CS nanoparticles showed 16.9 ± 5.8% lipid deposition ratio in animal model which was comparable to orally administered simvastatin (15.6 ± 4.1%) indicating their efficient behavior	Hong et al., 2014
PLGA, polyethyleneglycol (PEG), collagen IV	Annexin A1/ lipocortin 1-mimetic peptide (Ac2–26)	90% / 77.15 ± 1.1 nm / −15.49 ± 0.84 mV	C57BL/6J mice/ intravenous	PLGA-PEG- collagen IV nanoparticles loaded with Ac2–26 showed blockage of tissue damage up to thirty percent in ischemia-reperfusion injury area *in-vivo* and found more effective than free Ac2–26	Kamaly et al., 2013
Chitosan	Plasmid Cholesteryl ester transfer protein (pCETP)	95.2 ± 0.7% / 340.2 ± 14.6 nm / 22.9 ± 1.3 mV	New Zealand white rabbits/ intranasal	Chitosan nanoparticles containing plasmid cholesteryl ester transfer protein administered through intranasal route showed 59.2% reduction in aortic lesions *in-vivo* which was very high compared to saline (29.0%) and comparable to pCETP solution (71.0%) taken through intramuscular route	Yuan et al., 2008

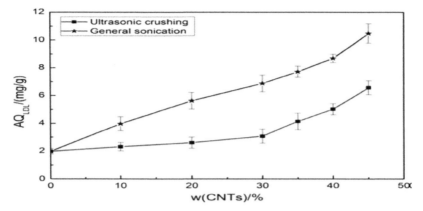

FIGURE 8.10 Adsorption amount of LDL [AQ$_{LDL}$/(mg/g)] on the S-CNTs/AC beads with different CNT contents and by different CNT dispersion methods. (From Lu et al., 2011, With permission. © 2011 Springer.).

GNPs are novel colloidal nanomedicines which are implemented for the treatment of various diseases like cancer (Haume et al., 2016). GNPs were firstly developed by Faraday in 1857 through phosphorous mediated reduction of gold chloride followed by stabilization with carbon disulfide (Faraday, 1857). GNPs are recently used in various forms like gold nanorods, nanospheres, nanoshells, and nanostars (Pietro et al., 2016). Ease of surface modification of GNPs makes them efficient targeting nanomedicines (Dreaden et al., 2012). GNPs have specific optical properties like scattering and light absorption at a wavelength of 650–900 nm (near infrared region). When exposed to NIR wavelength GNPs produce heat through surface Plasmon resonance effect. This generation of hyperthermic effect in GNPs may be useful in anticancer therapy (Arvizo et al., 2010). Recently, Rizwan et al., (2017) evaluated the capability of GNPs to reduce high glucose-induced oxidative-nitrosative stress-regulated inflammation *in-vitro* for the treatment of atherosclerosis. The results of the study revealed blockage of NF-κB activation through ERK1/2MAPK/Akt/tuberin-mTOR pathways regulated inflammatory gene expression by implementation of GNPs which may be beneficial to prevent atherosclerosis. In one more study, de Oliveira Gonçalves et al., (2015) investigated 5-aminolevulinic acid (ALA) conjugated GNPs functioned with PEG for effective treatment of atherosclerosis in white New Zealand rabbit model. Results of study showed that 5-ALA retained its structure in nanoparticles after *in-vivo* administration. Furthermore, 5-ALA got converted into protoporphyrin IX in GNPs followed by its accumulation in atherosclerotic plaque indicating their efficacy to treat atherosclerosis. Furthermore, the capability of GNPs to inhibit D-ribose glycation of bovine serum albumin was investigated by Liu et al., (2014) to prevent atherosclerosis. Results showed effective inhibition of D-ribose glycation by GNPs when used in the size range of 2–20 nm *in-vitro*. It was also reported that maximum D-ribose glycation inhibition occurred by using GNPs of the maximum total surface area. The role of other nanomedicines for the treatment of atherosclerosis is explained in Table 8.6.

8.4 NANOMEDICINES AND MYOCARDIAL INFARCTION (MI)

MI is caused due to insufficient blood supply to a part of the heart muscle for a specific time interval leading to death of heart cells (Valensi et al., 2011). This is resulted due to coronary artery blockage which occurs because of thrombus formation or atherosclerotic plaque rupture (Boateng and Sanborn, 2013). MI is categorized under major health problem in the USA, and approximately 1.1 million MI cases are reported there annually (Reed et al., 2017). Figure 8.11 describes the different classes of MI on the basis of coronary arteries condition (Thygesen et al., 2012). MI is usually diagnosed through the presence of immediate release of cardiac-specific troponin isoforms I (cTnI) and T (cTnT) from the myocardium (Wilson et al., 2009). Occurrence of the oxygen deficiency during MI leads to death of cardiomyocyte. This dead cardiomyocyte later on coverts into fibrotic scar and replaces the muscle tissue. This fibrotic scar reduces contractility of tissues generating cardiac dysfunction or heart failure on the last stage (Buxton, 2012). Currently used therapeutic agents against MI show beneficial effects like reduction in hypertrophy, patient mortality rate, and myocardial remodeling. However, available strategies show a poor prognosis of MI and deleterious effects on normal tissues of a patient's body. Therefore, novel nanomedicines are urgently needed to reduce these side effects of conventional strategies.

8.4.1 *LIPOSOMES AS EFFECTIVE NANOMEDICINES TO TREAT MYOCARDIAL INFARCTION (MI)*

Harel-Adar et al., (2011) evaluated phosphatidylserine presenting liposomes to improve the infarct repair in a rat model. Results of magnetic

TABLE 8.6 Solid Lipid Nanoparticles (SLNs) and Dendrimers as Effective Nanomedicines in Treatment of Atherosclerosis

Nanocarrier/ Composition	Drug	Entrapment/ size/zeta potential	Animal model/Route of Administration	Key Findings	References
SLNs/thymidine 3'-(1,2-dipalmitoyl-sn-glycero-3-phosphate), N-[5'-(2',3'-dioleoyl)uridine]-N',N',N'-trimethyl ammonium, polyethylene glycol	α-tocopherol, prostacyclin PGI2	—/92 nm/−23.6 mV	—	Developed SLNs loaded with iron oxide particles, α-tocopherol, and prostacyclin PGI2 showed inhibition of *in-vitro* platelet aggregation and higher sensitivity towards magnetic resonance imaging (MRI) compared to Feridex (a clinically used contrast agent)	Oumzil et al., 2016
SLNs/ Compritol, Tween 80	Carmustine	—	New Zealand White rabbits/ intravenous	Carmustine loaded SLNs were able to reduce the size of *in-vivo* atherosclerotic lesion up to 90% in comparison to control formulation and were found non-toxic to other tissues	Daminelli et al., 2016
Dendrimers/ 5-aminolevulinic acid (ALA),	Proto porphyrin IX	—	BALB/c mice/ topical	ALA dendrimers showed 4.6 fold higher uptake of the protoporphyrin IX in macrophages compared to endothelial cells, indicating their effectiveness in photodynamic therapy of atherosclerosis	Rodriguez et al., 2015
Dendrimers/ Poly (amidoamine) (PAMAM), manganese diethylene triamine pentaacetic acid (Mn-DTPA)	—	—/13.3 ± 1.2 nm/−18.7 ± 2.1 mV	Apoe−/− mice/ Intravenous	Manganese (Mn) G8 dendrimers produced enlarged magnetic resonance image of atherosclerotic lesions after 3 days of administration in animals at a dose of 0.05 mmol Mn/Kg	Nguyen et al., 2015
Dendrimers/ Poly (amidoamine) (PAMAM), ethylenediamine, galactose	Cholesteryl ester hydrolase (CEH) expression vector	—	C57BL/6 Mice/ intravenous	Galactose functionalized PAMAM dendrimers increased CEH expression in hepatocytes *in-vivo* and promoted high-density lipoprotein-associated cholesteryl ester hydrolysis which may be beneficial for atherosclerosis elimination	He et al., 2017

resonance imaging (MRI) confirmed the presence of liposomes at infarct macrophages *in-vivo*. Developed liposomes promoted the release of anti-inflammatory cytokines from macrophages and enhanced angiogenesis at infarct area enhancing its repairment. Furthermore, vascular endothelial growth factor entrapped immunoliposomes (VEGF) were investigated by Tang et al., (2014) for improved stem cell therapy in a MI rat model. VEGF, when used in combination with mesenchymal stem cells (MSCs), produced maximum attenuation in cardiac function loss compared to individual immunoliposomes and MSCs (Figure 8.12). Blood vessel density was enhanced up to 80% by using immunoliposomes and the MSCs combination *in-vivo*.

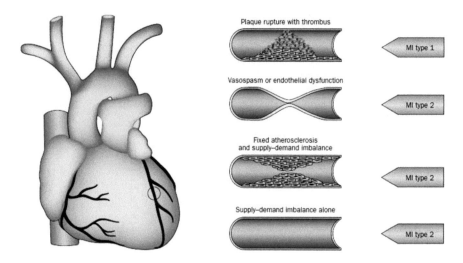

FIGURE 8.11   Differentiation between MI types 1 and 2 according to the condition of the coronary arteries. Abbreviation: MI, myocardial infarction. (From Thygesen et al., 2012, With permission. © 2012 Nature Publishing Group.).

Scott et al., (2007) investigated long-circulating liposomes conjugated with IgG2a mAb RMP–1 for effective treatment of MI in a rat model of MI. Radiolabeled immunoliposomes showed 83% increased targeting effect to site of MI compared non-infracted area and theses novel carriers also enhanced *in-vivo* blood circulation time up to 48 hours. Overview of liposomes as effective nanomedicines to treat MI is described in Table 8.7.

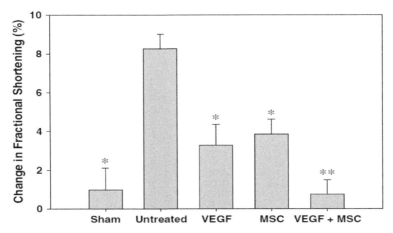

FIGURE 8.12 Cardiac function loss over time represented by FS change. Y-axis is calculated by subtracting the FS at 4 weeks post-MI from the FS at 1-week post-MI. *Significant difference compared to the "untreated," by ANOVA, P < 0.001. **Significant difference compared to "untreated," "VEGF" and "MSC," by ANOVA, P < 0.001. (From Tang et al., 2014, With permission. © 2014 Elsevier.).

8.4.2 ROLE OF POLYMERIC NANOPARTICLES IN TREATMENT OF MYOCARDIAL INFARCTION (MI)

Lu et al., (2014) investigated the therapeutic potential of placental growth factor (PGF) loaded PLGA nanoparticles in acute MI rat model for cardioprotective effect. The biodegradable nanoparticles injected into myocardium showed extended *in-vivo* release of PGF up to 15 days followed by reduction in infarct size and improvement in cardiac function compared to solution form of PGF. Furthermore, insulin-like growth factor–1 (IGF–1) conjugated PLGA nanoparticles were evaluated by Chang et al., (2013) to induce a cardioprotective effect after acute MI in FVB mice. The developed nanoparticulate system showed higher retention of IGF–1 in heart muscles compared to the plain IGF–1 solution and nanoparticles were capable of reducing infarct size and cardiomyocyte apoptosis up to 21 days upon administration of a single injection. In one more study, Binsalamah et al., (2011) reported efficacy of CS-alginate nanoparticles loaded with PGF to prevent MI in the infarct rat model through intramyocardial delivery. Animal hearts treated with the developed nanoparticulate system showed increased vascular density, ventricular function, and enhanced serum level of interleukin–10 after 8 weeks of coronary ligation. Table 8.8 gives an overview of research work done on polymeric nanoparticles to eliminate MI.

TABLE 8.7 Role of Liposomes in Effective Treatment of Myocardial Infarction (MI)

Composition	Drug	Entrapment/ size /zeta potential	Animal model/ Route of Administration	Key Findings	References
Monomethoxy poly(ethylene glycol) succinimidyl succinate, Dioleoyl phosphatidyl ethanolamine, mouse monoclonal antibody 2G4-2D7 specific for cardiac myosin	—	— / 250 nm/ —	New Zealand White rabbits/ intravenous	Developed long circulating immunoliposomes showed better accumulation at infarct area and increased blood circulation time up to two days as revealed in radiolabeling studies	Torchilin et al., 1996
Egg phosphatidyl choline, cholesterol, 1,2-distearoyl-sn-glycero–3-phospho ethanolamine-N-[methoxy (polyethylene glycol)–2000]	Coenzyme Q10	12 mol% / 167.3 ± 45.8 nm/ —	New Zealand White rabbits/ intracardiac	Coenzyme Q10 loaded liposomes showed 30% irreversible damage to the myocardium in animals which was very low compared to animals treated with empty liposomes (60%) indicating their protective effect against MI	Verma et al., 2007
Dipalmitoyl phosphatidyl choline, distearoyl phosphor ethanolamine-poly ethylene glycol 2000	Berberine chloride	10.4% / 0.11 μm/ —	C57BL/6J mice/ intravenous	Berberine chloride loaded liposomes showed preservation of cardiac ejection fraction up to 64% compared to empty liposomes or free berberine solution at 28th day after myocardial infarction in experimental animals	Allijn et al., 2017
Phosphatidylcholine, cholesterol, N-glutaryl phosphatidylethanolamine modified antimyosin	—	— / 196 ± 11 nm/ —	New Zealand White rabbits/ intracardiac	Rabbits treated with developed immunoliposomes showed reduction of myocardial infarct size to five-fold smaller compared to control liposomes administered in equivalent amounts	Khaw et al., 2007
Hydrogenated soy phosphatidylcholine, 1,2-distearoyl-sn-glycero-3-phospho ethanolamine-[methoxy (polyethylene glycol)–2000], cholesterol	Adenosine	— / 134 ± 21 nm/ −2.3 ± 1.1 mV	Wistar rats/ intravenous	Adenosine loaded liposomes showed effective accumulation in the infarct area as revealed in bioluminescence imaging, and they also reduced infarct size to a very low value (29.5 ± 6.5%) compared to control solution (53.2 ± 3.5%) after 30 min of administration	Takahama et al., 2009

TABLE 8.8 Polymeric Nanoparticles as Effective Nanomedicines in Treatment of Myocardial Infarction (MI)

Composition	Drug	Entrapment/size/zeta potential	Animal model/Route of Administration	Key Findings	References
Pluronic F–127 [a poly(ethylene oxide)-poly(propylene oxide)-poly(ethylene oxide) triblock copolymer]	Recombinant human VEGF (Vascular endothelial growth factor)	80% / 270 nm/ —	Sprague–Dawley rats/ intravenous	Core/shell nanoparticles and their gel loaded with VEGF showed improved cardiac output and ejection fraction in experimental animals even after 4th day of administration	Oh et al., 2010
Polyethylenimine–22	Hypoxia-regulated VEGF plasmid	95% / 76.90 ± 6.62 nm / 26.17 ± 1.3 mV	New Zealand White rabbits/ intramyocardial	The developed nanoparticulate system showed increased blood vessel density to 20.1 ± 1.3 and improved blood flow up to 1.28 ± 0.09 ml/min/g to the left ventricle of experimental animals compared to control nanoparticles after 14 days of treatment	Ye et al., 2011
Poly (lactic-co-glycolic acid) (PLGA)	Cyclosporine A (CsA)	—/111 ± 3 nm/ 20.3 ± 2.4 mV	Cyclophilin D–/– mice/ intravenous	PLGA nanoparticles loaded with CsA effectively inhibited opening of mitochondrial permeability transition pore in animal myocardium and protected the heart of animal from MI up to 4 weeks	Ikeda et al., 2016
PLGA, Polyethylene glycol (PEG), mouse monoclonal anti-rat CD68 antibody	VEGF (Vascular endothelial growth factor	83% / 4.1 ± 0.7 μm / –30 mV	Sprague–Dawley rats/ intramyocardial	Developed microparticles loaded with VEGF showed significantly higher myocardial retention up to 3 months after administration and least uptake below 4% by macrophages *in-vivo* indicating their efficacy in MI treatment	Formiga et al., 2013

8.4.3   ROLE OF OTHER NANOMEDICINE IN EFFECTIVE ELIMINATION OF MYOCARDIAL INFARCTION (MI)

Tan et al., (2017) investigated SLNs loaded with total flavonoid extract from *Dracocephalum moldavica* L. (TFDM) for cardioprotection against MI in Sprague–Dawley rats. Results of pharmacodynamic evaluation revealed better efficacy of TFDM loaded SLNs to reduce *in-vivo* infarct area, inflammation associated factors, and cardiac enzymes in serum compared to free TFDM. Furthermore, RGD peptide conjugated PEGylated SLNs loaded with puerarin was developed by Dong et al., (2017) and evaluated in acute MI rat model for the cardioprotective effect against MI. Developed SLNs extended the *in-vivo* release of puerarin up to 12 hours, followed by the presence of high puerarin concentration in plasma and myocardium compared to conventional SLNs formulations. Puerarin loaded SLNs also showed the highest reduction in infarct size *in-vivo*. Description of other nanomedicines used for MI eradication is given in Table 8.9.

8.5   INTELLECTUAL PROPERTY RIGHTS (IPR) REGARDING USE OF NANOMEDICINES IN TREATMENT OF CARDIOVASCULAR DISORDERS

Description of various patents regarding the use of nanomedicine in the treatment of cardiovascular disorders is summarized in Table 8.10.

8.6   CONCLUSIONS

Cardiovascular disorders are considered as a huge burden on healthcare departments in the developed nations due to high mortality and morbidity caused by them. The occurrence rate of cardiovascular disorders is high, especially in an older aged and obese population of any country. Detection and treatment patterns of cardiovascular disorders, however, have been improved due to development in genetics, molecular biology, bioengineering, and nanotechnology. Nanomedicines have provided a new exploration to treat cardiovascular disorders despite of their position in early development phase. Compared to conventional strategies, nanomedicines show characteristics of multitasking and multi-modular agents which are capable for simultaneous detection and treatment of cardiovascular disorders. Their targeted delivery has capability to minimize adverse effects and toxicity of

TABLE 8.9 Role of Various Nanomedicines in Treatment of Myocardial Infarction (MI)

Nanocarrier/Composition	Drug	Entrapment/size /zeta potential	Animal model/Route of Administration	Key Findings	References
Nanostructured lipid carriers (NLCs)/polyethylene glycol monostearate, oleic acid, glycerol monostearate	Baicalin	83.5%/83.9 nm /–32.1 mV	Sprague-Dawley (SD) rats/intravenous	PEGylated NLCs loaded with baicalin showed 7.2 fold increased AUC and 3 fold increased concentration of baicalin in animal myocardium compared to simple NLCs	Zhang et al., 2016
Dendrimers/Arginine-grafted bioreducible poly (disulfide amine) (ABP)-conjugated polyamidoamine (PAMAM)	Human Relaxin 1-expressing plasmid DNA with hypoxia response element (HRE) 12 copies (HR1)	—/121 nm/—	Sprague-Dawley (SD) rats/intramyocardial	Developed dendrimers were capable of reversing adverse cardiac defects to normal caused after MI development and they completely removed *in-vivo* infarct after 4 weeks of treatment	Lee et al., 2016
Carbon nanotubes/polyethylene glycol diacrylate, 1-(2-aminoethyl) piperazine, dodecyl amine, poly (N-isopropyl acrylamide)	Brown adipose-derived stem cells (BASCs)	—	Sprague-Dawley (SD) rats/intravenous	Poly (N-isopropylacrylamide)/single-wall carbon nanotubes based hydrogel containing BASCs showed increased seeding cells engraftment in infarcted myocardium of animals after 2 days of administration and was found effective compared to plain poly (N-isopropylacrylamide) hydrogel containing BASCs	Li et al., 2014
Gold nanoparticles (GNPs)/hydrogen tetra-chloroaurate (III), sodium citrate tribasic dihydrate,	Deoxy ribozyme (DNAzyme)	—/14 ± 3 nm/—	Sprague-Dawley (SD) rats/intramyocardial	DNAzyme conjugated GNPs showed TNF-α knockdown up to 50%, significant anti-inflammatory effects, and improved cardiac function in the MI animal model	Somasuntharam et al., 2016

TABLE 8.10 List of Patents Regarding the Use of Nanomedicine in Treatment of Cardiovascular Disorders

Title of patent	Brief description	Inventors	Patent Number	Reference
Liposomes compositions and methods for the treatment of atherosclerosis	This invention describes a method of preparation of unilamellar liposomes of size 100–150 nm and their role in atherosclerosis treatment	Michael J. Hope, Wendi Rodrigueza	US6139871 A	Hope and Rodrigueza, 2000
Liposomal prostaglandin formulations	This patent gives a method of formulation of prostaglandin loaded liposomes and their efficacy to treat cardiovascular disorders like myocardial infarction and systemic vasculitis	Marc J. Ostro	WO1992019243 A1	Ostro, 1992
Edaravone liposome injection and new application thereof	This invention explains the preparation procedure of edaravone loaded liposomes by using ingredients like phospholipid, cholesterol, polysorbate 80 and their utility to treat thrombophlebitis	Wang Ming	CN101601656 B	Ming, 2010
Method of making anti-thrombotic conjugate nanoparticles	This invention deals with a manufacturing method of polyvinyl alcohol-polylactic acid copolymer nanoparticles and their evaluation for anti-thrombotic effect	Jonathan Zhao	EP2269666 B1	Zhao, 2013
Functionalized nanoparticles, a method for preparing the same and application thereof	This patent describes the development procedure of poly(D,L-lactide-co-glycolide) nanoparticles coupled with insulin growth factor −1 (IGF−1) and their role in the eradication of myocardial infarction (MI)	Patrick C.H. Hsieh, Min-Feng Cheng, Chih Han Chang, Wei-Yin Liao	US9198873 B2	Hsieh et al., 2015
Daidzein solid lipid nanoparticles and preparation method thereof	This invention deals with the encapsulation of a hydrophobic drug daidzein inside solid lipid nanoparticles made up of glyceryl monostearate and their role in effective treatment of cardiovascular disorders like atherosclerosis and myocardial infarction (MI)	Li Yaping, Gao Fang, Huang Yan, Zhang Zhi-wen, Gu Wangwen	CN102258475 B	Yaping et al., 2013

TABLE 8.10 *(Continued)*

Title of patent	Brief description	Inventors	Patent Number	Reference
Use of solid lipid nanoparticles comprising cholesteryl propionate and/or cholesteryl butyrate	This patent discloses a development method of solid lipid nanoparticles using warm microemulsions containing cholesteryl propionate and/or cholesteryl butyrate for treatment of atherosclerosis and vascular thrombosis	Maria Rosa Gasco	WO2006128888 A1	Gasco, 2006
Pamam dendrimer derivatives for antithrombotic therapy	This invention describes the utility of polyamidoamine (PAMAM) dendrimer G4 and G5 derivatives for treatment of thrombosis with the generation of minimum toxicity after administration	Lara Esteban Duran, Luis Guzman, John Amalraj, Santos Leonardo Silva	WO2015008195 A1	Duran et al., 2015
Carbon nanotubes for imaging and drug delivery	This patent deals with method of preparation single-wall carbon nanotubes conjugated to RGD peptide to target monocytes of vascular endothelium for treatment of atherosclerosis	Bryan R. Smith, Eliver Ghosn	US20140079630 A1	Smith and Ghosn, 2014
Preparation and application of nanoparticles for thrombus-targeting and thermal-ablation	This invention describes the development of gold nanoparticles in size range 2–500 nm for thrombus targeting and its removal through thermal ablation using near-infrared light	Yin Le, Chen Jian, Huoda, Wang Weicheng, Hu Yong	CN105460976 A	Le et al., 2016

conventional therapeutic agents. However, the future scale-up technology and clinical exploration of nanomedicines will govern their position in the pharmaceutical market.

KEYWORDS

- **atherosclerosis**
- **cardiovascular diseases**
- **myocardial infarction**
- **nanomedicine**
- **receptor targeting**

REFERENCES

Absar, S., Nahar, K., Kwon, Y. M., & Ahsan, F., (2013). Thrombus-targeted nanocarrier attenuates bleeding complications associated with conventional thrombolytic therapy. *Pharm. Res.*, *30*, 1663–1676.

Agyare, E., & Kandimalla, K., (2014). Delivery of polymeric nanoparticles to target vascular diseases. *J. Biomol. Res. Ther.*, *3*(1), pp. S1–001.

Alaarg, A., Jordan, N. Y., Verhoef, J. J., Metselaar, J. M., Storm, G., & Kok, R. J., (2016). Docosahexaenoic acid liposomes for targeting chronic inflammatory diseases and cancer: An in vitro assessment. *Int. J. Nanomedicine.*, *11*, 5027–5040.

Alaarg, A., Senders, M. L., Varela-Moreira, A., Pérez-Medina, C., Zhao, Y., Tang, J., et al., (2017). A systematic comparison of clinically viable nanomedicines targeting HMG-CoA reductase in inflammatory atherosclerosis. *J. Control. Rel.*, *262*, 47–57.

Allijn, I. E., Czarny, B. M., Wang, X., Chong, S. Y., Weiler, M., Da Silva, A. E., et al., (2017). Liposome-encapsulated berberine treatment attenuates cardiac dysfunction after myocardial infarction. *J. Control. Release.*, *247*, 127–133.

Almer, G., Frascione, D., Pali-Schöll, I., Vonach, C., Lukschal, A., Stremnitzer, C., et al., (2013). Interleukin–10: An anti-inflammatory marker to target atherosclerotic lesions via PEGylated liposomes. *Mol. Pharm.*, *10*(1), 175–186.

Andrés, V., Pello, O. M., & Silvestre-Roig, C., (2012). Macrophage proliferation and apoptosis in atherosclerosis. *Curr. Opin. Lipidol.*, *23*(5), 429–438.

Antoniades, C., Psarros, C., Tousoulis, D., Bakogiannis, C., Shirodaria, C., & Stefanadis, C., (2010). Nanoparticles a promising therapeutic approach in atherosclerosis. *Cancer Drug Deliv.*, *7*, 303–311.

Arvizo, R., Bhattacharya, R., & Mukherjee, P., (2010). Gold nanoparticles: Opportunities and challenges in nanomedicine. *Exp. Opin. Drug Deliv.*, *7*(6), 753–763.

Bagre, A. P., Jain, K., & Jain, N. K., (2013). Alginate coated chitosan core-shell nanoparticles for oral delivery of enoxaparin: *In vitro* and *in vivo* assessment. *Int. J. Pharm.*, *456*(1), 31–40.

Bai, S., & Ahsan, F., (2009). Synthesis and evaluation of pegylated dendrimeric nanocarrier for pulmonary delivery of low molecular weight heparin. *Pharm. Res.*, *26*(3), 539–548.

Bai, S., & Ahsan, F., (2010). Inhalable liposomes of low molecular weight heparin for the treatment of venous thromboembolism. *J. Pharm. Sci.*, *99*(11), 4554–4564.

Bai, S., Thomas, C., & Ahsan, F., (2007). Dendrimers as a carrier for pulmonary delivery of enoxaparin, a low-molecular-weight heparin. *J. Pharm. Sci.*, *96*(8), 2090–2106.

Bai, S., Thomas, C., Rawat, A., & Ahsan, F., (2006). Recent progress in dendrimer-based nanocarriers. *Crit. Rev. Ther. Drug Carrier Syst.*, *23*(6), 437–495.

Bakker-Woudenberg, I. A., (2002). Long-circulating sterically stabilized liposomes as carriers of agents for treatment of infection or for imaging infectious foci. *Int. J. Antimicrob. Agents.*, *19*(4), 299–311.

Bardania, H., Shojaosadati, S. A., Kobarfard, F., Dorkoosh, F., Zadeh, M. E., Naraki, M., & Faizi, M., (2017). Encapsulation of eptifibatide in RGD-modified nanoliposomes improves platelet aggregation inhibitory activity. *J. Thromb. Thrombolysis*, *43*(2), 184–193.

Bartneck, M., Peters, F. M., Warzecha, K. T., Bienert, M., Van Bloois, L., Trautwein, C., Lammers, T., & Tacke, F., (2014). Liposomal encapsulation of dexamethasone modulates cytotoxicity, inflammatory cytokine response, and migratory properties of primary human macrophages. *Nanomedicine*, *10*(6), 1209–1220.

Bennett, J. S., (2001). Novel platelet inhibitors. *Annu. Rev. Med.*, *52*, 16–84.

Binsalamah, Z. M., Paul, A., Khan, A. A., Prakash, S., & Shum-Tim, D., (2011). Intramyocardial sustained delivery of placental growth factor using nanoparticles as a vehicle for delivery in the rat infarct model. *Int. J. Nanomedicine.*, *6*, 2667–2678.

Blasi, P., Giovagnoli, S., Schoubben, A., Ricci, M., & Rossi, C., (2007). Solid lipid nanoparticles for targeted brain drug delivery. *Adv. Drug Deliv. Rev.*, *59*(6), 454–477.

Boateng, S., & Sanborn, T., (2013). Acute myocardial infarction. *Dis. Mon.*, *59*(3), 83–96.

Brochu, H., Polidori, A., Pucci, B., & Vermette, P., (2004). Drug delivery systems using immo-bilized intact liposomes: A comparative and critical review. *Curr. Drug Deliv.*, *1*(3), 299–312.

Buxton, D. B., (2012). The impact of nanotechnology on myocardial infarction treatment. Nanomedicine *(Lond).*, *7*(2), 173–175.

Chacko, A. M., Hood, E. D., Zern, B. J., & Muzykantov, V. R., (2011). Targeted nanocarriers for imaging and therapy of vascular inflammation. *Curr. Opin. Colloid. Interface Sci.*, *16*, 215–227.

Chang, M. Y., Yang, Y. J., Chang, C. H., Tang, A. C., Liao, W. Y., Cheng, F. Y., et al., (2013). Functionalized nanoparticles provide early cardioprotection after acute myocardial infarction. *J. Control. Release.*, *170*(2), 287–294.

Chen, J. P., Yang, P. C., Ma, Y. H., & Wu, T., (2011). Characterization of chitosan magnetic nanoparticles for in situ delivery of tissue plasminogen activator. *Carbohyd. Pol.*, *84*(1), 364–372.

Cheng, Y., Wang, J., Rao, T., He, X., & Xu, T., (2008). Pharmaceutical applications of dendrimers: Promising nanocarriers for drug delivery. *Front. Biosci.*, *13*, 1447–1471.

Chowdhury, R., Khan, H., Heydon, E., Shroufi, A., Fahimi, S., Moore, C., et al., (2013). Adherence to cardiovascular therapy: A meta-analysis of prevalence and clinical consequences. *Eur. Heart J.*, *34*(38), 2940–2948.

Chung, B. L., Toth, M. J., Kamaly, N., Sei, Y. J., Becraft, J., Mulder, W. J., et al., (2015). Nanomedicines for endothelial disorders. *Nano Today*, *10*(6), 759–776.

Chung, T. W., Wang, S. S., & Tsai, W. J., (2008). Accelerating thrombolysis with chitosan-coated plasminogen activators encapsulated in poly-(lactide-co-glycolide) (PLGA) nanoparticles. *Biomaterials*, *29*(2), 228–237.

Collen, D., Stump, D. C., & Gold, H. K., (1988). Thrombolytic therapy. *Ann. Rev. Med.*, *39*, 405–423.

Cormode, D. P., Skajaa, T., Fayad, Z. A., & Mulder, W. J., (2009). Nanotechnology in medical imaging: Probe design and applications. *Arterioscler. Thromb. Vasc. Biol.*, *29*, 992–1000.

Coutre, S., & Leung, L., (1995). Novel antithrombotic therapeutics targeted against platelet glycoprotein IIb/IIIa. *Annu. Rev. Med.*, *46*, 257–265.

Crommelin, D. J. A., Scherphof, G., & Storm, G., (1995). Active targeting with particulate carrier systems in the blood compartment, *Adv. Drug Deliv. Rev.*, *17*, 49–60.

Daminelli, E. N., Martinelli, A. E., Bulgarelli, A., Freitas, F. R., & Maranhão, R. C., (2016). Reduction of atherosclerotic lesions by the chemotherapeutic agent carmustine associated to lipid nanoparticles. *Cardiovasc. Drugs Ther.*, *30*(5), 433–443.

De Meyer, S. F., Vanhoorelbeke, K., Broos, K., et al., (2008). Antiplatelet drugs. *Br. J. Haematol.*, *142*, 515–528.

De Oliveira, G. K., Da Silva, M. N., Sicchieri, L. B., De Oliveira, S. F. R., De Matos, R. A., & Courrol, L. C., (2015). Aminolevulinic acid with gold nanoparticles: A novel theranostic agent for atherosclerosis. *Analyst.*, *140*(6), 1974–1980.

De Vries, M. R., & Quax, P. H., (2016). Plaque angiogenesis and its relation to inflammation and atherosclerotic plaque destabilization. *Curr. Opin. Lipidol.*, *27*(5), 499–506.

Ding, B. S., Dziubla, T., Shuvaev, V. V., Muro, S., & Muzykantov, V. R., (2006). Advanced drug delivery systems that target the vascular endothelium. *Mol. Interv.*, *6*(2), 98–112.

Dong, Z., Guo, J., Xing, X., Zhang, X., Du, Y., & Lu, Q., (2017). RGD modified and PEGylated lipid nanoparticles loaded with puerarin: Formulation, characterization and protective effects on acute myocardial ischemia model. *Biomed. Pharmacother.*, *89*, 297–304.

Dreaden, E. C., Alkilany, A. M., Huang, X., Murphy, C. J., & El-Sayed, M. A., (2012). The golden age: Gold nanoparticles for biomedicine. *Chem. Soc. Rev.*, *41*(7), 2740–2779.

Duran, L. E., Guzman, L., Amalraj, J., & Silva, S. L., (2015). *Pamam Dendrimer Derivatives for Antithrombotic Therapy*. W. O. Patent 2015008195 A1.

Durán-Lara, E., Guzmán, L., John, A., Fuentes, E., Alarcón, M., Palomo, I., & Santos, L. S., (2013). PAMAM dendrimer derivatives as a potential drug for antithrombotic therapy. *Eur. J. Med. Chem.*, *69*, 601–608.

Faraday, M., (1857). The Bakerian lecture: Experimental relations of gold (and other metals) to light. *Philos. T. R. Soc. A.*, *147*, 145–181.

Feng, L., & Mumper, R. J., (2013). A critical review of lipid-based nanoparticles for taxane delivery. *Cancer Lett.*, *334*(2), 157–175.

Formiga, F. R., Garbayo, E., Díaz-Herráez, P., Abizanda, G., Simón-Yarza, T., Tamayo, E., Prósper, F., & Blanco-Prieto, M. J., (2013). Biodegradation and heart retention of polymeric microparticles in a rat model of myocardial ischemia. *Eur. J. Pharm. Biopharm.*, *85*(3 Pt A), 665–672.

Gasco, M. R., (2015). *Use of Solid Lipid Nanoparticles Comprising Cholesteryl Propionate and/or Cholesteryl Butyrate*. W. O. Patent 2006128888 A1.

Gong, H., Peng, R., & Liu, Z., (2013). Carbon nanotubes for biomedical imaging: The recent advances. *Adv. Drug Deliv. Rev.*, *65*(15), 1951–1963.

Harel-Adar, T., Ben Mordechai, T., Amsalem, Y., Feinberg, M. S., Leor, J., & Cohen, S., (2011). Modulation of cardiac macrophages by phosphatidylserine-presenting liposomes improves infarct repair. *Proc. Natl. Acad. Sci. USA.*, *108*(5), 1827–1832.

Haume, K., Rosa, S., Grellet, S., Śmiałek, M. A., Butterworth, K. T., Solov'yov, A. V., et al., (2016). Gold nanoparticles for cancer radiotherapy: A review. *Cancer Nanotechnol.*, *7*(1), 8–15.

He, H., Lancina, M. G., Wang, J., Korzun, W. J., Yang, H., & Ghosh, S., (2017). Bolstering cholesteryl ester hydrolysis in liver: A hepatocyte-targeting gene delivery strategy for potential alleviation of atherosclerosis. *Biomaterials, 130*, 1–13.

Hong, Z., Xu, Y., Yin, J. F., Jin, J., Jiang, Y., & Du, Q., (2014). Improving the effectiveness of (-)-epigallocatechin gallate (EGCG) against rabbit atherosclerosis by EGCG-loaded nanoparticles prepared from chitosan and polyaspartic acid. *J. Agric. Food Chem., 62*(52), 12603–12609.

Hope, M. J., & Rodrigueza, W., (2000). *Liposomes Compositions and Methods for the Treatment of Atherosclerosis*. U. S. Patent 6139871 A.

Hosseini, H., Li, Y., Kanellakis, P., Tay, C., Cao, A., Tipping, P., Bobik, A., Toh, B. H., & Kyaw, T., (2015). Phosphatidylserine liposomes mimic apoptotic cells to attenuate atherosclerosis by expanding polyreactive IgM producing B1a lymphocytes. *Cardiovasc. Res., 106*(3), 443–452.

Hsieh, P. C. H., Cheng, M. F., Chang, C. H., & Liao, W. Y., (2015). *Functionalized Nanoparticles, Method for Preparing the Same and Application Thereof*. U. S. Patent 9198873 B2.

Iijima, S., (1991). Helical microtubules of graphitic carbon. *Nature, 354*, 56–58.

Ikeda, G., Matoba, T., Nakano, Y., Nagaoka, K., Ishikita, A., Nakano, K., et al., (2016). Nanoparticle-mediated targeting of cyclosporine A enhances cardioprotection against ischemia-reperfusion injury through inhibition of mitochondrial permeability transition pore opening. *Sci. Rep., 6*, 20467.

Iqbal, M. A., Md, S., Sahni, J. K., Baboota, S., Dang, S., & Ali, J., (2012). Nanostructured lipid carriers system: Recent advances in drug delivery. *J. Drug Target., 20*(10), 813–830.

Jain, A., Mehra, N. K., Nahar, M., & Jain, N. K., (2013). Topical delivery of enoxaparin using nanostructured lipid carrier. *J. Microencapsul., 30*(7), 709–715.

Jain, N. K., & Asthana, A., (2007). Dendritic systems in drug delivery applications. *Expert Opin. Drug Deliv., 4*(5), 495–512.

Jaiswal, P., Gidwani, B., & Vyas, A., (2016). Nanostructured lipid carriers and their current application in targeted drug delivery. *Artif. Cells Nanomed. Biotechnol., 44*(1), 27–40.

Ji, S. R., Liu, C., Zhang, B., Yang, F., Xu, J., Long, J., Jin, C., Fu, D. L., Ni, Q. X., & Yu, X. J., (2010). Carbon nanotubes in cancer diagnosis and therapy. *Biochim. Biophys. Acta., 1806*(1), 29–35.

Jogala, S., Rachamalla, S. S., & Aukunuru, J., (2016). Development of PEG-PLGA based intravenous low molecular weight heparin (LMWH) nanoparticles intended to treat venous thrombosis. *Curr. Drug Deliv., 13*(5), 698–710.

Kamaly, N., Fredman, G., Subramanian, M., Gadde, S., Pesic, A., Cheung, L., et al., (2013). Development and in vivo efficacy of targeted polymeric inflammation-resolving nanoparticles. *Proc. Natl. Acad. Sci. USA., 110*(16), 6506–6511.

Khatak, S., & Dureja, H., (2015). Recent techniques and patents on solid lipid nanoparticles as novel carrier for drug delivery. *Recent Pat. Nanotechnol., 9*(3), 150–177.

Khaw, B. A., DaSilva, J., & Hartner, W. C., (2007). Cytoskeletal-antigen specific immuno-liposome-targeted *in vivo* preservation of myocardial viability. *J. Control. Release., 120*(1 & 2), 35–40.

Kim, H., Kee, P. H., Rim, Y., Moody, M. R., Klegerman, M. E., Vela, D., et al., (2013). Nitric oxide improves molecular imaging of inflammatory atheroma using targeted echogenic immunoliposomes. Atherosclerosis., *231*(2), 252–260.

Kim, I. S., Choi, H. G., Choi, H. S., Kim, B. K., & Kim, C. K., (1998). Prolonged systemic delivery of streptokinase using liposome. *Arch. Pharm. Res., 21*(3), 248–252.

Koudelka, S., Mikulik, R., Mašek, J., Raška, M., Turánek, K. P., Miller, A. D., & Turánek, J., (2016). Liposomal nanocarriers for plasminogen activators. *J. Control. Release.*, *227*, 45–57.

Kunamneni, A., Taleb, T., Abdelghani A., et al., (2007). Streptokinase–the drug of choice for thrombolytic therapy. *J. Thromb. Thrombolysis.*, *23*, 9–23.

Kunjachan, S., Jayapaul, J., Mertens, M. E., Storm, G., Kiessling, F., & Lammers, T., (2012). Theranostic systems and strategies for monitoring nanomedicine-mediated drug targeting. *Curr. Pharm. Biotechnol.*, *13*(4), 609–622.

Laing, S. T., Kim, H., Kopechek, J. A., Parikh, D., Huang, S., Klegerman, M. E., et al., (2010). Ultrasound-mediated delivery of echogenic immunoliposomes to porcine vascular smooth muscle cells *in vivo*. *J. Liposome Res.*, *20*(2), 160–167.

Laing, S. T., Moody, M. R., Kim, H., Smulevitz, B., Huang, S. L., Holland, C. K., et al., (2012). Thrombolytic efficacy of tissue plasminogen activator-loaded echogenic liposomes in a rabbit thrombus model. *Thromb. Res.*, *130*(4), 629–635.

Lammers, T., Aime, S., Hennink, W. E., Storm, G., & Kiessling, F., (2011). Theranostic nanomedicine. *Acc. Chem. Res.*, *44*(10), 1029–1038.

Le, Y., Jian, C., Huoda, W. W., & Yong, H., (2016). *Preparation and Application of Nanoparticles for Thrombus-Targeting and Thermal-Ablation.* C. N. Patent 105460976 A.

Lee, G. Y., Kim, J. H., Choi, K. Y., Yoon, H. Y., Kim, K., Kwon, I. C., et al., (2015). Hyaluronic acid nanoparticles for active targeting atherosclerosis. *Biomaterials*, *53*, 341–348.

Lee, Y. S., Choi, J. W., Oh, J. E., Yun, C. O., & Kim, S. W., (2016). Human relaxin gene expression delivered by bioreducible dendrimer polymer for post-infarct cardiac remodeling in rats. *Biomaterials*, *97*, 164–175.

Li, X., Radomski, A., Corrigan, O. I., Tajber, L., De Sousa, M. F., Endter, S., Medina, C., & Radomski, M. W., (2009). Platelet compatibility of PLGA, chitosan, and PLGA-chitosan nanoparticles. *Nanomedicine (Lond).*, *4*(7), 735–746.

Li, X., Zhou, J., Liu, Z., Chen, J., Lü, S., Sun, H., et al., (2014). A PNIPAAm-based thermosensitive hydrogel containing SWCNTs for stem cell transplantation in myocardial repair. *Biomaterials*, *35*(22), 5679–5688.

Liang, F., & Chen, B., (2010). A review on biomedical applications of single-walled carbon nanotubes. *Curr. Med. Chem.*, *17*(1), 10–24.

Liu, W., Cohenford, M. A., Frost, L., Seneviratne, C., & Dain, J. A., (2014). Inhibitory effect of gold nanoparticles on the D-ribose glycation of bovine serum albumin. *Int. J. Nanomedicine.*, *9*, 5461–5469.

Lobatto, M. E., Fuster, V., Fayad, Z. A., & Mulder, W. J., (2011). Perspectives and opportunities for nanomedicine in the management of atherosclerosis. *Nat. Rev. Drug Discov.*, *10*(11), 835–852.

Lu, Y., Gong, Q., Lu, F., Liang, J., Ji, L., Nie, Q., & Zhang, X., (2011). Preparation of sulfonated porous carbon nanotubes/activated carbon composite beads and their adsorption of low-density lipoprotein. *J. Mater. Sci. Mater. Med.*, *22*(8), 1855–1862.

Lu, Z. X., Mao, L. L., Lian, F., He, J., Zhang, W. T., Dai, C. Y., et al., (2014). Cardioprotective activity of placental growth factor in a rat model of acute myocardial infarction: Nanoparticle-based delivery versus direct myocardial injection. *BMC. Cardiovasc. Disord.*, *14*, 53–66.

Miao, Z. L., Deng, Y. J., DU, H. Y., Suo, X. B., Wang, X. Y., Wang, X., Wang, L., Cui, L. J., & Duan, N., (2015). Preparation of a liposomal delivery system and it's *in vitro* release of rapamycin. *Exp. Ther. Med.*, *9*(3), 941–946.

Ming, W., (2010). *Edaravone Liposome Injection and New Application Thereof.* C. N. Patent 101601656 B.

Mishra, S., Bedja, D., Amuzie, C., Foss, C. A., Pomper, M. G., Bhattacharya, R., Yarema, K. J., & Chatterjee, S., (2015). Improved intervention of atherosclerosis and cardiac hypertrophy through biodegradable polymer-encapsulated delivery of glycosphingolipid inhibitor. *Biomaterials*, *64*, 125–135.

Naseri, N., Valizadeh, H., & Zakeri-Milani, P., (2015). Solid lipid nanoparticles and nanostructured lipid carriers: Structure, preparation, and application. *Adv. Pharm. Bull.*, *5*(3), 305–313.

Nguyen, P. D., O'Rear, E. A., Johnson, A. E., Lu, R., & Fung, B. M., (1989). Thrombolysis using liposomal-encapsulated streptokinase: An *in vitro* study. *Proc. Soc. Exp. Biol. Med.*, *192*(3), 261–269.

Nguyen, T. H., Bryant, H., Shapsa, A., Street, H., Mani, V., Fayad, Z. A., Frank, J. A., Tsimikas, S., & Briley-Saebo, K. C., (2015). Manganese G8 dendrimers targeted to oxidation-specific epitopes: *In vivo* MR imaging of atherosclerosis. *J. Magn. Reson. Imaging.*, *41*(3), 797–805.

Oh, K. S., Song, J. Y., Yoon, S. J., Park, Y., Kim, D., & Yuk, S. H., (2010). Temperature-induced gel formation of core/shell nanoparticles for the regeneration of ischemic heart. *J. Control. Release.*, *146*(2), 207–211.

Ostro, M. J., (1992). *Liposomal Prostaglandin Formulations*. W.O. Patent 1992019243 A1.

Oumzil, K., Ramin, M. A., Lorenzato, C., Hémadou, A., Laroche, J., Jacobin-Valat, M. J., et al., (2016). Solid lipid nanoparticles for image-guided therapy of atherosclerosis. *Bioconjug. Chem.*, *27*(3), 569–575.

Palella, F. J. Jr., & Phair, J. P., (2011). Cardiovascular disease in HIV infection. *Curr. Opin. HIV. AIDS.*, *6*(4), 266–271.

Paliwal, R., Paliwal, S. R., Agrawal, G. P., & Vyas, S. P., (2011). Biomimetic solid lipid nanoparticles for oral bioavailability enhancement of low molecular weight heparin and its lipid conjugates: *In vitro* and *in vivo* evaluation. *Mol. Pharm.*, *8*(4), 1314–1321.

Pardeshi, C., Rajput, P., Belgamwar, V., Tekade, A., Patil, G., Chaudhary, K., & Sonje, A., (2012). Solid lipid-based nanocarriers: An overview. *Acta Pharm.*, *62*(4), 433–472.

Park, J. H., Ye, M., & Park, K., (2005). Biodegradable polymers for microencapsulation of drugs. *Molecules*, *10*(1), 146–161.

Patidar, A., Thakur, D. S., Kumar, P., & Verma, J., (2010). A review on novel lipid-based nanocarriers. *Int. J. Pharm. Pharm. Sci.*, *2*(4), 30–35.

Pazzini, C., Marcato, P. D., Prado, L. B., Alessio, A. M., Höehr, N. F., Montalvão, S., et al., (2015). Polymeric nanoparticles of enoxaparin as a delivery system: *In vivo* evaluation in normal rats and in a venous thrombosis rat model. *J. Nanosci. Nanotechnol.*, *15*(7), 4837–4843.

Perkins, W. R., Vaughan, D. E., Plavin, S. R., Daley, W. L., Rauch, J., Lee, L., & Janoff, A. S., (1997). Streptokinase entrapment in interdigitation-fusion liposomes improves thrombolysis in an experimental rabbit model. *Thromb. Haemost.*, *77*(6), 1174–1178.

Pietro, P. D., Strano, G., Zuccarello, L., & Satriano, C., (2016). Gold and silver nanoparticles for applications in theranostics. *Curr. Top. Med. Chem.*, *16*(27), 3069–3102.

Psarros, C., Lee, R., Margaritis, M., & Antoniades, C., (2012). Nanomedicine for the prevention, treatment, and imaging of atherosclerosis. *Maturitas.*, *73*(1), 52–60.

Reed, G. W., Rossi, J. E., & Cannon, C. P., (2017). Acute myocardial infarction. *Lancet.*, *389*(10065), 197–210.

Rizwan, H., Mohanta, J., Si, S., & Pal, A., (2017). Gold nanoparticles reduce high glucose-induced oxidative-nitrosative stress-regulated inflammation and apoptosis via tuberin-mTOR/NF-κB pathways in macrophages. *Int. J. Nanomedicine.*, *12*, 5841–5862.

Rizzo, L. Y., Theek, B., Storm, G., Kiessling, F., & Lammers, T., (2013). Recent progress in nanomedicine: Therapeutic, diagnostic and theranostic applications. *Curr. Opin. Biotechnol., 24*(6), 1159–1166.

Rodriguez, L., Vallecorsa, P., Battah, S., Di Venosa, G., Calvo, G., Mamone, L., et al., (2015). Aminolevulinic acid dendrimers in photodynamic treatment of cancer and atheromatous disease. *Photochem. Photobiol. Sci., 14*(9), 1617–1627.

Sanchez-Gaytan, B. L., Fay, F., Lobatto, M. E., Tang, J., Ouimet, M., et al., (2015). HDL-mimetic PLGA nanoparticle to target atherosclerosis plaque macrophages. *Bioconjug. Chem., 26*(3), 443–451.

Schiener, M., Hossann, M., Viola, J. R., Ortega-Gomez, A., Weber, C., Lauber, K., Lindner, L. H., & Soehnlein, O., (2014). Nanomedicine-based strategies for treatment of atherosclerosis. *Trends Mol. Med., 20*(5), 271–281.

Scott, R. C., Wang, B., Nallamothu, R., Pattillo, C. B., Perez-Liz, G., Issekutz, A., et al., (2007). Targeted delivery of antibody conjugated liposomal drug carriers to rat myocardial infarction. *Biotechnol. Bioeng., 96*(4), 795–802.

Sheikhpour, M., Golbabaie, A., & Kasaeian, A., (2017). Carbon nanotubes: A review of novel strategies for cancer diagnosis and treatment. *Mater. Sci. Eng. C Mater. Biol. Appl., 76*, 1289–1304.

Shidhaye, S. S., Vaidya, R., Sutar, S., Patwardhan, A., & Kadam, V. J., (2008). Solid lipid nanoparticles and nanostructured lipid carriers-innovative generations of solid lipid carriers. *Curr. Drug Deliv., 5*(4), 324–331.

Smith, B. R., & Ghosn, E., (2014). *Carbon Nanotubes for Imaging and Drug Delivery*. U. S. Patent 20140079630 A1.

Somasuntharam, I., Yehl, K., Carroll, S. L., Maxwell, J. T., Martinez, M. D., Che, P. L., et al., (2016). Knockdown of TNF-α by DNAzyme gold nanoparticles as an anti-inflammatory therapy for myocardial infarction. *Biomaterials, 83*, 12–22.

Takahama, H., Minamino, T., Asanuma, H., Fujita, M., Asai, T., Wakeno, M., et al., (2009). Prolonged targeting of ischemic/reperfused myocardium by liposomal adenosine augments cardioprotection in rats. *J. Am. Coll. Cardiol., 53*(8), 709–717.

Tan, M. E., He, C. H., Jiang, W., Zeng, C., Yu, N., Huang, W., Gao, Z. G., & Xing, J. G., (2017). Development of solid lipid nanoparticles containing total flavonoid extract from Dracocephalum moldavica L. and their therapeutic effect against myocardial ischemia-reperfusion injury in rats. *Int. J. Nanomedicine., 12*, 3253–3265.

Tan, M. L., Choong, P. F. M., & Dass, C. R., (2010). Recent developments in liposomes, microparticles, and nanoparticles for protein and peptide drug delivery. *Peptides, 31*, 184–193.

Tang, Y., Gan, X., Cheheltani, R., Curran, E., Lamberti, G., Krynska, B., Kiani, M. F., & Wang, B., (2014). Targeted delivery of vascular endothelial growth factor improves stem cell therapy in a rat myocardial infarction model. *Nanomedicine, 10*(8), 1711–1178.

Thatipamula, R., Palem, C., Gannu, R., Mudragada, S., & Yamsani, M., (2011). Formulation and *in vitro* characterization of domperidone loaded solid lipid nanoparticles and nanostructured lipid carriers. *Daru., 19*(1), 23–32.

Thygesen, K., Alpert, J. S., Jaffe, A. S., Simoons, M. L., Chaitman, B. R., White, H. D., et al., (2012). Third universal definition of myocardial infarction. *Nat. Rev. Cardiol., 9*(11), 620–633.

Tinkle, S., McNeil, S. E., Mühlebach, S., Bawa, R., Borchard, G., Barenholz, Y. C., Tamarkin, L., & Desai, N., (2014). Nanomedicines: Addressing the scientific and regulatory gap. *Ann. N Y Acad. Sci., 1313*, 35–56.

Torchilin, V. P., Narula, J., Halpern, E., & Khaw, B. A., (1996). Poly(ethylene glycol)-coated anti-cardiac myosin immunoliposomes: Factors influencing targeted accumulation in the infarcted myocardium. *Biochim. Biophys. Acta.*, *1279*(1), 75–83.

Undas, A., (2017). Prothrombotic fibrin clot phenotype in patients with deep vein thrombosis and pulmonary embolism: A new risk factor for recurrence. *Biomed. Res. Int.*, *2017*, 8196256.

Vaidya, B., Nayak, M. K., Dash, D., Agrawal, G. P., & Vyas, S. P., (2016). Development and characterization of highly selective target-sensitive liposomes for the delivery of streptokinase: *In vitro/in vivo* studies. *Drug Deliv.*, *23*(3), 801–807.

Vaidya, B., Nayak, M. K., Dash, D., Agrawal, G. P., & Vyas, S. P., (2011). Development and characterization of site-specific target sensitive liposomes for the delivery of thrombolytic agents. *Int. J. Pharm.*, *403*(1 & 2), 254–261.

Valensi, P., Lorgis, L., & Cottin, Y., (2011). Prevalence, incidence, predictive factors and prognosis of silent myocardial infarction: A review of the literature. *Arch. Cardiovasc. Dis.*, *104*(3), 178–188.

Vander Valk, F. M., Van Wijk, D. F., Lobatto, M. E., et al., (2015). Prednisolone-containing liposomes accumulate in human atherosclerotic macrophages upon intravenous administration. *Nanomedicine*, *11*(5), 1039–1046.

Verma, D. D., Hartner, W. C., Thakkar, V., Levchenko, T. S., & Torchilin, V. P., (2007). Protective effect of coenzyme Q10-loaded liposomes on the myocardium in rabbits with an acute experimental myocardial infarction. *Pharm. Res.*, *24*(11), 2131–2137.

Vyas, S. P., & Vaidya, B., (2009). Targeted delivery of thrombolytic agents: Role of integrin receptors. *Expert Opin. Drug Deliv.*, *6*(5), 499–508.

Wadajkar, A. S., Santimano, S., Rahimi, M., Yuan, B., Banerjee, S., & Nguyen, K. T., (2013). Deep vein thrombosis: Current status and nanotechnology advances. *Biotechnol. Adv.*, *31*(5), 504–513.

Wang, C., Ji, B., Cao, Y., Sun, B., & Liu, X., (2010). Evaluating thrombolytic efficacy and thrombus targetability of RGDS-liposomes encapsulating subtilisin FS33 *in vivo*. *Sheng Wu Yi Xue Gong Cheng Xue Za Zhi.*, *27*(2), 332–336.

Weber, C., & Noels, H., (2011). Atherosclerosis: Current pathogenesis and therapeutic options. *Nat. Med.*, *17*(11), 1410–1422.

Wilson, S. R., Sabatine, M. S., Braunwald, E., Sloan, S., Murphy, S. A., & Morrow, D. A., (2009). Detection of myocardial injury in patients with unstable angina using a novel nanoparticle cardiac troponin I assay: Observations from the PROTECT-TIMI 30 trial. *Am. Heart J.*, *158*, 386–391.

Wong, B. S., Yoong, S. L., Jagusiak, A., Panczyk, T., Ho, H. K., Ang, W. H., & Pastorin, G., (2013). Carbon nanotubes for delivery of small molecule drugs. *Adv. Drug Deliv. Rev.*, *65*(15), 1964–2015.

Xu, Y, Y., Yang, J., Shen, T., Zhou, F., Xia, Y., Fu, J. Y., et al., (2012). Intravenous administration of multi-walled carbon nanotubes affects the formation of atherosclerosis in Sprague-Dawley rats. *J. Occup. Health.*, *54*(5), 361–369.

Yaghoobi, N., Faridi, M. R., Faramarzi, M. A., Baharifar, H., & Amani, A., (2017). Preparation, optimization and activity evaluation of PLGA/Streptokinase nanoparticles using electrospray. *Adv. Pharm. Bull.*, *7*(1), 131–139.

Yang, H., & Kao, W. J., (2006). Dendrimers for pharmaceutical and biomedical applications. *J. Biomater. Sci. Polym. Ed.*, *17*(1 & 2), 3–19.

Yaping, L., Fang, G., Yan, H., Zhiwen, Z., & Wangwen, G., (2013). *Daidzein Solid Lipid Nanoparticles and Preparation Method Thereof.* C. N. Patent 102258475 B.

Ye, L., Zhang, W., Su, L. P., Haider, H. K., Poh, K. K., Galupo, M. J., Songco, G., Ge, R. W., Tan, H. C., & Sim, E. K., (2011). Nanoparticle-based delivery of hypoxia-regulated VEGF transgene system combined with myoblast engraftment for myocardial repair. *Biomaterials, 32*(9), 2424–2431.

Yoon, G., Park, J. W., & Yoon, I. S., (2013). Solid lipid nanoparticles (SLNs) and nanostructured lipid carriers (NLCs): Recent advances in drug delivery. *Int. J. Pharm. Investig., 43*(5), 353–356.

Yu, M., Jie, X., Xu, L., Chen, C., Shen, W., Cao, Y., Lian, G., & Qi, R., (2015). Recent advances in dendrimer research for cardiovascular diseases. *Biomacromolecules, 16*(9), 2588–2598.

Yuan, X., Yang, X., Cai, D., Mao, D., Wu, J., Zong, L., & Liu, J., (2008). Intranasal immunization with chitosan/pCETP nanoparticles inhibits atherosclerosis in a rabbit model of atherosclerosis. *Vaccine, 26*(29 & 30), 3727–3734.

Zernecke, A., & Weber, C., (2005). Inflammatory mediators in atherosclerotic vascular disease. *Basic Res. Cardiol., 100*(2), 93–101.

Zernecke, A., Shagdarsuren, E., & Weber, C., (2008). Chemokines in atherosclerosis: An update. *Arterioscler. Thromb. Vasc. Biol., 28*(11), 1897–908.

Zhang, M., He, J., Jiang, C., Zhang, W., Yang, Y., Wang, Z., & Liu, J., (2017). Plaque-hyaluronidase-responsive high-density-lipoprotein-mimetic nanoparticles for multistage intimal-macrophage-targeted drug delivery and enhanced anti-atherosclerotic therapy. *Int. J. Nanomedicine., 12*, 533–558.

Zhang, S., Wang, J., & Pan, J., (2016). Baicalin-loaded PEGylated lipid nanoparticles: Characterization, pharmacokinetics, and protective effects on acute myocardial ischemia in rats. *Drug Deliv., 23*(9), 3696–3703.

Zhang, Y., Zhou, J., Guo, D., Ao, M., Zheng, Y., & Wang, Z., (2013). Preparation and characterization of gadolinium-loaded PLGA particles surface modified with RGDS for the detection of thrombus. *Int. J. Nanomedicine., 8*, 3745–3756.

Zhao, J., (2013). *Method of Making Anti-Thrombotic Conjugate Nanoparticles.* E. P. Patent 2269666 B1.

Zhao, K., Li, D., Shi, C., Ma, X., Rong, G., Kang, H., Wang, X., & Sun, B., (2016). Biodegradable polymeric nanoparticles as the delivery carrier for drug. *Curr. Drug Deliv., 13*(4), 494–499.

CHAPTER 9

Nanomedicines for the Treatment of Gastric and Colonic Diseases

MD. ADIL SHAHARYAR[1,2], MAHFOOZUR RAHMAN[3], KAINAT ALAM[4], SARWAR BEG[5], KUMAR ANAND[2], CHOWDHURY MOBASWAR HOSSAIN[1], ARIJIT GUHA[1], MUHAMMAD AFZAL[6], IMRAN KAZMI[7], REHAN ABDUR RUB[8] and SANMOY KARMAKAR[2*]

[1] *Bengal School of Technology, Chinsurah, Hooghly, West Bengal, India, Tel.: +91-9748902723,*
Email: adil503@yahoo.co.in

[2] *Department of Pharmaceutical Technology, Jadavpur University, Kolkata,West Bengal, India. Tel.: +91-8017136385*
E-mail: sanmoykarmakar@gmail.com

[3] *Department of Pharmaceutical Sciences, Faculty of Health Sciences, Sam Higginbottom Institute of Agriculture, Technology, and Sciences, Allahabad, Uttar Pradesh, India, Tel.:+91-8627985598,*

Email: mahfoozkaifi@gmail.com

[4] *Christian College of Nursing, Faculty of Health Sciences, Sam Higginbottom Institute of Agriculture, Technology, and Sciences, Allahabad, Uttar Pradesh, India*

[5]*Product Nanomedicine Research Lab, Department of Pharmaceutics, School of Pharmaceutical Education and Research, Jamia Hamdard (Hamdard University), New Delhi, India*

[6] *Department of Pharmacology, College of Pharmacy, Aljouf University, Sakaka, KSA*

[7] *Glocal School of Pharmacy, Glocal University, Mirzapur Pole, Saharanpur, Uttar Pradesh, India*

[8] *Phmaceutics Research Lab. SPER, Jamia Hamdard, New Delhi-62, India*

Corresponding author. E-mail: sanmoykarmakar@gmail.com

ABSTRACT

Nanomedicine is a rapidly growing field in the area of diagnostics, imaging, and targeted therapeutic approaches for the basis of the pathophysiology of gastrointestinal diseases. Nanocarriers capable of delivering of anti-inflammatory drugs specifically to gastrointestinal tract affected region for a prolonged period of time reduces the side effects of encapsulated drugs. Therefore, the present book chapter provides in details on various gastric disorders and their pathophysiology on the recent development of nano-medicines for the detection and treatment of gastrointestinal diseases.

9.1   INTRODUCTION

Nano is a prefix, which means dwarf in Greek. To better equip with the understanding of the prefix "nano," some examples can be presented like one-billionth of a meter or the width of 6-carbon atom, considered as one nanometer (nm). The width of a human hair and red blood cell is approximately 80,000 nm and 7000 nm, respectively. Size of atoms is less than 1 nm whereas molecules are larger (Whiteside et al., 2003). The historical observation and size measurement of nanoparticles took place in the early 20th century by Richard Adolf Zsigmondy using gold colloids and other nanoparticles with the help of ultra-microscope. In 1959, a physicist and noble laureate by the name Richard Feynman presented a title "There is plenty of room at the bottom" at an American Physical Society meeting in which he pointed out that atoms and molecules can be manipulated and such systems should be developed for wider application (Feynman et al., 1959). Nanotechnology is a marriage between science and engineering, involving design, synthesis, characterization, and application of materials and devices having a functional organization in at least one dimension on the nanometer scale (Emerich and Sahoo et al., 2003). Medical application of nanotechnology is recognized as nanomedicine, which includes diagnosis and therapy using nanomaterials. It also includes methods to evaluate nanomaterial for its potential toxicity and also takes into account the development of nanosensors and nanodevices for drug delivery and bioimaging (Kim et al., 2011). Diseased or inflamed tissues undergo pathophysiologically as well as anatomical changes, giving way for the development of various nanotechnology (Vasir et al., 2005) based products. The above facts are exploited in the treatment of gastrointestinal and colonic diseases both in terms of therapeutic and diagnostics.

The gastrointestinal system of humans is also referred to as the digestive system and is basically classified into two parts namely the gastro-intestinal tract or alimentary canal and accessory organs associated with it. Physical and mechanical processes are involved in digestion. Digestion involves the breakdown of food, absorption of nutrients and waste elimination. The accessory organs are mainly liver, pancreas, and gall bladder. From mastication by teeth to emulsification by bile of liver, the GI covers a wide range of functions (Cheng et al., 2010). Diseases of the gastrointestinal system are a menace for public health, and its burden is increasing day by day. In the United States alone it affects 60 to 70 million people (Cheng et al., 2010) of which 48.3 million people required Ambulatory care visits (2010), 21.7 million people were hospitalized reported deaths were, 5.4 million people required diagnostic and therapeutic inpatient procedures and 20.4 million people underwent ambulatory surgical procedures (Cheng et al., 2010). These mentioned data are troubling and calls for steps, to be taken in the area of gastroenterology.

9.2 GASTROINTESTINAL DISORDERS

The saying "Happiness for me is largely a matter of digestion" by Lin Yutang in his book "The Importance of Living" highlights the importance of digestion in a rhetoric manner. A healthy gastrointestinal system is an indication of well-being. When any one of the components of the GI system is disturbed, it brings down with it other components of the body. For maintaining the well-being of the body and mind, the GI system should be in good condition. Many diseases occur due to malfunctioning of the GI system. These malfunctions, manifested in the form of major diseases along with their treatment is discussed below (Gabe, 1998).

9.2.1 EXCESSIVE INTESTINAL GAS

The total volume of intra-luminal gas normally is found to be 100–200 ml volume of gas in the stomach increases due to food intake (air entering with food), chemical reactions, the diffusivity of gas from blood and bacterial fermentation. The absorbed excess gas in the GI is removed by blood, anal evacuation, and bacterial consumption. This balance in the volume of gas between the intraluminal and venous blood is vital (Azpiroz et al., 2004). CO_2, O_2 is large while N_2 is poorly absorbed in the gut wall. Excess gas is removed by belching. The impairment of output due to different factors results in the increase in intraluminal gas volume. It involves three categories of symptoms, namely: (a) Aerophagia which arises due to dyspeptic type symptoms with excessive eructation. (b) Flatulence with odor from the anal passage is another symptom due to gases like hydrogen sulfides, methanethiol, and dimethylsulphide produced by Sulphur reducing bacteria. There are 20 evacuations in healthy individual constituting 200–700 ml of gas (Azpiroz et al., 2004). The 3rd category of symptom is attributed to bloating and meteorism. Bloating is due to impairment of anal evacuation resulting in an increase in fecal transit time thus causing colonic fermentation. Diet control is an integral part of excessive intra-luminal gas treatment. A diet producing low flatulence is recommended which includes fish and eggs. In the case of carbohydrate-containing foods like gluten-free bread, rice bread, rice is preferred. Vegetables like tomatoes and lettuce while low flatulence fruits include cherries and grapes. High flatulence foods include onion, beans, bananas, etc. Recent studies suggest exercises increases intestinal gas clearance. Patients suffering from lactose intolerance are given lactose-free diet, but presently some studies suggest if lactose taken in the normal range does not cause abdominal symptoms (Azpiroz et al., 2004).

9.2.2 REFLUX DISEASE

Gastroesophageal reflux disease is a chronic disease characterized by reflux of stomach acid into the esophagus with symptoms of heartburn and regurgitation. The regurgitation takes place usually after a heavy meal. Other symptoms include dysphagia, chest pain, and coughing. In this disease, 20% of American and 15% of the worldwide population face weekly symptoms of heartburn and regurgitation. Endoscopy of the upper part of GI is commonly used to detect GERD. It is of two types erosive and non-erosive. On-erosive reflux disease (NERD) is characterized by the absence of esophageal mucosal erosion in GERD during upper Endoscopy findings while in erosive esophagitis there is esophageal mucosal inflammation and swelling. The erosive type is known as Barrett's esophagus, which often succumbs to esophageal cancer. Remedial measures to enhance the performance of GI system include taking digestive aids in short durations, not lying down immediately after taking food, taking food with low rented sugar (De Vault et al., 1999).

9.2.3 IRRITABLE BOWEL SYNDROME

Irritable bowel syndrome is a disease of gastrointestinal tract characterized by abdominal pain, irregularities in stool and bloating. Its global prevalence is 11.2% with 5–10% reported for Europe, China, and the United States. Pain syndromes, overactive bladder, and migraine are the somatic comorbidities linked to IBS. Psychiatric conditions include depression and anxiety. Visceral sensitivity has also been found to be associated with IBS. According to population-oriented studies, it is highly prevalent in female sex with a ratio of 1.67. Risk factors include sex (female), Age (>50 years), Birth cohort, Breastfeeding (<6 months), Herbivore pet in childhood, birth weight (low), and body mass index (low). It has been found that there is an increase in colon muscle contraction in IBS as compared to non-IBS patients. Diet rich in fibers is a good way to ward of IBS. Similarly, colostrum has been found to be of significant importance in the treatment of IBS (Malagelada et al., 2006).

9.2.4 GASTROINTESTINAL DISEASE DUE TO DISRUPTION OF INTESTINAL PERMEABILITY

Digestion and absorption are not the only roles played by GI system, but it also shields the internal human body from external pathogens, toxins, and antigens (Farrell et al., 2012). This protective shield has two aspects namely intestinal barrier and intestinal permeability. The intestinal wall is composed of 4 layers namely the mucosa, submucosa, muscularis, and the serosa. The line of distinguishment between the intestinal barrier and permeability lies in the fact that the former is a separation of the gut lumen from the inner host by a functional entity comprising of mechanical elements (mucous and epithelial layer), humoral elements (defensins, IgA), immunological elements (lymphocyte, innate immune cells), muscular, and neurological elements; while the latter is a function of the barrier at specific sites on the intestinal wall. Permeability is measured by flux rate across the intestinal barrier. Impairment of intestinal permeability accounts for the imbalance in intestinal homeostasis (Eberl et al., 2005). Protection of the GI system is maintained by low pH of the gastric acid and enzymes produced by the pancreas. Low pH prevents the bacterial growth, and pancreatic enzyme disrupts the bacterial cell wall. Neutral intragastric pH causes increased the vulnerability of lumen to pathogens like *shigella and cholera*. Decrease in pancreatic enzymes in pancreatitis has been linked to bacterial growth. Some non-specific barriers involved in the defense mechanism are epithelial crypt cells and villi. The microbes are washed away by the fluid secreted by the epithelial cells along with the help of peristaltic movement. To assist the epithelial cells, the villi secrete mucous which prevent the bacterial adhesion to the receptors on the surface of the epithelial cells. Immunoglobin-A(IgA) and defensins (anti-microbial peptides) are the other non-specific barriers. IgA binds to antigens and increases their intestinal clearance (Ouellette et al., 1997). To add to this, there are some dynamic barriers like cytoskeleton and tight intercellular junctions which allow bidirectional passage of different types of inflammatory cells and substances. Inflammatory cells equip the mucosal defense system by presenting antigen to immune cells. ATP is an essential component of the epithelial system which helps it maintain its integrity and dynamism. Interference in ATP usage causes disruption in the epithelial system. NSAID and Tacrolimus (FK506) interfere with the ATP usage (Gabe et al., 1998; Bjarnason et al., 1993). The immune system safeguards the lumen by taking a sample of the antigen present in that particular area (Mckay et al., 1999). The epithelial barrier disruption is attributed to many inflammatory diseases, weakness of genetic function regulating the barrier (Demeo et al., 2002). Some of the major GI disorders are discussed in the following sections.

9.2.4.1 CROHN'S DISEASE

The Crohn's disease affects the intestinal lumen (colon and small intestine) and comes under the purview of inflammatory bowel disease. It is a chronic inflammatory disease having dubious pathogenesis. Several postulates, involving its mechanism, have been proposed but the one which is acclaimed figure has been mentioned. In Crohn's disease, the factors responsible are environmental (antigen in food, bacterial flora), increased gut permeability due to antigen from the external environment. Individuals who are susceptible have increased immune response due to environmental antigens. It has been found that susceptible animal models lacking Interleukin–10 and microbe-free environment fails to succumb to the disease (Kennedy et al., 2000). Though both Crohn's disease and ulcerative colitis share the same symptoms of watery or bloody diarrhea and abdominal pain or cramps yet they have some differences. Ulcerative colitis affects the upper layer of the intestine forming sores or ulcers (Hermiston et al., 1995) but the Crohn's disease starts from the upper layer and finally damages all the four layers making a hole in the intestine known as Stella and reaches the tissues (Ukabam, 1983; Baumgart et al., 2012). In Crohn's disease, a patchy appearance with healthy tissues in between is found while in Ulcerative colitis the ulcer is uniformly distributed. Finally, in Crohn's disease, the intestine wall thickens and get blocked, requiring surgery. To some extent, Crohn's disease can be treated, but the complications involved have a wide range of impact. The formation of Stula results in decreased absorption of nutrients causing symptoms like (a) persistent low grade fever, (b) loss of appetite and weight (c) fatigue, (d) anemia due to blood loss and or poor iron absorption, (e) skin and eye infection, (f) arthritis and sore joints involving knees or hip, (g) osteoporosis from due to calcium and vitamin D malabsorption, (h) inadequate vitamin K absorption, causing poor blood clotting, (i) stunted growth in children, (j) delayed pupates, and (k) stone in kidney and gallbladder (Mueller, 2002; Farmer, 1975; Zieve, 2009; Panes et al., 2007).

Since malabsorption of nutrients is one of the reasons for the above symptoms, taking nutrient supplement helps in improving the nutrient absorption problem from the intestine. The dietary remedy includes taking whole food and digestive aids (Izzo et al., 2005). Colostrum has shown to decrease the level of toxins from the body in Crohn's disease.

9.2.4.2   CELIAC DISEASE (CS)

Celiac disease (CS) is a pathological condition of the digestive system, characterized by malabsorption of nutrients present in the food as well as abnormal immune action towards gluten (Hall et al., 1991), a type of protein present in barley, wheat, and rye (Van, 2005; Hollander et al., 1992) causes allergy and finally destroying villi in the CS. It has been found that patient with CS has abnormal intestinal permeation. It is unclear whether abnormal intestinal permeation initiates (Van et al., 1993) the disease or the disease itself along with inflammation of the tissues. The disease manifests itself into symptoms prevalent usually in infants and children and can be categorized into following two categories: (1) Abdominal bleeding and pain, chronic diarrhea, vomiting, constipation, stool either fatty or pale and foul smelling, loss of weight, irritabilities. (2) Symptoms associated with the digestive symptoms occur less, but other symptoms may involve like unexplained iron deficiency causing anemia; fatigue, arthritis, osteoporosis, depression or anxiety, numbness in the hands with tingling effect and feet, seizure, missed menstrual period, infections, recurrent miscarriage; canker sores inside the mouth, dermatitis herpetiformis (itchy skin rash with itchiness). If the patient remains untreated with persistent complications, may cause liver disease, type–1 diabetes, intestinal cancer, Addison's disease (adrenal insufficiency and hypercortisolism) (Marsh et al., 1990), Sjogren's syndrome (an auto-immune disease affecting salivary glands). For the diagnosis of the disease anti-tissue transglutaminase antibodies (+TGA) or anti-endomysium antibodies (EMA) is used or biopsy of the small intestine is done. Ontogenic development in humans concerned with intestine takes place around 38th week and continuous till infancy covering the neonatal stage. Luminal maturation takes place around this period range and is influenced by factors like breastfeeding during which IgA is released thus helping the immune system in identifying luminal antigens (Reinhardt, 1984; Roberts, 1977; Falth-Magnusson et al., 1988). Several studies suggest, altered luminal permeability especially in children due to release of inflammatory cytokines. Eczema and asthma are two chronic disease caused due to food allergy. Food allergy can be life-threatening if it causes anaphylactic shock (Jalonen, 1991; Heyman et al., 1994). It is conventionally treated by epinephrine or steroids and is characterized by symptoms like hives, swelling, itchy skin, coughing, tingling or swelling in the mouth, breathing trouble, and diarrhea.

9.2.4.3   PANCREATITIS

Inflammation of the pancreas gland, medically, is termed as pancreatitis (Izrailov, 2018). In pancreatitis, during inflammation of pancreases, the pancreatic enzymes instead of getting activated in the intestines attack the tissues of the pancreas. It is more prevalent in men than women and is of two types: acute and chronic. Acute pancreatitis is life-threatening involving many complications and characterized by pain, that first starts in the upper abdomen and later extends to the back. This pain gets augmented after eating, but while in resting condition, it is mild. There are other symptoms which include a swollen and tender abdomen; nausea and vomiting, fever, and a rapid pulse (Noel, 2009; Sakorafas et al., 2000). Though permeability doesn't seem to be the culprit in this disease yet both morbidity and mortality is increased due to its influence (Russo et al., 2004). The linkage between permeability and acute pancreatitis is currently investigated, and further study is required to be conducted.

9.2.4.3.1   Chronic Pancreatitis

Chronic pancreatitis is an inflammation (Kumar et al., 2005) of the pancreas which gets worse with the passage of time, permanently damaging the pancreas. The reason behind damaging of the pancreas is same as that of acute pancreatitis. Alcohol is the main rogue behind chronic pancreatitis as it slowly damages the pancreatic duct. Other reasons involved are hereditary disorders related to pancreas, cystic fibrosis, hypercalcemia, hyperlipidemia, autoimmune disease, unknown causes and some medicines (Kumar et al., 2005). In most of the patient, there is an emergence of upper abdominal pain which may affect the back and the pain worsens with eating and drinking (Warshaw et al., 1998). The pathway of the pain is the same as that of acute pancreatitis. The pain subsides with the progression of the disease as enzymes secretion stops. Oily stool, weight loss, and nausea are other symptoms associated with chronic pancreatitis (Warshaw et al., 1998).

9.2.4.4   LIVER DISEASE

The liver is the laboratory of the body performing essential functions. Dysfunction of the liver is a major issue caused by various reasons that include viruses, medicine, genetics, immune system disorder, and toxicity.

9.2.4.4.1   Intestinal Permeability

Alcoholic and fatty liver are the two diseases where intestinal permeability is involved. Fatty acid of liver does not itself progresses and succumbs to advanced liver disease, but combination of both can result in liver failure. This occurrence of the two diseases is the most important factor behind cryptogenic cirrhosis. Studies suggest that some cofactor is responsible for the conversion of fatty acid liver to advanced liver disease and alcoholic liver disease. Endotoxin is believed to be this cofactor. Steatohepatitis, severe fatty liver, alcoholic liver is some of the liver-oriented disease believed to be caused by intestinal permeability (Yang, 1997; Wigg, 2001; Bode et al., 1987). So, the two-major cause for initiation or progression of liver disease is endotoxins and intestinal permeability.

9.2.5   H. PYLORI *INFECTION AND RELATED GASTROINTESTINAL DISORDERS*

The most common gastrointestinal infection is caused by Helicobacter pylori propounded more than 25 years ago (Marshall, 1985; Yamaoka et al., 2008). In the case of middle-aged adults, its prevalence all over the globe is more than 80%. Socioeconomic conditions like income, lower educational level, etc. are considered to be one of the reasons. The infection mainly attacks during childhood and remains throughout the lifetime. Some of the other modes of infection includes fecal-oral, oral-oral, iatrogenic, due to maternal child contact. There is a 6-fold increase in infection due to a sexual relationship with an infected partner.

Peptic ulcer disease (PUD) affects millions of people all over the world resulting in the rise in economic burden. The main reason behind PUD and chronic gastric carcinoma is *Helicobacter pylori* infection. *Gastric carcinoma* is a lethal disease with millions of people dying. The rate of occurrence of gastric carcinoma and *Helicobacter pylori* is same (Suerbaum et al., 2002). *Helicobacter pylori* infection affects the gastrointestinal system by increasing gastrin (Schubert et al., 2008) and causing hyposecretion as well as transient hypochlorhydria, duodenal ulcer, and gastric ulcer. On the contrary, low acid secretion might cause gastric ulcers and gastric carcinoma (Suerbaum et al., 2002). Chemokines and cytokines like interleukin–8, tumor necrosis factor-α (TNF-α) and interleukin 1β (Selgrad et al., 2008) are induced by *Helicobacter pylori* and cause mucosal inflammation. The inflammation is enhanced by the involvement of neutrophils, lymphocytes, and plasma cells. The adherence of *Helicobacter pylori* to gastric epithelium results in disruption of the mucosa and causes the host cell to release toxic proteins, cytotoxins, platelet activating factor and lipopolysaccharide, which augments the damage. PUD(Marshall et al., 2005) involves a high economic burden (Pezzi et al., 1995). NSAIDs indirectly damage the mucosa. About 90% of the ulcers cases are due to either non-steroidal anti-inflammatory drugs (NSAID) or *Helicobacter pylori* or both. Non-steroidal anti-inflammatory drugs inhibit the intestinal antimicrobial resistance (Tsai, 2005; Kivi et al., 2008). Novel strategies should be developed to treat Helicobacter infection so as to take clinical practice to new Heights.

9.2.6 GASTROINTESTINAL STROMAL TUMOR (GIST)

Gastrointestinal stromal tumor (GIST) is unusual cancer that affects the esophagus, stomach, small intestine and colon of the gastrointestinal system. Cancer initiates from connective tissues, fats, muscles, nerves, cartilage, etc. and finally attacks the system that affects the movement of food. Mainly stomach is affected by GIST and is prevalent in adults between ages 40–80 years but can occur in people of all ages. GIST can occur in the gastrointestinal tract and outside of it of varying sizes and occurring at various locations. The tumor has the potential to rebound after its removal. GIST, in its early stages, is difficult to diagnose due to lack of symptoms. Vague abdominal pain, abdominal bleeding found in stool or vomit, anemia due to low blood count, feeling of fullness causing a decrease in appetite are some of the symptoms associated with gastrointestinal stroma (Miettinen, 2006; Pidhorecky, 2006; Miettinen, 2006; Lehnert et al., 1998).

9.2.7   GASTROPARESIS

The movement of food through the gastrointestinal tract is controlled by the vagus nerve. When the vagus nerve is damaged due to prolonged high blood sugar as in the case of type–1 and type–2 diabetes, is referred to as gastroparesis (Park et al., 2006). Retention of food for a longer period in the GI takes place (Park et al., 2006). Different symptoms that occurring this disease are loss of control over blood sugar, stomach spasm, abdominal pain, early fullness of stomach, nausea, vomiting, bloating, heartburn, gastro-esophageal Reflux, weight loss, lack of appetite (Park et al., 2006). X-rays, manometer, and gastric emptying scans are the diagnostic tools for gastroparesis. Adjustment in insulin injection dose is required for patients with diabetes, dietary fiber, medications, jejunostomy tube, parenteral nutrition, botulinum toxins, gastric neurostimulators.

9.3   NANOMATERIALS

Though the scientific world considers nanomaterial to be an invention of the modern time yet history has a different narration. Nanoparticles were founded and used in the 9th century in mesopotamia. It was employed to give blistering effect on the surface of pots. The first glimpse of nanomaterial metals was given by Michael Faraday (Faraday et al., 1847). Visualization and characterization are two aspects of chemical and physical properties required for its development. It is the gift from modern technology that has created the possibility of accurately measuring and visualizing the particles size distribution with different techniques like transmission directly and scanning electron microscope as well as scanning tunneling microscopy (Horber, 2003; Karoutsos et al., 2009).

Flying on the wings of these advances, researchers have developed, advanced methods to generate particles of different sizes, geometries, and compositions. Nanoscale materials are extensively used in a wide range of fields, from medicine to engineering. The discovery of new properties of these nanoparticles has led to advancement especially in the field of nanomedicine. Great heights have been scaled with respect to synthesizing different types of new materials that can be employed as nanovehicles (Kroto, 1985; Li et al., 2005). In the area of imaging and diagnostics, the discovery of quantum effects (size-dependent properties) of nanomaterials, the entire scenario in the field of nanomedicine has changed. In quantum effect, nanoparticles produce specific emission, absorption or scattering that is harnessed for application in the imaging and diagnostic domain (Chan, 2002; Chen; 2008; Hikage et al., 2010). Nanomaterial is employed in drug delivery where it acts as drug carriers. Single-cell targeting has changed the innovative scenario all over the globe. Nanomaterials are characterized by small size (1 to 1000 nm), stability, large surface area to volume ratio, long shelf life and the potential to carry drugs in high concentration as compared to another molecule carrier (Rahman et al., 2017). The nanomaterial may be designed in such a way so as to skip the digestive process and perform site-specific delivery of drugs. The kinetics of drug release can be controlled and may be attached to the tailored surfaces to perform site-specific delivery (Beg S 2017). These properties equip the researchers to fabricate and employ nanomaterial in bio-imaging, diagnostic, and therapeutic drug delivery in the field of gastroenterology.

9.3.1   NANOCARRIERS USING IN TREATMENT OF GI DISORDERS

Nanocarrier systems are particulate dispersion or solid particles in which the drug is dissolved, attached or entrapped (Rahman, 2016). The carrier systems are generally polymers and are of two types namely natural and synthetic, with particulate vector nature (Ahmad, 2015a). These polymers must be non-toxic, biocompatible, non-carcinogenic, and non-immuno-genic (Ahmad, 2015b; Ahmad, 2010; Rahman, 2012c). They must possess biodegradable character, and the degraded product must not be toxic to the body. Synthetic degradable polymers used are lactides, glycolactides, poly-caprolactones, polyesters, polyorthoesters, polyanhydrides, polyamides, polyalkylcyanoacrylates, pseudo poly amino acids or polyphosphazenes (Aneja, 2014; Ahmad, 2013; Rahman, 2012a, 2012b, 2015). Non-degrad-able includes acrylic polymers, polyethylene glycol, polyethylene oxide, silicon, and elastomers (Ahmad, 2011; Akhter, 2011; Rahman, 2013, 2015; Kumar, 2017). Naturally occurring proteins such as gelatin or albumin, and polysaccharides like alginate, chitosan, amylopectin, or pectin are usually employed and find leverage over others as they can be degraded by colonic bacteria thus providing opportunity for colonic drug delivery (Pinhassil, 2010; Shukla et al., 2004). There are some polysaccharides like glucose, galactose, and acid derivatives that are easily degraded. The degraded product easily undergoes assimilation or elimination. Synthetic polymers have advantage over natural polymers that the size of average molecular weight of the former can be controlled which is not possible for the latter. Properly designed method for the synthesis of macromolecules can regulate size, molecular weight distribution, specific architecture, mechanical, and viscoelastic properties, and surface energy. There have been extensive studies on aliphatic polyesters with respect to their biocompatibility and biodegradability (Laroui, 2011; Ravi, 2008; Zambaux et al., 1999). Co-polymers like poly(D, L-Lactide), poly(glycolide), PGA are easily degraded into glycolic and lactic acid. They are natural metabolites, finally forming carbon dioxide and water. Similarly, lipase degrades polycaprolactone, which is more hydrophobic than PLA or PGA. Due to the hydrolysis of acrylate function, poly alkyl cyanoacrylates can easily dissolve in biological fluids (Brigger et al., 2002). Nanocarriers drug delivery for treatments of gastric disorders are illustrated in Figure 9.1 and Table 9.1.

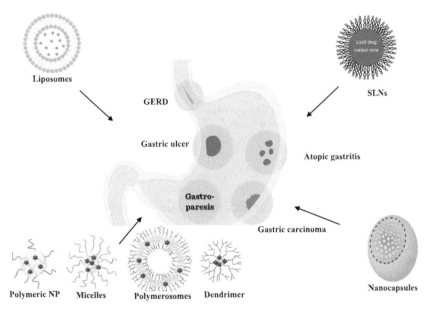

FIGURE 9.1 Illustrations of nanocarriers as drug delivery, diagnostics, and imaging in treatment of gastric disorders.

TABLE 9.1   Several Nanocarriers Employed in Gastro-Inflammatory Disorders (Elinav, 2013; Tahara, 2011; Van Deventer, 1997)

Drugs name	Nanocarrier	Advantages
Tacrolimus	PLGA- NPs	Significant reduction in inflammation and minimize adverse effects
Dexamethasone + butyrate	Solid lipid matrix (stearic acid–butyrate)	Reduced inflammation via reduction of pro-inflammatory cytokines, IL–1β and TNF–1α).
Dexamethasone	Liposomes	Alleviates Inflammation
5-ASA	Chitosan-NP	It founds 3.5-fold higher adhesion to inflamed tissue than chitosan free pellets.
Antisense-DNA nucleotide	Chitosan-PLGA	Specific uptake by inflamed mucosa and significant, amelioration of inflammation
m IL–10	Gelatin Microspheres	Amelioration of inflammation in DSS colitis; prevention of the onset of colitis
Dexamethasone	PLA-NP	Amelioration of inflammation in colitis
5-ASA	PCL-NP	NP–5-ASA 60 times more effective than 5-ASA solution and received anti-inflammatory effect.
5-ASA	Silica-NPs	It founds 6-fold higher adhesion to inflamed tissue; higher efficacy and prolonged effects
PLGA NPs versus Eudragit NP	Tacrolimus	PLGA NPs founds prolonged sustained drug release and significant higher reductions in inflammations.

5-ASA: 5-Aminosalicylic acid, PCL: poly-caprolactone, IL: interleukin, PLA: Polylactic acid, PLGA: poly (lactic-co-glycolic acid).

9.3.2   NANOPARTICLES

Drug carrier systems are called particulate vectors and are grouped according to (a) vector size, (b) targeting mechanisms, (c) nanomaterials like microparticles, micelles or nanoparticles. The particle size of nanoparticles and microparticles range from $1A^0$ to 1 μm and so bear similarity in size with most of the cells and bacteria (Rahman, 2012a, 2012b, 2012c). This unique size helps the micro and nanoparticles to be uptaken for interaction with the cell. On the basis of constituent elements, they are classified as follows.

9.3.2.1 LIPOSOMES

Liposomes (De Lattre et al., 1993) are spherical vesicles (Nano or micro) with phospholipid bilayer walls and an aqueous central layer. The aqueous layer may incorporate hydrophilic drugs while the phospholipid layer may contain hydrophobic drugs. Liposomes are made up of natural phospholipids, sterols or glycerol-lipids that imparts good biocompatibility and modified release (Rahman, 2015, 2016). The complex vesicular structure of the liposomes is chemically and physically unstable.

9.3.2.2 DRY EMULSIONS

Dry emulsions are spherical structures with peripheral part composed of surfactant and an oily inner core part for lipophilic drugs and can be prepared by using different types of amphiphilic macromolecules like protein, modified polysaccharide, and stabilizers (Pedersen et al., 1998). They are available as a freeze-dried powder which can be easily reconstituted in water for oil in water emulsion.

9.3.2.3 NANOCAPSULES

Nanocapsules or nanospheres are polymeric colloidal, 3rd generation vectors having a size ranging between 100 nm to 1 μm. Nanospheres are spherical in structure with shell and core made up of polymer and oil respectively. The drug can be attached to the network or adsorbed on the surface of the nanoparticle through covalent hydrophobic or electrostatic attraction. The affinity of the Nanospheres towards the cell or tissue can be enhanced by adding cell or tissue specific ligand during its synthesis (Chan et al., 2009).

9.4 STRATEGIES IN THE NANOPARTICLE DELIVERY IN DIGESTIVE SYSTEM

In Inflammatory bowel disease and colon cancer mainly two sites are affected namely colon and ileum. There is a need to target these sites by employing targeted Drug Delivery System. The different strategies developed for drug delivery in nanomedicine are transit time in the gastrointestinal tract, enzyme environment, ambient pressure and pH (Rahman, 2017).

9.4.1   TIME-DEPENDENT STRATEGY

The strategy developed, targets transit time of the drug in the gastrointestinal tract. Different coatings are applied, and the drug release feature is programmed in such a way so as to release the drug at the target site. In this, the outer layer is destroyed by releasing the drug. The drug release is affected by factors like degree of swelling of polysaccharides and pH change. Dorkoosh et al., (2001) took an effort to modify time release profile of the drug by changing the thickness of the layer of polymer to target any part of intestine and colon. Laroui et al., (2010) used two polymers namely chitosan and alginate to study their release profile at different concentrations. The release kinetic is affected by osmotic pressure of the salt or charged polymers or erosion rate of the coating by the polymer.

9.4.2   PH-DEPENDENT STRATEGY

The pH varies considerably in the GI tract. The pH in the stomach is 2–3, small intestine 5–6 and the colon 7. Eudragit poly (meth) acrylate is one such polymer containing an acidic or alkaline group that causes the pH-dependent release of the drug. Another example is that of polyacrylamide which is highly stable at acidic pH and degrades in neutral pH, thus protecting the drug from the harsh environment of the stomach. Certain polysaccharides have pH-dependent stability like chitosan, pectin or alginates (Crociani, 1994; Onishi, 2000; Sajeesh et al., 2010). Lamprecht et al. (2003) showed that with a small change in pH from 6.8 to 7.4, the release profile of 5-fluorouracil coated with Eudragit P–4581F microspheres can be modified. The colon drug delivery system is highly sensitive to weak pH changes which are harnessed for targeted drug delivery system. One of the strategies is to encapsulate a prodrug which gets converted to the active moiety due to specific pH and target the specific area.

9.4.3   PRESSURE DEPENDENT STRATEGY

The colon is characterized by a strong peristaltic movement which finally leads to an increase in viscosity and pressure of luminal content causing evacuation of the feces. This concept is utilized in the development of intestinal pressure-controlled colon delivery capsules (PCDC). It employs water insoluble Polymer ethyl cellulose which swells up (Hu, 2000; Shibata,

2001; Takaya et al., 1995). This PCDC does not rupture in the stomach and small intestine due to the fluid present in it. In the colon, due to increasing peristaltic pressure, it ruptures and releases the drug at the site.

9.4.4 ENZYME-BASED STRATEGY

In this strategy enzymes released by colonic bacteria is utilized. These enzymes cause degradation of nanomaterial and release of drug at the target site. Laroui et al.; (Laroui et al., 2010) fabricated polylactic acid nanoparticles containing an anti-inflammatory tripeptide encapsulated in a hydrogel. The hydrogel is composed of alginate and chitosan. The capsule is taken orally and is degraded in the colon by bacterial enzymes. Amylase, amylopectin, and chitosan are the polysaccharides that can be conveniently employed as they are bacterial enzyme sensitive (Laroui, 2010; Crociani, 1994; Laroui, 2011; Mc Connell et al., 2008).

9.4.5 THERAGNOSTIC APPROACH USING NANOMATERIAL IN GI DISEASES

Theragnostic deals with molecules for the purpose of both therapeutic and diagnostic using a single carrier. Nanomaterials are being fabricated and loaded with drugs for the purpose of therapy and imaging in Cancer (Rahman, 2017). It employee's molecules that can respond to both biological and chemical stimuli in a particular area targeted for treatment. A nanomaterial system may be linked to stimuli that indicate information like the elevation of temperature, hypoxia, pH, specific binding to inflammatory ligands. In imaging, the signaling molecules are encapsulated or attached to the carrier by covalent or non-covalent interactions (Veiseh et al., 2009) or by modifying the surface (Fang et al., 2009). The binding or surface modification depends upon two factors namely application and area that is being targeted for treatment. Functional groups like carbonyl, amines or silane (Fang, 2009; Li, 2005; Sun et al., 2008) are covalently linked and provide various functions under different pH or oxidative conditions. For flexibility, hydrophobic, electrostatic or hydrogen bonding interactions can be selected (Choi, 2008; Sun, 2010; Veiseh, 2009; Yang et al., 2007). Different types of polymers are selected to link the diagnostic and therapeutic molecule with the nanocarrier so as to produce a large number of interactions. For obtaining covalent, hydrophobic or ionic interaction, generally, Amides,

Ester (Veiseh, 2010 Xiong et al., 2010), disulfides, hydrazone or thioether linkages are used. Yang et al., (2009) in his study showed the inhibition of tumor growth and angiogenesis in various animal tumor models by blocking the binding between urokinase plasminogen activator (UPA) receptor and its natural ligands, UPA, ATF peptides. As a target ligand, he used UPA and amino-terminal fragments (ATF) peptide which binds with its receptor UPAR. He coated ATF nanoparticles which were actively taken up by cancer cells in tumor resulting in both tumor imaging and therapeutic delivery (Bu, 2004; Li et al., 2005).

9.4.5.1 TISSUE ENGINEERING

Damaged tissue and its repairing is a challenge to nanomedicine that necessitated the growth of tissue engineering. In this artificially stimulated cell proliferation is put on a scaffold made up of suitable material along with growth factor. An ideal scaffold is biodegradable and biocompatible at the same time should possess adequate interconnected pores and high surface area for the migration or diffusion of medium and distribution of nutrients, thus maintaining electrical and chemical communication among the cells (Serrano et al., 2007). Hassani et al. (2009) showed the treatment of gastric ulcers using nanotechnology. He explained that polystyrene made up of microspheres and nanospheres adhered to the gastric ulcer preferentially as compared to healthy tissues. For nanoparticles, a minimum size of 50 nm showed maximum adherence to the affected tissues.

9.4.5.2 SPECIFIC IMAGING AND THERAPEUTIC IN COLON CANCER

Quantum dots have revolutionized the field of imaging in cancer treatment. These are nanoparticles which emit light on changing size. They are more brighter and cheaper than organic dyes and can be excited by any light source which is blue shift to the emission spectrum. On conjugation with MRI can produce excellent 3D images of tumor. The main disadvantage with quantum dots is toxicity in human. There are many nanotechnological based products such as liposome (Doxil), nanoparticles (Abraxane) which have been approved and are presently in the market. The property of nano-carrier, exploited in this field is their size (10–100 nm) due to which they get accumulated within tumor. Photodynamic therapy is a new and non-invasive technique for treating diseases like tumors. It involves the illumination of particles which is being targeted for a specific area of a body by an external light source. The light can be used for different actions. It can be used to heat the metal particle and destroy surrounding molecules as well as generate high energy reactive oxygen species. The therapy is advantageous in the sense that no toxicity can be found in the body as the light only targets the specific area where the particle is confined. So, it counters the toxicity caused by quantum dots. Kirui et al. (Kirui et al., 2010) fabricated gold-iron oxide nanoparticles and strategically employed them for targeting, imaging, and laser photothermal therapy of cancerous cells at 800 nm of irradiation. Gold nanoparticles were functionalized with the terminal carboxy part of phospholipid and was conjugated with single chain antibody. This functionalized and conjugated nanoparticle binds selectively to overexpressed A33 antigen present on the surface of SW 122 colorectal cancer cells. The nanoparticles are uptaken by cancer cells and absorb light at 808 nm finally leading to the destruction of the cancerous cell.

9.4.5.3 GENE DELIVERY USING NANOPARTICLE

Gene therapy is basically mediated through transfection. Vectors used in it are virus and non-virus. The transfection through virus has high success rate, but oncogenesis and immunogenicity are its lacunae. Non-Virus based gene therapy includes formation of complexes of DNA or siRNA (small interference RNA) with cationic polymer chitosan. DNA or siRNA therapy poses the problem of low penetration across cell membrane (Meade et al., 2007). Lately, efforts have been made in developing tissue targeted nucleic acid delivery system using synthetic reagents (Farokhzad, 2004; Gunther, 2005; Hood, 2002; Kim, 2004; Pardridge, 2004; Ravi, 2008; Schiffelers, 2004; Toub et al., 2006). Polymeric nanoparticles loaded with siRNA have their surface attached with peptide, chemoattractant, and antibodies. Nanoparticles protect and augment the pharmacological action not only invitro but also in vivo (Fattal et al., 1998). In case of irritable bowel disease, biodegradable Polymers help reach the siRNA to cytosol for effective transfection (Toub et al., 2006).

9.4.5.4 NANOMEDICINE IN INFLAMMATORY BOWEL DISEASE

It is a chronic debilitating inflammatory disease covering both Crohn's disease and ulcerative colitis. Recent studies have shed light on the pathogenesis of IBD which is caused due to factors like T cells of immune system, human genes, bacteria, environment (Fiochhi et al., 2015), TNF–α. Saleem pointed out the involvement of NOD–2 genes in Crohn's disease (Saleem et al., 2015). Singh et al. (2015) showed the oral and topical limitation of 5-aminosalicylic acid and corticosteroid due to higher doses and low permeation. Liposomes have tremendous potential in drug delivery due to controlled rate specific target and low dose (Kesisoglue et al2005). IBD is characterized by increased permeability leading to accumulation of liposome, macrophages, and dendritic cells. These events result in greater uptake of the liposome by macrophages and dendritic cells as compared to tablet and solutions. The size of liposomes plays an active role in these uptakes (Kesisoglue et al., 2015) when therapy with corticosteroid fails then tacrolimus is used, but is accompanied by side effects (Singh et al2015). To overcome this side effect, it is loaded in a cationic as well as anionic liposome and administered to colitis as well as healthy rats (Jubeeh et al., 2004). Anionic liposomes adhered two times more than the cationic one, to the inflamed mucosa. Cationic liposomes adhered to the healthy mucosa three times more than anionic one (Jubeeh et al., 2004). In a study, anionic liposome were loaded with superoxide dismutase, 4-aminotenypol, and catalase and administered in colitis-induced rats. The liposome displayed higher uptake, increased residence time and higher anti-inflammatory effect in comparison to the drugs alone (Jubeeh et al., 2005). Nakase et al. formulated microparticles comprising of poly-D-lactic acid loaded with dexamethasone for oral administration to DSS induced colitis in rats (Nakase et al., 2003). These particles were preferentially taken up by the phagocytic cells which crowd the inflammatory site. Various macroscopic drug delivery vehicles have been designed to limit drug absorption in the upper intestine and cause release of the encapsulated drug at the specific area of colon in response to pH, colonic enzymes, transit time of colon, mechanical pressure. IBD is distributed in specific areas throughout the intestine but tablet, capsules, pellets, etc. deliver the encapsulated drug to healthy tissues also (Nakase et al., 2003). Limitation of macroscopic system is lack of specificity. They are also cleared by the diarrheal symptoms from the intestine, experienced by 92% of patients suffering from irritable bowel disease. Thus, these shortcomings in the treatment of IBD lead to introduction of nanoparticle-loaded drug delivery system which has the advantage of site specificity in response to inflamed issues with increase mucus production, accumulation of phagocytic immune cells and disruption of barrier owing to inflammation. A deep sincere and authentic study on micro and nanoparticles was conducted by Lamprecht et al. (2001), where he showed that the nanoparticles of (polylactic-co-glycolic acid) accumulated more in the mucosa of colitic ulcer than healthy tissues. The binding was highest with particle size less than 0.1 μm, proving that there exists an inverse proportionality between binding affinity to inflamed tissue and particle size. Another challenge which possess to the targeted nanoparticle delivery system is the uncontrolled release of drug molecule in the upper gastrointestinal tract due to high surface area. To counter the challenge authors linked anti-inflammatory drug 5-aminosalicylic acid to nanoparticles made of poly(caprolactone) (Pertuit et al., 2007). The upper gastrointestinal tract, outburst of the drug decreased significantly, and the 5-aminosalicylic acid solution was 60 times more effective in the IBD treatment. In another work, Laroui et al. loaded a tri-peptide in a polysaccharide made hydrogel which degraded in response to specific pH at the inflamed colonic site releasing the nanoparticle. The combination is delivered at 1000 times of lower dose than the peptide alone. The advantage of this system is protection and prevention of burst of drugs in the upper gastrointestinal tract and site specificity (Laroui et al., 2010).

In the domain of biological therapeutics delivery of siRNA to colon has been a major challenge. These siRNA decreases the expression of TNFα, pro-inflammatory cytokines responsible for IBD. With great effort Laroui et al., loaded the TNFα- siRNA nanoparticles and showed its effect on the macrophages using hydrogel. The formulation was administered in dextran sodium sulfate-induced colitis in mice model (Laroui et al., 2010, 2011).

Using polymer Wilson et al. (2010) formulated a combination which degraded in the presence of reactive oxygen species. He loaded the polymer nanoparticle with siRNA which finally depressed TNF α-mRNA, thus, changing the course of the inflammation in the colon (Wilson et al., 2010). According to Kountours et al. (2001), the release of siRNA is a ROS dependent mechanism. Biopsy of inflamed intestinal part reveals a 10 to 100-fold increase in reactive oxygen species. He developed nanoparticles loaded with siRNA and named as thioketal nanoparticle (tkns) which can release encapsulated agent in reactive oxygen species environment be formulated from poly1,4- phenylene acetone dimethyl thioketal). These TKNS do not degrade in acid, base, and protease catalyzed medium (Shukla, 2004; Colonna et al., 1996).

9.4.5.5 NANOPOWDER AS HEMOSTATIC AGENT IN ULCER BLEEDING USED THROUGH GI ENDOSCOPY

Nanoparticles find immense use in ulcer bleeding. Endoscopic hemostasis is achieved by injection tamponade, electrocautery, and hemoclips in Ulcer bleeding. A novel, safe, and effective method using nanopowder named TC 325 was used by Giday et al. in an anticoagulant induced severe gastrointestinal bleeding in an animal model. In a clinical study 20 patients was administered in a hospital with peptic ulcer bleeding (forecast score 1a or 1b) and within 24 hours underwent endoscopic hemospray, with around 95% patient achieving hemostasis. The Lone patient not achieving hemostasis had forest 1A ulcer bleeding. Bleeding started in 2 patients within 72 hours. A 30-day follow up resulted in no major complication (Giday and Sung et al., 2011).

9.4.5.6   *PREVENTION OF CLOGGING OF PLASTIC STENTS USING NANOPARTICLE*

For addressing malignant and biliary disorders endoscopically, stents are used to remove the blockage. The plastic stents get occluded by the sludge due to adherence of protein glycoprotein or bacteria. To solve the issue nanotechnology provides solution by modifying the surface of the plastics stent through soil release phenomenon. The modification of biliary plastic stent includes a thin coating of stable abrasion material using SOLGEL technology. A reduction in sludge adherence has been observed when plastic stent made of Teon with soft gel technology was used as compared to the uncoated one. These coating comprised of organic epoxides 190 gram per mole or 500 grams per mole and propylamine silane. Thus, nanocoating of this material prevents clogging of biliary plastic stent (Sietz et al., 2007).

9.4.5.7   *NANO-BASED CAPSULE ENDOSCOPY*

Capsule endoscopy is indispensable and widely used technology in bowel and colonic diseases. The drawback of this technique is failure to detect deep tissue disorders and lack of therapeutic activity. European Union is funding projects like NEMO for the integration of optical Technology, nanotechnology, biosensing, and maneuvering technology so as to create capsule endoscopy with molecular imaging and secretion analysis capability. It can also detect marked and deep tissues disorder. It is advantageous in detecting cancerous and pancreas tissue in the GI tract. Robotic beetles, capsule endoscopy is a step-in innovation consisting of ultrasound, transducers, bioanalytical, and mechanical sensors along with robotic arms for obtaining tissue samples, releasing drug and thermally destroying tissue

(Laroui et al., 2011). Currently, the researchers are trying to fabricate intelligent endoscopic capsules with both diagnostic and therapeutic ability. The vector capsule consists of locomotion, diagnostic, and therapeutic system (Laroui et al., 2011).

9.5 CONCLUSION

The present book chapter summarized the infestations of gastrointestinal disorders, their pathophysiology and applications of nanomedicine in its effective treatment. The application of nanomedicines is a rapidly developing area of investigation, and in the near future, it will play a key role in the effective treatment of gastric disorders as diagnostics, imaging, and drug delivery. Indeed, some examples of nanocarriers based therapies showed best therapeutics over conventional materials in terms of efficacy. There are needs of optimization of nanomaterial, which will determine the applicability of these methods in modern care healthcare practice. Nanorobots and capsule endoscopy current innovation for surgical or endoscopic procedures and detection of small bowel lesions.

KEYWORDS

- colon cancer
- gastric and celiac diseases
- gastroenterology
- nanomedicines
- nanotechnology
- theranostics

ACKNOWLEDGEMENT

We express our sincere thanks to DST-INSPIRE fellowship program, DST-SERB, UGC-UPE II and AICTE-RPS schemes of Govt. of India for providing financial support which was utilized for the present study.

REFERENCES

Ahmad, J., Akhter, S., Rizwanullah, M., et al., (2015a). Nanotechnology-based inhalation treatments for lung cancer: State of the art. *Nanotechnol Sci. Appl.*, *19*(8), 55–66.

Ahmad, J., Amin, S., Rahman, M., et al., (2015b). Solid matrix based lipidic nanoparticles in oral cancer chemotherapy: Applications and pharmacokinetics. *Curr. Drug. Metab.*, *16*(8), 633–644.

Ahmad, M. Z., Akhter, S., Ahmad, I., et al., (2011). Development of polysaccharide-based colon targeted drug delivery system: Design and evaluation of Assam Bora rice starch-based matrix tablet. *Curr. Drug Deliv.*, *8*(5), 575–581.

Ahmad, M. Z., Akhter, S., Anwar, M., et al., (2013). Colorectal cancer targeted Irinotecan-Assam Bora rice starch-based microspheres: A mechanistic, pharmacokinetic and biochemical investigation. *Drug Dev. Ind. Pharm.*, *39*(12), 1936–1943.

Ahmad, M. Z., Akhter, S., Jain, G. K., et al., (2010). Metallic nanoparticles: Technology overview & drug delivery applications in oncology. *Expert. Opin. Drug Deliv.*, *7*(8), 927–942.

Akhter, S., Ahmad, Z., Singh, A., et al., (2011). Cancer targeted metallic nanoparticle: Targeting overview, recent advancement, and toxicity concern. *Curr. Pharm. Des.*, *17*(18), 1834–1850.

Aneja, P., Rahman, M., Beg, S., et al., (2014). Cancer targeted magic bullets for effective treatment of cancer. *Recent Pat. Antiinfect. Drug. Discov.*, *2*, 121–135.

Azpiroz, F., & Serra, J., (2004). Treatment of excessive intestinal gas. *Current Treatment Options in Gastroenterology*, *7*(4), 299–305.

Baumgart, D. C., & Sandborn, W. J., (2012). Crohn's disease. *Lancet*, *380*(9853), 1590–1605.

Beg, S., Rahman, M., Jain, A., et al., (2017). Nanoporous metal-organic frameworks as hybrid polymer-metal composites for drug delivery and biomedical applications. *Drug. Discov. Today.*, *22*(4), 625–637.

Bjarnason, I., Hayllar, J., MacPherson, A. J., et al., (1993). Side effects of non-steroidal anti-inflammatory drugs on the small and large intestine in humans. *Gastroenterology*, *104*(6), 1832–1847.

Bode, C., Kugler, V., & Bode, J. C., (1985). Pyloric Campylobacter infection and gastroduodenal disease. *Med. J. Aust.*, *142*(8), 439–444.

Bode, C., Kugler, V., & Bode, J. C., (1987). Endotoxemia in patients with alcoholic and nonalcoholic cirrhosis and in subjects with no evidence of chronic liver disease following acute alcohol excess. *J. Hepatol*, *4*(1), 8–14.

Brigger, I., Dubernet, C., & Couvreur, P., (2002). Nanoparticles in cancer therapy and diagnosis. *Adv. Drug. Deliv. Rev.*, *54*(5), 631–651.

Bu, X., Khankaldyyan, V., Gonzales-Gomez, I., et al., (2004). Species-specific urokinase receptor ligands reduce glioma growth and increase survival primarily by an antiangiogenesis mechanism. *Lab. Invest.*, *84*(6), 667–678.

Chan, L. M., Lowes, S., Hirst, B. H., et al., (2004). The ABCs of drug transport in intestine and liver: Efflux proteins limiting drug absorption and bioavailability. *Eur. J. Pharm. Sci.*, *21*(1), 25–51.

Chan, W. C., Maxwell, D. J., Gao, X., et al., (2002). Luminescent quantum dots for multiplexed biological detection and imaging. *Curr. Opin. Biotechno.*, *13*(1), 40–46.

Chen, L. D., Liu, J., Yu, X. F., et al., (2008). The biocompatibility of quantum dot probes used for the targeted imaging of hepatocellular carcinoma metastasis. *Biomaterials*, *29*(31), 4170–4176.

Cheng, L. K., O'Grady, G., Du, P., et al., (2010). Gastrointestinal system. *Wiley Interdiscip. Rev. Syst. Biol. Med.*, *2*(1), 65–79.

Choi, J. S., Park, J. C., Nah, H., et al., (2008). A hybrid nanoparticle probe for dual-modality positron emission tomography and magnetic resonance imaging. *Angew. Chem. Int. Ed. Engl.*, *47*(33), 6259–6262.

Colonna, S., Gaggero, N., Carrea, G., et al., (1996). Enantio and diastereoselectivity of cyclohexanone monooxygenase catalyzed oxidation of 1, 3-dithioacetals. *Tetrahedron*, *7*(2), 565–570.

Crociani, F., Alessandrini, A., Mucci, M. M., et al., (1994). Degradation of complex carbohydrates by Bifidobacterium spp. *Int. J. Food Microbiol.*, *24*(1 & 2), 199–210.

De Vault, K. R., & Castell, D. O., (1999). Updated guidelines for the diagnosis and treatment of gastroesophageal reflux disease. The Practice Parameters Committee of the American College of Gastroenterology. *Am.J.Gastroenterology*, *94*(6), 1434–1442.

Delattre, J., Couvreur, P., Puisieux, F., et al., (1993). In: Doc-Lavoisier, T., (ed.), *JR Les Liposomes: Aspects Technologiques, Biologiques et Pharmacologiques*. Paris: Editions INSERM, pp. 266. 310 F ISBN 2-85206-891-5. Biochemical Education. 23. 78–78.

DeMeo, M. T., Mutlu, E. A., Keshavarzian, A., et al., (2002). Intestinal permeation and gastrointestinal disease. *J. Clin. Gastroenterol*, *34*(4), 385–396.

Dorkoosh, F. A., Verhoef, J. C., Borchard, G., et al., (2001). Development and characterization of a novel peroral peptide drug delivery system. *J. Control. Release*, *71*(3), 307–318.

Eberl, G., (2005). Inducible lymphoid tissues in the adult gut: Recapitulation of a fetal developmental pathway. *Nat. Rev. Immunol.*, *5*, 413–420.

Elinav, E., & Peer, D., (2013). Harnessing nanomedicine for mucosal theranostics: A silver bullet at last? *ACS Nano.*, *7*, 2883–2890.

Emerich, D. F., & Thanos, C. G., (2003). Nanotechnology and medicine. *Expert. Opin. Biol. Ther.*, *3*(4), 655–663.

Falth-Magnusson, K., Kjellman, N. I., & Magnusson, K. E., (1988). Antibodies IgG, IgA, and IgM to food antigens during the first 18 months of life in relation to feeding and development of atopic disease. *J. Allergy Clin. Immunol.*, *81*(4), 743–749.

Fang, C., Bhattarai, N., Sun, C., et al., (2009). Functionalized nanoparticles with long-term stability in biological media. *Small*, *5*(14), 1637–1641.

Faraday, M., (1847). The Bakerian lecture: Experimental relations of gold (and other metals) to light. *Phil. Trans. R. Soc. Lond.*, *147*, 145–181.

Farmer, R. G., Hawk, W. A., & Turnbull, R. B., (1975). Clinical patterns in Crohn's disease: A statistical study of 615 cases. *Gastroenterology*, *68*(4), 627–635.

Farokhzad, O. C., Jon, S., Khademhosseini, A., et al., (2004). Nanoparticle-aptamer bioconjugates: A new approach for targeting prostate cancer cells. *Cancer Res.*, *64*(21), 7668–7672.

Farrell, C. P., Barr, M., Mullin, J. M., et al., (2012). Epithelial barrier leak in gastrointestinal disease and multiorgan failure. *J. Epith. Biol. Pharmacol.*, *5*, 13–18.

Fattal, E., Vauthier, C., Aynie, I., et al., (1998). Biodegradable polyalkylcyanoacrylate nanoparticles for the delivery of oligonucleotides. *J. Control Release.*, *53*(1 & 3), 137–143.

Feynman, R. P., (1959). *There's Plenty of Room at the Bottom, Annual Meeting of American Physiological Society*. California Institute of Technology, Pasadena, CA. 23 (5). pp. 22-36.

Fiocchi, C., (2015). Inflammatory bowel disease pathogenesis: Where are we? *Journal of Gastroenterology and Hepatology*, *30*(S1), 12–18.

Gabe, S. M., Bjarnason, I., & Tolou-Ghamari, Z., (1998). The effect of tacrolimus (FK506) on intestinal barrier function and cellular energy production in humans. *Gastroenterology, 115*(1), 67–74.

Giday, S. A., Kim, Y., Krishnamurty, D. M., et al., (2011). Long-term randomized controlled trial of a novel nanopowder hemostatic agent (TC–325) for control of severe arterial upper gastrointestinal bleeding in a porcine model. *Endoscopy, 43*(4), 296–299.

Guha, A., Biswas, N., Bhattacharjee, K., et al., (2016). pH-responsive cylindrical MSN for oral delivery of insulin-design, fabrication, and evaluation. *Drug Delivery, 23*(9), 3552–3561.

Gunther, M., Wagner, E., Ogris, M., et al., (2005). Specific targets in tumor tissue for the delivery of therapeutic genes. *Curr. Med. Chem. Anticancer Agents, 5*(2), 157–171.

Hall, E. J., & Batt, R. M., (1991). Abnormal permeability precedes the development of a gluten sensitive enteropathy in Irish setter dogs. *Gut., 32*(7), 749–753.

Hassani, S., Pellequer, Y., & Lamprecht, A., (2009). Selective adhesion of nanoparticles to inflamed tissue in gastric ulcers. *Pharm. Res., 26*(5), 1149–1154.

Hermiston, M. L., & Gordon, J. I., (1995). Inflammatory bowel disease and adenomas in mice expressing a dominant negative N-cadherin. *Science, 270*(5239), 1203–1207.

Heyman, M., Darmon, N., & Dupont, C., (1994). Mononuclear cells from infants allergic to cow's milk secrete tumor necrosis factor alpha, altering intestinal function. *Gastroenterology, 106*(6), 1514–1523.

Hikage, M., Gonda, K., Takeda, M., et al., (2010). Nanoimaging of the lymph network structure with quantum dots. *Nanotechnology, 21*(18), 185103.

Hollander, D., (1992). The intestinal permeability barrier: A hypothesis as to its regulation and involvement in Crohn's disease. *Scand. J. Gastroenterol, 27*(9), 721–726.

Hood, J. D., Bednarski, M., Frausto, R., et al., (2002). Tumor regression by targeted gene delivery to the neovasculature. *Science, 296*(5577), 2404–2407.

Horber, J. K., & Miles, M. J., (2003). Scanning probe evolution in biology. *Science, 302*(5647), 1002–1005.

Hu, Z., Mawatari, S., Shibata, N., et al., (2000a). Application of a biomagnetic measurement system (BMS) to the evaluation of gastrointestinal transit of intestinal pressure-controlled colon delivery capsules (PCDCs) in human subjects. *Pharm. Res., 17*(2), 160–167.

Hu, Z., Mawatari, S., Shimokawa, T., et al., (2000b). Colon delivery efficiencies of intestinal pressure-controlled colon delivery capsules prepared by a coating machine in human subjects. *J. Pharm. Pharmacol., 52*(10), 1187–1193.

Izrailov, R. E., Tsvirkun, V. V., Alikhanov, R. B., et al., (2018). Laparoscopic pancreatic head resection. *Khirurgiia (Mosk), 2*, 45–51.

Izzo, A. A., & Coutts, A. A., (2005). Cannabinoids and the digestive tract. *Handb. Exp. Pharmacol., 168*, 573–598.

Jalonen, T., (1991). Identical intestinal permeability changes in children with different clinical manifestations of cow's milk allergy. *J. Allergy. Clin. Immunol., 88*(5), 737–742.

Karoutsos, V., (2009). Scanning probe microscopy: Instrumentation and applications on thin films and magnetic multilayers. *J. Nanosci. Nanotechnol., 9*(12), 6783–6793.

Kennedy, R. J., Hoper, M., & Deodhar, K., (2000). Interleukin 10-deficient colitis: New similarities to human inflammatory bowel disease. *Br. J. Surg., 87*(10), 1346–1351.

Kim, B., Tang, Q., Biswas, P. S., et al., (2004). Inhibition of ocular angiogenesis by siRNA targeting vascular endothelial growth factor pathway genes: Therapeutic strategy for herpetic stromal keratitis. *Am. J. Pathol., 165*(6), 2177–2185.

Kim, K. S., Khang, G., & Lee, D., (2011). Application of nanomedicine in cardiovascular diseases and stroke. *Current Pharmaceutical Design, 17*, 1825–1833.

Kirui, D. K., Rey, D. A., & Batt, C. A., (2010). Gold hybrid nanoparticles for targeted phototherapy and cancer imaging. *Nanotechnology, 21*(10), 105105.

Kivi, M., & Tindberg, Y., (2006). Helicobacter pylori occurrence and transmission: A family affair? *Scand. J. Infect. Dis., 38*(6 & 7), 407–417.

Kountouras, J., Chatzopoulos, D., & Zavos, C., (2001). Reactive oxygen metabolites and upper gastrointestinal diseases. *Hepatogastroenterology, 48*(39), 743–751.

Kroto, H. W., Heath, J. R., O'Brien, S. C., et al., (1985). C_{60}: Buckminsterfullerene. *Nature, 318*(6042), 162–163.

Kumar, P., & Clark, M., (2005). *Clinical Medicine* (6th edn.). Edinburgh: WB Saunders xviii, 1508 p.

Kumar, V., Bhatt, P. C., Rahman, M., et al., (2017). Fabrication, optimization, and characterization of umbelliferone β-D-galactopyranoside-loaded PLGA nanoparticles in treatment of hepatocellular carcinoma: *In vitro* and *in vivo* studies. *Int. J. Nanomedicine., 12*, 6747–6758.

Lamprecht, A., Ubrich, N., Yamamoto, H., et al., (2001). Biodegradable nanoparticles for targeted drug delivery in treatment of inflammatory bowel disease. *J. Pharmacol. Exp. Ther., 299*(2), 775–781.

Lamprecht, A., Yamamoto, H., Takeuchi, H., et al., (2003). Microsphere design for the colonic delivery of 5-fluorouracil. *J. Control Release, 90*(3), 313–322.

Laroui, H., Dalmasso, G., Nguyen, H. T., et al., (2010a). Drug-loaded nanoparticles targeted to the colon with polysaccharide hydrogel reduce colitis in a mouse model. *Gastroenterology, 138*(3), 843–853, 1–2.

Laroui, H., Wilson, D. S., Dalmasso, G., et al., (2010b). Nanomedicine in GI. *Am. J. Physiol. Gastrointest. Liver. Physiol., 300*(3), G371–G383.

Lehnert, T., (1998). Gastrointestinal sarcoma (GIST)—a review of surgical management. *Ann. Chir. Gynaecol., 87*(4), 297–305.

Li, H., Soria, C., Griscelli, F., et al., (2005). Amino-terminal fragment of urokinase inhibits tumor cell invasion *in vitro* and *in vivo*: Respective contribution of the urokinase plasminogen activator receptor-dependent or -independent pathway. *Hum. Gene Ther., 16*(10), 1157–1167.

Li, J., Ng, H. T., & Chen, H., (2005). Carbon nanotubes and nanowires for biological sensing. *Methods. Mol. Biol., 300*, 191–123.

Malagelada, J. R., (2006). A symptom-based approach to making a positive diagnosis of irritable bowel syndrome with constipation. *Int. J. Clin. Pract., 60*(1), 57–63.

Marsh, M. N., Bjarnason, I., & Shaw, J., (1990). Studies of intestinal lymphoid tissue: XIV–HLA status, mucosal morphology, permeability and epithelial lymphocyte populations in first degree relatives of patients with celiac disease. *Gut., 31*(1), 32–36.

Marshall, B. J., & Windsor, H. M., (2005). The relation of *Helicobacter pylori* to gastric adenocarcinoma and lymphoma: Pathophysiology, epidemiology, screening, clinical presentation, treatment, and prevention. *Med. Clin. North. Am., 89*(2), 313–344.

Marshall, B. J., McGechie, D. B., & Rogers, P. A., (1985). Pyloric Campylobacter infection and gastroduodenal disease. *Med. J. Aust., 142*(8), 439–444.

McConnell, E. L., Murdan, S., & Basit, A. W., (2008). An investigation into the digestion of chitosan (noncrosslinked and crosslinked) by human colonic bacteria. *J. Pharm. Sci., 97*(9), 3820–3829.

McKay, D. M., & Baird, A. W., (1999). Cytokine regulation of epithelial permeability and ion transport. *Gut, 44*(4), 283–289.

Meade, B. R., & Dowdy, S. F., (2007). Exogenous siRNA delivery using peptide transduction domains/cell penetrating peptides. *Adv. Drug Deliv. Rev., 59*(2 & 3), 134–140.

Miettinen, M., & Lasota, J., (2006a). Gastrointestinal stromal tumors: Review on morphology, molecular pathology, prognosis, and differential diagnosis. *Arch. Pathol. Lab. Med., 130*(10), 1466–478.

Miettinen, M., & Lasota, J., (2006b). Gastrointestinal stromal tumors-definition, clinical, histological, immunohistochemical, and molecular genetics features, and differential diagnosis. *Virchows. Arch., 438*(1), 1–12.

Mueller, M. H., Kreis, M. E., Gross, M. L., et al., (2002). Anorectal functional disorders in the absence of anorectal inflammation in patients with Crohn's disease. *Br. J. Surg., 89*(8), 1027–1031.

Noel, R. A., Braun, D. K., Patterson, R. E., et al., (2009). Increased risk of acute pancreatitis and biliary disease observed in patients with type 2 diabetes: A retrospective cohort study. *Diabetes Care, 32*(5), 834–838.

Onishi, H., Koyama, K., Sakata, O., et al., (1994). Preparation of chitosan/alginate/calcium complex microparticles loaded with lactoferrin and their efficacy on carrageenan-induced edema in rats. *Drug Dev. Ind. Pharm., 36*(8), 879–884.

Ouellette, A. J., (1997). Paneth cells and innate immunity in the crypt microenvironment. *Gastroenterology, 113*(5), 1779–1784.

Panes, J., Gomollon, F., Taxonera, C., et al., (2007). Crohn's disease: A review of current treatment with a focus on biologics. *Drugs, 67*(17), 2511–2537.

Pardridge, W. M., (2004). Intravenous, non-viral RNAi gene therapy of brain cancer. *Expert. Opin. Biol. Ther., 4*(7), 1103–1113.

Park, Moo-In., & Camilleri, M., (2006). Gastroparesis: Clinical update. *Am. J. Gastroenterol., 101*(5), 1129–1139.

Pedersen, G. P., Fäldt, P., Bergenstaåhl, B., et al., (1998). Solid state characterization of a dry emulsion: A potential drug delivery system. *Int. J. Pharm., 171*(2), 257–270.

Pertuit, D., Moulari, B., Betz, T., et al., (2007). 5-Aminosalicylic acid bound nanoparticles for the therapy of inflammatory bowel disease. *J. Control. Release, 123*(3), 211–218.

Pezzi, J. S., & Shiau, Y. F., (1995). Helicobacter pylori and gastrointestinal disease. *Am. Fam. Physician., 52*(6), 1717–1724.

Pidhorecky, I., Cheney, R. T., Kraybill, W. G., et al., (2006). Gastrointestinal stromal tumors: Current diagnosis, biologic behavior, and management. *Ann. Surg. Oncol., 7*(9), 705–712.

Pinhassi, R. I., Assaraf, Y. G., Farber, S., et al., (2010). Arabinogalactan-folic acid-drug conjugate for targeted delivery and target-activated release of anticancer drugs to folate receptor-overexpressing cells. *Biomacromolecules, 11*(1), 294–303.

Rahman, M., Ahmad, M. Z., Ahmad, J., et al., (2015a). Role of graphene nano-composites in cancer therapy: Theranostic applications, metabolic fate, and toxicity issues. *Curr. Drug Metab., 16*(5), 397–409.

Rahman, M., Ahmad, M. Z., Kazmi, I., et al., (2012a). Emergence of nanomedicine as cancer-targeted magic bullets: Recent development and need to address the toxicity apprehension. *Curr. Drug. Discov. Technol., 9*(4), 319–329.

Rahman, M., Ahmad, M. Z., Kazmi, I., et al., (2012b). Advancement in multifunctional nanoparticles for the effective treatment of cancer. *Expert. Opin. Drug. Deliv., 9*(4), 367–381.

Rahman, M., Ahmed, M. Z., Kazmi, I., et al., (2012c). Novel approach for the treatment of cancer: Theranostic nanomedicines. *Pharmacologia, 3,* 371–376.

Rahman, M., Akhter, S., Ahmad, J., et al., (2015b). Nanomedicine-based drug targeting for psoriasis: Potentials and emerging trends in nanoscale pharmacotherapy. *Expert. Opin. Drug. Deliv., 12*(4), 635–652.

Rahman, M., Akhter, S., Ahmad, M. Z., et al., (2015c). Emerging advances in cancer nanotheranostics with graphene nanocomposites: Opportunities and challenges. *Nanomedicine (Lond)., 10*(15), 2405–2422.

Rahman, M., Beg, S., Ahmed, A., et al., (2013). Emergence of functionalized nanomedicines in cancer chemotherapy: Recent advancements, current challenges, and toxicity considerations. *Recent Patent on Nanomedicine., 2,* 128–139.

Rahman, M., Beg, S., Verma, A., et al., (2017). Therapeutic applications of liposomal-based drug delivery and drug targeting for immune linked inflammatory maladies: A contemporary viewpoint. *Curr. Drug. Targets., 18*(13), 1558–1571.

Rahman, M., Kumar, V., Beg, S., et al., (2016). Emergence of liposome as targeted magic bullet for inflammatory disorders: Current state of the art. *Artif. Cells Nanomed. Biotechnol., 44*(7), 1597–1608.

Ravi, S., Peh, K. K., Darwis, Y., et al., (2008). Development and characterization of polymeric microspheres for controlled release protein loaded drug delivery system. *Indian J. Pharm. Sci., 70*(3), 303–309.

Reinhardt, M. C., (1984). Macromolecular absorption of food antigens in health and disease. *Ann. Allergy., 53*(6 pt 2), 597–601.

Roberts, S. A., & Freed, D. L., (1977). Neonatal IgA secretion enhanced by breastfeeding. *Letter Lancet., 2*(8048), 1131.

Russo, M. W., Wei, J. T., & Thiny, M. T., (2004). Digestive and liver disease statistics. *Gastroenterology, 126*(5), 1448–1453.

Sahoo, S. K., & Labhasetwar, V., (2003). Nanotech approaches to drug delivery and imaging. *Drug. Discov. Today., 8*(24), 1112–1120.

Sajeesh, S., Bouchemal, K., Sharma, C. P., et al., (2010). Surface-functionalized polymeth-acrylic acid-based hydrogel microparticles for oral drug delivery. *Eur. J. Pharm. Biopharm., 74*(2), 209–218.

Sakorafas, G. H., & Tsiotou, A., (2000). Etiology and pathogenesis of acute pancreatitis: Current concepts. *J. Clin. Gastroenterol., 30*(4), 343–356.

Salem, M., Seidelin, J., et al., (2014). Species-specific engagement of human NOD2 and TLR signaling upon intracellular bacterial infection: Role of Crohn's associated NOD2 gene variants. *Clinical & Experimental Immunology, 179*(3). P. 426-434.

Schiffelers, R. M., Ansari, A., Xu, J., et al., (2004). Cancer siRNA therapy by tumor selective delivery with ligand-targeted sterically stabilized nanoparticle. *Nucleic Acids Res., 32*(19), e149.

Schubert, M. L., & Peura, D. A., (2008). Control of gastric acid secretion in health and disease. *Gastroenterology, 134*(7), 1842–1860.

Seitz, U., Block, A., Schaefer, A. C., et al., (2007). Biliary stent clogging solved by nanotechnology? *In vitro* study of inorganic-organic sol-gel coatings for Teflon stents. *Gastroenterology, 133*(1), 65–71.

Selgrad, M., & Malfertheiner, P., (2008). New strategies for Helicobacter pylori eradication. *Curr. Opin. Pharmacol., 8*(5), 593–597.

Serrano, F., Lopez, G. L., Jadraque, M., et al., (2007). A Nd: YAG laser-microperforated poly(3-hydroxybutyrate-co–3-hydroxyvalerate)-basal membrane matrix composite film as substrate for keratinocytes. *Biomaterials, 28*(4), 650–660.

Shibata, N., Ohno, T., Shimokawa, T., et al., (2001). Application of pressure-controlled colon delivery capsule to oral administration of glycyrrhizin in dogs. *J. Pharm. Pharmacol., 53*(4), 441–447.

Shukla, A. K., Verma, M., Singh, K. N., et al., (2004). Superoxide induced deprotection of 1, 3-dithiolanes: A convenient method of dedithioacetalization. *Indian. J. Chem-Sect B., 43*(08), 1748–1752.

Suerbaum, S., & Michetti, P., (2002). *Helicobacter pylori* infection. *N. Engl. J. Med., 347*(15), 1175–1186.

Sun, C., Du, K., Fang, C., et al., (2010). PEG-mediated synthesis of highly dispersive multifunctional superparamagnetic nanoparticles: Their physicochemical properties and function *in vivo. ACS Nano., 4*(4), 2402–2410.

Sun, C., Lee, J. S., & Zhang, M., (2008). Magnetic nanoparticles in MR imaging and drug delivery. *Adv. Drug. Deliv. Rev., 60*(11), 1252–1265.

Sung, J. J., Luo, D., Wu, J. C., et al., (2011). Early clinical experience of the safety and effectiveness of hemospray in achieving hemostasis in patients with acute peptic ulcer bleeding. *Endoscopy, 43*(4), 291–295.

Tahara, K., Samura, S., Tsuji, K., Yamamoto, H., et al., (2011). Oral nuclear factor-κB decoy oligonucleotides delivery system with chitosan modified poly(D, L-lactide-co- glycolide) nanospheres for inflammatory bowel disease. *Biomaterials, 32*, 870–878.

Takaya, T., Ikeda, C., Imagawa, N., et al., (1995). Development of a colon delivery capsule and the pharmacological activity of recombinant human granulocyte colony-stimulating factor (rhG-CSF) in beagle dogs. *J. Pharm. Pharmacol., 47*(6), 474–478.

Toub, N., Bertrand, J. R., Tamaddon, A., et al., (2006). Efficacy of siRNA nanocapsules targeted against the EWS-Fli1 oncogene in Ewing sarcoma. *Pharm. Res., 23*(5), 892–900.

Tsai, C. J., Perry, S., & Sanchez, L., (2005). Helicobacter pylori infection in different generations of Hispanics in the San Francisco Bay Area. *Am. J. Epidemiol., 162*(4), 351–357.

Ukabam, S. O., Clamp, J. R., Cooper, B. T., et al., (1983). Abnormal small intestinal permeability to sugars in patients with Crohn's disease of the terminal ileum and colon. *Digestion, 27*(2), 70–74.

Van Deventer, S. J. H., Elson, C. O., & Fedorak, R. N., (1997). Multiple doses of intravenous interleukin 10 in steroid-refractory Crohns disease. *Gastroenterology, 113*, 383–389.

Van, D. H., (2005). Recent advances in coeliac disease. *Gut., 55*(7), 1037–1046.

Van, R. M. E., Uil, J. J., & Mulder, C. J., (1993). Intestinal permeability in patients with celiac disease and relatives of patients with celiac disease. *Gut., 34*(3), 354–357.

Vasir, J. K., & Labhasetwar, V., (2005). Targeted drug delivery in cancer therapy. *Technol. Cancer Res. Treat., 4*(4), 363–374.

Veiseh, O., Gunn, J. W., & Zhang, M., (2010). Design and fabrication of magnetic nanoparticles for targeted drug delivery and imaging. *Adv. Drug Deliv. Rev., 62*(3), 284–304.

Veiseh, O., Sun, C., Fang, C., et al., (2009). Specific targeting of brain tumors with an optical/ magnetic resonance imaging nanoprobe across the blood-brain barrier. *Cancer Res., 69*(15), 6200–6207.

Warshaw, A. L., Banks, P. A., & Fernández-Del, C. C., (1998). AGA technical review: Treatment of pain in chronic pancreatitis. *Gastroenterology, 115*(3), 765–776.

Whitesides, G. M., (2003). The 'right size' in nanobiotechnology. *Nat. Biotechnol.*, *21*(10), 1161–1165.

Wigg, A. J., Roberts-Thomson, I. C., & Dymock, R. B., (2001). The role of small intestinal bacterial overgrowth, intestinal permeability, endotoxemia, and tumor necrosis factor alpha in the pathogenesis of non-alcoholic steatohepatitis. *Gut*, *48*(2), 206–211.

Wilson, D. S., Dalmasso, G., Wang, L., et al., (2010). Orally delivered thioketal nanoparticles loaded with TNF-alpha-siRNA target inflammation and inhibit gene expression in the intestines. *Nat. Mater.*, *9*(11), 923–928.

Xiong, X. B., Ma, Z., Lai, R., et al., (2010). The therapeutic response to multifunctional polymeric nanoconjugates in the targeted cellular and subcellular delivery of doxorubicin. *Biomaterials*, *31*(4), 757–768.

Yamaoka, Y., (2008). *Helicobacter Pylori: Molecular Genetics and Cellular Biology*. Caister Academic. Pr., ix + 261 p.; ill.; index. ISBN: 978-1-904455-31-8 . The Quarterly Review of Biology, 85, 110-110.

Yang, L., Mao, H., Cao, Z., et al., (2009). Molecular imaging of pancreatic cancer in an animal model using targeted multifunctional nanoparticles. *Gastroenterology*, *136*(5), 1514–1525, e2.

Yang, S. Q., Lin, H. Z., & Lane, M. D., (1997). Obesity increases sensitivity to endotoxin liver injury: Implications for the pathogenesis of steatohepatitis. *Proceedings of the National Academy of Sciences of the United States of America*, *94*(6), 2557–2562.

Yang, Z., Zheng, S., Harrison, W. J., et al., (2007). Long-circulating near-infrared fluorescence core-cross-linked polymeric micelles: Synthesis, characterization, and dual nuclear/optical imaging. *Biomacromolecules*, *8*(11), 3422–3428.

Zambaux, M. F., Bonneaux, F., Gref, R., et al., (1999). Preparation and characterization of protein C-loaded PLA nanoparticles. *J. Control. Release*, *60*(2 & 3), 179–188.

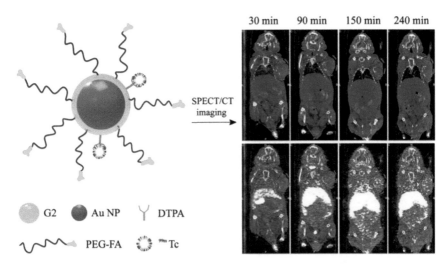

FIGURE 2.4 Development of folate receptor targeting dendrimer-entrapped gold nanoparticles [{(Au0)6-G2-DTPA(99mTc)-PEG-FA}] for targeted SPECT/CT dual-mode imaging of tumors. This delivery system showed promise for efficient, low-cost tumor diagnosis. Images are reprinted with permission from Li et al. (2016).

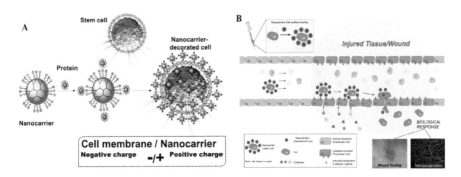

FIGURE 2.5 (A) Schematic design of dendrimer coated stem cells through electrostatic interaction. (B) The data indicates a wound healing response in the mice model, and their biological response study indicates the nanoparticle can generate neo-vascularization. Source: Used with permission from Li et al. (2016). © 2016 American Chemical Society.

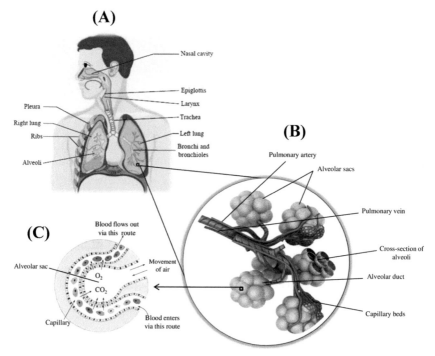

FIGURE 6.1   Human respiratory organ (A) Structure of the human respiratory system; (B) An enlarged view of terminal alveolar region and dense capillary network; (C) Close up view of gaseous exchange of carbon dioxide and oxygen between the capillaries and alveoli.

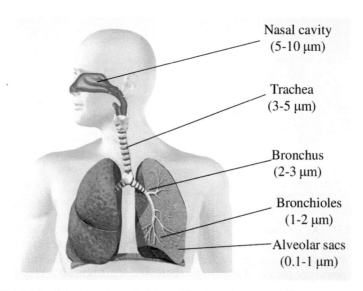

FIGURE 6.2   Particles size dependent deposition in various parts of lungs.

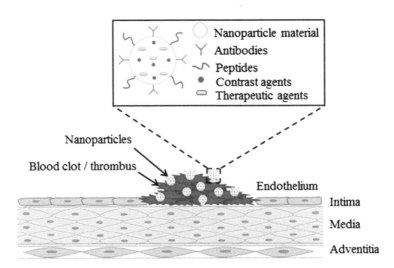

FIGURE 8.1 Multifunctional nanoparticle system with target specific, diagnostic, and therapeutic capabilities for deep vein thrombosis (DVT) management (From Wadajkar et al., 2013, With permission.).

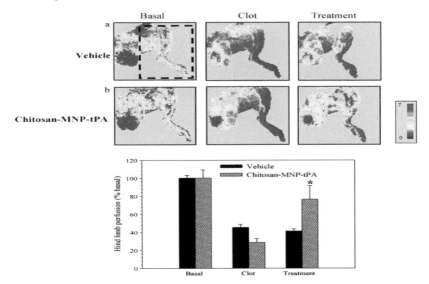

FIGURE 8.3 Representative results of the effects of tPA (Actilyse®) immobilized to the chitosan-coated magnetic nanoparticle (chitosan-MNP–tPA) on tissue perfusion in a rat iliac embolic model. Tissue perfusion of hind limb was measured with a laser Doppler perfusion imager. Five minutes after clot lodging, vehicle (a) or tPA covalently bound to chitosan-coated magnetic particle C250 (b) was administered from the right iliac artery under magnet guidance and treated for 120 min. Laser Doppler signals in designated areas, as illustrated with the square in (a), were acquired for quantitative analysis of hindlimb perfusion. *$p < 0.05$ compared with vehicle. (From Chen et al., 2011, With permission).

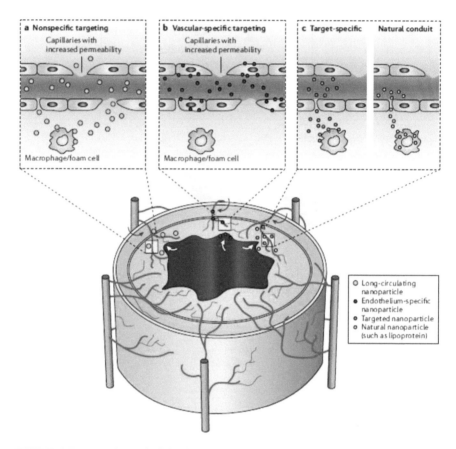

FIGURE 8.7 Targeting principles in atherosclerotic plaques. The vessel walls of larger arteries are supplied with nutrients by the lumen and the vasa vasorum—a network of small microvessels. In the lesioned vessel wall the vasa vasorum undergoes angiogenic expansion, with neovessels reaching into the base of the plaque, which is accompanied by the upregulation of cell-surface receptors and increased permeability of the endothelium. The upregulation of receptors and the increased permeability also affect the endothelium on the luminal side of the plaque. The main targeting principles can be classified into nonspecific targeting of the plaque (part a), specific targeting of the vasculature (part b) and specific targeting of components (part c) of the plaque (for example, the extracellular matrix or macrophages) with either synthetic nanoparticles or via interaction through a natural conduit. The targeting of the plaque occurs via both the vasa vasorum and the main lumen at lesioned sites and is exemplified on the figure with corresponding arrows. Depending on the targeting principle applied, the cellular distribution of nanoparticles in the plaque will vary considerably. (From Lobatto et al., 2011, With permission).

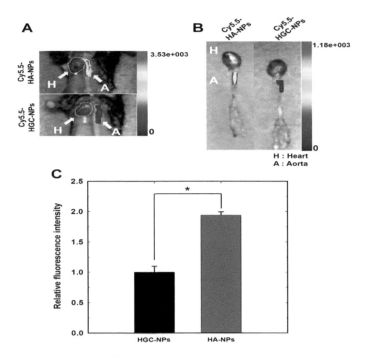

FIGURE 8.9 *In-vivo* near-infrared fluorescence (NIRF) imaging of atherosclerotic lesion using cyanine (Cy) 5.5 - labeled hyaluronic acid nanoparticles (HA-NPs) and Cy5.5 - labeled hydrophobically modified glycol chitosan nanoparticles (HGC-NPs). (A) Fluorescence images of atherosclerotic lesion in ApoE KO mice. (B) Fluorescence images of isolated aorta from ApoE KO mice. (C) Relative fluorescence intensity of atherosclerotic lesion in (B). Asterisk (*) indicates difference at the $p < 0.05$ significance level. (From Lee et al., 2015, With permission.)

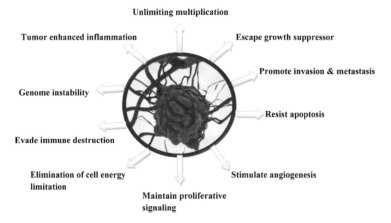

FIGURE 10.1 The tumor microenvironment and characteristics of cancer.

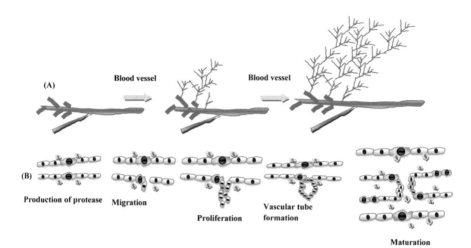

FIGURE 10.2 (A) Schematics showing the angiogenesis process from pre-existing blood vessels, (B) various steps occur in tumor angiogenesis.

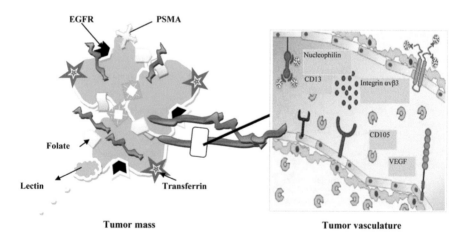

FIGURE 10.3 Schematic representation of tumor mass with their receptors along with tumor vasculature is showing their receptors.

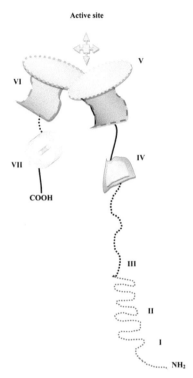

FIGURE 10.4   Schematic representation of the CD13 receptor as a model.

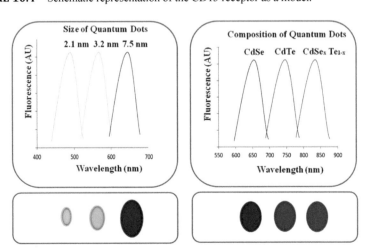

FIGURE 14.4   Effects of the size and composition on the emission wavelength of QDs. (A) Change of the spectrum from 450 nm to 650 nm with a change of nanoparticle diameter from 2 nm to 7.5 nm. Below the spectrum size of the nanoparticles of the same composition is presented. (B) By changing the composition of QDs, while retain the constant size (5 nm diameter), emission vary be vary in the range of 610 nm to 800 nm (Bailey and Nie, 2003).

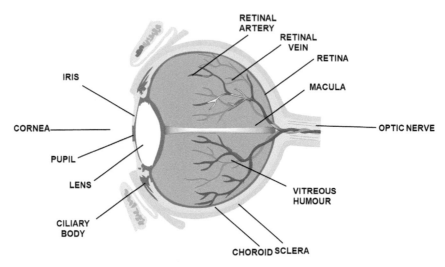

FIGURE 16.1   Structure of the eye.

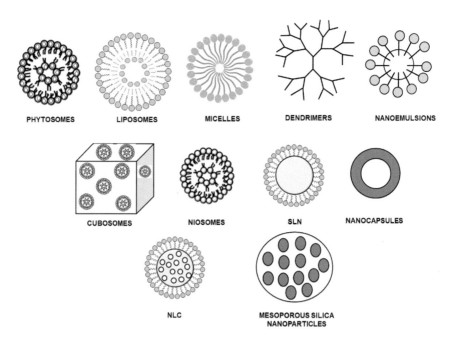

FIGURE 16.2   Schematic representation of various nanocarriers explored for ocular delivery.

CHAPTER 10

Angiogenesis Treatment with CD13 Targeting Nanomedicines

MADHU GUPTA[1*], RAMESH K. GOYAL[1], and VIKAS SHARMA[2]

[1]*Delhi Pharmaceutical Science & Research University, Pushp Vihar Sector-3, MB Road, New Delhi-110017, India, Tel.: +91-9205916054, E-mail: madhugupta98@gmail.com, vikassharma15@gmail.com*

[2]*Shri Rawatpura Sarkar Institute of Pharmacy, NH-75, Jhansi Road, Datia, MP, India*

*Corresponding author. E-mail: madhugupta98@gmail.com

ABSTRACT

Tumor targeting through angiogenesis such as tumorigenesis and metastatic processes has received considerable interest nowadays for the progression and development of efficient guided nanomedicines. Scientific community explored the various probes, which can target and bind the specific molecular addresses over-expressed in the tumor vasculature. In this respect, CD13 is a metalloprotease ectoenzyme and plays a significant role in tumor angiogenesis and metastasis and is over-expressed in many angiogenic processes. Hence, these molecules may be considered as specific biomarkers not only to monitor disease progression but also to rationally design targeted nanomedicines. A peptide-based ligand as asparagine-glycine-arginine (NGR) peptides are the most efficient approach with specific binding to CD13 protein had been widely accepted to achieve the ligand-mediated targeting to small molecules, peptides, proteins as well as to drug/imaging agent-containing nanomedicines, with maximizing their therapeutic index. At present most of them entered and occupied their position in the clinical trials, or other having promising reports on the therapeutic outcome in animal models. This book chapter cover the brief addressing on the structure and possible function of CD13 protein in development of malignant tumors, with recent and relevant examples of various CD13-assisted nanomedicines such as liposomes, polymeric nanoparticles, quantum dots, drug conjugates, and many more applied in drug/gene therapy as well as in imaging and theragnostics with discussing a brief overview of current clinical trials focused on more rational design of CD13-targeting agents.

10.1 TUMOR ANGIOGENESIS

Cancer is a very complex, dynamic, and versatile nature of diseases having a crucial challenge for developing efficient and safe therapies. According to the World Health Organization (WHO) survey, new global cancer incidences are expected to increase from 14 million in 2012 to as many as 22 million. The main reason is the lack of selective delivery of anti-cancer agents to timorous tissue. The higher systemic exposure to anti-cancerous agents frequently results in dose-limiting toxicity (Wicki et al., 2015). Cancer is the ultimate surplus combination of several characteristics such as unlimited multiplication, evasion from growth suppressors, promoting invasion and metastasis, resisting apoptosis, stimulating angiogenesis, maintaining proliferative signaling, elimination of cell energy limitation, evading immune destruction, genome instability and mutation, and tumor enhanced inflammation (Wang et al., 2017; Rahman, 2017; Beg, 2017) (Figure 10.1).

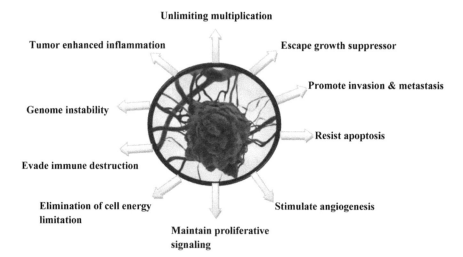

FIGURE 10.1 (See color insert.) The tumor microenvironment and characteristics of cancer.

Angiogenesis is the process in which the sprouting and growth of new vessels from existing vasculature (Ribatti and Djonov, 2012; Rahman, 2016; Ahmad, 2015a, 2015b). The most common way for newer blood vessel formation in malignancy is angiogenesis; hence this process is termed as tumor angiogenesis. In this process, endothelial cells may be transformed from a resting level to a rapid growth phase by a diffusible chemical signal originating from the tumor cells. However, the endothelial cells transformation depends on increased production of one or more positive regulators of angiogenesis, namely vascular endothelial growth factor (VEGF), fibroblast growth factor–2 (FGF–2), interleukin–8 (IL–8), placental growth factor (PlGF), transforming growth factor-beta (TGF-beta), platelet-derived growth factor (PDGF), angiopoietins (Angs) and others. These factors can be migrated from tumor cells to the extracellular matrix, or released from host cells recruited to the tumor. This switch may also cause the involve down-regulation of endogenous inhibitors of angiogenesis such as endostatin, angiostatin or thrombospondin and has thus been regarded as the result of tipping the net balance between positive and negative regulators (Ribatti, 2009). Angiogenic signals may be able to the preferential differentiation of endothelial cells into tip cells that initiate to migrate and exist at the leading front of the growing vessels. A number of factors including VEGF receptor (VEGFR)–3 (for lymphatic endothelial cells), VEGFR–1 and–2 (for blood endothelial cells), PDGF-B, and the Notch ligand delta-like ligand (Dll)–4 have been shown to contribute to the endothelial tip cell phenotype (Hellstrom et al., 2007). Endothelial cells resided just back to the tip cells and called stalk cells, release other factors such as VEGFR–1 and Notch–1 and –4 which are important for inducing a quiescent state of these cells (Jakobsson et al., 2010), maturation of the vascular wall, lumen formation and to support perfusion. However, this process may be usually disrupted by either overproduction of pro-angiogenic signals, lack of angiogenesis inhibitors, path-finding signals or maturation factors, and finally producing the excessive tip-cell formation and migration of endothelial cells, which do not assume a quiescent phenotype associated with a healthy vasculature (Figure 10.2) (Ridgway et al., 2006; Noguera-Troise et al., 2006).

10.2 RECEPTOR OVER-EXPRESSED ON TUMOR CELLS AND ANGIOGENESIS

Scientist continually focused on understanding the cancer pathophysiology at the cell and molecular level, and all the pathophysiological variants recognized and executed for drug delivery at the site-specific molecular level (Vasir et al., 2005). Ultimately, the site-specific biotargets relating to cancer are approached through a bio-receptors and drug interaction. The main reason for exploiting these cellular receptors is their altered density at the normal and diseased condition at million levels. The diverse pathophysiological conditions responsible for disease progression and the role of the receptor(s) as molecular modulator has opened, new vistas for cellular or intracellular drug targeting. These ligand-receptor interactions mediate the cellular events as they may be up or down regulated or may appear in the isoforms variant for specific biofunctionalization which may be determinant of a disease (Gupta et al., 2014).

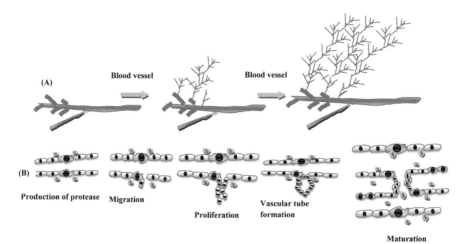

FIGURE 10.2   (See color insert.) (A) Schematics showing the angiogenesis process from pre-existing blood vessels, (B) various steps occur in tumor angiogenesis.

Several bioreceptors are recognized till date and successfully employed for cancer targeting at specific site namely; transferrin receptors, EGFR, VEGF, integrin receptors, PSMA, hormone receptors, CD44, CD13, integrin, and receptors/epitopes expressed on tumor cells (Figure 10.3). The receptors which are either expressed on malignant cells are intensely being implemented for their possible utilization as handles for nonimmunogenic, site-specific targeting to the ligand-specific bio-sites (Gupta et al., 2014). The receptors mainly present on the cell surface biomembrane and give support for cancer cells by providing nutrients and other important things that are important for tumor growth and neovascularization (Gupta et al., 2014). The receptors have two essential components, i.e., extra-cellular ligand binding domain and an intracellular signal-transmitting domain. In these receptors, specific binding sites are present so that, only the specific ligand of given orientation can be recognized and specific ligand-carrier complex attached to the specific receptor. Then the system is internalized or liberates the therapeutic moiety at the extra-cellular site, and the biological response is produced (Agarwal et al., 2008).

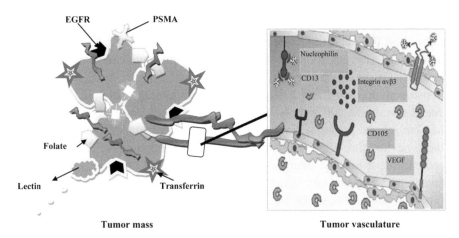

FIGURE 10.3 (See color insert.) Schematic representation of tumor mass with their receptors along with tumor vasculature is showing their receptors.

10.3 CD13 RECEPTOR: STRUCTURAL INTEGRITY

Research data supported that hydrolytic enzyme, as peptidases forms are present in various malignancies. In regard to peptidases, one of the most studied candidates is aminopeptidase N (APN; EC 3.4.11.2, also referred as CD13, microsomal aminopeptidase, aminopeptidase M, alanine aminopeptidase, particle-bound aminopeptidase, p146, p161 or gp150). It is a Zn^{2+} dependent membrane-bound ectopeptidase and controls several functions such enzymatic regulation of peptides as well as characteristics associated with malignant cells, like tumor cell invasion, differentiation, proliferation, and apoptosis, motility, and angiogenesis. The CD13 also regarded as a viral receptor and implicated in cholesterol turnover. All these properties make them attractive and fascinating and designated as moonlighting ectoenzyme (Wickstrom et al., 2011).

CD13 is a membrane-spanning, type-II cell surface protein with 160 kDa with heavily glycosylated, and coding part of the human gene (ANPEP), located in chromosome 15 (q25-q26). Its presence may be observed in the renal and intestinal epithelial cells, in the nervous system (synaptic membranes and pericytes), in myeloid cells (monocytes, macrophages, and DCs) and in fibroblast-like cells, such as synoviocytes. It is overexpressed on the activated or angiogenic endothelial cells but in a normal cell in a very lesser amount. The sequential immunoprecipitation and deglycosylation techniques confirmed the five different forms of CD13 owing to the variation in oligosaccharide composition in endothelial cells and monocytes (O'Connell et al., 1991). CD13 is located on the cell membrane with its N-terminus part and facing catalytic domain exterior environment to the cell. The full-length APN contains 967 amino acids, in which having a short N-terminal cytoplasmic domain, single transmembrane part, and a large cellular ectodomain containing the active site (Luan et al., 2007). It is zinc-dependent metalloprotease, related to the M1 family and having a single zinc ion in the structure. APN firmly affixed the zinc ion through highly conserved HELAH amino acid motif. After proteolysis, it may be degraded into two parts, with range between 90,000 and 45,000, respectively. Structural investigation revealed that APN is a dimer type that is coordinated by noncovalent bond (Hans et al., 2000). Basically, APN contains seven domains naming as domain I to domain VII. Domain I is the cytosolic part of APN having 9-amino acids, that is responsible and assisting for the anchorage to the cell. Domain II is the membrane-spanning domain and resembling alpha-helix type and appearing the structure of this domain to be highly conserved, varying in different type II aminopeptidases. Domain III presents exterior environment to the cell, which resembles a stalk supporting the catalytic part of the APN. Domain IV consisted of amino acids 70 to 252 and had not played a significant role in the enzymatic activity. One of the two-conserved N-glycosylation sites occurs in this domain. Domain V and VI consisted of the amino acid 253–580, which is the catalytic site of the APN and ligand of the zinc-ion associated with the critical roles for the enzymatic activity. Domain VII is the last domain containing 581 to 967 amino acids. It is the C-terminal of the APN with a high content of alpha-helix. In this domain, the dimer is combined through the noncovalent bond (Anthony et al., 2006). The activity of the enzyme is based on the three-dimensional structure of APN, but it has not been determined, which still appears to be perplexing enough to get a better solution in the future.

APN has performed a function with three possible mechanisms of action. When the ligands bind with APN, it works as (i) as an enzyme, (ii) as a receptor, and/or (iii) as a signaling molecule. All performance related at least one of the possible acts likely (i) peptide cleavage, (ii) endocytosis, and (iii) signal transduction. The biological response produced by these mechanisms, although some complicated phenomena such as angiogenesis, invasion, and chemotaxis, might be the outcome of the interplay between enzymatic activity and signaling functions. While, the virus or even cholesterol and NGR-peptide binding therapeutics (NGR-targeted moiety conjugated with nanosystem), the receptor functions could mediate the signal transduction that is required for endocytosis (Mina-Osorio, 2008) (Figure 10.4).

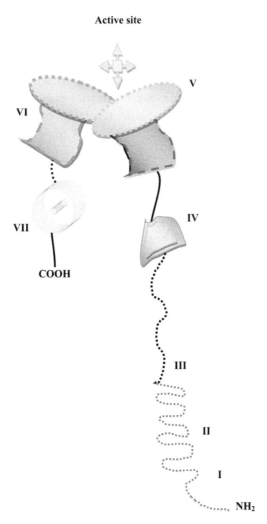

FIGURE 10.4 **(See color insert.)** Schematic representation of the CD13 receptor as a model.

10.4   CD13: CONTRIBUTION TO TUMOR PROGRESSION

CD13 is a key component in tumor vasculature and is over-expressed in the membranes of endothelial tumor cells. NGR works as ligand that can be integrated with CD13. CD13 receptors play significant role in chemokine processing and tumor invasion. Research studies demonstrated that APN/CD13 receptor is heavily over-expressed on several angiogenic blood vessels and other types of vessels undergoing angiogenesis but not on normal vasculature. All these facts confirmed the critical role of CD13 in tumor progression and in angiogenesis. The higher level of APN expression is specifically to solid tumor such as melanoma, liver, renal, pancreas, breast, gastric, thyroid ovarian, and prostate cancer. The CD13 biomarker is present on the myeloid lineage in the hematopoietic compartment at early phase and specifically located on the surface of myeloid progenitors and their differentiated offsprings. Its presence is also confirmed by nonhematopoietic tissue or organs and selectively overexpressed on vascular endothelial cells undergoing angiogenesis (Luan and Xu, 2007).

Angiogenesis is the formation of new blood vessels, and tumor growth depends on a steady blood supply with nutrients and oxygen for survival and tumor growth and ultimately for spreading the tumor cells into other tissue or site. These tumor blood vessels over-expressed certain markers that are either present at very low levels or are entirely absent in normal blood vessels like APN or CD13 marker. In tumor pathophysiology, APN levels raised due to hypoxia, angiogenic growth factors and others signals regulating capillary tube formation during angiogenesis (Bhagwat et al., 2001). Indeed, vascular endothelial growth factor (VEGF), a key angiogenesis regulator, induced the increased expression of APN at an early stage of tumor growth, which is crucial for sustained growth of most solid tumors. The endothelial cells produced several enzymes that are responsible for degradation of basement membrane components. These enzymes are type IV collagenases, matrix metalloproteases, and aminopeptidases. The studies supported that type IV collagen is substrate of the APN/CD13 and enhance the migration of endothelial cells into the surrounding tissue and contribute to the mechanism of tumor metastasis (Pasqualini et al., 2000). Finally, all the addressing issue turned the benign tumor to malignant tumor with raising the severity of the disease. Hence, the level of APN expression is inhibited then the angiogenesis and metastasis of the tumor can be inhibited, and growth of a tumor can be inhibited.

10.5 TARGETING THE CD13 RECEPTORS BY VARIOUS CANCER THERAPEUTICS

The CD13 is present in human malignancy and contributing to neoplastic features. By disturbing the CD13 expression, function or their signaling, may lead to the progression of various novel anticancer targets. The more intensive research was experimented into therapeutic options, biological agents and some kinds of carriers that can combine with NGR-peptide to offer a variety of novel tumor-targeted drug delivery systems. NGR-containing peptides implicated for CD13 targeting and lead to better therapeutic management in tumor diagnosis and tumor blood vessel imaging. The targets can be utilized in several ways to exert their anticancer potential, which are discussed below in text and summarized in Table 10.1.

10.5.1 NATURAL INHIBITORS

Numerous APN inhibitors act directly on tumor cells and play important in the anticancer field. Bestatin (ubenimex): Initially, it was explored as an immunomodulant however presently various clinical trials were conducted with this agent for treating the hematological malignancy. The clinical level has been performed in carcinoma of the lung, bladder, stomach, head, and neck, esophagus, and skin malignant melanoma (Bauvois and Dauzonne, 2006).

10.5.2 CHR-2797 (TOSEDOSTAT)

A novel metalloenzyme inhibitor, CHR–2797 which has potential to convert into a pharmacologically active acid product (CHR–79888) inside cells. It is a potent inhibitor of a number of intracellular aminopeptidases and shows their antineoplastic property against a range of tumor cell lines in vitro and in vivo (Krige et al., 2008). CHR–2797 is orally bioavailable agent, and in combination with paclitaxel it can be employed as undergoing phase Ib clinical investigation in the treatment of solid tumors (van Herpen et al., 2010). According to the reported phase I trial with accelerated titration design, 40 patients were selected with advanced solid tumors. CHR–2797 was taken once a time daily, but it showed toxicities such as fatigue, diarrhea, peripheral edema, nausea, dizziness, and constipation. One patient prompted partial response in case of renal cell carcinoma, and

TABLE 10.1 Recent Literature Depicted the CD13 Based Targeting Through Various Nanocarriers

Target moiety	Delivery system	Bioactive molecule	Disease/cell line/ animal model	Concluding remarks	Ref.
NGR	CdSe/ZnS Quantum dots	Green Fluorescent	rat BCECs cells	The NGR-PEG-QDs for glioma/tumor vasculature-targeted fluorescence imaging. This nanoprobe was competent to migrate across the BBB/BTB, which contributed to a prominent achievement in the specific targeting of glioma cells and tumor vasculatures in vitro and in vivo.	Huang et al., 2017
NGR	Liposomes	Bortezomib	(NB) cell line, GI-LI-N	Novel vascular targeted BTZ formulation is endowed with high therapeutic index and low toxicity, providing a new tool for future applications in neuroblastoma clinical studies.	Zuccari et al., 2017
NGR	Multifunctional (PEI–PEG)-based nanoparticles	DOX and pDNA	HUVEC cells	The results indicated that designed nanosystem influenced the distribution of CD13 on HUVEC through the interaction between CD13 and NGR, causing CD13 clustering, leading to the co-localization of CD13 and CAV1 and internalization via the CvME.	Liu et al., 2017
NGR	Poly (ethyleneimine) poly (ethylene glycol) (PEI–PEG)-based nanoparticles (TPIC)	Co-delivery of DNA and doxorubicin (DOX)	HUVEC cells	The results indicated that TPIC influenced the distribution of CD13 on HUVEC through the interaction between CD13 and NGR, causing CD13 clustering, leading to the co-localization of CD13 and CAV1 and internalization via the CvME. Additionally, this internalization was not dependent on the enzyme activity of CD13 but was inhibited by cholesterol depletion.	Liu et al., 2016
NGR	PAMCN	Cisplatin	Angiogenesis/ HUVEC cells/ KM Mice	The in-vivo study indicated that PAMCN group based treated mice survived 100% and reduced the hepatic and renal toxicity significantly compared to free drug-treated group. All the results revealed that this novel drug delivery system could be a potential approach for antiangiogenic chemotherapy.	Zhang et al., 2015

TABLE 10.1 *(Continued)*

Target moiety	Delivery system	Bioactive molecule	Disease/cell line/ animal model	Concluding remarks	Ref.
NGR	pcCPP/NGR-LP (Liposomes)	siRNA	HT–1080 cells and MCF–7 cells	pcCPP/NGR-LP is expected to be a promising delivery system for oncotherapy. The dual-modified liposomes may be considered as alternatives to photodynamic therapy in which NIR light is used to locally generate cytotoxic reactive oxygen species at the tumor site that cause cancer cell death, but not to induce intracellular delivery of macromolecular therapeutics.	Yang et al., 2015
cKNGRE	68Ga-NOTA-c (NGR)	----	Fischer-344 rats	In NeDe tumor-bearing rats higher 68Ga-NOTA-c (NGR) accumulation was found in the tumors than that of the 68Ga-NODAGA-[c(RGD)]2. Due to its high selectivity and strong binding affinity to APN/CD13, favorable biodistribution and pharmacokinetics, 68Ga-NOTA-c(NGR) might be a potential molecular probe for the noninvasive detection of CD13/ APN-positive tumors, metastases, and neovasculature.	Mate et al., 2015
Asn–Gly–Arg (NO2) COOCH3	13F–1 is a 5-fluorouracil prodrug	5- FU	Colo205 xenografts	The 13F–1 administered in mice by injection, significantly inhibited the growth of Colo205 xenografts without sign of toxicity. Hence, it can be used as a promising agent for treatment of cancers with high expression of APN/CD13	Cui et al., 2014.
CNGRCK	cNGR-DNB-NPs	DTX	HT–1080, HUVEC/ Balb/c mice	The results suggest that cNGR peptide-conjugated PLGA-PEG-NPs can be successfully explored for the design of dual-targeted delivery systems for targeting the tumor cells and vasculature and have a potential approach for CD13-overexpressed specific tumors.	Gupta et al., 2014
iNGR	Poly(ethylene glycol)-poly (L-lactic-co-glycolic acid) nanoparticles	Paclitaxel	Human umbilical vein endothelial cells	The anti-glioma efficacy of paclitaxel-loaded iNGR-NP is verified by its improved anti-angiogenesis activity and the significantly prolonged survival time in mice bearing intracranial glioma.	Kang et al., 2014

TABLE 10.1 (Continued)

Target moiety	Delivery system	Bioactive molecule	Disease/cell line/ animal model	Concluding remarks	Ref.
CNGRCK2H-K3HK11	SWNTs	2-me-thoxyestradiol	MCF–7 cells/ female BALB/c mice/	The inhibition ratio of this SWNTs drug delivery system at 24, 48, and 72 hr was about 57.7%, 83.6%, and 88.2%. Compared with normal saline group, the relative tumor volumes in the 2ME, SWNTs–2ME, and NGR-SWNTs–2ME groups were decreased 1 week after administration This novel neovascularity targeting drug delivery system containing NGR-SWNTs–2ME may be beneficial to improve treatment efficacy and minimize side effects in future cancer therapy.	Chen et al., 2013
NGR tripep-tide ligand	Monodisperse spherical micelles	---	---	These results suggest that enhanced delivery to solid tumors can be achieved by targeting upregulated receptors in the tumor vasculature with multivalent ligand-presenting nanoparticles, but additional work is required to optimize such systems for multivalent targeting.	Simnick et al., 2011
CNGRCG	cNGR–PEG–PLA-NPs	pDNA	HUVEC and HepG2 cells	The unique mechanism of caveolae-mediated endocytosis was indeed mainly involved in the internalization of cNGR–PEG–PLA-NPs into HUVEC and led to significant gene transfection efficiency. Such favorable properties including high transfer efficiency, low cytotoxicity, and fast uptake by nondestructive endocytic pathways make cNGR–PEG–PLANPs to be a promising carrier for the intracellular delivery of therapeutic agents.	Liu et al., 2011
CNGRCK	BPEI-SS-PEG-cNGR/pDNA polyplexes	pDNA	HT–1080, MDA-MB–231 cells, HEK293, B16F1	The results implied that multifunctional polymer-mediated gene transfection took place tumor-specifically and via GSH-dependent pathway.	Son et al., 2010

TABLE 10.1 *(Continued)*

Target moiety	Delivery system	Bioactive molecule	Disease/cell line/ animal model	Concluding remarks	Ref.
CYGGRGNG and STR-R4	Dual ligand modified liposomes	Rho-DOPE	MS–1 and TE cells	The dual-modification of liposomes with NGR and STR-R4 resulted in an enhanced uptake due to synergistic effects in CD13 positive cells. The proposed strategy holds considerable promise for the further development of in vivo applicable systems.	Takara et al., 2010
GNGRG-GVRSS-SRTPSDKYC	LPD-PEG-NGR (liposome-polycation-DNA) nanoparticle	DOX and siRNA	HT–1080 and HT–29 cells	The tumor-homing peptide (NGR) modified nanoparticle provides an enhancement of drug potency and may potentially be a therapeutic agent against drug-resistant tumors when combined with siRNA against drug-resistant genes such as p-glycoprotein.	Chen et al., 2010
GNGRG	Liposomes	DOX	Colon26 NL–17 cells	The dual-targeting liposomes containing DOX strongly suppressed tumor growth in Colon26 NL–17 carcinoma-bearing mice.	Murase et al., 2010
cKNGRE	Temperature sensitive liposomes	fluorescent reporter Oregon Green	HT–1080, MCF–7 cells	The results of this study are significant because they demonstrate improved avidity of an NGR-targeted LTSL without the limitation of a disulfide bridge.	Negussie et al., 2010
cNGR	cNGR-pQDs	--	--	The results showed that cNGR-pQDs were specifically accumulated on the tumor endothelial cells surface and to a lesser extent in the vessel lumen, supporting a higher specificity of cNGR-pQDs for angiogenic tumor vasculature.	Oostendorp et al., 2008
CNGRC	Stealth NGR/ PEG/PDBA-coupled-SHA PEI/ pDNA targeting polyplex loaded	pDNA	DCs	All the results suggest that PLGA-PEG-PLGA encapsulation of this stealth targeting polyplex has no negative effects on key properties of immature DCs and should pave the way for targeting DCs for vaccination purposes.	Moffatt et al., 2006

TABLE 10.1 *(Continued)*

Target moiety	Delivery system	Bioactive molecule	Disease/cell line/ animal model	Concluding remarks	Ref.
	with PLGA-PEG-PLGA tri-block copolymer				
CNGRC	CNGRC/PEG/PEI/DNA	DNA	HUVEC and HT–1080/nude mice	*In vivo* transduction analysis using the CNGRC/PEG/PEI/DNA vector to target the intravenous delivery of a yellow fluorescence protein (YFP)-expressing plasmid to subcutaneous H1299 tumors confirmed delivery of YFP to both tumor cells and tumor endothelial cells. The use of this peptide to further increase tumor-specific delivery mediated by our novel PEI/DNA vector now provides a basis for developing tumor-targeted gene therapies for use in the clinical treatment of cancer.	Moffatt et al., 2005

four patients had stable disease for >6 months (Reid et al., 2009). Tosedostat is currently in a registration study in patients with relapsed/ refractory acute Macleod leukemia.

10.5.3 APN/CD13 RECEPTOR-BASED TARGETED DELIVERY SYSTEM

The NGR peptide is the most affinity based peptide for targeting the CD13 receptors that have been sieved with *in vivo* phage display library and may be prepared for various sequences according to requirement. It is mainly composed of asparagine-glycine-arginine (NGR) peptides are the most efficient approach with their targeting receptors well characterized. NGR peptide is specifically homing the APN receptors and binds them (Wang et al., 2011). Now date, numerous bioactive had been conjugated to NGR peptides to enhance their therapeutic output namely cytotoxic agents, cytokines, antiangiogenic compounds, viral particles, fluorescent compounds, contrast agents, DNA complexes, and other biologic response modifiers (Corti et al., 2008).

The NGR based targeted drug delivery displayed greatest tumor selectivity and under rapid progression over the last few years. The very first time, doxorubicin (DOX) is conjugated to NGR peptide and presented improved therapeutic outcome with even lesser toxicity than the free drug itself. This conjugate has potential to reduce the toxic nature as well as improve efficacy against human cancer xenografts in nude mice, compared with free doxorubicin (Arap et al., 1998). The APN/CD13 is widely up-regulated in various cell lines including epithelial cells, mast cells, fibroblasts, and muscle cells with different locations such as cell membranes, cytoplasm, plasma, and stromal fibrillar components of some connective tissues, but the detailed mechanism of NGR based tumor-targeting still unclear. It is believed that rapid deamidation of asparagines is possible in tumor homing NGR peptide. So, NGR could be converted to isoaspartate-glycine-arginine (isoDGR), and isoDGR binding might be inhibited the αvβ3 integrin-mediated endothelial cell adhesion, proliferation, and related tumor development (Curnis et al., 2006). Both types of NGR explored as linear and cyclic forms, among which the cyclized form (cNGR) were designed by the disulfide bonding of the two cysteines. Literature abound study supposed that linear NGR peptides have remarkable biodistribution and efficacy, however, the antitumor activity was increased up to 10 fold higher in case of cyclic form (Colombo et al., 2002). Despite the greater activity of cyclic one, the preference had been given to use linear NGR-containing motifs to avoid the formation of disulfide bridges between adjacent peptides on the liposome surface that may render the ligand ineffective (Corti and Ponzoni, 2004). Up to now, a diverse set of NGR structures has been designed and explored for drug delivery and tumor imaging, enlisted in Table 10.1.

a. Biopolymer: Numerous newer devices can be executed for targeted purpose to reduce the side effects, to improve the bioavailability, or to enhance the therapeutic effects at target specific sites. In present scenario, most of these specifically designed systems for tumor treatment are based on polymeric materials to manage the drug release at its programmed level (Peng et al., 2009; Rahman, 2015a). Hence, recent development basically exploits for synthesis of biopolymer-based nanosystems particles and their application as drug/gene carriers. The biopolymer offers several favorable properties in regard to synthetic polymers used in drug/gene delivery, such as biocompatibility, biodegradability, and abundant renewable sources as well as biocompatibility, biodegradability, and low immunogenicity. These biopolymers are group of polymers that originated from living organisms having in three groups: polysaccharides, proteins, and nucleic acids. However, the sizes, charge, morphology of surface and release pattern are important influential parameters that affect the efficacy of biopolymer-based nanoparticles used as drug/gene delivery carriers. Biopolymer-based nanosystem mainly fabricated with such as protein (silk, collagen, gelatin, β-casein, zein, and albumin), protein-mimicked polypeptides and polysaccharides (chitosan, alginate, pullulan, starch, and heparin). Todays, biopolymers can be considered as suitable materials as nanoparticles for clinical application (Nitta and Numata, 2013; Aneja, 2014).

Biopolymer can be efficiently used as novel drug delivery systems for APN targeted systems. Recent research explored that poly(γ-glutamic acid)-based targeted drug delivery system (PAMCN) could be efficiently targeted the transmembrane metalloprotease aminopeptidase-N (APN/CD13). This system was loaded with cisplatin and presenting the sustainable release profile with a half-maximal release time ($t_{1/2}$) of 23 hr. The flow cytometry and fluorescence microscopic analysis concluded that drug nanosystem might be specifically bound to human umbilical vein endothelial cells (HUVEC). PAMCN system improved the therapeutic efficacy to HUVEC cells with lesser IC_{50} value decreased to 90.83 ± 33.00 μg/ml than free CDDP and presented less tube formation amounts (p< 0.01) than free CDDP in matrigel angiogenesis inhibition assay. The *in-vivo* study indicated that PAMCN group based treated mice survived 100% and reduced the hepatic and renal toxicity significantly compared to the free drug-treated group. All the results revealed that this novel drug delivery system could be a potential approach for antiangiogenic chemotherapy (Zhang et al., 2015; Rahman, 2015b; Ahmad, 2013).

b. Polymeric Micelles: Polymeric micelles are generally prepared by amphiphilic block copolymers such as poly (ethylene oxide)-poly(β-benzyl-L-aspartate) and poly (N-isopropylacrylamide)-polystyrene. These are in size range of less than 100 nm and assembled with a hydrophobic core and hydrophilic shell. The hydrophobic micelle core acts as a drug reservoir for water-insoluble drug and bounded by hydrophilic corona that provides a protective interface to the core, thus making the particles an appropriate candidate for i.v., administration (Adams et al., 2003; Rahman, 2012a; Akhtar, 2011). The most significant aspect of micelles is that the drug released can be controlled by an external stimulus like pH, temperature, ultrasound or certain enzymes. The polymeric micelles are easily modulated by small functional groups that ultimately enhance their targeting potential (Rapoport, 2007; Rahman, 2012b; Ahmad, 2010, 2011). In a similar fashion, Wang, and co-workers have designed NGR-modified docetaxel (DTX)-loaded PEG-b-PLA polymeric micelles (NGR-PM-DTX). The quantitatively and qualitatively analysis showed that NGR enhances the uptake of micelles by CD13-overexpressed tumor cells (fibrosarcoma, HT1080) and endothelial cells (HUVEC), and this uptake could be inhibited by free NGR demonstrated that NGR ligand plays an important role for CD13 receptor-mediated targeting (Wang et al., 2009). The targeted site was selected for brain tumor angiogenic blood vessels and targeted by NGR-modified DSPE–PEG micelles containing paclitaxel (NGR-M-PTX). The study supported that the designed formulation bind to and kill brain tumor angiogenic blood vessels and also penetrating the brain tumor interstitial space, resulting in direct cell death and ultimately produce antiangiogenic therapy using NGR-M-PTX exhibits potent *in vivo* antitumor activity in a C6 glioma–bearing animal model (Zhao et al., 2011).

c. Carbon Nanotubes: Recent research focused attention on the new family of nanomaterials termed as single-walled carbon nanotubes (SWNTs) and considered as therapeutic molecules delivery into cells. SWNTs have ultrahigh area owing to all atomies exposed on the surface which allows efficient loading of multiple molecules (Gomez-Gualdron et al., 2012). The supramolecular binding of aromatic molecules can be easily achieved by π-π bonding of those molecules onto the polyaromatic surface of nanotubes (Wang et al., 2012). SWNTs could be efficiently loaded various biomolecules including drugs, peptides, proteins, plasmid DNA and small interfering RNA into cells via endocytosis (Grigoryan et al., 2011). In one study, the neovascularity-targeting drug delivery system containing NGR-SWNTs–2ME presents higher tumor inhibition effect and actively target than 2ME alone and SWNTs–2ME without NGR peptide ligand. These entire multifactor influences the tumor neovascularity and as a viable opportunity for future cancer therapy (Chen et al., 2013).

d. Quantum Dots: Quantum dots (QD) are small nanoparticulate size ranged (2–10 nm) colloidal fluorescent semiconductor nanocrystals consisted from 10–50 atoms of groups II–IV or III–V of the periodic table (Qi and Gao, 2008). Their structure mainly composed of a metalloid crystalline core and a shell that provide protection for core and renders the QD available for *in vivo* applications (Hardman, 2006). The properties that determine their absorption and light emission ultimately controlled the size and shape of quantum dots. The most specific features of QD is their fluorescence spectrum, which make them optimal fluorophores and suitable for biomedical imaging. Fluorescent QD combined with bioactive moieties or specific ligands (e.g., ligands, and antibodies) and easily traced out the targeted site and play an important role for targeting applications. QD are stable nanocrystals for several months without degradation or alteration. QD are mostly exploited as long-term, multi-contrast imaging agents for detection and diagnosis of cancer *in-vivo* (Díaz and Vivas-Mejia, 2013; Rahman, 2012c, 2013). Oostendorp et al., in 2008, prepared cNGR-labeled paramagnetic quantum dots (cNGR-pQDs) for tumor angiogenic activity via the noninvasive assessment using quantitative in-vivo molecular magnetic resonance imaging (MRI). The MRI results were validated using ex-vivo two-photon laser scanning microscopy (TPLSM). The results showed that cNGR-pQDs were specifically accumulated on the tumor endothelial cells surface and to a lesser extent in the vessel lumen, supporting a higher specificity of cNGR-pQDs for angiogenic tumor vasculature (Oostendorp et al., 2008).

e. Gene Delivery System: NGR peptides offer excellent therapeutic properties in case of gene therapy. The recombinant adeno-associated virus having linear and cyclic versions of NGR specially localize transduce cells expressing CD13 (Grifman et al., 2001). Several studies demonstrated that recombinant adenoviral vectors with NGR set up in the fiber knob infect human glioma cells 100 to 1000 times more effective than the virus containing wild-type fiber. Other reports published that NGR incorporated into Moloney murine leukemia virus envelope escort proteins enhanced retrovirus binding and transduction of endothelial cells (Mizuguchi et al., 2001).

In order to successful non-viral cancer gene therapy, multifunctional polymeric gene delivery systems have draw attention. In the effective polymeric gene delivery vector, polymers can facilitate efficient DNA binding and release, endosomal buffering, and tissue or intracellular targeting with minimal toxicity. To proof the concept, Son et al., 2010, used biological functionalities were tailored with high level of sophistication in vector design. The simultaneous installations of various types of functionalities were achieved in one-pot reaction under mild conditions through pre-thiolation strategy incorporating thiol-containing components. This designed system contained linear polyethylene glycol (PEG), cyclic NGR (CNGRCK, cNGR) tumor-targeted peptide and bioreducible branched polyethyleneimine (BPEI). They used low molecular weight BPEI that was thiolated with propylene sulfide and mixed with α-Maleimide-ω-N-hydroxysuccinimide ester polyethyleneglycol (MAL-PEG-NHS, MW:5000) and cNGR peptide. The study output suggests that multifunctional polymers consisting of some key domains facilitated enhanced transfection to tumor cells. The disulfide linkage cause the reductive cleavage triggered by GSH in cytoplasm and led to efficient release of pDNA. BPEI was allowed for effective compaction and buffering effect. One of the most significant features is achieved by cNGR peptide within the polymeric framework that is for tumor-targeted gene delivery, and tumor-specificity was established by competitive inhibition assay with free cNGR peptide. In 2005, Moffatt et al., described the novel CD13 targeted polyethyleneimine (PEI)–DNA vector formulation that is capable of efficient tumor-specific delivery after intravenous administration to nude mice. The employed the CNGRC peptide to PEI for involving the strong affinity between phenyl(di) boronic acid (PDBA) and salicylhydroxamic acid (SHA) as well as a polyethylene glycol (PEG) linker to reduce steric hindrance between the vector and the peptide. The *in-vitro* targeting assessment implicated that CNGRC/PEG/PEI/DNA vector carrying a β-galactosidase (β-Gal)-expressing plasmid achieved 5-fold higher transduction than untargeted PEG/PEI/DNA-β-gal vector of CD13-positive lung cancer, fibrosarcoma, bladder cancer, and human umbilical vein endothelial cells. The gene delivery was especially used for CD13-positive cells and results confirmed by competition study. The *in-vivo* study was performed on nude mice and showed that 12-fold increase in β-Gal expression in tumors as compared with expression in either lungs or tumors from animals treated with the original PEI/DNA- β-gal vector (Moffatt et al., 2005). The functional siRNA delivery to cells by loading siRNA into cationic liposomes bearing a photolabile-caged cell-penetrating peptide (pcCPP) and asparagine-glycine-arginine peptide (NGR) molecules attached to the liposome surface (pcCPP/NGR-LP). pcCPP/NGR-LP is expected to be a promising delivery system for oncotherapy. The dual-modified liposomes may be considered as alternatives to photodynamic therapy in which NIR light is used to locally generate cytotoxic reactive oxygen species at the tumor site that cause cancer cell death, but not to induce intracellular delivery of macromolecular therapeutics (Yang et al., 2015). Liu and his team (2016) designed the (NGR) peptide-modified multifunctional poly(ethyleneimine)– poly(ethylene glycol) (PEI–PEG)-based nanoparticles (TPIC) for the co-delivery of DNA and doxorubicin (DOX). The results indicated that TPIC influenced the distribution of CD13 on HUVEC through the interaction between CD13 and NGR, causing CD13 clustering, leading to the co-localization of CD13 and CAV1 and internalization via the CvME. Additionally, this internalization was not dependent on the enzyme activity of CD13 but was inhibited by cholesterol depletion (Liu et al., 2016).

f. Liposomes: In the last decades, liposome-encapsulated therapeutic agents have been considered to increase the selective toxicity of chemo-therapeutics in cancer, resulting in higher therapeutic efficacy and or mini-mized damage to normal tissues such as heart or bone marrow (Allen, 2002; Rahman, 2016). Poly(ethylene glycol) (PEG) grafted liposomes termed as long-circulating liposomes are able to passively accumulated in tumors via the enhanced permeability and retention (EPR) effect. To attain greater chemotherapeutic efficacy, the PEGylated liposomes have been conjugated with attaching the specific targeting-ligands to the distal end of the PEG chain (Fondell et al., 2010). The specific ligand-receptor interaction induced the receptor-mediated endocytosis of liposomes. Previous studies docu-mented that doxorubicin-based liposomal formulations targeted the tumors by coupling with linear peptides containing the GNGRG sequence (Pastorino et al., 2006; Pastorino et al., 2003; Garde et al., 2007). These formulations successfully caused tumor endothelial cells and tumor cells apoptosis with antitumor and antimetastatic activity observed in an orthotopic model of human neuroblastoma (Pastorino et al., 2003).

In one study, it was reported that NGR peptide targeted liposomal doxo-rubicin might be sufficiently bind and kill the angiogenic blood vessels and furthermore indirectly, the tumor cells that vessels support, mainly in the tumor core. The anti-GD2-targeted liposomes would be employed for direct cell killing, including cytotoxicity against cells that are at the tumor periphery and independent of the tumor vasculature. All the results promised that this combination approach is a search for more effective and less toxic cancer treatments (Pastorino et al., 2006). In the same series, dual-ligand liposomal system is prepared with specific ligand and a cell penetrating peptide (CPP) that could be employed to enhance selectivity and cellular uptake. The dual-ligand based PEGylated liposomes were able to recognize CD13 due to NGR motif. The RX unit was R4, and it can be masked by the PEG aqueous layer. It was observed that single NGR- or STR-R4 ligand liposomes showed no enhanced cellular uptake; however, the dual-modified liposome presents a synergistic effect on cellular uptake. Hence, the dual-ligand system had shown to be

TABLE 10.2 Recent Literature Depicted the CD13 Based Targeting Through Various Nanocarriers

Target moiety	Delivery system	Bioactive molecule	Disease /cell line/ animal model	Concluding remarks	Ref.
NGR	CdSe/ZnS Quantum dots	Green Fluorescent	rat BCECs cells	the NGR-PEG-QDs for glioma/tumor vasculature-targeted fluorescence imaging. This nanoprobe was competent to migrate across the BBB/BTB, which contributed to a prominent achievement in the specific targeting of glioma cells and tumor vasculatures in vitro and in vivo.	Huang et al., 2017
NGR	Liposomes	Bortezomib	(NB) cell line, GI-LI-N	Novel vascular targeted BTZ formulation is endowed with high therapeutic index and low toxicity, providing a new tool for future applications in neuroblastoma clinical studies.	Zuccari et al., 2017
NGR	Multifunctional (PEI–PEG)-based nanoparticles	DOX and pDNA	HUVEC cells	The results indicated that designed nanosystem influenced the distribution of CD13 on HUVEC through the interaction between CD13 and NGR, causing CD13 clustering, leading to the co-localization of CD13 and CAV1 and internalization via the CvME.	Liu et al., 2017
NGR	Poly(ethyleneimine) poly(ethylene glycol) (PEI–PEG)-based nanoparticles (TPIC)	Co-delivery of DNA and doxorubicin (DOX)	HUVEC cells	The results indicated that TPIC influenced the distribution of CD13 on HUVEC through the interaction between CD13 and NGR, causing CD13 clustering, leading to the co-localization of CD13 and CAV1 and internalization via the CvME. Additionally, this internalization was not dependent on the enzyme activity of CD13 but was inhibited by cholesterol depletion.	Liu et al., 2016
NGR	PAMCN	Cisplatin	Angiogenesis/ HUVEC cells/ KM Mice	The in-vivo study indicated that PAMCN group based treated mice survived 100% and reduced the hepatic and renal toxicity significantly compared to free drug-treated group. All the results revealed that this novel drug delivery system could be a potential approach for antiangiogenic chemotherapy.	Zhang et al., 2015

TABLE 10.2 (*Continued*)

Target moiety	Delivery system	Bioactive molecule	Disease /cell line/ animal model	Concluding remarks	Ref.
NGR	pcCPP/NGR-LP (Liposomes)	siRNA	HT-1080 cells and MCF-7 cells	pcCPP/NGR-LP is expected to be a promising delivery system for oncotherapy. The dual-modified liposomes may be considered as alternatives to photodynamic therapy in which NIR light is used to locally generate cytotoxic reactive oxygen species at the tumor site that cause cancer cell death, but not to induce intracellular delivery of macromolecular therapeutics.	Yang et al., 2015
cKNGRE	68Ga-NOTA-c(NGR)	—	Fischer–344 rats	In NeDe tumor-bearing rats higher 68Ga-NOTA-c(NGR) accumulation was found in the tumors than that of the 68Ga-NODAGA-[c(RGD)]2. Due to its high selectivity and strong binding affinity to APN/CD13, favorable biodistribution and pharmacokinetics, 68Ga-NOTA-c(NGR) might be a potential molecular probe for the noninvasive detection of CD13/APN-positive tumors, metastases, and neovasculature.	Mate et al., 2015
Asn–Gly–Arg (NO$_2$) COOCH3	13F–1 is a 5-fluorouracil prodrug	5-FU	Colo205 xenografts	The 13F–1 administered in mice by injection, significantly inhibited the growth of Colo205 xenografts without sign of toxicity. Hence, it can be used as a promising agent for treatment of cancers with high expression of APN/CD13	Cui et al., 2014.
CNGRCK	cNGR-DNB-NPs	DTX	HT–1080, HU-VEC/ Balb/c mice	The results suggest that cNGR peptide-conjugated PLGA-PEG-NPs can be successfully explored for the design of dual-targeted delivery systems for targeting the tumor cells and vasculature and have a potential approach for CD13-overexpressed specific tumors.	Gupta et al., 2014

TABLE 10.2 (Continued)

Target moiety	Delivery system	Bioactive molecule	Disease /cell line/ animal model	Concluding remarks	Ref.
iNGR	Poly(ethylene glycol)-poly (L-lactic-co-glycolic acid) nanoparticles	Paclitaxel	Human umbilical vein endothelial cells	The anti-glioma efficacy of paclitaxel-loaded iNGR-NP is verified by its improved anti-angiogenesis activity and the significantly prolonged survival time in mice bearing intracranial glioma.	Kang et al., 2014
CNGRCK-2HK3HK11	SWNTs	2-methoxyestra-diol	MCF–7 cells/ female BALB/c mice/	The inhibition ratio of this SWNTs drug delivery system at 24, 48, and 72 hr was about 57.7%, 83.6%, and 88.2%. Compared with normal saline group, the relative tumor volumes in the 2ME, SWNTs–2ME, and NGR-SWNTs–2ME groups were decreased 1 week after administration This novel neovascularity targeting drug delivery system containing NGR-SWNTs–2ME may be beneficial to improve treatment efficacy and minimize side effects in future cancer therapy.	Chen et al., 2013
NGR tripep-tide ligand	Monodisperse spherical micelles	—	—	These results suggest that enhanced delivery to solid tumors can be achieved by targeting upregulated receptors in the tumor vasculature with multivalent ligand-presenting nanoparticles, but additional work is required to optimize such systems for multivalent targeting.	Simnick et al., 2011
CNGRCG	cNGR–PEG–PLA-NPs	pDNA	HUVEC and HepG2 cells	The unique mechanism of caveolae-mediated endocytosis was indeed mainly involved in the internalization of cNGR–PEG–PLA-NPs into HUVEC and led to significant gene transfection efficiency. Such favorable properties including high transfer efficiency, low cytotoxicity, and fast uptake by nondestructive endocytic pathways make cNGR–PEG–PLANPs to be a promising carrier for the intracellular delivery of therapeutic agents.	Liu et al., 2011

TABLE 10.2 *(Continued)*

Target moiety	Delivery system	Bioactive molecule	Disease /cell line/ animal model	Concluding remarks	Ref.
CNGRCK	BPEI-SS-PEG-cNGR/pDNA polyplexes	pDNA	HT–1080, MDA-MB–231cells, HEK293, B16F1	The results implied that multifunctional polymer-mediated gene transfection took place tumor-specifically and via GSH-dependent pathway.	Son et al., 2010
CYGGRGNG and STR-R4	Dual ligand modified liposomes	Rho-DOPE	MS–1 and TE cells	The dual-modification of liposomes with NGR and STR-R4 resulted in an enhanced uptake due to synergistic effects in CD13 positive cells. The proposed strategy holds considerable promise for the further development of in vivo applicable systems.	Takara et al., 2010
GNGRG-GVRSS-SRTPSDKYC	LPD-PEG-NGR (liposome- polycation-DNA) nanoparticle	DOX and siRNA	HT–1080 and HT–29 cells	The tumor-homing peptide (NGR) modified nanoparticle provides an enhancement of drug potency and may potentially be a therapeutic agent against drug-resistant tumors when combined with siRNA against drug-resistant genes such as p-glycoprotein.	Chen et al., 2010
GNGRG	Liposomes	DOX	Colon26 NL–17 cells	The dual-targeting liposomes containing DOX strongly suppressed tumor growth in Colon26 NL–17 carcinoma-bearing mice.	Murase et al., 2010
cKNGRE	Temperature sensitive liposomes	Fluorescent reporter Oregon Green	HT–1080, MCF–7 cells	The results of this study are significant because they demonstrate improved avidity of an NGR-targeted LTSL without the limitation of a disulfide bridge.	Negussie et al., 2010
cNGR	cNGR-pQDs	—	—	The results showed that cNGR-pQDs were specifically accumulated on the tumor endothelial cells surface and to a lesser extent in the vessel lumen, supporting a higher specificity of cNGR-pQDs for angiogenic tumor vasculature.	Oostendorp et al., 2008

TABLE 10.2 *(Continued)*

Target moiety	Delivery system	Bioactive molecule	Disease /cell line/ animal model	Concluding remarks	Ref.
CNGRC	Stealth NGR/PEG/ PDBA-coupled-SHA PEI/pDNA targeting polyplex loaded with PLGA-PEG-PLGA tri-block copolymer	pDNA	DCs	All the results suggest that PLGA-PEG-PLGA encapsulation of this stealth targeting polyplex has no negative effects on key properties of immature DCs and should pave the way for targeting DCs for vaccination purposes.	Moffatt et al., 2006
CNGRC	CNGRC/PEG/PEI/ DNA	DNA	HUVEC and HT–1080/nude mice	*In vivo* transduction analysis using the CNGRC/PEG/ PEI/DNA vector to target the intravenous delivery of a yellow fluorescence protein (YFP)-expressing plasmid to subcutaneous H1299 tumors confirmed delivery of YFP to both tumor cells and tumor endothelial cells. The use of this peptide to further increase tumor-specific delivery mediated by our novel PEI/DNA vector now provides a basis for developing tumor-targeted gene therapies for use in the clinical treatment of cancer.	Moffatt et al., 2005

potential options in the development of efficient and specific drug delivery systems (Takara et al., 2010). The other study explored the importance of novel cyclic NGR containing peptide, cKNGRE, which does not contain a disulfide bridge. The lysolipid containing temperature sensitive liposomes (LTSLs) was prepared and center attached to the surface of specific designed liposomes. The in-vitro fluorescence microscopy presented the comparative evaluation of cKNGRE-OG demonstrated binding and active uptakes by CD13+ cancer cells and minimal binding to CD13-cancer cells and have 3.6-fold greater affinity for CD13+ cancer cells than a linear NGR-containing peptide. The designed system showed rapidly released (>75% in <4 s) drug at 41.3°C with minimal release at 37°C. All the study concluded that cKNGRE-targeted temperature sensitive liposome that lacks a disulfide bridge and has sufficient binding affinity for biological applications (Negussie et al., 2010).

g. Polymeric Nanoparticles: The polymeric NPs are colloidal particulate systems that have self-assembled in case of amphiphilic block copolymers when exposed to an aqueous media. They provide several merits such as a higher drug payload, prolonged blood circulation with controlled release profiles. The polymeric NPs can easily accommodate the cytotoxic drugs into the hydrophobic core (Kumar, 2017; Pandey, 2018). The outer exposed hydrophilic part provided the stable dispersion by imparting a steric stabilization effect that ultimately enhanced its blood residence time following intravenous injection. The recent study established by Liu et al., 2011, that peptide-mediated polymeric nanoparticles targeted gene vector for highly efficient receptor-mediated intracellular delivery. Their research group implicated the cyclic Asn-Gly-Arg (cNGR) to target gene-loaded poly (lactic acid)–poly (ethylene glycol) nanoparticles (PLA–PEG NPs) to HUVEC over-expressing CD13. DNA was complexed with cNGR–PEG–PLA-NPs by using 6-lauroxyhexyl lysinate (LHLN) as cationic surfactant with homogeneous small-sized complexes (200 nm) with positive charge (~10 mV). The cNGR peptide coupled nanoparticles facilitated the fast and efficient internalization of cNGR–PEG–PLA NPs into HUVEC cells. Furthermore, the results of the mechanism studies and transfection assays presented the caveolae-mediated endocytosis was used in the internalization of cNGR–PEG–PLA NPs into HUVEC and led to significant gene transfection efficiency. Our research group Gupta et al., 2014 validated the cNGR peptide conjugated to the PEG terminal end in the PLGA-PEG block copolymer. The ligand-conjugated nanoparticles (cNGR-DNB-NPs) encapsulating docetaxel (DTX) were used for in vitro cytotoxicity, cell apoptosis, and cell cycle analysis. The results revealed the enhanced therapeutic potential of cNGR-DNB-NPs. The higher cellular uptake was also found in cNGR peptide anchored NPs into HUVEC and HT–1080 cells. The *in-vivo* biodistribution and antitumor efficacy studies confirmed that targeted NPs presenting the higher therapeutic outcome through targeting the tumor-specific site. Hence, the cNGR functionalized PEG-PLGA-NPs might be opportunistic approach for therapeutic applications to efficient antitumor drug delivery. Recent study concluded that multifunctional PEI-PEG based nanoparticles containing DOX and pDNA, influenced the distribution of CD13 on HUVEC through the interaction between CD13 and NGR, causing CD13 clustering, leading to the co-localization of CD13 and CAV1 and internalization via the CvME (Liu et al., 2017).

h. Other Targeted Systems: Previous research stated that NGR based CNGRC peptide, when conjugated with the tumor necrosis factor-α (TNF), a cytokine possessing potent vascular damaging properties and antitumor activity, directed the generation of a new therapeutic moiety with higher anti-tumor efficacy, termed NGR-TNF (Corti, 2004). The agent reduced the dose at picogram level of NGR-TNF, provided the synergistic antitumor effects with various chemotherapeutic drugs (such as doxorubicin, melphalan, cisplatin, paclitaxel, and gemcitabine), by modulating drug-penetration barriers (Corti et al., 2008). Presently, NGR-hTNF is currently entered in Phase III clinical studies for mesothelioma is ongoing in Austria, Italy, Canada as well as in four randomized Phase II trials in four types of solid tumors, as monotherapy or in combination with chemotherapeutic regimens (www.molmed.com). The other studies showed that an APN isoform that was upregulated on tumor blood vessels, NGR-TNF motif presented synergistic antitumor activity. On the other hand, APN expressed in normal kidney and in myeloid cells failed to bind to NGR-TNF. All these observations displayed the selectivity and tumor-homing properties of NGR-drug conjugates and may be a significant achievement for the progression of vascular-targeted therapies based on the NGR/CD13 system (Curnis et al., 2002). Other cytokines and antiangiogenic molecules, namely IFN- α, IFN-α 2a, endostatin, and tumstatin fragment also attached with NGR peptides for enhancing the antitumor effect (Meng et al., 2007; Yokoyama and Ramakrishnan, 2005; Curnis et al., 2005). Minute amount at pictogram level of recombinant IFN-CNGRC conjugate targeted the tumor vasculature as well as delayed tumor growth in mice (Curnis et al., 2005).

The first clinical trial of intravenously administered NGR-hTNF was done at four doubling-dose levels ($0.2–0.4–0.8–1.6$ μg/m^2). The frequent treatment-related toxicity was up to grade 1–2 chills (69%) occurring during the first infusions. 75% of patients respond with DCE-MRI and showed decrease over time of K (trans), which was more pronounced at 0.8 μg/m^2. During therapy, 44% patient presented stable disease for 5.9 months, including a colon cancer patient who experienced an 18-month progression-free time (Gregorc et al., 2010). Another phase 1b study was performed on 15 patients with NGR-hTNF-doxorubicin (Dox). The results findings supported that no toxicity was observed and it is well tolerated, only 11% of the side effects were shown in regard to NGR-hTNF. The dose level of 0.8 µg/m² NGR-hTNF plus Dox 75 mg/m² was adopted for phase II development (Gregorc et al., 2009).

Fusion proteins having an extracellular domain of tissue factor as truncated tissue factor, tTF, and the NGR motif were expressed in Escherichia coli BL21 (DE3). The *in-vivo* studies were performed on xenografted nude mice and concluded that tTF-NGR induced partial or complete thrombotic occlusion of tumor vessels, followed by histological examination and significant tumor growth inhibition (Kessler et al., 2008; Bieker et al., 2009). The human fibrosarcoma xenograft model presented response by after administration of tTF-NGR and magnetic resonance imaging (MRI) study revealed a significant reduction of tumor perfusion. This is the clinical first-in-man application of low dose of this targeted coagulation factor displayed good tolerability and decreased tumor perfusion as measured by MRI (Bieker et al., 2009). 13F–1 is a 5-fluoro-uracil prodrug, mainly consisted of Asn–Gly–Arg (NO2) COOCH3 tripeptide and may be exploited for cancer growth targeting via CD13 receptors. The results displayed the activity against human colonic carcinoma growth. The 13F–1 administered in mice by injection, significantly inhibited the growth of Colo205 xenografts without sign of toxicity. Hence, it can be used as a promising agent for treatment of cancers with high expression of APN/CD13 (Cui et al., 2014). In one study, the In vivo imaging property of a novel 68Ga-labeled NOTA-c (NGR) molecule in vivo using mini PET was investigated. The results showed that owing to its higher selectivity and strong binding affinity to APN/CD13, favorable biodistribution, and pharmacokinetics, 68Ga-NOTA-c (NGR) might be a potential molecular probe for the noninvasive detection of CD13/APN-positive tumors, metastases, and neovasculature (Mate et al., 2015). Recent study suggested that long circulatory CdSe/ZnS quantum dots (QDs)-based nanoscale and fluorescence NGR-based specific CD13 peptides were able to recognize the BBB and target CD13-overexpressing glioma and tumor vasculature in vitro and in vivo, contributing to fluorescence imaging of this brain malignancy (Huang et al., 2017).

10.6 CONCLUSION AND FUTURE PERSPECTIVE

Tumor angiogenesis had a very critical role in development and progression of tumor. Various effective tumor inhibition agents might check the tumor progression but would not eradicate the tumor as a stand-alone therapy. Cancer can be effectively treated by inhibition and targeting of tumor angiogenesis. Drug targeting or cytokine delivery to tumor neovasculature is an appealing option for tumor inhibition and ultimately destroying the tumor vessels and hence, suppression of tumor cells.

CD13 is a key component in tumor vasculature and is over-expressed in the membranes of endothelial tumor cells. CD13 receptors play significant role in chemokine processing and tumor invasion. The binding affinity of NGR peptide motif is CD13 isoform, which is expressed on the endothelial cells of tumor blood vessels but not on the normal vasculature. Till date, several research work experimented to explore the new moiety to target the CD13 receptor such as enzyme inhibitors as well as APN-targeted carrier constructs and proof the concept. The therapeutic outcome of CD13 receptor targeting via the NGR peptide anchored to chemotherapeutic drug modulated by some of the functions of CD13 that influence either the positive or negative way. Nanoformulations have occupied their place and very emerging therapy options in the future oncotherapy. Till date, this addressing issue would be required further investigation. However, some of the therapeutic entered or are currently being evaluated in clinical trials. The other innovation may be the therapeutic and imaging-based approach that is termed as nanotheragnostic way. Several examples are cited in the text so that readers can easily understand the approach.

All together these elements forecast a bright future for targeted therapeutic NGR based approach by making sure of proper localization of drugs at their desired site. These breakthrough aspects might shift the paradigm of treatment for many devastating diseases, including cancer, and contribute to tackling a myriad of important human conditions.

KEYWORDS

- **CD13 receptor**
- **nanomedicine**
- **targeting strategies**
- **theranostic**
- **tumor malignancy**

REFERENCES

Adams, M. L., Lavasanifar, A., & Kwon, G. S., (2003). Amphiphilic block copolymers for drug delivery. *J. Pharm. Sci.*, *92*(7), 1343–1355.

Agarwal, A., et al., (2008). Ligand-based dendritic systems for tumor targeting. *Int. J. Pharm.*, *350*(1 & 2), 3–13.

Ahmad, J., Akhter, S., Rizwanullah, M., et al., (2015a). Nanotechnology-based inhalation treatments for lung cancer: State of the art. *Nanotechnol. Sci. Appl.*, *19*(8), 55–66.

Ahmad, J., Amin, S., Rahman, M., et al., (2015b). Solid matrix based lipidic nanoparticles in oral cancer chemotherapy: Applications and pharmacokinetics. *Curr. Drug. Metab.*, *16*(8), 633–644.

Ahmad, M. Z., Akhter, S., Ahmad, I., et al., (2011). Development of polysaccharide-based colon targeted drug delivery system: Design and evaluation of Assam Bora rice starch-based matrix tablet. *Curr. Drug. Deliv.*, *8*(5), 575–581.

Ahmad, M. Z., Akhter, S., Anwar, M., et al., (2013). Colorectal cancer targeted Irinotecan-Assam Bora rice starch-based microspheres: A mechanistic, pharmacokinetic and biochemical investigation. *Drug. Dev. Ind. Pharm.*, *39*(12), 1936–1943.

Ahmad, M. Z., Akhter, S., Jain, G. K., et al., (2010). Metallic nanoparticles: Technology overview & drug delivery applications in oncology. *Expert. Opin. Drug. Deliv.*, *7*(8), 927–942.

Akhter, S., Ahmad, Z., Singh, A., et al., (2011). Cancer targeted metallic nanoparticle: Targeting overview, recent advancement, and toxicity concern. *Curr. Pharm. Des.*, *17*(18), 1834–1850.

Allen, T. M., (2002). Ligand-targeted therapeutics in anticancer therapy. *Nat. Rev. Cancer*, *2*(10), 750–763.

Aneja, P., Rahman, M., Beg, S., et al., (2014). Cancer targeted magic bullets for effective treatment of cancer. *Recent. Pat. Antiinfect. Drug. Discov.*, *2*, 121–135.

Anthony, A., Leslie, G., & Brian, W. M., (2006). Structure of aminopeptidase N from *Escherichia coli* suggests a compartmentalized, gated active site. *Proc. Natl. Acad. Sci. USA*, *103*(36), 13339–13344.

Arap, W., Pasqualini, R., & Ruoslahti, E., (1998). Cancer treatment by targeted drug delivery to tumor vasculature in a mouse model. *Science*, *279*(5349), 377–380.

Bauvois, B., & Dauzonne, D., (2006). Aminopeptidase-N/CD13 (EC 3.4.11.2) inhibitors: Chemistry, biological evaluations, and therapeutic prospects. *Med. Res. Rev.*, *26*(1), 88–130.

Beg, S., Rahman, M., Jain, A., et al., (2017). Nanoporous metal-organic frameworks as hybrid polymer-metal composites for drug delivery and biomedical applications. *Drug. Discov. Today.*, *22*(4), 625–637.

Bhagwat, S. V., et al., (2001). CD13/APN is activated by angiogenic signals and is essential for capillary tube formation. *Blood*, *97*(3), 652–659.

Bieker, R., et al., (2009). Infarction of tumor vessels by NGR peptide-directed targeting of tissue factor: Experimental results and first-in-man experience. *Blood*, *113*(20), 5019–5027.

Chen, C., et al., (2013). Single-walled carbon nanotubes mediated neovascularity targeted antitumor drug delivery system. *J. Pharm. Pharmaceut. Sci.*, *16*(1), 40–51.

Chen, Y., Wu, J. J., & Huang, L., (2010). Nanoparticles targeted with NGR motif deliver c-myc siRNA and doxorubicin for anticancer therapy. *The Amer. Soc. Gene cell Ther*, *18*(4), 828–834.

Colombo, G., et al., (2002). Structure-activity relationships of linear and cyclic peptides containing the NGR tumor homing motif. *J. Biol. Chem.*, *277*(49), 47891–47897.

Corti, A., & Ponzoni, M., (2004). Tumor vascular targeting with tumor necrosis factor alpha and chemotherapeutic drugs. *Signal Transd. Comm. Cancer Cells*, *1028*, 104–112.

Corti, A., (2004). Strategies for improving the antineoplastic activity of TNF by tumor targeting. *Methods Mol. Med.*, *98*, 247–264.

Corti, A., et al., (2008). The neovasculature homing motif NGR: More than meets the eye. *Blood*, *112*(7), 2628–2635.

Curnis, F., et al., (2002). Differential binding of drugs containing the NGR motif to CD13 isoforms in tumor vessels, epithelial, and myeloid cells. *Cancer Res.*, *62*(3), 867–874.

Curnis, F., et al., (2005). Targeted delivery of IFN-gamma to tumor vessels uncouples anti-tumor from counter-regulatory mechanisms. *Cancer Res.*, *65*(7), 2906–2913.

Curnis, F., et al., (2006). Spontaneous formation of L-isoaspartate and gain of function in fibronectin. *J. Biol. Chem.*, *281*(47), 36466–36476.

Díaz, M. R., & Vivas-Mejia, P. E., (2013). Nanoparticles as drug delivery systems in cancer medicine: Emphasis on RNAi-containing nanoliposomes. *Pharmaceut.*, *6*(11), 1361–1380.

Fondell, A., et al., (2010). Nuclisome: A novel concept for radionuclide therapy using targeting liposomes. *Eur. J. Med. Mol. Imaging*, *37*(1), 114–123.

Garde, S. V., et al., (2007). Binding and internalization of NGR-peptide-targeted liposomal doxorubicin (TVT-DOX) in CD13-expressing cells and its antitumor effects. *Anticancer Drugs*, *18*, 1189–1200.

Gomez-Gualdron, D. A., et al., (2012). Dynamic evolution of supported metal nanocatalyst/carbon structure during single-walled carbon nanotube growth. *ACS Nano*, *6*(1), 720–735.

Gregorc, V., et al., (2009). Phase Ib study of NGR-hTNF, a selective vascular targeting agent, administered at low doses in combination with doxorubicin to patients with advanced solid tumors. *Br. J. Cancer*, *101*(2), 219–224.

Gregorc, V., et al., (2010). Defining the optimal biological dose of NGR-hTNF, a selective vascular targeting agent, in advanced solid tumors. *Eur. J. Cancer*, *46*(1), 198–206.

Grifman, M., et al., (2001). Incorporation of tumor-targeting peptides into recombinant adeno-associated virus capsids. *Mol Ther.*, *3*(6), 964–975.

Grigoryan, G., et al., (2011). Computational design of virus-like protein assemblies on carbon nanotube surfaces. *Sci.*, *332*(6033), 1071–1076.

Gupta, M., Agrawal, U., & Vyas, S. P., (2014). Receptor-mediated targeting and their intracellular trafficking pathway: Importance in drug delivery. In: Bhupinder, S. B., (ed.), *Nanobiomedicine*. M/s Studium Press LLC, USA. Volume 4, (ISBN: 978-1-626990-54-8).

Gupta, M., et al., (2014). Dual-targeted polymeric nanoparticles based on tumor endothelium and tumor cells for enhanced antitumor drug delivery. *Mol. Pharm.*, *11*(3), 697–715.

Hans, S., Oye, N., & Jorgen, O., (2000). Structure and function of aminopeptidase N. *Adv. Exp. Med. Biol.*, *477*, 25–34.

Hardman, R. A., (2006). Toxicologic review of quantum dots: Toxicity depends on physicochemical and environmental factors. *Environ. Health Perspect.*, *114*(2), 165–172.

Hellstrom, M., et al., (2007). Dll4 signaling through Notch1 regulates formation of tip cells during angiogenesis. *Nature*, *445*(7129), 776–780.

Huang, N., et al., (2017). Efficacy of NGR peptide-modified PEGylated quantum dots for crossing the blood-brain barrier and targeted fluorescence imaging of glioma and tumor vasculature. *Nanomed.: Nanotechno. Bio. Med.*, *13*, 83–93.

Jakobsson, L., et al., (2010). Endothelial cells dynamically compete for the tip cell position during angiogenic sprouting. *Nat. Cell. Biol.*, *12*(10), 943–953.

Kang, T., et al., (2014). iNGR-modified PEG-PLGA nanoparticles that recognize tumor vasculature and penetrate gliomas. *Biomaterials*, *35*, 4319–4332.

Kessler, T., et al., (2008). Generation of fusion proteins for selective occlusion of tumor vessels. *Curr. Drug Discov. Technol.*, *5*(1), 1–8.

Krige, D., et al., (2008). CHR–2797: An antiproliferative aminopeptidase inhibitor that leads to amino acid deprivation in human leukemic cells. *Cancer Res.*, *68*(16), 6669–6679.

Kumar, V., Bhatt, P. C., Rahman, M., et al., (2017). Fabrication, optimization, and characterization of umbelliferone β-D-galactopyranoside-loaded PLGA nanoparticles in treatment of hepatocellular carcinoma: *In vitro* and *in vivo* studies. *Int. J. Nanomedicine*, *12*, 6747–6758.

Liu, C., et al., (2011). Enhanced gene transfection efficiency in CD13-positive vascular endothelial cells with targeted poly (lactic acid)–poly (ethylene glycol) nanoparticles through caveolae-mediated endocytosis. *J. Control. Rel.*, *151*, 162–175.

Liu, C., et al., (2017). A preliminary study on the interaction between Asn-Gly-Arg (NGR)-modified multifunctional nanoparticles and vascular epithelial cells. *Acta Pharmaceutica Sinica B*, *7*(3), 361–372.

Luan, Y., & Xu, W., (2007). The structure and main functions of aminopeptidase N. *Curr. Med. Chem.*, *14*(6), 639–647.

Mate, G., et al., (2015). *In vivo* imaging of Aminopeptidase N (CD13) receptors in experimental renal tumors using the novel radiotracer 68Ga-NOTA-c(NGR). *Eur. J. Pharm. Sci.*, *69*, 61–71.

Meng, J., et al., (2009). High-yield expression, purification, and characterization of tumor-targeted IFN-alpha2a. *Cytotherapy*, *9*(1), 60–68.

Mina-Osorio, P., (2008). The moonlighting enzyme CD13: Old and new functions to target. *Trends Mol. Med.*, *14*(8), 361–371.

Mizuguchi, H., et al., (2001). A simplified system for constructing recombinant adenoviral vectors containing heterologous peptides in the HI loop of their fiber knob. *Gene Ther.*, *8*(9), 730–735.

Moffatt, S., & Cristiano, R. J., (2006). Uptake characteristics of NGR-coupled stealth PEI/pDNA nanoparticles loaded with PLGA-PEG-PLGA tri-block copolymer for targeted delivery to human monocyte-derived dendritic cells. *Int. J. Pharm.*, *321*, 143–154.

Moffatt, S., Wiehle, S., & Cristiano, R. J., (2005). Tumor-specific gene delivery mediated by a novel peptide–polyethyleneimine–DNA polyplex targeting aminopeptidase N/CD13. *Human Gene Ther.*, *16*(1), 57–67.

Murase, Y., et al., (2010). A novel DDS strategy, dual-targeting, and its application for antineovascular therapy. *Cancer Lett.*, *287*(2), 165–171.

Negussie, A. H., et al., (2010). Synthesis and in vitro evaluation of cyclic NGR peptide targeted thermally sensitive liposome. *J. Control. Rel.*, *143*, 265–273.

Nitta, S. K., & Numata, K., (2013). Biopolymer-based nanoparticles for drug/gene delivery and tissue engineering. *Int. J. Mol. Sci.*, *14*(1), 1629–1654.

Noguera-Troise, I., et al., (2006). Blockade of Dll4 inhibits tumor growth by promoting non-productive angiogenesis. *Nature*, *444*(7122), 1032–1037.

O'Connell, P. J., Gerkis, V., & Dapice, A. J., (1991). Variable O-glycosylation of CD13 (aminopeptidase N). *J. Biol. Chem.*, *266*(7), 4593–4597.

Oostendorp, M., et al., (2008). Quantitative molecular magnetic resonance imaging of tumor angiogenesis using cNGR-labeled paramagnetic quantum dots. *Cancer Res.*, *68*(18), 7676–7683.

Oral presentation at ASCO (2015), of survival data in poor prognosis patients achieved in the Phase III trial on NGR-hTNF treatment in pleural mesothelioma. www.molmed.com.

Pandey, P., Rahman, M., Bhatt, P. C., et al., (2018). Implication of nanoantioxidant therapy for treatment of hepatocellular carcinoma using PLGA nanoparticles of rutin. *Nanomedicine (Lond).* doi: 10. 2217/nnm–2017–0306.

Pasqualini, R., et al., (2000). Aminopeptidase N is a receptor for tumor-homing peptides and a target for inhibiting angiogenesis. *Cancer Res.*, *60*(3), 722–727.

Pastorino, F., et al., (2003). Vascular damage and anti-angiogenic effects of tumor vessel-targeted liposomal chemotherapy. *Cancer Res.*, *63*(21), 7400–7409.

Pastorino, F., et al., (2006). Targeting liposomal chemotherapy via both tumor cell specific and tumor vasculature-specific ligands potentiates therapeutic efficacy. *Cancer Res.*, *66*(20), 10073–10082.

Peng, L., et al., (2009). Novel gene-activated matrix with embedded chitosan/plasmid DNA nanoparticles encoding PDGF for periodontal tissue engineering. *J. Biomed. Mater. Res. Part A*, *90*, 564–576.

Qi, L., & Gao, X., (2008). Emerging application of quantum dots for drug delivery and therapy. *Expert Opin. Drug Deliv.*, *5*(3), 263–267.

Rahman, M., Ahmad, M. Z., Ahmad, J., et al., (2015b). Role of graphene nano-composites in cancer therapy: Theranostic applications, metabolic fate, and toxicity issues. *Curr. Drug Metab.*, *16*(5), 397–409.

Rahman, M., Ahmad, M. Z., Kazmi, I., et al., (2012a). Emergence of nanomedicine as cancer-targeted magic bullets: Recent development and need to address the toxicity apprehension. *Curr. Drug. Discov. Technol.*, *9*(4), 319–329.

Rahman, M., Ahmad, M. Z., Kazmi, I., et al., (2012b). Advancement in multifunctional nanoparticles for the effective treatment of cancer. *Expert. Opin. Drug. Deliv.*, *9*(4), 367–381.

Rahman, M., Ahmed, M. Z., Kazmi, I., et al., (2012c). Novel approach for the treatment of cancer: Theranostic nanomedicines. *Pharmacologia*, *3*, 371–376.

Rahman, M., Akhter, S., Ahmad, M. Z., et al., (2015a). Emerging advances in cancer nanotheranostics with graphene nanocomposites: Opportunities and challenges. *Nanomedicine (Lond)*, *10*(15), 2405–2422.

Rahman, M., Beg, S., Ahmed, A., et al., (2013). Emergence of functionalized nanomedicines in cancer chemotherapy: Recent advancements, current challenges, and toxicity considerations. *Recent Patent on Nanomedicine*, *2*, 128–139.

Rahman, M., Beg, S., Verma, A., et al., (2017). Therapeutic applications of liposomal-based drug delivery and drug targeting for immune linked inflammatory maladies: A contemporary viewpoint. *Curr. Drug. Targets.*, *18*(13), 1558–1571.

Rahman, M., Kumar, V., Beg, S., et al., (2016). Emergence of liposome as targeted magic bullet for inflammatory disorders: Current state of the art. *Artif. Cells. Nanomed. Biotechnol.*, *44*(7), 1597–1608.

Rapoport, N., (2007). Physical stimuli-responsive polymeric micelles for anti-cancer drug delivery. *Prog. Polym. Sci.*, *32*, 962–990.

Reid, A. H., et al., (2009). A first-in-man phase I and pharmacokinetic study on CHR–2797 (Tosedostat), an inhibitor of M1 aminopeptidases, in patients with advanced solid tumors. *Clin. Cancer Res.*, *15*(15), 4978–4985.

Ribatti, D., & Djonov, V., (2012). Intussusceptive microvascular growth in tumors. *Cancer Lett.*, *316*(2), 126–131.

Ribatti, D., (2009). Endogenous inhibitors of angiogenesis: A historical review. *Leuk. Res.*, *33*(5), 638–644.

Ridgway, J., et al., (2006). Inhibition of Dll4 signaling inhibits tumor growth by deregulating angiogenesis. *Nature*, *444*, 1083–1087.

Shu-Xiang, C., et al., (2014). 13F–1, a novel 5 fluorouracil prodrug containing an Asn–Gly–Arg (NO_2) COOCH3 tripeptide, inhibits human colonic carcinoma growth by targeting Aminopeptidase N (APN/CD13). *Eur. J. Pharmacol.*, *734*, 50–59.

Simnick, A. J., et al., (2011). *In vivo* tumor targeting by a NGR-decorated micelle of a recombinant diblock copolypeptide. *J. Control. Rel.*, *155*, 144–151.

Son, S., Singha, K., & Kim, W.J., (2010). Bioreducible BPEI-SS-PEG-cNGR polymer as a tumor-targeted nonviral gene carrier. *Biomaterials, 31*, 6344–6354.

Takara, K., et al., (2010). Design of a dual-ligand system using a specific ligand and cell penetrating peptide, resulting in a synergistic effect on selectivity and cellular uptake. *Int. J. Pharm., 396*, 143–148.

Van Herpen, C. M. L., et al., (2010). A phase Ib dose-escalation study to evaluate safety and tolerability of the addition of the aminopeptidase inhibitor tosedostat (CHR–2797) to paclitaxel in patients with advanced solid tumors. *British J. Cancer, 103*(9), 1362–1368.

Vasir, J. K., Reddy, M. K., & Labhasetwar, V. D., (2005). Nanosystems in drug targeting: Opportunities and challenges. *Curr. Nanosci., 1*(1), 47–64.

Wang, C., et al., (2012). Adsorption and properties of aromatic amino acids on single-walled carbon nanotubes. *Nanoscale, 4*(4), 1146–1153.

Wang, M., et al., (2017). Role of tumor microenvironment in tumorigenesis. *J. Cancer, 8*(5), 761–773.

Wang, R. E., et al., (2011). Development of NGR peptide-based agents for tumor imaging. *Am. J. Nucl. Med. Mol. Imaging, 1*(1), 36–46.

Wang, X., et al., (2009). NGR-modified micelles enhance their interaction with CD13-overexpressing tumor and endothelial cells. *J. Control. Rel., 139*, 56–62.

Wicki, A., et al., (2015). Nanomedicine in cancer therapy: Challenges, opportunities, and clinical applications. *J. Control. Rel., 200*, 138–157.

Wickstrom, M., et al., (2011). Aminopeptidase N (CD13) as a target for cancer chemotherapy. *Cancer Sci., 102*(5), 501–508.

Yang, Y., et al., (2015). Dual-modified liposomes with a two-photon-sensitive cell penetrating peptide and NGR ligand for siRNA targeting delivery. *Biomaterials, 48*, 84–96.

Yokoyama, Y., & Ramakrishnan, S., (2005). Addition of an aminopeptidase N-binding sequence to human endostatin improves inhibition of ovarian carcinoma growth. *Cancer, 104*(2), 321–331.

Zhang, L., et al., (2015). Fabrication of poly(γ-glutamic acid)-based biopolymer as the targeted drug delivery system with enhanced cytotoxicity to APN/CD13 over-expressed cells. *J. Drug Target., 23*(5), 453–461.

Zhao, B. J., et al., (2011). The antiangiogenic efficacy of NGR-modified PEG–DSPE micelles containing paclitaxel (NGR-M-PTX) for the treatment of glioma in rats. *J. Drug Target., 19*(5), 382–390.

Zuccari, G., et al., (2015). Tumor vascular targeted liposomal-bortezomib minimizes side effects and increases therapeutic activity in human neuroblastoma. *J. Control. Rel., 211*, 44–52.

CHAPTER 11

Liposomal Nanomedicines as State-of-the-Art in the Treatment of Skin Disorders

MAHFOOZUR RAHMAN[1*], KAINAT ALAM[2], MD. ADIL SHAHARYAR[3], SARWAR BEG[4], ABDUL HAFEEZ[5], REHAN ABDUR RUB[6], and VIKAS KUMAR[7]

[1]Department of Pharmaceutical Sciences, Shalom Institute of Health andAllied Sciences (SIHAS), Sam Higginbottom University of Agriculture,Technology & Sciences (SHUATS), Allahabad, India

[2]Christian College of Nursing, Shalom Institute of Health and AlliedSciences (SIHAS), Sam Higginbottom University of Agriculture,Technology & Sciences (SHUATS), Allahabad, India

[3]Department of Pharmaceutical Technology, Jadavpur University, Kolkata, India

[4]Product Nanomedicine Research Lab, Department of Pharmaceutics, School of Pharmaceutical Education and Research, Jamia Hamdard (Hamdard University), New Delhi, India

[5]Glocal School of Pharmacy, Glocal University, Saharanpur, U.P., India

[6]Pharmaceutics Research Lab, SPER, Jamia Hamdard, New Delhi-62, India

[7]Natural Product Drug Discovery Laboratory, Department of Pharmaceutical Sciences, Shalom Institute of Health and Allied Sciences(SIHAS), Sam Higginbottom University of Agriculture, Technology & Sciences (SHUATS), Allahabad, India.

*Corresponding author. E-mail: mahfoozkaifi@gmail.com

ABSTRACT

In 1965, the first lipid-based vesicular systems were described which called as liposomes and started from biophysics to paved new applications as drug delivery systems. Conventional topical medication in skin diseases is more effective and has lesser side effects than the systemic application. Yet, physicochemical properties of drugs and skin anatomy particularly the stratum corneum (SC) are the major barrier in the effective topical pharmacotherapy. Exploiting of the fact that beneath the SC and epidermis of the skin is composed lipids and lipoproteins. Strategies to overcome the limitations, several enhancement strategies have been made including penetration enhancer, iontophoresis, electroporation, supersaturated solutions, and novel drug delivery systems. The emergence of Liposomal systems produces a several-fold higher therapeutic concentration in the various layers of skin and lower systemic concentrations as compared to conventional dosage forms and has improved topical drug delivery in the realm of dermatology. Over the last 5-decade umpteen research papers on liposomal topical drug delivery has been published and technical advances such as remote drug loading, extrusion, etc., enhance the encapsulation of therapeutic molecules.

Furthermore, developments lead to more advanced liposomes such as stealth liposome, stimuli-responsive liposome, cationic liposome and immuno-liposomes investigated for the topical disease. These advances have led to many clinical trials in various areas as topical delivery for psoriasis, skin cancer, acne, atopic dermatitis (AD), antifungal, and anti-inflammatory drugs. A number of liposomal formulations are on the market, and many are under clinical trial, and many more will receive approval in the near future.

11.1   INTRODUCTION

Conventional-based topical drug delivery system has various setbacks such as limited skin permeation and more time taken for therapeutic responses (Rahman et al., 2015). In the past few decades more demanding and increasing interest in the exploration of novel techniques to enhance higher permeation of drug through the skin (Rahman et al., 2016a). Enhanced percutaneous delivery results by adequate reduction of the barrier properties of the SC, which renders its main skin barrier (Rahman et al., 2016b). Employment of high voltage and laser light pulse sources are attempted to create permeable windows, which promotes percutaneous absorption (Zhang et al., 1997). Iontophoresis uses electric potential for delivery of ionizable drugs through the skin (Tavakoli et al., 2015). Electroporation employed short duration high voltage to enhance skin permeability by the creation of pores into the SC lipids (Tavakoli et al., 2015). Another is sonophoresis which employed ultrasonic frequencies from KHz to MHz to enhance permeability of the skin (Herwadkar et al., 2012). Penetration enhancer like dimethyl sulfoxide, lecithin (Xie F., 2016), cyclodextrin (Xie, 2016), glycerol, azone, highly concentrated surfactant solutions or supersaturated solutions (Xie, 2016), also makes higher drug permeation by virtue of raising the fluidity of lipids in the SC layer (Xie, 2016). The entire attempt has potential for enhanced drug permeability into the skin. Apart from these, emergence of novel carriers especially liposomes in dermatology gained wider attention as drug delivery in treatment of various skin disorders (Xie, 2016). Liposomes also used in combination with penetration enhancer, which results to change drug disposition in the skin (Rahman et al., 2016a). Liposomes are a microscopic vesicular structure composed of amphiphilic lipid arranged in concentric bilayer and aqueous compartments (Rahman et al., 2016a). The phospholipids resemble to natural lipid which revealed their unique biocompatibility and other characteristics too. It forms a vesicular structure when comes to contact with aqueous phase (Jóhannesson et al., 2016). The drug molecules are encapsulated as according to nature of drugs, lipophilic drug entrap in bilayer, and hydrophilic drug entrap in aqueous compartment (Wadhwa, 2016). The various types of liposomes depend on the lipid composition, method of preparation and the nature of encapsulated agents. Furthermore, single bilayer is referred to as unilamellar lipid vesicle; depending on its size it is further divided into small unilamellar vesicles (SUV) or large unilamellar vesicles (LUV) (Akhtar, 2014). Whereas, if more number of bilayers are present they called multilamellar vesicles (MLV) (Akhtar, 2014). The liposomes are prepared from single lipid or a mixture with water, cholesterol, and electrolytes. Cholesterol provides lamellar stability and flexibility, whereas an electrolyte provides isotonicity and enhances the lipid bilayer formation (Budai et al., 2013). Liposomal bilayer formation is due to the posses of amphiphilic nature, the polar region of head attracts water and lipophilic will forms tail, which repels water (Budai et al., 2013). Now a day, there are various phospholipids, which used in the preparation of liposomes such as natural phospholipids, i.e., lecithin; it is a major component of biological membranes (Budai et al., 2013).

Liposomal properties depend upon size and morphology, which ultimately depends on the method of preparation and nature of lipids, which may of cationic, anionic, and neutral in nature (De Leeuw, 2009). Size is important for circulation times in blood plasma and disposition *in vivo*. Therefore, controlling of size is important in this regard and also provides integrity, stability, and especially higher skin permeation reported with decreased particle size (De leeuw, 2009). At initial, liposomes as attractive model systems for biological membranes, due to similar in lipid composition and structure (Matsuzaki et al., 1993). First time as a drug carrier, discovered by Sessa and Weissman (Sessa G &Weissman G., 1970), which reported that encapsulation of lysozyme in MLV. It gained diversity in drug delivery realm because possess of entrapment ability, biodegradable, and non-toxicity (Rahman et al., 2016c). Despite of these, liposomes well investigated by the various route of administration and accepted as potential carriers for various drugs such as low molecular weight compounds, therapeutic protein, and diagnostic agents to deliver drug efficiently and to the desired site (Rahman et al., 2016c). Conventional formulation several demerits like rapid clearance from the circulation or site, hindrance by biological barriers, and altered biodistribution of entrapped drugs. To overcome such problems, liposomes loaded drugs provides an optimum action at lower dosage (Rahman et al., 2016c). Literatures revealed that toxicity of various categories of drugs like antimicrobial, antiviral, and chemotherapeutic agents (Pinto-Alphandary, 2000) have to reduce by liposome drug delivery. Liposomal delivery is also effective in potentiating the immunogenicity of antigenic substances (Mastrobattista et al., 1999). Apart from that liposomes also have been demonstrated as efficient vehicle for gene delivery as gene therapy (Zhang et al., 2010). In 2010, investigations were reported on the clinical use of liposome on human diseases. Thus, its several examples are discussed in miscellaneous dermatological section.

11.2   LIPOSOMES AS TOPICAL/TRANSDERMAL DRUG DELIVERY

Conventional based topical formulations are poor in delivery of the ingredients into the dermal layers, because of fail to penetrate the horny layer. Not only intravenously, liposomes have diverse potential as topical drug delivery systems (Akhtar, 2014). Their topical delivery characteristics depend on size, surface charge and chemical compositions (Akhtar, 2014). Liposomal-based topical drug delivery has some advantages over to systemic application. Their similar composition to biological membranes and biocompatible to this (Rahman, 2016a). Moreover, the liposomes may serve as local depots for sustained and targeted delivery of several dermatological compounds such as antibiotics, corticosteroids, retinoic acid into the SC and epidermis, by possessing of phospholipid into the lipid layers of liposomes (Foldvari et al., 1990). Furthermore, also minimizes systemic absorption (El Maghraby,

2008). Thus liposomes may act as penetration enhancers and facilitate topical delivery. Whereas interacting with SC and destabilizes the lipid matrix, which results to enhance drug flux through skin (Kirjavainen, 1996). Initially, dermatological applications of liposomes were due to moisturizing and restorative functions only (Magdassi, 1997). Whereas further it permeation into epidermal or even up to deeper layer of cells was reported by many literatures. Drug-loaded liposomes when have to applied on the skin, it fuses with the cellular membranes and active molecules is released into the cells. The drugs may also be released at the target sites via skin appendages and enhanced systemic absorption (Magdassi, 1997). Liposomal drug delivery has found better results in the treatment of acne vulgaris over conventional ones by virtue of skin appendages (Garg, 2016). The liquid state of liposomes for finasteride to deliver in the pilosebaceous unit using of hamster Xank and ear models (Garg, 2016). Liposomes as a topical carrier depending on the type and composition (Rahman, 2016a). There to now liposomes as a convenience to second generation liposomes gained wide attention and alternatives to improving dermal drug delivery (Rahman, 2016a). Moreover, its therapeutic application is illustrated in Figure 11.1. There are various types of lipids and surfactant, which utilizes in the preparation of vesicle formation. Liposomes have been widely employed to enhance percutaneous absorption of several therapeutic molecules like as diclofenac, beta histidine, tetracaine, and triamcinolone, etc. (Zylberberg & Matosevic, 2016). The proposed mechanism is follicular uptake of liposomes, which permits delivery of drugs into the deeper skin layer. Although, liposomes drug delivery through skin is dependent on size. As such, Schramlova et al. (1997) reported the liposomes 600 nm in size; easily penetrate into skin, whereas 1000 nm size more remains into upper surface of SC. Thus SC is the main barrier to the percutaneous absorption and also considered as main route for penetration. However, literatures also reported that hair follicles and sebaceous glands also play a key role for transdermal delivery. In earlier time, hair follicles route was not well accepted because of possessing of 0.1% of the total surface area of the skin. But to date, it is well accepted and to provide a large space for easy transport of vesicles into the skin (Kajimoto et al., 2011). Literatures also reported that liposomal formulation provides beneficial effects in the treatment of hair follicle-related disorders such as acne and alopecia (Fang, 2014). Classical liposomes have certain limitation like limited penetration and vesicles confined to SC. To overcome this problem especial vesicular carrier is designed, which permits transdermal drug delivery through the deeper skin layer. This showed better stability and higher entrapment efficiency (Fang, 2014).

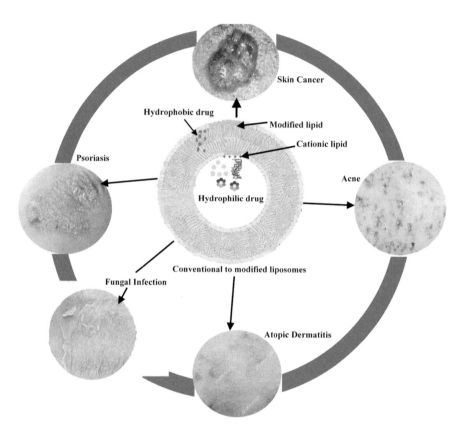

FIGURE 11.1   Illustration of liposomes and its topical applications.

11.3   THERAPEUTIC APPLICATIONS OF LIPOSOMES IN VARIOUS DREADFUL SKIN DISORDERS

11.3.1   *PSORIASIS*

Psoriasis is immune-based skin disorders, characterized by keratinocytes hyperproliferation. Nearly 2–3% of the world population affected. Psoriasis is available in different forms such as plaque psoriasis, guttate psoriasis, pustular, and erythrodermic psoriasis. Among this, the most common is plaque psoriasis (Rahman, 2012a). It forms red scaly plaques over the scalp, lower back and extended up to the limbs. There are so many causing factors such as streptococcal infection, cutaneous trauma, drugs, alcohol, cigarette smoking, stress, and ultraviolet exposure (Rahman, 2012a). But relapsing and spontaneous remission is a great challenge in psoriasis treatment (Rahman, 2012b). An abnormal T cell activation is principle pathway in pathogenesis of psoriasis. Furthermore, binding of antigen on to surface of antigen presenting cells in skin layers to forms major histocompatibility complex (MHC), which is moving into lymph node to bind to T lymphocyte receptors, their activation leads to forms cluster differentiation 2 (CD2) and lymphocyte functional antigen 3 (Rahman et al., 2012a; 2012b). Whereas these clusters cells enter into circulatory system to different skin layers produce cytokines, chemokines, and various growth factors, which causes hyperkeratosis in psoriasis (Rahman et al., 2012a, 2012b). Furthermore, all these inflammatory markers regulate the production and expansion of T helper (T1) and $T_H 17/T_H 22$ cells, which further produces IL–4, IL–13 and IL–5 in psoriasis cascade (Van den Berg, 2013). Moreover, $T_H 17$ cells produce IL–17A and IL–23R, which are strongest evidenced in pathogenesis of psoriasis to date (Van den Berg, 2013). Other than this, the abnormal alteration in eicosanoid and polyunsaturated fatty acids (PUFA) metabolism, and free radical generation also involved in pathogenesis of psoriasis (Rahman et al., 2013). For prevention of psoriasis, various pharmacotherapy's available, which provides short term suboptimal therapeutic effects with various adverse effects such as body weight gain, hypertension, and osteoporosis, etc. (Rahman, 2012b; Rahman et al., 2013). Therefore requires cost-effective and safe therapeutic drug delivery system. Among novel drug delivery, emergence of liposomes gained unique identity in novel drug delivery area (Rahman, 2012b). It is biocompatible and enhanced patient compliance by possessing of phospholipids, minimization of dose and lesser adverse effects in antipsoriatic drug delivery (Rahman, 2012b). Dithranol is the oldest therapy for psoriasis; it acts on interleukin receptor of keratinocyte cell to get beneficial effect. Skin irritations, staining, and restricted skin permeation are of a bigger challenge. Dithranol loaded liposomes are developed, which shows higher skin permeation, minimizes skin irritations and prevents staining (Myśliwiec et al., 2016). Another researcher developed 0.5% dithranol loaded liposomes, which shows better skin permeation over conventional ointment (Saraswat, 2007). Methotrexate (MTX) is an immunosuppressant and widely use in psoriasis treatment. It acts via blocking the conversion of dihydrofolate to tetrahydrofolate (THF). THF is essential for the formation of De-Nova synthesis (Gladman, 2017). MTX is administered in psoriatic patient preferably by oral and parenteral routes. Long term administration causes significant adverse effects including mucosal ulceration, stomatitis, bone marrow suppression, cirrhosis, and loss of appetites (Gladman, 2017). Topical delivery of MTX is limited skin permeation, which results to suboptimal therapeutic effects in psoriasis (Gladman, 2017). To minimize these drawbacks, MTX loaded liposomal formulations has been developed and applied on six patients founds higher clearance rate of psoriasis, and one patient had complete recovery (Ali et al., 2008). Trotta et al., developed MTX loaded liposomes using dipotassium glycyrrhizinate (KG), when applied on porcine skin results to show 3–4 times higher skin permeation as compared to conventional liposomes (Trotta et al., 2002). Another researcher revealed the presence of higher concentration of cyclic 3, 5-adenosine monophosphate (cAMP) and guanosine 3, 5-cyclic guanine monophosphate (cGMP) causes keratinocyte hyperproliferation in psoriasis (Touitou et al., 1992). Dyphylline is a theophylline derivative, which inhibits cAMP phosphodiesterase inhibitor and readily inactivates cyclic AMP or cGMP. Limited skin permeation of dyphylline is a challenge for its effective delivery (Touitou et al., 1992). To overcome such problems, it is loaded into PEG-containing liposomes, which enhances circulation time and passive targeting as compared to a non-PEGylated liposome containing dyphylline (Touitou et al., 1992). Tacrolimus (FK506) is another immunosuppressant employed for psoriasis treatment; it is obtained from the fermentation broth of *Streptomyces tsukubaensis*. It causes T lymphocyte inactivation by acting on to cytoplasmic receptor such as FKBP immunophilin (Pople & Singh, 2011). Tacrolimus have several demerits like burning sensation, pruritus, and risk of cutaneous infections. It is delivered by topical formulation, but limited skin permeation in psoriatic skin is a challenge (Pople & Singh, 2011). Tacrolimus loaded into liposomes founds greater entrapment efficiency and higher localization of said drug in dermis (Pople & Singh, 2011). Tamoxifen is a prodrug employed for treatment of psoriasis. They are metabolized into 4-hydroxytamoxifen, which ultimately binds to DNA and hindered their synthesis. Moreover, also arrest the keratinocyte cell cycle (Remitz et al., 1999). Inappropriate skin permeation is a limitation of tamoxifen in psoriasis (Pople and Singh, 2011). To overcome this, it is loaded into multilamellar liposomes, which found 59.87 $\mu g/cm^2/h$ permeation of tamoxifen into skin, whereas its solution shows only 21.65 $\mu g/cm^2/h$ (Katare, 2009). Another different researcher develops tamoxifen loaded flexible vesicle, and it founds higher skin permeation and inhibition of epidermal keratinocytes cells in psoriasis (Pople & Singh, 2011). Moreover, it is also summarized in Table 11.1.

TABLE 11.1 Therapeutic Application of Liposomes in Topical/Transdermal Drug Delivery

Drug-loaded Carrier	Remarks	Reference
Skin Cancer		
5-aminolevulinic acid (5-ALA) loaded Liposomes	It showed 7.3 times higher dermal uptake over drug loaded solution and received larger photodynamic action in non-melanoma skin cancers as basal cell (BCC) and squamous cell carcinomas (SCC)	Jain et al., 2016
Resveratrol and paclitaxel co-encapsulated in a PEGylated liposomes	It received greater efficacy of in combination against drug sensitive (MCF–7) and drug-resistant (MCF–7/Adr) tumors	Meng, 2016
Tamoxifen loaded cationic liposome-PEG-PEI complex	It received greater anticancer efficacy (86% tumor growth inhibition) against breast cancer cells line in mice bearing BT474 tumors.	Jain et al., 2014
Mitoxantrone-peptide-targeted liposomes	It causes 20 fold reduction in IC 50 over to conventional liposomes	Rahman et al., 2012b
Drug-loaded Carrier	Outcomes	Reference
Psoriasis		
Dithranol loaded liposomes	Administration of 0.5% dithranol loaded liposomes on human, which shows higher therapeutic efficacy, nonstaining to skin and cloth compared to 1.15% dithranol ointment.	Myśliwiec et al., 2016
Dyphylline loaded liposomal gel	Mice skin treated dyphylline loaded liposomal gel, which causes higher skin permeation and targeted action as compared to Dyphylline loaded solution.	Touitou et al., 1992
MTX loaded liposomes	Porcine skin treated with MTX loaded liposomes founds 3–4 times higher skin permeation as compared to conventional cream.	Ali et al., 2008
MTX loaded EV	Clinical study revealed that 0.25% MTX loaded EV gel on 25 patients, received higher permeation and reduction of PASI score from 6.2 to 2.9 in the study period of 10 weeks over 1% MTX gel.	Trotta et al., 2002
Tretinoin (TRE) loaded liposomes	It founds greater efficacy with better patient compliance.	Raza et al., 2013
Resveratrol loaded liposomes	It founds 92–96% encapsulation efficiency and better-sustained release found.	JORAHOLMEN et al., 2015
Resveratrol loaded EV	It founds higher drug permeation in porcine skin and inhibited ROS production and peroxidation in psoriasis.	Joraholmen et al., 2015

TABLE 11.1 (*Continued*)

Drug-loaded Carrier	Remarks	Reference
Loaded drugs carrier	*Outcomes*	*Reference*
Acne		
Clindamycin loaded multilamellar liposomes	Clinically applied in 30 patient for up to 4-week duration, it showed much more effective action to reduces comedones, papules, and pustules.	Johnson, 2004; Skalko et al., 1992
Tea tree oil loaded liposomes	It is effective against *P. acne* in mild to moderate acne.	Biju et al., 2005
Salicylic acid loaded liposomes	It founds prolonged release of salicylic acid and shows 10 times higher skin retention in the porcine skin over to free salicylic acid dispersion.	Bhalerao, & Raje, 2003
Finasteride loaded liposomes	It is loaded into liposomes suspension and gel, which results to significantly enhance the skin permeation in mice skin by 54 and 52.4%, respectively. Whereas solution showed 24% and conventional gel only 29%.	Kumar et al., 2007
Loaded drugs	*Outcomes*	*Ref*
Atopic Dermatitis (AD)		
IL–13 antisense oligonucleotides (IL–13 ASO loaded liposomes	It founds several significant effects in animal model in reducing the severity of AD.	Kim et al., 2009
Adenosylcobalamin (Vitamin B12 derivative) loaded liposomal hydrogel	It administered transcutaneously, results to higher drug skin permeation and showed significant reduction in dermatitis score of AD in term of dorsal skin thickness, parakeratosis, etc.	Jung et al., 2011
Hirsutenone (HST) loaded EV	It showed significantly higher permeation flux of said drug across the skin as compared to conventional cream. The same formulation showed greater improvement in the severity index of AD.	Kang et al., 2011

11.3.2   SKIN CANCER

Cancer is the largest cause of mortalities in various countries. Multidrug resistance (MDR) is a major obstacle in cancer treatment because of most death occurs due to metastatic tumor resistant to cancer chemotherapy (Apalla et al., 2017). MDR makes failure of cancer chemotherapy in breast, lung, and ovarian cancer, etc. Skin cancer is most common in humans. There are several therapies available for skin cancer treatment it includes surgery, radiation, immunotherapy, and chemotherapy, which provides suboptimal therapeutic effects with higher dose-related toxicity and high-risk damage of vital organs (Rahman, 2012a, 2012b). Thus, emergence of liposomal system loaded anticancer drugs gained great privilege in cancer therapy. They offer lots of advantages include as biocompatibility, higher drug encapsulation, prevent from premature drug release, targeted drug delivery and optimum biodistribution (Tavano, 2016). Other than this, certain challenges possess like high cost of production, phospholipid oxidation, rapid eliminations from systemic circulation and instability due to aggregation, sedimentation, and hydrolysis (Rahman, 2012c). In skin cancer therapy, topical liposomal system gained promising interest (Apalla et al., 2017). Yarosh et al. uses liposome system containing DNA repair enzymes for the treatment of *Xeroderma pigmentosum* or skin cancer (Yarosh et al., 2001). T4 endonuclease V, is a bacteriophage T4 produced DNA repair enzyme, which is substitute of the UV damaged enzyme complex, and it is essential for the correction of defect in DNA. Thus, it is loaded into liposomes and available as T4NS liposome lotion to treat skin disease (Yarosh et al., 2001). Moreover, it enhances removal of damaged DNA, which occurs in the first few hours after the treatment (Yarosh et al., 2001). In another different study revealed that daily applications of T4N5 liposomes lotion to sun damaged skin in 30 *Xeroderma pigmentosum* patients (with a history of skin cancer or actinic keratosis), results to lower the rate of production of new cationic keratosis and basal cells (BCCs) carcinomas (Yarosh et al., 2001).

Non-melanoma skin cancers as BCC and squamous cell carcinomas (SCC) is increasing day to day (Madan, 2010). A novel, non-invasive treatment is Photodynamic Therapy (PDT) has been widely employed in skin cancer therapy. In this, administration of tumor-localized photosensitizers is followed by applying of non-ionizing laser light, which leads to cellular destruction in the neoplastic area (Cai et al., 2017). 5-aminolevulinic acid (5-ALA) is a prodrug and mainly employed as PDT in the treatment of BCC and SCC by systemic or topical application to skin tumors (Inoue, 2017), whereas 5-ALA application produces porphyrin accumulations mainly protoporphyrin IX (PpIX), which called as photosensitizer (Inoue, 2017). 5-ALA is a hydrophilic drug; it poorly permeates SC of skin and sometimes remains some kind of neoplasms. Therefore, improving the dermal uptake of 5-ALA founds 7.3 times higher as compared to a solution using of liposomes to enhance the photodynamic efficiency of PDT in skin cancer treatment (Jain et al., 2016) via improving of biodistribution and local concentration of the drug. However, the skin retention was increased by 4.5 times (Jain et al., 2016). Another different researcher developed 5-ALA loaded liposomes, composed of dimyristoyl-PC or glycerol dilaurate (LIN et al., 2016). Both the formulations enhance PpIX expression in dorsal rat skin and pilosebaceous units as compared to alone 5-ALA. On the other hand, hydrophobic 5-ALA derivatives also investigated for PDT in skin cancer (LIN et al., 2016). Lipophilic nature makes better diffusing properties because of possessing enhanced lipophilicity. Furthermore, its derivatives are encapsulated into liposomal formulations into lipid layer of liposomes domain, makes it higher drug incorporation and enhances chemical and physical stability of such novel system (Raza et al., 2013). Multidrug resistance is a major challenge toward cancer treatment. Therefore, recent advancement made co-delivery of multiple anticancer drugs using nanocarriers gained a promising therapeutic strategy against drug-resistant tumor (Wang et al., 2016). Researcher develops resveratrol and paclitaxel co-encapsulated in a PEGylated liposome, which shows higher efficacy of both drugs against drug sensitive (MCF–7) and drug-resistant (MCF–7/Adr) tumors *in vivo* (Meng, 2016). Tamoxifen is effective against breast cancer in pre and post-menopausal women. Its limited permeation through the skin is a major problem. To overcome these, it is loaded into cationic liposome-PEG-PEI complex, which showed 86% tumor growth inhibition in estrogen receptor-positive breast cancer cells in mice bearing BT474 tumors (Jain et al., 2014). Furthermore, the said drug delivery may provide a useful transdermal drug delivery tool for breast cancer therapy (Jain et al., 2014). The above discussion is summarized in Table 11.1.

11.3.3   ACNE

Acne is severe inflammatory dermatosis of the pilosebaceous unit, which are characterized by various abnormalities in sebum production, follicular epithelial desquamation, bacterial proliferation, inflammation, and abnormal immunological reaction (Jalian et al., 2016). It affects nearly about 40–50 million people in the US only. Whereas adolescents people are affected by up to 80%, which creates a negative impact on the quality of personnel life

(Zouboliscc et al., 2005). Propionibacterium acnes cause abnormal sebum production as well as immunological host reaction (Gollnick, 2003). Acne can be mild, moderate or severe. In mild to moderate acne, topical therapy gained first option, whereas in systemic therapy used for severe and moderate cases (Gollnick, 2003). Pathophysiologically, the aim of acne treatment includes reduction in keratinization, reduction of interfollicular P. acnes, inflammation reduction and normalization of sebaceous gland activity (Strauss et al., 2007). There are several keratolytic agents such as retinoids, retinoids like drugs, benzoyl peroxide, salicylic acid, and azelaic acid; and antibiotics such as clindamycin, erythromycin, and erythromycin-zinc complex (Strauss et al., 2007). Retinoids and similar drugs which include tretinoin (TRE), adapalene, and tazarotene are most common topical agents used (Nyirady et al., 2001). Some of these agents produce high incidence of side effects such as skin dryness, and skin irritations (Nyirady et al., 2001). These side effects diminish patient compliance and efficacy of the therapy (Nyirady et al., 2001). To overcome such problems, use of a lower dose of antiacne-agents as well as employed novel drug delivery systems including liposomes (Ellis et al., 1998). Emergence of liposomes shows great potentials to reduce the side effects without reducing the efficacy (Ellis et al., 1998). Studied have reported the potential of liposomal systems, which are widely available in literature (Ellis et al., 1998). Clindamycin is used as most widely anti-biotic therapy for the management of acne vulgaris (Johnson, 2004). In order to improve the effectiveness of clindamycin in this, it is loaded into multilamellar liposomes (Johnson, 2004; Skalko et al., 1992). Furthermore, it is compared with free clindamycin lotion and clinically applied in acne patient in a "n=30, 4 week" treatment. Thus, the liposomes showed much more effective in reducing the total number of comedones, papules, and pustules (Skalko et al., 1992). In another different study, a significant result was founding with different liposomes as skin penetration, such as hydrogenated soya lecithin/cholesterol showed highest value of penetration among liposomal formulations (Honzak and Sentjurc, 2000). When it applied clinically on 76 patients for six weeks of duration, it founds 33.3% reduction in open comedones, whereas clindamycin solution showed only 8.3% for same (Honzak, 2000). Tea tree oil showed efficacy against P. acnes in mild to moderate acne (Enshaieh et al., 2007). Various colloidal formulations such as microemulsions, liposomes, and multiple emulsions and their follicular uptake reported on bovine udder skin. Liposomes were found 0.41±0.009% mg oil/gram of sebum plug, whereas multiple emulsions showed only 0.21±0.006% mg oil/gram of sebum plug follicular uptake (Biju et al., 2005). Salicylic acid is efficiently used in treatment of acne (Lee, 2003). Application of salicylic acid loaded liposomes results to prolonged the release of salicylic acid and 10 times higher skin retention in the porcine skin as compared to free salicylic acid dispersion (Bhalerao and Raje, 2003). Finasteride is an anti-androgen, which restricts the productions of dihydrotesterone from testosterone by competitively inhibiting the type II 5-α reductase isoenzyme (Bhalerao, Raje, 2003). It is loaded into liposomes suspension and gel, which results to significantly enhance the skin permeation in mice skin by 54 and 52.4%, respectively (Kumar et al., 2007). Whereas solution showed 24% and conventional gel only 29% skin permeation. In another way, negative charged containing liposomes showed higher permeation over neutral liposomes. Overall the finasteride loaded liposome results to five-fold enhancement of drugs into the skin (31.6 ± 3) than drug solution (2.6 ± 1) and conventional gel with (5.0 ± 1) (Kumar et al., 2007). RU58841 is a nonsteroidal antiandrogen. It has strong topical efficacy against sebaceous gland on animal models (Battmann et al., 1994). It's *in vitro* application on human skin, and liposomes loaded RU get optimum delivery of RU 58841 (Bernard et al., 1995). Whereas liposomal formulation *in vivo* cutaneous distribution studies on rat skin revealed that better permeation and retention of said drug over longer duration. Further-more, liposomal enhances the drug concentration up to 30 to 150 μm, which showed accumulation in the sebaceous ducts and upper portion of the gland (Bernard et al., 1995). Benzoyl peroxide is most commonly used drugs in treatment of acne via mild keratolytic as well as bactericidal effect. Its main site of applications is skin. Skin irritation is common, and dose-related effects seem (Fulton et al., 1974). Encapsulation into liposomes is an interesting approach to enhance the therapeutic efficacy and minimizes skin irritation over conventional benzoyl peroxide topical application (Fluhr et al., 1999). Whereas 30 acne patients receive liposomal loaded benzoyl peroxide gel for duration of 3 months, founds two times improved therapeutic response as compared to conventional gel for same (Fluhr et al., 1999). Apart from this, get much less skin irritation in the first 2 weeks of therapy and complete disap-pearance thereafter. Whereas no burning sensation was founding throughout the studied duration (Fluhr et al., 1999). Apart from this, benzoyl peroxide loaded liposomes showed strong, effective action against *Propionibacteria* and *Micrococcaceae* in acne patients over conventional formulation for same (Fluhr et al., 1999). All-trans retinoic acid or TRE is effective against mild to moderate against acne. It acts through comedolysis and maturation of the follicular epithelium (Webster, 1998; Lavker et al., 1994). Its topical applica-tion causes local problem such as irritation, mild to severe erythema, dryness, peeling and scaling (Lavker et al., 1994). To overcome such problems, retinoic acid-loaded liposomes developed and show improved effectiveness of retinoic acid after topical application (Schaller et al., 1997). Adapalene is widely used topical anti-acne drug; it has many side effects. To overcome side effects, adapalene loaded liposomes developed using of film hydration method (Kumar, Banga, 2016). The optimized liposomal formulation for said drug founds 86.66 nm in diameter, 97.01%w/w encapsulation efficiency. The *in vitro* skin permeation studies of liposomal formulation delivered drug skin permeation of 6.72 ± 0.83 µg/cm^2 (Kumar, Banga, 2016). Furthermore, confocal microscopic studies also proved higher drug penetration in the hair follicles, which showed desired targeted action for acne (Kumar, Banga, 2016). The above discussion is shown in Table 11.1.

11.3.4 ATOPIC DERMATITIS (AD)

Atopic dermatitis (AD) is a dreadful eczematous skin inflammation, which characterized by eczematous skin inflammation by virtue of excessive infiltration of IgE, T-lymphocytes, and mast cells. Its common symptoms are dry inflamed skin, pruritus, itching, sleep disturbances and emotional distress (Lipozenci, 2007). It is a common and long-lasting skin disease, which affects large world population (Cork, 2009). AD can occur at of any age. But usually appears at early childhood and periodically relapses throughout the patient life (Cork, 2009). AD is multifactorial originated pathogenesis along with genetic defects (Cork, 2009) and immune imbalance. Abnormal T-helper cells (TH2) is major responsible in the development of AD (Lugovic et al., 2005). Presently, there is no absolute therapy for the treatment of AD, because possesses of complexity interplay between patient susceptible gene, skin barrier abnormalities and immune imbalance (Lugovic et al., 2005). Apart from this various approach use for their treatment such as pharmacological and non-pharmacological. They include avoidance of causative allergens, skin hydration, topical anti-inflammatory or immunosuppressant therapies, antipruritic medication and antiseptic medication (Lugovic et al., 2005). Mild case can be potentially controlled by emollient therapy only, but moderate to severe patient requires intensive therapy. Whereas, these therapies can achieve control over AD with varying degree (Brown et al., 2006). Thus, many challenges associated with these therapies such as nonspecific targeted delivery (Brown et al., 2006), systemic toxicity and patient compliance. In the last two-decade, various scientific research leads to develop novel target-specific drug delivery systems via optimize therapeutic effect and minimizes off-target effects (Romero, 2013). In this particular, liposomes gained wider attention in AD therapy (Elsayed et al., 2007). Many studies reported the significance of liposomal drug delivery systems in term of drug permeation into the target tissue and improved therapeutic efficacy in reducing the severity of AD (Elsayed et al., 2007). As such, development of elastic liposomes for the topical delivery of IL–13 antisense oligonucleotides (IL–13 ASO). This founds several significant effects in animal model in reducing the severity of AD (Kim et al., 2009). Moreover, it produces dose and ratio dependent of said formulation results to inhibits of IL–13 in treated animal as compared to control groups (Kim et al., 2009). Furthermore, greater degree of inhibition of IL–4 and IL–5 and significant decrease in infiltrated inflammatory cell was founds (Kim et al., 2009). Another researcher developed adenosylcobalamin (Vitamin B12 derivative) loaded liposomal hydrogel, administered transcutaneously, which results to higher drug skin permeation and showed significant reduction in dermatitis score of AD in term of dorsal skin thickness, parakeratosis, etc. (Jung et al., 2011). Hirsutenone (HST) is a naturally occurring immune regulator. Further, it is loaded into elastic liposomes showed significantly higher permeation flux of said drug across the skin as compared to conventional cream (Kang et al., 2011). The same formulation showed greater improvement in the severity index of AD-like skin lesions and to normalize other immune-related inflammatory markers such as cyclo-oxygenase–2, nitric oxide synthase, IL–4, IL–13, IgE, and eosinophils (Kang et al., 2011). Moreover, liposomal application in AD is summarized in Table 11.1.

11.3.5   MISCELLANEOUS DERMATOLOGICAL APPLICATIONS

Topical liposomal formulations containing encapsulated drugs recently received commercial viability. In 1988, the first antifungal liposomal topical product containing 1% econazole was marketed named as PevarylLipogel® (Nae, 1996). Melanin is available in liposome loaded formulation named as lipoxomes®, which delivered the melanin to the hair follicles and hair shaft to hair follicles of peoples who have bold, white or grey hair (Sand et al., 2007). Glycolic acid is used as exfoliative agent and moisturizer; it causes an irritant effect when it applied alone as on to the skin (Perugini et al., 2000). Glycolic acid is loaded into liposomes, and they showed superior effects over alone use of the said drug (Perugini et al., 2000). Retinoid is widely employed in the treatment of various skin-based disorders like as acne, psoriasis, and skin cancer (Trapasso et al., 2009). But this drug has demerits like skin burning and limited dermal permeation. These demerits are minimized by applying of liposomal drug delivery systems (Trapasso et al., 2009). A number of drugs like progesterone, betamethasone dipropionate, and dexamethasone are employed for treatment of atopic eczema, psoriasis, mycosis, idiopathic hirsutism and cutaneous infection (Sunders and Shek, 2000). The better treatment of such said disease occurs through liposomal drug delivery systems (Shao et al., 2016). It produces several benefits such as higher recovery rates, higher drug skin permeation, and lesser adverse effect. Liposomal systems also utilizing for loading of wide variety of peptides to treat several skin diseases by topical application (Shao et al., 2016), as well as also recover the barrier functions of SC in wounds and burns (Shao et al., 2016). On the other hand, studied reported that liposomal formulations having higher recovery rates to injures like skin barrier as compared to mechanical mixture of lipids (Shao et al., 2016).

11.4 LIPOSOMAL FORMULATIONS FACE CHALLENGES TOWARDS TOPICAL/TRANSDERMAL DRUG DELIVERY

To date, there are many liposomal formulations available in the market, which is far behind the expectations. It might due to high cost of the products and physical instability. The liposomal products need to lyophilized requires reconstitutions prior to administration (Pierre, 2011). The liposomal formulation should have higher entrapment efficiencies, narrow size distributions, long-term stabilities, and optimum release profile. After all, the production methods require a sound knowledge about the ingredient used in manufacturing of the said formulations, i.e., lipids/phospholipid/cholesterol, organic solvents, etc. to enhance liposomes stability. Whereas the formulations should be free from toxic solvents and detergents (Pierre, 2011). The industrial application of liposomes just began at the mid of 1980, because of delay in solving of technological problems and quality control of production at large scale. In this regard, requires high-quality lipids and phospholipids, reproducibility at the large-scale production, and quality control test to validate the process and final product in term of physiochemical stability of liposomal production and sterile liposomal products (Pierre, 2011). However, such problems are needed to solve before they can have commercialized. Direct mixing and homogenization are the commonest methods for scale-up of liposomes for micro size range. But for nanosize range, this method is not suited. There are variations in the vesicle size of liposomes and special characteristics from lab scale to large-scale production (Pierre, 2011). The big disadvantages are the high cost of phospholipids, which restricts the emergence of liposomes for clinical uses. Thus, therefore resolving such costs and procedure for large-scale productions to expand such delivery systems, these all are still challenges. Other than this, many sensitive substances are exposed to mechanical stresses, extreme pH values, and volatile organic solvents during this preparation (Pierre, 2011). In spites of these, conventional, and novel preparation techniques have own merits and demerits. Further requires more research on this area to provide a more in-depth in new liposomal product development and its patents too.

11.5 CONCLUSION

Among many drug delivery systems, liposomes drug delivery most extensively investigated in treatment of topical disorders such as psoriasis, skin cancer, acne, AD, and other topical disorders too. This delivery approach provided enhanced efficacy and lowered the incidence of side effects as compared to conventional formulations. Whereas it also reduces skin irritation, staining, and attributes controlled release. Apart from this, their high cost of production and physical stability problems still is a barrier in the development of a commercial product of liposomes in the treatment of said skin disorders. Moreover, many cases of clinical studies evaluated in skin diseases, which founds higher efficacy, better tolerability as compared to conventional formulations. Now, furthermore, an investigation is needed and to allow the large-scale production of liposomes at low costs.

KEYWORDS

- **cationic liposome**
- **immuno liposomes**
- **liposome**
- **topical diseases**
- **topical drug delivery**

REFERENCES

Akhtar, N., (2014). Vesicles: A recently developed novel carrier for enhanced topical drug delivery. *Curr. Drug. Deliv.*, *11*(1), 87–97.

Ali, M. F., Salah, M., Rafea, M., & Saleh, N., (2008). Liposomal methotrexate hydrogel for treatment of localized psoriasis: Preparation, characterization and laser targeting. *Med. Sci. Monit.*, *12*, 166–74.

Apalla, Z., Nashan, D., Weller, R. B., & Castellsagué, X., (2017). Skin cancer: Epidemiology, disease burden, pathophysiology, diagnosis, and therapeutic approaches. *Dermatol Ther (Heidelb)*, *7*(1), 5–19.

Battmann, T., Bonfils, A., Branche, C., Humbert, J., Goubet, F., Teutsch, G., & Philibert, D., (1994). RU 58841, a new specific topical antiandrogen: A candidate of choice for treatment of acne, androgenetic alopecia and hirsutism. *J. Steroid. Biochem. Mol. Biol.*, *48*, 55–60.

Bernard, E., Dubois, J. L., & Wepierre, J., (1995). Percutaneous absorption of a new antiandrogen included in liposomes or in solution. *Int. J. Pharm.*, *126*, 235–243.

Bhalerao, S. S., & Raje, H. A., (2003). Preparation, optimization, characterization and stability studies of salicylic acid liposomes. *Drug. Dev. Ind. Pharm.*, *29*, 451–467.

Brown, M. B., Martin, G. P., Jones, S. A., et al., (2006). Dermal and transdermal drug delivery systems: Current and future prospects. *Drug Deliv.*, *13*(3), 175–187.

Budai, L., Kaszás, N., Gróf, P., et al., (2013). Liposomes for topical use: A physicochemical comparison of vesicles prepared from egg or soy lecithin. *Sci. Pharm.*, *20181*(4), 1151–1166.

Cai, W., Gao, H., Chu, C., et al., (2017). Engineering phototheranostic nanoscale metal-organic frameworks for multimodal imaging-guided cancer therapy. *ACS. Appl. Mater. Interfaces*, *9*(3), 2040–2051.

Cork, M. J., Danby, S. G., Vasilopoulos, Y., et al., (2009). Vasilopoulos, epidermal barrier dysfunction inatopic dermatitis. *J. Invest. Dermatol.*, *1299*(8), 1892–1908.

De Leeuw, J., De Vijlder, H. C., Bjerring, P., et al., (2009). Liposomes in dermatology today. *Eur. Acad. Dermatol. Venereol*, *23*, 505–516.

El Maghraby, G. M., Barry, B. W., & Williams, A. C., (2008). Liposomes and skin: From drug delivery to model membranes. *Eur. J. Pharm. Sci. Pharm.*, *34*, 203–222.

Ellis, C. N., Millikan, L. E., Smith, E. B., et al., (1998). Comparison of adapalene 0.1% solution and tretinoin 0.025% gel in topical treatment of acne vulgaris. *Br. J. Dermatol.*, *139*(S52), 41–47.

Elsayed, M. M., Abdallah, O. Y., Naggar, V. F., et al., (2007). Deformable liposomes and ethosomes as carriers for skin delivery of ketotifen. *Pharmazie.*, *62*(2), 133–137.

Enshaieh, S., Jooya, A., Siadat, A. H., et al., (2007). The efficacy of 5% topical tea tree oil gel in mild to moderate acne vulgaris: A randomized, double-blind placebo-controlled study. *Indian J. Dermatol. Venereol. Leprol.*, *73*, 22–25.

Fang, C. L., Aljuffali, I. A., Li, Y. C., et al., (2014). Delivery and targeting of nanoparticles into hair follicles. *Ther. Deliv.*, *5*(9), 991–1006.

Fluhr, J. W., Barsom, O., Gehring, W., et al., (1999). Antibacterial efficacy of benzoyl peroxide in phospholipid liposomes. A vehicle-controlled, comparative study in patients with papulopustular acne. *Dermatology*, *198*, 273–277.

Foldvari, M., Gesztes, A., & Mezei, M., (1990). Dermal drug delivery by liposome encapsulation: Clinical and electron microscopic studies. *J. Microencapsul.*, *7*, 479–489.

Fulton, J. E., Farzad-Bakshandeh, A., & Bradley, S., (1974). Studies on the mechanism of action to topical peroxide and vitamin A acid in acne vulgaris. *J. Cutan. Pathol.*, *1*, 191–200.

Garg, T., (2016). Current nanotechnological approaches for an effective delivery of bio-active drug molecules in the treatment of acne. *Artif. Cells. Nanomed. Biotechnol.*, *44*(1), 98–105.

Gladman, D. D., (2017). Should methotrexate remain the first-line drug for psoriasis? *Lancet*, *389*(10068), 482–483.

Gollnick, H., (2003). Current concepts of the pathogenesis of acne. *Drugs, 63*, 1579–1596.

Herwadkar, A., Sachdeva, V., Taylor, L. F., et al., (2012). Low-frequency sonophoresis mediated transdermal and intradermal delivery of ketoprofen. *Int. J. Pharm., 423*(2), 289–296.

Honzak, L., & Sentjurc, M., (2000). Development of liposome encapsulated clindamycin for treatment of acne vulgaris. *P. Fügers. Arch., 440*, 44, 45.

Inoue, K., (2017). 5-Aminolevulinic acid-mediated photodynamic therapy for bladder cancer. *Int. J. Urol., 24*(2), 97–101.

Jain, A. K., Lee, C. H., & Gill, H. S., (2016). 5-Aminolevulinic acid coated microneedles for photodynamic therapy of skin tumors. *J. Control. Release, 239*, 72–81.

Jain, A. S., Goel, P. N., Shah, S. M., et al., (2014). Tamoxifen guided liposomes for targeting encapsulated anticancer agent to estrogen receptor-positive breast cancer cells: *In vitro* and *in vivo* evaluation. *Biomed Pharmacother., 68*(4), 429–438.

Jalian, H. R., Levin, Y., & Wanner, M., (2016). Physical modalities for treating acne and rosacea. *Semin. Cutan. Med. Surg., 35*(2), 96–102.

Jóhannesson, G., Stefánsson, E., & Loftsson, T., (2016). Microspheres and nanotechnology for drug delivery. *Dev. Ophthalmol., 55*, 93–103.

Johnson, B. A., & Nunley, J. R., (2000). Topical therapy for acne vulgaris. How do you choose the best drug for each patient? *Postgrad. Med., 107*, 69–70, 73–76, 79, 80.

Jøraholmen, M. W., Škalko-Basnet, N., Acharya, G., et al., (2015). Resveratrol-loaded liposomes for topical treatment of the vaginal inflammation and infections. *Eur. J. Pharm. Sci. Pharm., 79*, 12–21.

Jung, S. H., Cho, Y. S., Jun, S. S., et al., (2011). Topical application of liposomal cobalamin hydrogel for atopic dermatitis therapy. *Pharmazie., 66*(6), 430–435.

Kajimoto, K., Yamamoto, M., Watanabe, M., et al., (2011). Noninvasive and persistent transfollicular drug delivery system using a combination of liposomes and iontophoresis. *Int. J. Pharm., 403*(1 & 2), 57–65.

Kang, M. J., Eum, J. Y., Jeong, M. S., et al., (2011). Tat peptide-admixed elastic liposomal formulation of hirsutenone for the treatment of atopic dermatitis in NC/Nga mice. *Int. J. Nanomedicine., 26*, 2459–2467.

Katare, O. P., Kumar, R., Singh, B., et al., (2009). A multi-compartmental liposomal system for topical drug delivery. *Patent. Office. J.*, 6210.

Kim, S. T., Lee, K. M., Park, H. J., et al., (2009). Topical delivery of interleukin–13 antisense oligonucleotides with cationic elastic liposome for the treatment of atopic dermatitis. *J. Gene. Med., 11*(1), 26–37.

Kirjavainen, M., Urtti, A., Jääskeläinen, I., et al., (1996). Interactions of liposomes with human skin in vitro the influence of lipid composition and structure. *Biochim. Biophys. Acta., 1304*, 179–189.

Kumar, R., Singh, B., Bakshi, G., & Katare, O. P., (2007). Development of liposomal systems of finasteride for topical applications: Design, characterization, and *in vitro* evaluation. *Pharm. Dev. Technol., 12*, 591–601.

Kumar, V., & Banga, A. K., (2016). Intradermal and follicular delivery of adapalene liposomes. *Drug. Dev. Ind. Pharm., 42*(6), 871–879.

Lavker, R. M., Leyden, J. J., & Thorne, E. G., (1994). An ultrastructural study of the effects of topical tretinoin on microcomedones. *Clin. Ther., 14*, 773–780.

Lee, H. S., & Kim, I. H., (2003). Salicylic acid peels for the treatment of acne vulgaris in Asian patients. *Dermatol. Surg., 29*, 1196–1199.

Lin, M. W., Huang, Y. B., Chen, C. L., et al., (2016). A formulation study of 5-aminolevulinic encapsulated in DPPC liposomes in melanoma treatment. *Int. J. Med. Sci., 13*(7), 483–489.

Lipozenci, J. C., & Wolf, R., (2007). Atopic dermatitis: An update and review of the literature. *Dermatol. Clin.*, *25*, 605–612.

Lugovic, L., Lipozencic, J., & Jakicrazumovic, J., (2005). Prominent involvement of activated Th1subset of T-cells and increased expression of receptor for IFN-on keratinocytes in atopic dermatitis acute skin lesions. *Int. Arch. Allergy. Immunol.*, *137*, 125–133.

Madan, V., Lear, J. T., & Szeimies, R. M., (2010). Non-melanoma skin cancer. *Lancet*, *375*, 673–685.

Magdassi, S., (1997). Delivery systems in cosmetics. *Colloids Surf A Physico. Chem. Eng Aspects.*, *123, 124*, 671–679.

Mastrobattista, E., Koning, G. A., & Storm, G., (1999). Immuno-liposomes for the targeted delivery of antitumor drugs. *Adv. Drug. Deliv. Rev.*, *40*, 103–127.

Matsuzaki, K., Imaoka, T., Asano, M., et al., (1993). Development of a model membrane system using stratum corneum lipids for estimation of drug skin permeability. *Chem. Pharm. Bull. (Tokyo)*, *41*, 575–579.

Meng, J., Guo, F., Xu, H., et al., (2016). Combination therapy using co-encapsulated resveratrol and paclitaxel in liposomes for drug resistance reversal in breast cancer cells *in vivo*. *Sci. Rep.*, *6*, 22390.

Myśliwiec, H., Myśliwiec, P., Baran, A., et al., (2016). Dithranol treatment of plaque-type psoriasis increases serum TNF-like weak inducer of apoptosis (TWEAK). *Adv. Med. Sci.*, *61*(2), 207–211.

Naev, R., (1996). Feasibility of topical liposome drugs produced on an industrial scale. *Adv. Drug. Deliv. Rev.*, *18*, 343–347.

Nyirady, J., Grossman, R. M., Nighland, M., et al., (2001). A comparative trial of two retinoids commonly used in the treatment of acne vulgaris. *J. Dermatolog. Treat.*, *12*, 149–157.

Perugini, P., Genta, I., Pavanetto, F., et al., (2000). Study on glycolic acid delivery by liposomes and microspheres. *Int. J. Pharm.*, *196*, 51–61.

Pierre, M. B., & Dos, S. M. C. I., (2011). Liposomal systems as drug delivery vehicles for dermal and transdermal applications. *Arch. Dermatol. Res.*, *303*(9), 607–621.

Pinto-Alphandary, H., Andremont, A., & Couvreur, P., (2000). Targeted delivery of antibiotics using liposomes and nanoparticles: Research and applications. *Int. J. Antimicrob. Agents.*, *13*, 155–168.

Pople, P. V., & Singh, K. K., (2011). Development and evaluation of colloidal modified nanolipid carrier: Application to topical delivery of tacrolimus. *Eur. J. Pharm. Biopharm*, *79*, 82–94.

Rahman, M., Ahmad, M. Z., Kazmi, I., et al., (2012c). Advancement in multifunctional nanoparticles for the effective treatment of cancer. *Expert Opin. Drug Deliv.*, *9*(4), 367–381.

Rahman, M., Ahmed, M. Z., Kazmi, I., et al., (2012a). Insight into the biomarkers as the novel anti-psoriatic drug discovery tool: A contemporary viewpoint. *Curr. Drug Discov. Technol.*, *9*(1), 48–62.

Rahman, M., Akhter, S., Ahmad, J., et al., (2015). Nanomedicine-based drug targeting for psoriasis: Potentials and emerging trends in nanoscale pharmacotherapy. *Expert. Opin. Drug. Deliv.*, *12*(4), 635–52.

Rahman, M., Alam, K., Ahmad, M. Z., et al., (2012b). Classical to current approach for treatment of psoriasis: A review. *Endocr. Metab. Immune. Disord. Drug. Targets, 12*(3), 287–302.

Rahman, M., Beg, S., Ahmad, M. Z., et al., (2013). Omega–3 fatty acids as pharmacotherapeutics in psoriasis: Current status and scope of nanomedicine in its effective delivery. *Curr. Drug. Targets*, *14*, 708–722.

Rahman, M., Beg, S., Anwar, F., et al., (2016c). *Nanotechnology-Based Nano-Bullets in Antipsoriatic Drug Delivery: State of the Art: Nanoscience in Dermatology* (pp. 157–166). Elsevier.

Rahman, M., Beg, S., Sharma, G., et al., (2016b). Lipid-based vesicular nanocargoes as nanotherapeutic targets for the effective management of rheumatoid arthritis. *Recent. Pat. Antiinfect. Drug. Discov.*, *11*(1), 3–15.

Rahman, M., Kumar, V., Beg, S., et al., (2016a). Emergence of liposome as targeted magic bullet for inflammatory disorders: Current state of the art. *Artif. Cells Nanomed. Biotechnol.*, *44*(7), 1597–1608.

Raza, K., Singh, B., Lohan, S., et al., (2013). Nano-lipoidal carriers of tretinoin with enhanced percutaneous absorption, photostability, biocompatibility, and anti-psoriatic activity. *Int. J. Pharm.*, *456*(1), 65–72.

Remitz, A., Reitamo, S., Erkko, P., et al., (1999). Tacrolimus ointment improves psoriasis in a microplaque assay. *Br. J. Dermatol.*, *141*, 103–107.

Romero, E., & Morilla, M. J., (2013). Highly deformable and highly fluid vesicles as potential drug delivery systems: Theoretical and practical considerations. *Int. J. Nanomed.*, *8*, 3171–3186.

Sand, M., Bechara, F. G., Sand, D., et al., (2007). A randomized, controlled, double-blind study evaluating melanin-encapsulated liposomes as a chromophore for laser hair removal of blond, white, and gray hair. *Ann. Plast. Surg.*, *58*(5), 551–554.

Saraswat, A., Agarwal, R., Katare, O. P., et al., (2007). A randomized, double-blind, vehicle-controlled study of a novel liposomal dithranol formulation in psoriasis. *J. Dermatolog. Treat.*, *18*(1), 40–45.

Schaller, M., Steinle, R., & Korting, H. C., (1997). Light and electron microscopic findings in human epidermis reconstructed *in vitro* upon topical application of liposomal tretinoin. *Acta. Derm. Venereol.*, *77*, 122–126.

Schramlová, J., Blazek, K., Bartácková, M., et al., (1997). Electron microscopic demonstration of the penetration of liposomes through skin. *Folia Biol (Praha)*, *43*, 165–169.

Sessa, G., & Weissmann, G., (1970). Incorporation of lysozyme into liposomes. *J. Biol. Chem.*, *245*, 3295–3301.

Shao, M., Hussain, Z., Thu, H. E., et al., (2016). Drug nanocarrier, the future of atopic diseases: Advanced drug delivery systems and smart management of disease. *Colloids. Surf. B. Biointerfaces.*, *147*, 475–491.

Skalko, N., Cajkovac, M., & Jalšenjak, I., (1992). Liposomes with clindamycin hydrochloride in the therapy of acne vulgaris. *Int. J. Pharm.*, *85*, 97–101.

Strauss, J. S., Krowchuk, D. P., Leyden, J. J., et al., (2007). Guidelines of care for acne vulgaris management. *J. Am. Acad. Dermatol.*, *56*, 651–663.

Sunders, Z. E., & Shek, P. N., (2000). Prophylaxis against lipo-polysaccharide-induced lung injuries by liposome-entrapped dexamethasone in rats. *Biochemical. Pharmacology*, *59*, 1155–1161.

Tavakoli, N., Minaiyan, M., Heshmatipour, M., & Musavinasab, R., (2015). Transdermal iontophoretic delivery of celecoxib from gel formulation. *Res. Pharm. Sci.*, *10*(5), 419–428.

Tavano, L., & Muzzalupo, R., (2016). Multi-functional vesicles for cancer therapy: The ultimate magic bullet. *Colloids. Surf. B. Biointerfaces.*, *147*, 161–171.

Touitou, E., Shaco-Ezra, N., Dayan, N., et al., (1992). Dyphylline liposomes for delivery to the skin. *J. Pharm. Sci.*, *81*, 131–134.

Trapasso, E., Cosco, D., Celia, C., et al., (2009). Retinoids: New use by innovative drug-delivery systems. *Expert Opin. Drug Deliv.*, *6*, 465–483.

Trotta, M., Peira, E., Debernardi, F., et al., (2002). Elastic liposomes for skin delivery of dipotassium glycyrrhizinate. *Int. J. Pharm.*, *241*, 319–327.

Van den Berg, W. B., & McInnes, I. B., (2013). Th17 cells and IL–17 a -focus on immune-pathogenesis and immune-therapeutics. *Semin. Arthritis. Rheum.*, *43*, 158–170.

Wadhwa, S., Singh, B., & Sharma, G., (2016). Liposomal fusidic acid as a potential delivery system: A new paradigm in the treatment of chronic plaque psoriasis. *Drug. Deliv.*, *23*(4), 1204–1213.

Wang, K., Kievit, F. M., & Zhang, M., (2016). Nanoparticles for cancer gene therapy: Recent advances, challenges, and strategies. *Pharmacol. Res.*, *114*, 56–66.

Webster, G. F., (1998). Topical tretinoin in acne therapy. *J. Am. Acad. Dermatol.*, *39*, 38–44.

Xie, F., Chai, J. K., Hu, Q., et al., (2016). Transdermal permeation of drugs with differing lipophilicity: Effect of penetration enhancer camphor. *Int. J. Pharm.*, *507*(1 & 2), 90–101.

Yarosh, D. L., Klein, J., O'Connor, A., et al., (2001). Effect of topically applied T4 endonuclease V in liposomes on skin cancer in xeroderma pigmentosum: A randomized study. *Lancet*, *357*, 926–929.

Zhang, L., Li, L., & An, Z., (1997). *In vivo* transdermal delivery of large molecules by pressure mediated electro incorporation and electroporation: A novel method for drug and gene delivery. *Bioelectrochem. Bioenerg.*, *42*, 283–292.

Zhang, Y., Li, H., Sun, J., et al., (2010). DC-Chol/DOPE cationic liposomes: A comparative study of the influence factors on plasmid pDNA and siRNA gene delivery. *Int. J. Pharmaceutics.*, *390*, 198–207.

Zouboulis, C. C., Eady, A., Philpott, M., et al., (2005). What is the pathogenesis of acne? *Exp. Dermatol.*, *14*, 143–152.

Zylberberg, C., & Matosevic, S., (2016). Pharmaceutical liposomal drug delivery: A review of new delivery systems and a look at the regulatory landscape. *Drug. Deliv.*, *23*(9), 3319–3329.

CHAPTER 12

Conventional to Novel Targeted Approaches for Brain Tumors: The Role of Nanomedicines for Effective Treatment

MAHFOOZUR RAHMAN[1*], KAINAT ALAM[2], MD. ADIL SHAHARYAR[3], SARWAR BEG[4], ABDUL HAFEEZ[5], and VIKAS KUMAR[6]

[1]*Department of Pharmaceutical Sciences, Shalom Institute of Health and Allied Sciences (SIHAS), Sam Higginbottom University of Agriculture, Technology & Sciences (SHUATS), Allahabad, India, E-mail: mahfoozkaifi@gmail.com*

[2]*Christian College of Nursing, Shalom Institute of Health and Allied Sciences (SIHAS), Sam Higginbottom University of Agriculture, Technology & Sciences (SHUATS), Allahabad, India*

[3]*Department of Pharmaceutical Technology, Jadavpur University, Kolkata, India*

[4]*Product Nanomedicine Research Lab, Department of Pharmaceutics, School of Pharmaceutical Education and Research, Jamia Hamdard (Hamdard University), New Delhi, India*

[5]*Glocal School of Pharmacy, Glocal University, Saharanpur, U.P., India*

[6]*Natural Product Drug Discovery Laboratory, Department of Pharmaceutical Sciences, Shalom Institute of Health and Allied Sciences (SIHAS), Sam Higginbottom University of Agriculture, Technology & Sciences (SHUATS), Allahabad, India*

Corresponding author. E-mail: mahfoozkaifi@gmail.com

ABSTRACT

Brain cancer treatment is restricted by various barriers, including the blood-brain barrier, difficulties transport within the brain interstitium and faces problem in delivering of drugs especially to tumor cells and highly invasive quality of gliomas and drug resistance. In this context, the main rogue is the blood-brain barrier, which restricts the therapeutic agents to cross the blood-brain barrier. Another way, shorter half-life of anticancer drugs, the low molecular weight of the drug, and poor steady-state concentration in the malignant glioma cells is a big problem. Therefore, continuous demanding of optimal anticancer drugs and drug delivery systems to the CNS for effective treatment of brain tumor. Literatures revealed that tumor itself develops its own vascular network and have greater endothelial gaps compared to normal vessels. This prominent feature is exploited in the development of anticancer drug delivery by targeting the BBB. Hence, the various anticancer drugs are using to treat primary CNS tumors. Furthermore, recent development made in targeting and crossing the blood-brain barrier (BBB) in a controlled and non-invasive manner by efficiently targeted agents such as antibodies or protein carriers and nanomedicines. Therefore, recent advancement in nanotechnology has provided promising results to this challenge. Thus, several nano-carriers such as solid lipid nanoparticles (NPs), polymeric NPs, liposomes, micelles, nanoemulsions are reported for the anticancer drug delivery to the brain tumor. Many of these nanomedicines are effectively transported across various *in vitro* and *in vivo* BBB models via endocytosis and or transcytosis mechanism. In future need to development of nanomedicine which focuses on increasing their drug-trafficking performance and specificity for a brain tumor, improving their BBB permeability and reducing their neurotoxicity.

12.1   INTRODUCTION

In oncology domain, brain cancer treatment still remains one of the biggest challenges. There are varieties of neoplasm in brain tumors namely primary and metastatic glial cells or their progenitors or gliomas. Systemic malignancies give rise to the metastatic, which ultimately develops within the parenchyma of the brain (Newton, 1994; Davis, 2001). WHO classified mainly three types of brain tumors as astrocytomas, oligodendrogliomas, and oligoastrocytomas (Kleihues, 1995). Astrocytomas further classified into grade I to IV. Among these the most threatening is grade 4, which known as glioblastoma multiform (GBM). Moreover, still the most standard way of classifying gliomas is the histopathological analysis. With the advancement made in the technology, is becoming increasingly clear that the new genetic subtype of these gliomas maybe present and their molecular changes are an important tool in prognosis of diseases (Kleihues, 1995). According to molecular analysis gain and loss of Several literatures revealed that frequently loss and addition of chromosomes named as 7p & 10q are responsible for GBM, whereas oligodendrogliomas are frequently dominated by loss of 1p & 19q chromosomes. That is, 5–8 per 100,000 inhabitants (Shai, 2008). Malignant Astrocytomas makeup for about 50 to 60% of primary brain tumors. Environmental or genetic factors seem to be ambiguous in the increasing incidence of brain tumors (Ohgaki, 2005). Mostly, surgical resection followed by radiotherapy and chemotherapy are the standard treatment for patients suffering from malignant brain tumors, but regardless of the continuous research and new approaches, the prognosis is still lacking (DeAngelis, 2001) in case of GBM. The median survival of patients is only 20 weeks by surgical resection, surgery, and radiation for 36 weeks and if a standard cytotoxic chemotherapy is included, there is a rise in median survival of patients to the tune of 40 to 50 weeks (Brandes, 1996). There has been an enormous advancement in the anticancer drug discovery and development in the last decades, which showed promising *in vitro* results, but failed miserably in the clinical trials (Omuro, 2007). In this content, the main rogue is the blood-brain barrier which prevents the therapeutic agents to cross it and reaches to the desired target in the CNS (Groothuis, 2000; Huynh, 2006). Maintenance of steady-state concentration in the malignant glioma cells is a big problem, which possesses low molecular weight and shorter half-life of chemotherapeutic agents in the blood (Sarin, 2008). There is continuously demanding of optimal agents and valuable system for the delivery of anticancer drug to the CNS for effective treatment of brain tumor (Friedman, 2000). Now from the current research, it is cleared that a tumor develops its own vascular network. Thus, new vascular within the tumors have greater endothelial gaps compared to normal vessels (Provenzale, 2005) and this feature is exploited in the development of anticancer drug delivery with targeting the BBB (Provenzale, 2005). Therefore, the present chapter is to deal with various approaches that have been designed to treat primary CNS tumors after resection usually reappears within 2 cm of the tumor margin, and so have high infiltration capacity. Already a number of various review articles have been published on this specific topic and summarizes the various studies taken in this area (Beduneau, 2007; Juillerat-Jeanneret, 2008; Rahman, 2012a; 2012b). The main aim of this paper is to highlight the recent findings and developments made in targeting and crossing the BBB in a controlled and non-invasive manner through designing efficiently targeted vectors (antibodies and protein carriers) or nanosystems (colloidal carriers).

12.2   CONVENTIONAL AND TARGETED CHEMOTHERAPEUTICS

Currently, many chemotherapeutic agents are already in clinical use or in trials which used for the treatment of primary brain tumors (Muldoon, 2008). Laquintana et al. used alkylating agents such as carmustine, lomustine, nimustine, which belongs from the category of nitrosoureas for treating patients with malignant astrocytoma. They act by methylating the O6 position of guanine in DNA (Musacchio, 2009). But they are also accompanied by systemic toxicity like myelosuppression, gastrointestinal effect, and nephrotoxicity. The therapy provides average benefit in survival of patients is due to shorter half-life of therapeutic agent(s) in blood (Sarin, 2008) and poor BBB permeability, thus causing its limited distribution in CNS. A comparative study was conducted between radiotherapy alone and radiation followed by procarbazine, lomustine, and vincristine in the treatment of high-grade astrocytoma, but overall survival benefits for patients was insignificant (Medical Research Council2001). The newest alkylating agent is temozolomide with short blood half-life of about 1.8 hours, so it requires high systemic dose. It has high absolute oral bioavailability and good BBB penetration (Baker, 1999). Further, it has been approved by FDA for the treatment of GBM and AAs. Whereas on prolong systemic administration, it shows side effects such as thrombocytopenia, nausea, and vomiting. According to FDA, this drug has a median survival time of 5.8 months (Bower, 1997). The current standard treatment for GBM is temozolomide, which is to be administered during 6 weeks of radiation therapy at a dose of 75 mg/2 minutes and followed by a 5-day regimen over the following month (Stupp, 2005). A comparative study was conducted between temozolomide, and no differences were found statistically in terms of survival rates among patient of both groups. Temozolomide gets leverage over procarbazine in terms of progression-free survival (12.4 vs. 8.3 weeks) (Yung, 1999). In the quest to determine the most efficient combination for treatment of patients suffering from malignant gliomas, a randomized phase III clinical trial was conducted between temozolomide alone and procarbazine, lomustine, vincristine taken together (Parasramka, 2017). From the result, it seems that temozolomide was better in terms of improvement of quality of life of patients. Survival rates were also evaluated (Osoba, 2000). For newly diagnosed gliomas temozolomide in combination with radiotherapy is the standard treatment which improves not only the median survival rates but also quality of life (Stupp, 2007). Antineoplastic agents like anthracyclines, platinum (II) complexes, etoposide, paclitaxel (PCL) to produce effective concentration in the brain in treating CNS tumors. As a result, high concentration is required which leads to systemic toxicity, side effects and finally compromising the quality of life of the patients (Stupp, 2007). With the adventure of targeted therapy, new dimensions in drug delivery system have opened which focus on the treatment of Glioma cells, with reduced systemic toxicity. It involves binding of the drug to the selective target on the glioma cells, which overexpressed receptors on the tumor cells or expression of receptors are present on normal brain cells (Stupp, 2007). At the molecular level, there are several cancers causing pathways have been detected in gliomas, which involve EGFR (Epidermal growth factor receptor), VEGFR (vascular endothelial growth factor receptor), PI3K (phosphoinositide 3' kinase), PDGFR (platelet-derived growth factor receptor), Ras/Raf/MAPK and mToR (mammalian target of Rhabdomycin). Whereas PI3K is upregulated by EGFR stimulation which also includes PDGFR, EGFR, VEGFR, fibroblast growth factor receptor (FGFR), Insulin Growth factor I Receptor and Ras. Through P70S6k and 4E-BP1 proteins, mToR regulates homeostasis and growth of cells (Newton, 2004). Small molecules to monoclonal antibodies have become an integral part of the normal therapy which acts by inhibiting these factors. Many clinical trials are being conducted using inhibitors alone or combination of inhibitors or combining inhibitors with anti-cancer drugs for the treatment of gliomas (Omuro, 2007). New chemotherapeutic agents which showed specific targeted action such as Gefitinib, Lapatinib, Erlotinib (EGFR inhibitors), Vatalanib (protein kinase C-beta and other angiogenesis pathway inhibitor), Imatinib (PDGR inhibitor), Temsirolimus (mToR inhibitor), Bortezomib (VEGFR inhibitor) and the recombinant humanized monoclonal antibody (Omuro, 2007).

12.3 DRUG TRANSPORT MECHANISM AT THE BBB

The homeostasis of the brain is maintained by the selective nature of the BBB, which allows certain beneficial endogenous substances and nutrients to pass through it and at the same time restricting the passage of xenobiotics or toxic metabolites, which are harmful for the brain. Monolayer brain capillary endothelial cells fused by tight junctions form the main component of the BBB (Lai, 2005). Apart from this, there are other components of BBB like astrocytic foot process, pericytes, and perivascular macrophages with the basal lamina, which regulates and strengthens it (Abbott, 2006; Lai, 2005). The barrier aspect of the BBB is attributed to the absence of fenestrations and presence of tight junctions in the endothelial cells. The endothelial cells of the BBB, when compared to vascular endothelial cells, shows low pinocytic activity and vesicular traffic resulting in limiting non-specific transendothelial transport. The barrier shows exception to the passage of small lipid-soluble molecules (Reese, 1967). The mechanisms which are involved in the transport of drug or endogenous substance(s) across BBB is of two types: passive diffusion and endogenous carrier-mediated transport. Passive diffusion deals with the passage of drug or endogenous substances depending upon the concentration gradient across BBB and the physicochemical properties of the drug. Drugs that qualify to pass through the BBB are lipophilic in nature with a molecular weight of 400–500 Da. The lipophilicity of the drug is associated with the octanol/water partition coefficient. There have been many quantitative approaches to establish relationship between BBB penetration, lipophilicity, molecular weight and various chemical features (Levin, 1980; Di, 2009). The endogenous carrier-mediated transport can further be classified into: (a) receptor-mediated endocytosis which involves three steps, namely, (1) receptor-mediated endocytosis at the luminal blood side, (2) intracellular movement, (3) exocytosis at the luminal side of brain endothelial. (b) carrier-mediated facilitated transport or active transport which dependent on ATP (Pardridge, 1999). The process involves many receptors like transferring receptors, insulin receptors, lipoprotein-related protein–1 and lipoprotein-related protein 2 receptors. There is also involvement of diphtheria toxin receptor (Gaillard, 2005), large anti-cancer biomolecules that can be transported across BBB via receptor-mediated endocytosis (Pardridge, 2003). Forty-three families of receptors constitute the solute carrier (SLC), which known as carrier-mediated facilitated transport system (Hediger, 2004). Specific substrates such as sugars, amino acids, oligopeptides, organic anions, and cations are transported by each SLC member (Huang, 2006). The two most important SLC superfamily members are organic cation transporter (SLC 21) and organic anion-cation transport system (SLC 22). ATP binding cassette (ABC) are ATP dependent while organic anion-cation transport system works by exchanging cations and anions thereby producing an electrochemical gradient. The organic cation transporters (OCT) are ATP independent transport system. These transporters are present apically or basolateral on the BBB. The location of the anti-cancer drug or endogenous substances decides whether it will be pumped in or out (Ito, 2005). Protein families like P-glycoprotein (Pgp/MDR1), Breast Cancer Resistance Protein (BCRP /ABCG2), multi-resistance drug associated protein (MRP) works through ATP dependent active transport system. Among the ABC transporter super-families, pgp is an efflux pump present at the apical side of BBB with molecular weight 170 kDA. Anti-cancer drugs like anthracyclines, topotecan, and PCL are pumped back into the systemic circulation by p-gp (Demeule, 2002). It is thought that the main reason behind the development of resistance in tumors against anticancer drugs is due to active transport. The presence of Pgp in resistant glioblastoma is one of the good examples, though its vasculature is leaky (Becker, 1991). A study was conducted in which syngeneic intracerebral B–16 brain tumor model was selected. It was found that the concentration of PCL in normal brain was greater than the Pgp knocked out mice compared to the wild type of mice (Gallo, 2003). Strategies have been developed in clinical and preclinical trials to inhibit the Pgp to enhance the penetration of anti-cancer drugs (Kemper, 2004). With the adventure of new Pgp inhibitors like valspodar, elacridar, zosuquidar (a ray of hope can be seen in brain cancer treatment (Kemper, 2004). The substrate for BCRP which is present on the apical side of BBB are flavopiridol, mitoxantrone, topotecan, methotrexate (MTX), sulfated conjugates of therapeutic agents, hormones (estrogen sulfate) (Ito, 2005). Presence of both p-gp and BCRP on the same side of BBB provides evidence that BCRP prevents penetration of substrates across BBB. There are nine types of MRP which make up a family called MRP family. All of them being efflux pumps possessing capacity to pump out different types of lipophilic anions (Kruh, 2003). Since MRP–1, MRP2, MRP–4 and MRP–5 are present on the apical side of the BBB, so they have an inhibitory role in the transport of anti-cancer drugs (Loscher, 2005). ChemicallyMRP–1 is a glutathione and glucuronate conjugate pump which inhibits the transport of anticancer drugs like anthracyclines, MTX, vinca alkaloids, epipodophyllotoxins, and camptothecins (Kruh, 2003). The nucleotide analog drugs are inhibited from crossing BBB by MRP–1 prototypical substrate such as E217BG, reduced folates, MTX, MRP–4 and MRP–5. These acts through transportation of cAMP and cGMP (Kruh, 2003). The BBB has certain characteristic features which remain present in the tumor vasculature while others get changed. This affects the crossing of anti-cancer drugs. There are varieties of primary brain tumor capillary endothelial cells consisting of tight junction proteins. These include zona occludens–1 (Sawada, 2000) Claudin–1, Claudin–5 and occluding (Liebner, 2000), which are either downregulated or lost. The disruption or leaky nature of the BBB of the brain tumor vasculature turns out to be a boon in disguise, thus enhancing anticancer drug penetration. Still, the outer rim of the BBB may have a normal structure that throws a challenge to the drug delivery system in achieving proper drug concentration in the tumor (Levin, 1980; de Vries, 2006). Drugs that are highly permeable depend on the flow of blood for achieving the required concentration. These drugs are blood flow rate limited. Another factor which is to be taken into account is the exudation of the interstitial fluid which leads to the decrease in penetration of the drug. The angiogenic and proangiogenic balance, as well as tumor growth, are the factors which decide whether the tumor vasculature will be normal or abnormal (Jain, 2007). It is observed that when angiogenic inhibitors are administered, it affects the structure and function of the tumor vasculature. Under certain conditions, it normalizes leading to improved penetration of BBB (Zhou, 2009; Zhou, 2008). Further investigation and study are required to appreciate the anti-angiogenic effect on the vasculature of the brain tumor as it involves both dose and time dependency. It has been seen that anti-angiogenic therapy affects ABC transporters like up-regulation of MRP–1 and MRP–3 and downregulation of Pgp (Becker, 1991). Up-regulation of Pgp expression was found in both glioblastoma and vascular endothelial cells after Doxorubicin (DOX) treatment (Rittierodt, 2003), which indicates that Pgp is involved in (DOX) resistance. The expression of ABC transporter keeps on changing which throws a doubtful light on the exploitation of efflux pumps present on the BBB vasculature. A 1.7-fold increase in the concentration of PCL was found in gene disrupted Pgp deficient mice as compared to wild type (Gallo, 2003). A conclusion can be drawn that Pgp has a greater role which is shown by the concentration of PCL. More studies are required to establish the role of efflux pumps present at the BBB of a brain tumor.

12.4 DRUG DELIVERY THROUGH INVASIVE METHODS

12.4.1 BBB DISRUPTION THROUGH INTRA-ARTERIAL (IA) DELIVERY OF DRUGS

Intra-arterial (IA) drug administration supersedes intravenous drug administration as the former bypasses the first pass metabolism, yet this factor alone is not sufficient enough to give good clinical results. Temporary disruption of the BBB through osmosis followed by IA drug administration increases the concentration of chemotherapeutic agent in the CNS without affecting neurocognitive functions. IA decreases systemic toxicity (Neuwelt, 2004). This disruption of BBB is carried out by infusing pre-warmed mannitol (25%), it administered through the internal carotid artery at a flow rate of 3–12 mL/s for a short duration of 10 mins and monitoring of the tumor through computed tomography or MRI is carried out. Disruption of BBB can also be carried out using ultrasound which shows no tissue damage (Vykhodtseva, 2008). Amalgamation of IA therapy of anti-cancer drugs and BBB through osmosis gives good result especially in some tumors like germ cell tumors (Jahnke, 2008). Complexity of the procedure limits its usage. Radiotherapy along with temozolomide is the current and conventional treatment in primary and metastatic brain cancers. It is the standard treatment in newly diagnosed gliomas with an advantage of increased permeability of BBB.

12.4.2 CONVECTION ENHANCED DELIVERY (CED)

The convection-enhanced delivery (CED) was developed by Bob et al. to improve the drug distribution. The method involves infusion of anti-cancer drug in or around the tumor under hydrostatic pressure through a catheter (Bobo, 1994). In an animal model, the drug was passed through the catheter at high flow microinfusion resulting in an improved distribution of high molecular weight anti-cancer drugs (Bobo, 1994). The CED techniques are encouraging and show future prospect in the delivery of agents like conjugates, monoclonal antibodies, antisense oligonucleotides or viral vectors (Beduneau, 2007). A slight increase in the median survival time was found when an interleukin–13 was conjugated to cintredekinbesudotox (PE38QQR). The conjugate was delivered by CED (Yung, 1999). Real-time monitoring is a big challenge which can be encountered using current techniques like fluorodeoxyglucose-positron emission tomography (FDG-PET), single photo emission computed tomography (SPECT) and MRI. Other challenges include high infusion rates, surgery requirement, low distribution of the drug.

12.4.3 IMPLANTED THERAPIES

A highly invasive technique, used to deliver anticancer drug, is the implantation of the polymeric matrix or reservoir in the brain. An implant is governed by rate control, shape, size, biodegradability. Since, 1996, FDA has approved Gliadel® for the treatment of newly diagnosed and recurrent malignant gliomas. It has shown an increase of 2-month survival time. Gliadel® is a biodegradable poly anhydride polymer in the form of wafer containing BCNU (carmustine) (Westphal, 2003) complications and low diffusion of drug in the brain parenchyma are its limitations (Westphal, 2003). Other anticancer drugs like PCL, cisplatin alone or in combination with Gliadel® have been studied clinically (Vukelja, 2007; Sheleg, 2002). Patients suffering from high-grade, recurrent gliomas when treated with Gliadel®

in combination with temozolomide was found to be safe and tolerable (Gururangan, 2001). Osmotic minipumps are another innovative technique which allows drugs to be released locally and continuously, but its success is limited to experimental tumors.

12.5 NONINVASIVE DRUG DELIVERY STRATEGIES WITHOUT BBB DISRUPTION

Oral and systemic are the two conventional routes currently used in the delivery of anti-cancer drugs to the CNS, but its success is limited by low penetration of BBB. Drugs such as etoposide and carboplatin when administered systemically at high concentration, these not only results in enhancement of penetration across BBB but also helps in accompanied by systemic toxicities, thus putting the effort on the back foot. So here we discuss different non-invasive techniques to disrupt the BBB (Gururangan, 2001).

12.5.1 DIRECT CONJUGATION OF ANTITUMOR DRUGS

Lipophilicity of drug is brought about or improved by attaching a lipophilic moiety to the parent drug. The lipophilic group detaches itself from the parent after entering brain parenchyma, thus increasing CNS concentration of the drug. Apart from lipophilicity, efflux pumps also play a part. Though PCL has high lipophilicity yet it has low CNS concentration due to Pgp efflux pumps (Gururangan, 2001). Another strategy is to conjugate vector like antibodies, peptides, protein carriers, and viruses to the drug to increase permeation through BBB. The low-density lipoprotein receptor-related protein has been targeted with a new vector, which is a 19-amino acid called Angiopep–2. New investigations in the field of CNS drug delivery have led to the development of a conjugation between PCL and the peptide Angiopep–2 called ANG1005 (Regina, 2008). A comparative study was carried out in orthotopic brain tumor model between PCL and ANG1005. ANG1005 supersedes PCL in permeability as well as better *in vivo* anti-tumoral activity partially due to bypassing of the Pgp efflux pump. ANG1005 has an added advantage of being transported across BBB via receptor-mediated endocytosis. The efficacy ANG1005 in the treatment of recurrent primary or metastatic cancer is being investigated in a clinical trial known as ANG1005 phase I clinical trials. The iron binding protein P97 (Melanotransferrin) closely associated with Transferrin (Tf) has been conjugated with PCL and Adriamycin (ADR) (Jones, 2007; Karkan, 2008). A

10-fold increase in brain penetration was found with P97 drug conjugate than the free drug. When P97-ADR conjugate was administered to animals having intracranial gliomas, it increased their survival. This indicates that P97 may have some role in the treatment of brain tumors. Conjugation has the potential of increasing the efficacy of the parent drug. A covalent chemical conjugate of Temozolomide and a transmembrane transporter called TMZ-Bio-shuttle has been designed and explored. Its studies reveal a better *in vitro* activity in some glioma cell lines at low dose (Waldeck, 2008).

12.5.2   CO-ADMINISTRATION OF CHEMOTHERAPEUTIC AGENTS WITH INHIBITORS OF EFFLUX TRANSPORTERS

Pgp, BCRP, and MRP are the active efflux pumps which give resistance to anticancer drugs in BBB penetration. Administration of chemotherapeutic agents along with efflux pump inhibitors is a good way to increase CNS drug concentration. PCL, docetaxel, and imatinib were used in pre-clinical models. They should improve CNS concentration due to increased permeability (Blakeley, 2008; Bihorel, 2007). Verapamil and cyclosporine A are first generation P-gp inhibitors, which possess certain problems like as (i) Low binding affinities and high systemic toxicity due to high dose. (ii) The enzymes like cytochrome P450 which metabolizes drugs (CYP3A) are inhibited. Second generation Pgp inhibitors are better inhibitors. Cyclosporine A analog, Valspodar is a 2nd generation Pgp inhibitor. But it faces problems due to interaction with CYP3A enzymes. Newer Generation Pgp inhibitors supersede the 1st and the 2nd generation Pgp inhibitors in not having any interaction with CYP3A enzymes and low side effects. This group of newer generation Pgp inhibitors includes Elacridar 11, Zosuquidar 12 and Tariquidar 13 (Breedveld, 2006).

Since the distribution of BCRP is the same as that of the Pgp, it can be exploited. Elacridar 11 and the proton pump inhibitor pantoprazole are the renowned BCRP inhibitors. A comparative study should be carried out in selecting the inhibitors which include determination of substrate profile of anti-cancer drugs and its interaction with other drugs (Breedveld, 2006).

12.5.3   AVAILABLE STRATEGIES TO ACHIEVE TARGET THERAPY

Targeted therapy exploits the overexpression of receptor and receptors that are not present in normal Brain tumor cells. Specificity and selectivity are the two main aspects of targeted therapy. Among the receptors the Benzodi-azepine receptors (PBRs). Apoptosis and its presence in between the inner and outer mitochondrial membrane makes it a good target (Pilkington, 2008). Apoptosis in rats C6 glioma cells line have been studied, and it was found that PK11195, A peripheral benzodiazepine receptor (PBR) ligand, and isochinoline carboxamide derivative facilitates it (Pilkington, 2008). A comparative study between gemcitabine alone andPK11195-Gemcitabine conjugate was carried out. The conjugate showed better tumor penetration and 2-fold increment in brain tumor selectivity as compared to Gemcitabine alone (Guo, 2001). PBR expression has positive correlation with the grade of malignancy of tumors, proliferative, and apoptotic indices, while relation with survival in a group of 130 patients (Vlodavsky, 2007). Because of these facts about PBR expression, it is now being exploited in the diagnosis of astrocytomas (Sekimata, 2008). It has been found that tumor growth and metastasis depend upon signaling pathways like EGFR, PI3k, mTOR, and Ras/Raf/MAPK (Shai, 2008). These pathways have been targeted as in the case of EGFR, when it is overly expressed in 60% GBM, which is the chief feature of primary Glioblastoma. The 1st generation of EGFR inhibitors like gefitinib (2) and erlotinib (4) did not produce good results with malignant gliomas. To some extent, it is due to shortened EGFR (EGFR VIII) present in 50–70% of the overexpressing EGFR gliomas (Omuro, 2008). *In-vitro* studies have shown that EGFR VIII resisted the EGFR inhibitors gefitinib (2) and erlotinib (da Fonseca, 2008). Regression of gliomas in phase I/II of the clinical trial was seen, when perillyl alcohol (POH) was administered intranasally as an inhibitor for nucleotide binding protein Ras (da Fonseca, 2008). It was often observed that existing of an alteration in the PI3K pathway in cancer patients. This alteration is also found in glioblastoma. In a clinical trial, while focusing on primary gliomas, a negative correlation was observed between apoptosis and PI3K pathway activation. It shows by inhibiting PI3K pathway (Opel, 2008). When PI3k was inhibited by LY294002, a dead receptor, as well as chemotherapy-induced apoptosis, was revived synergistically in glioblastomas (Opel, 2008). Monoagent of 1st generation targeted therapy in malignant gliomas, multi-targeting drugs, and combinational targeted therapy have all shown discouraging clinical results (Nappe, 2008). Studies are done for the combination Bortezomib, and VEGF inhibitor bevacizumab in comparison with temozolomide reported for the treatment of recurrent malignant gliomas. In case of progressive, recurrent malignant brain tumors, combining bevacizumab 9 and a cytotoxic agent or an EGFR inhibitor like irinotecan is considered as safe (Dietrich, 2008; Poulsen, 2009). Targeted drug therapies such as Erlotinib and Temsirolimus

(EGFR and mToR inhibitors), Erlotinib 4 and bevacizumab 9 (EGFR and VEGF inhibitors) or Temozolomide (Omuro, 2008). Investigational approach has also been taken in the areas of immunotherapy, ribozymes, and RNA interferences (RNAi), as a novel targeted therapy (Sampson, 2008; Bao, 2008).

12.5.4 *NONINVASIVE TECHNIQUE INVOLVING NANOMEDICINE FOR DELIVERING DRUGS TO TREAT BRAIN TUMOR*

Another technique to enhance the permeability of anti-cancer drug across BBB is using of nanoparticulate carriers. Current studies (Sarin, 2008) have shown that the maximum pore size of the glioma's microvasculature is 20 nm. It can be put into a notion that nanoparticulate having longer circulation time in blood and size less than 12 nm can easily cross this microvasculature pore of the gliomas (Sarin, 2008).

The study of nanoparticulate carriers focuses on colloidal carriers like liposomes, polymeric nanoparticles (NPs), solid lipid NPs, polymeric micelles, and dendrimers. When the colloidal system is administered it extravagates less into the tumor than the normal tissue because of the disrupted BBB (Moghimi, 2005). This results in selectivity of drugs into the brain tumor. Enhanced Permeability and Retention (EPR) is a terminology born out of passive targeting of the tumor by nanoparticle drug delivery system. Liposomes are nanocarriers of particle size ranging between 50–150 nm (Moghimi, 2005). The liposomes remain in the microvasculature while the drug diffuses through the liposomal membrane and finally crosses the pores of BBB of malignant gliomas. The main reason behind using nanocarriers through the systemic circulation is long half-life in the blood, but the reticuloendothelial cells (RES) rapidly remove it (Moghimi, 2001), which in turn depends on its particle size, charge, and surface properties (Ogawara, 2001). Prolongation of the life of the drug in the plasma is achieved by coating it with PEG or chemically linking the drug surface with PEG. This minimizes the interaction of the drug with RES and prevents its removal. The crossing of the PEGylated nanocarrier becomes difficult because of its low affinity towards BBB (Provenzale, 2005). But still, it's an important tool only when it behaves as a substrate for active transport systems like carrier-mediated transport, receptor-mediated endocytosis and adsorptive endocytosis (Provenzale, 2005). A non-invasive imaging technique through drug delivery can be formulated by encapsulating the imaging agent along with the anticancer drug (Provenzale, 2005; Rahman, 2012a; 2012b). With the adventure of nanotechnology nanosystems such as liposomes and NPs are emerging drug delivery systems in pre-clinics.

12.5.4.1 LIPOSOMES

Long history is evident that liposomes have been used as drug delivery agents because of their easy preparation, good biocompatibility, low toxicity, and easy availability. But their blood residence time is decreased due to macrophages of RES (Rahman, 2016). The problem can be overcome by decreasing the particle size less than 100nm or by surface modification, i.e., PEGylation (Rahman, 2016). The PEGylated liposomes can be targeted to the brain by attaching a monoclonal antibody. This modification is against transferrin receptors (OX–26, glial fibrillary protein or human insulin receptors (Rahman, 2013; Akhter, 2011). 5-Fluorouracil is a poor brain penetrator. To enhance its penetration through the systemic route, it is conjugated with Transferrin. The conjugation takes place by coupling of the amine group present on the surface of the stearylamine of the liposome and carboxyl group of Transferrin. Rats were injected intravenously, and free 5-Flurouracilcoupled and non-coupled liposomes were evaluated (Soni, 2008). The results were found at an average of 10-fold increment in the concentration of liposomal–5 Fluorouracil conjugate and 17-fold increment in liposomal-transferrin conjugate was obtained, which points out to the fact that transferrin receptors are present on the BBB which acts through receptors mediated endocytosis (Soni, 2008). A combination of Boron capture neutron therapy (BCNT) and PEGylated liposome-transferrin conjugate has been proposed in the treatment of malignant gliomas. The therapy aims to achieve the delivery of Sodium Borocaptate (Na2 10B 12H 11 SH, BSH) in the targeted specific brain tumor (Doi, 2008). BCNT involves the bombardment of 10B with a neutron (1n) resulting in the formation of α-particle (4He) and 7 Li, which are high-energy linear particles with biological activity and low tissue distribution (5–9 mm). This shields the tissues from non-specific radiation damage. It is important to achieve a requisite number of 10B atoms within the tumor (Yanagie, 2008).

Liposomes have been modified for the delivery of gene in the brain tumors. Torchillin studied the delivery of gene and conjugated transactivating transcriptional peptide (TATp) with modified liposome. The conjugation was used to deliver a gene encoding green fluorescent protein (PEGFp-N1), which targets intra-cranial brain tumor cells called U87 MG cells in nude mice. *In vivo* studies show better selectivity and delivery of (PEGFp-N1) genes by the modified liposomes-(TATp) conjugation than the plasmid loaded liposome (Gupta, 2007). Thus, TATP opens a new chapter and raises hope in gene delivery system in tumors.

12.5.4.2   NANOPARTICLES (NPS)

NPs are colloidal carrier system in which the drugs are dissolved, trapped, encapsulated, absorbed or chemically linked to the surface. They are polymeric particles 1–1000 nm in size (Kumar, 2017; Rahman, 2013). They can further be classified into nanocapsules, which are reservoir systems and nanospheres constituting of a matrix system (Rahman, 2013; 2012b). Biodegradable polymers are used for manufacturing NPs of which PLGA (Polylactic acid co-glycolic Acid) deserves special mention, as it is approved by FDA and can be easily processed to a size of 200 nm. These NPs get leverage over other systems as because they have high drug loading capacity and protects it from enzymatic and chemical degradation. When intravenously administered it is cleared from the blood in the same way as liposomes, so its size is reduced to 100 nm. To inhibit its interaction with RES (Rahman, 2012a; 2012b). The endothelial uptake of NP affects its biodistribution. The uptake can be enhanced by covalently linking polyethylene glycol or polyoxide chains on their surface or by coating it with hydrophilic surfactants (Rahman, 2012a). Chertok et al. have devised a unique method using a magnet (iron oxide/magnetite) as the core of a nanoparticle. The shell of the NP has been made of biocompatible polymer (starch/ dextran) (Chertok, 2007). Chertok et al. used this magnetic NP for targeting brain tumors (Chertok, 2007; Chertok, 2008). Under the effect of magnetic field magnetic nanoparticle (12mg Fe/kg) were injected into rats possessing 9L-gliosarcoma. For evaluation, magnetic resonance images were taken before and after administration of nanoparticle at 1-hour interval for a period of 4 hours. A 5-fold increase in the concentration of the magnetic NPs were found in total glioma when compared to non-targeted tumors (Chertok, 2008). A 3–6-fold increase in the target selectivity index was found by magnetic NPs. Target selectivity index is defined as NPs concentration in gliomas when compared to normal brain cells. Thermotherapy with magnetic NPs (magnetic fluid hyperthermia) is another method developed that deals with injecting the magnetic fluid directly into the tumor and then producing heat by alternating the magnetic field. This causes heating of all the body parts. This method has shown good tolerability *in-vivo* in all the tumors including GBM (van Landeghem, 2009).

Bio-nanocapsules and its effect on brain tumors were studied by Tsutsui and coworkers (Tsutsui, 2007). These bio nanocapsules are made up of surface antigen of hepatitis-B, chemical compounds, proteins, gene, and small interference RNA (siRNA). They are conjugated with anti-human EGFR antibody which has affinity for EGFR VIII (Tsutsui, 2007). These EGFR VIII are overly expressed in various malignancies of epithelial nature especially gliomas. Gli36 (EGFR VIII) (but not the wild type one) and Gli35 cell lines in rats were subjected to BNC's. Both effective and selective targeting was achieved in the brain tumor study. Another method was fabricated by Schneider et al. (2008). The combined vaccines with anti-sense nucleotides deliver by using poly-butyl cyanoacrylate NPs (Schneider, 2008). The logic behind using this combination therapy is to activate the immune system by active specific immunization with the help of new castle disease virus affected tumor cells. In this, the transforming growth factor (TGF)-β production is blocked by (TGF)-β antisense oligonucleotide (Schneider, 2008). Polysorbate 80 was used to coat the poly-butyl acrylate NPs, so as to increase BBB penetration. NP-treated animals showed prolonged survival rates than untreated controls with low concentration of (TGF)-β and increased rates of CD 25+T_lymphocyte in blood. This duo combination can be effectively used to destroy resistant glioblastomas (Schneider, 2008).

12.5.4.3   *SOLID LIPID NANOPARTICLES (SLNS)*

Solid lipid NPs are solid lipid dispersions stabilized by emulsifier or co-emulsifier complex in water. SLN have been used in delivering drugs to CNS. The lipids in SLN are usually used as food and conventional emulsifiers such as poloxamer bile salts and polysorbates (Blasi, 2007). The biodistribution of SLNs can be affected by changing the physiochemical properties of its surface. This enhances the specific delivery of the drug. The delivery of the drug to the CNS using SLN have been improved and widely published as reviews (Blasi, 2007). Anti-cancer drugs like campto-thecin, DOX, and PCL were delivered to the brain tumor. They can be used as normal SLN's or PEGylated SLN (Wong, 2007) when these drugs were encapsulated and delivered as SLN: a higher concentration was achieved in the CNS indicating, its better BBB crossing (Wong, 2007). SLN's have leverage over polymeric coated NP in terms of low cytotoxicity, protection of the labile drug from degradation, physical stability, easy preparation, and controlled release. The low toxicity of SLN and presence of biodegradable lipids make it a suitable formulation for delivery of drugs to the brain tumors (Brioschi, 2007). SLN's have potential to carry various anticancer drugs like DOX, PCL, and cholesteryl butyrate to brain tumors. In a rat brain glioma model using DOX, it was shown that SLNS attain a 12-fold increment of the drug concentration in tumor within 30 minutes as compared to free solution. Similarly, after 24 hours, it showed A 50-fold increment in drug concentration in tumor. Without disrupting BBB, DOX-SLN conjugate showed sub-therapeutic concentration while the drug alone failed to match that concentration (Brioschi, 2007). A study was conducted in which rabbits were administrated PCL SLN's and PCL alone as controlled. A 10-fold increase in concentration of PCL–SLN conjugate was found then PCL alone. Thus, it is evident from these studies that SLN's have good role in brain tumor (Wong, 2007) (Figure 12.1).

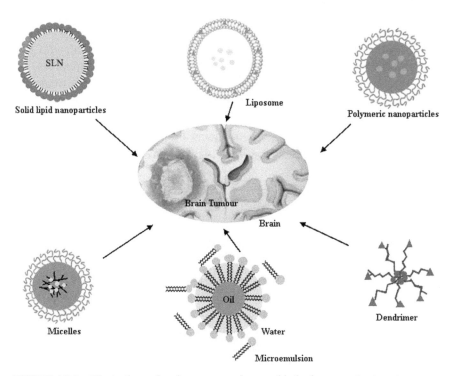

FIGURE 12.1 Illustrations of various nanocarriers used in brain cancer treatment.

12.5.4.4   POLYMERIC MICELLES AND DENDRIMERS

Amphiphilic copolymers are made into aqueous solution resulting polymeric micelles with size range of 10–100 nm. When the co-polymers concentration reaches threshold, value called CMC (critical micelle concentration) then self-assembling occurs (Rahman, 2012a). It is made up of inner core and outer shell. Hydrophobic polymer blocks [e.g., poly (propylene glycol) (PPG, poly (D,L-Lactide), poly (caprolactone), etc.] constitute the core while hydrophilic polymer (e.g., PEG) make up the outer shell (Rahman, 2012a). An interesting example is that of a copolymer called Pluronic block copolymer. It is made of two hydrophilic PEG and one hydrophobic PPG block (PEG-PPG-PEG) (Rahman, 2012a). It showed remarkable property of crossing the membrane of cultured brain microvessel endothelial cells and inhibiting P-gp (Batrakova, 2008). This inhibition of efflux pump by pluronic block co-polymer opens the window for the crossing of many therapeutic agents, not only across BBB but other biological membranes. Pluronic based micellar delivery system opens a new chapter in the novel delivery system for treating brain cancers. A novel technique was formulated by Musacchio et al. (Musacchio, 2009) in which PEG-PE micelle was loaded with PCL and a surface modified with a PBR ligand (Imidazopyridine derivative). The formulation was investigated for synergistic effect in anti-cancer treatment. LN18 human glioblastoma cell line was used for studying cytotoxic effect of these micelles. The synergistic effect of PCL-loaded PBR targeted micelles played a pivotal role in enhancing the cytotoxicity. In the field of nanomedicine, PBR-targeted nanoformulations have also shown promising therapeutic potential (Musacchio, 2009).

Dendrimers are other drug delivery system comprising of polymer branches attached to a central core. Point of attachment is called shell. The surfaces of the branches are known as termini. Their size is comparable to NPs or polymeric micelles and is less than 12 nm. Polythene-co-polyester (PEPE) dendrimers loaded with MTX and conjugated to D-Glucosamine was synthesized by Dhanikula et al. (Dhanikula, 2008). This conjugation with glucose not only enhanced the BBB penetration but also armed the formulation with tumor targeting specificity with the help of facilitated metabolism of glucose by Glucose transporter (GLUT) present within the tumors. The efficacy of MTX loaded dendrimer against tumor was evaluated against vascular human tumor spheroids and glioma cells (Dhanikula, 2008). In both the cell line, endocytosis of glycosylated dendrimers was more than non-glycosylated. Furthermore, IC50 of MTX in dendrimers was lower than the free MTX (Dhanikula, 2008) (Table 12.1).

TABLE 12.1 Application of Nanomedicines in Brain Cancer Treatment

Drug name and carrier	Advantages	References
5-Fluorouracil coupled transferrin conjugated stearyl amine surfaced liposomes	Rats were received intravenous injection of the said liposomes, results to enhances 10-fold higher concentration 0f 5-FU loaded liposome and 17-fold increment in liposomal transferrin conjugates.	Soni, 2008
Gene encoding green fluorescent protein (PEGFp-N1 is coupled with trans activating transcriptional peptide (TATp) conjugated liposome.	It provides better selectivity, and higher concentration of the said gene were achieved in intra-cranial brain tumor of U87 MG cells in nude mice.	Gupta, 2007
Doxorubicin-loaded solid lipid nanoparticles	Rat brain glioma cells were received 50-fold increment of drug concentration without disrupting BBB.	Brioschi, 2007
Paclitaxel-loaded SLN	Studies reported that 10-fold enhancement of paclitaxel-SLN was found in brain tumor as compared to paclitaxel alone.	Wong, 2007
Paclitaxel-loaded PEG-PE micelle and its surface modified with a PBR derivatives	It produced synergistic effect in cytotoxicity against LN18 human glioblastoma cell line as compared to free paclitaxel.	Musacchio, 2009
Polythene-co-polyester (PEPE) dendrimers loaded with methotrexate (MTX) conjugated to D-glucosamine	The said formulation enhanced the BBB penetration and tumor targeting specificity as compared to non-glycosylated dendrimers.	Dhanikula, 2008
Antisense nucleotides ((TGF) -β antisense oligonucleotide) anchored with poly-butyl cyanoacrylate nanoparticles	This combination provides the higher inhibition of transforming growth factor (TGF)-β by enhanced BBB penetration and showed prolonged survival rates than untreated controls. Moreover, they said combination is effectively used to destroy resistant glioblastomas.	Schneider, 2008

12.6 CONCLUSION

The blood-brain barrier, composed by the endothelial cells of the brain capillaries forms the tight junction, which restricts the entry of compounds into the brain. The various strategies have been employed to enhance the amount and concentration of anticancer drug molecules in the brain cancer including invasive and non-invasive techniques. Invasive techniques included surgical resection followed by radiotherapy and chemotherapy are the standard treatment for patients suffering from malignant brain tumors but still lacking specific specificity and higher toxicity. Continuous research in this particular area developed more specific targeted anticancer drug molecules, which targeted the brain tumors and it is delivered through using of nanomedicines included liposomes, NPs, solid lipid NPs, polymeric NPs, dendrimers, etc.

KEYWORDS

- **brain tumor**
- **cancer targeted therapy**
- **chemotherapy**
- **liposomes**
- **malignant tumor**
- **nanomedicines**
- **polymeric nanoparticles**

REFERENCES

Abbott, N. J., Ronnback, L., & Hansson, E., (2006). Astrocyte-endothelial interactions at the blood-brain barrier. *Nat. Rev. Neurosci.*, 41–53.

Akhter, S., Ahmad, Z., Singh, A., et al., (2011). Cancer targeted metallic nanoparticle: Targeting overview, recent advancement and toxicity concern. *Curr. Pharm. Des.*, *17*(18), 1834–1850.

Aneja, P., Rahman, M., Beg, S., et al., (2014). Cancer targeted magic bullets for effective treatment of cancer. *Recent. Pat. Antiinfect. Drug. Discov.*, *9*(2), 121–135.

Baker, S. D., Wirth, M., Statkevich, P., et al., (1999). Absorption, metabolism and excretion of 14C-temozolomide following oral administration to patients with advanced cancer. *Clin. Cancer. Res.*, 309–317.

Bao, S., Wu, Q., Li, Z., et al., (2008). Targeting cancer stem cells through L1CAM suppresses glioma growth. *Cancer. Res.*, 6043–6048.

Barth, R. F., Yang, W., Wu, G., et al., (2008). Thymidine kinase–1 as a molecular target for boron neutron capture therapy of brain tumors. *Proc. Natl. Acad. Sci.*, 17493–17497.

Batrakova, E. V., & Kabanov, A. V., (2008). Pluronic block copolymers: Evolution of drug delivery concept from inert nanocarriers to biological response modifiers. *J. Control. Release*, 98–106.

Becker, I., Becker, K. F., Meyermann, R., et al., (1991). The multidrug-resistance gene MDR1 is expressed in human glial tumors. *Acta. Neuropathol.*, 516–519.

Beduneau, A., Saulnier, P., & Benoit, J. P., (2007). Active targeting of brain tumors using nanocarriers. *Biomaterials*, 4947–4967.

Bihorel, S., Camenisch, G., Lemaire, M., et al., (2007). Modulation of the brain distribution of imatinib and its metabolites in mice by valspodar, zosuquidar and elacridar. *Pharm. Res.*, 1720–1728.

Blakeley, J., (2008). Drug delivery to brain tumors. *Current Neurology and Neuroscience Reports*, 235–241.

Blasi, P., Giovagnoli, S., Schoubben, A., et al., (2007). Solid lipid nanoparticles for targeted brain drug delivery. *Adv. Drug. Deliv. Rev.*, 454–477.

Bobo, R. H., Laske, D. W., Akbasak, A., et al., (1994). Convention-enhanced delivery of macromolecules in the brain. *Proc. Natl. Acad. Sci.*, 2076–2080.

Bower, M., Newlands, E. S., Bleehen, N. M., et al., (1997). Multicentre CRC phase II trial of temozolomide in recurrent or progressive high-grade glioma. *Cancer Chemother. Pharmacol.*, 484–488.

Brandes, A. A., & Fiorentino, M. V., (1996). The role of chemotherapy in recurrent malignant gliomas: An overview. *Cancer Invest.*, 551–559.

Breedveld, P., Beijnen, J. H., & Schellens, J. H., (2006). Use of P-glycoprotein and BCRP inhibitors to improve oral bioavailability and CNS penetration of anticancer drugs. *TRENDS in Pharmacological. Sciences*, 17–24.

Brioschi, A., Zenga, F., Zara, G. P., et al., (2007). Solid lipid nanoparticles: Could they help to improve the efficacy of pharmacologic treatments for brain tumors? *Neurol. Res.*, 324–330.

Cao, Y., Tsien, C. I., Shen, Z., et al., (2005). Use of magnetic resonance imaging to assess blood-brain/blood glioma barrier opening during conformal radiotherapy. *J. Clin. Oncol.*, 4127–4136.

Chertok, B., David, A. E., Huang, Y., et al., (2007). Glioma selectivity of magnetically targeted nanoparticles: A role of abnormal tumor hydrodynamics. *J. Control. Release*, 315–323.

Chertok, B., Moffat, B. A., David, A. E., et al., (2008). Iron oxide nanoparticles as a drug delivery vehicle for MRI monitored magnetic targeting of brain tumors. *Biomaterials*, 487–496.

Da Fonseca, C. O., Linden, R., Futuro, D., et al., (2008). Ras pathway activation in gliomas: A strategic target for intranasal administration of perillyl alcohol. *Arch. Immunol. Ther. Exp.*, 267–276.

Davis, F. G., & McCarthy, B. J., (2001). Current epidemiological trends and surveillance issues in brain tumors. *Expert. Rev. Anticancer. Ther.*, 395–401.

De Vries, N. A., Beijnen, J. H., Boogerd, W., et al., (2006). Blood-brain barrier and chemotherapeutic treatment of brain tumors. *Expert. Rev. Neurother.*, 1199–1209.

DeAngelis, L. M., (2001). Brain tumors. *N. Engl. J. Med.*, 114–123.

Decaudin, D., (2004). Peripheral benzodiazepine receptor and its clinical targeting. *Anticancer Drugs*, 737–745.

Demeule, M., Regina, A., Jodoin, J., et al., (2002). Drug transport to the brain: Key roles for the efflux pump P-glycoprotein in the blood-brain barrier. *Vascul. Pharmacol.*, 339–348.

Dhanikula, R. S., Argaw, A., Bouchard, J. F., et al., (2008). Methotrexate loaded polyether-copolyester dendrimers for the treatment of gliomas: Enhanced efficacy and intratumoral transport capability. *Mol. Pharm.*, 105–116.

Di, L., Kerns, E. H., Bezar, I. F., et al., (2009). Comparison of blood-brain barrier permeability assays: *In situ* brain perfusion, MDR1-MDCKII and PAMPA-BBB. *J. Pharm. Sci.*, 1980–1991.

Dietrich, J., Norden, A. D., & Wen, P. Y., (2008). Emerging antiangiogenic treatments for gliomas - efficacy and safety issues. *Curr. Opin. Neurol.*, 736–744.

Doi, A., Kawabata, S., Iida, K., et al., (2008). Tumor-specific targeting of sodium borocaptate (BSH) to malignant glioma by transferrin-PEG liposomes: a modality for boron neutron capture therapy. *J. Neurooncol.*, 287–294.

Friedman, H. S., Kerby, T., & Calvert, H., (2000). Temozolomide and treatment of malignant glioma. *J. Clin. Cancer. Res.*, 2585–2597.

Gaillard, P. J., Visser, C. C., & De Boer, A. G., (2005). Targeted delivery across the blood-brain barrier. *Expert. Opin. Drug. Deliv.*, 299–309.

Gallo, J. M., Li, S., Guo, P., et al., (2003). The effect of P-glycoprotein on paclitaxel brain and brain tumor distribution in mice. *Cancer. Res.*, 5114–5117.

Groothuis, D. R., (2000). The blood-brain and blood-tumor barriers: A review of strategies for increasing drug delivery. *Neuro. Oncology.*, 45–59.

Guo, P., Ma, J., Li, S., et al., (2001). Targeted delivery of a peripheral benzodiazepine receptor ligand gemcitabine conjugate to brain tumors in a xenograft model. *Cancer Chemother. Pharmacol.*, 169–176.

Gupta, B., Levchenko, T. S., & Torchilin, V. P., (2007). TAT Peptide-modified liposomes provide enhanced gene delivery intracranial human brain tumor xenografts in nude mice. *Oncol. Res.*, 351–359.

Gururangan, S., Cokor, L., & Rich, J. N., (2001). Phase I study of Gliadel wafer plus temozolomide in adults with recurrent supratentorial high-grade gliomas. *Neuro-oncol.*, 246–250.

Haga, S., Hinoshita, E., Ikezaki, K., et al., (2001). Involvement of the multidrug resistance protein 3 in drug sensitivity and its expression in human glioma. *Jpn. J. Cancer. Res.*, 211–219.

Halatsch, M. E., Schmidt, U., Behnke-Mursch, J., et al., (2006). Epidermal growth factor receptor inhibition for the treatment of glioblastoma multiforme and other malignant brain tumors. *Cancer. Treat. Rev.*, 74–89.

Hediger, M. A., Romero, M. F., Peng, J. P., et al., (2004). The ABCs of solute carriers: Physiological, pathological and therapeutic implications of human membrane transport proteins introduction. *P. Flugers. Arch.*, 465–468.

Huang, Y., & Sadee, W., (2006). Membrane transporters and channels in chemoresistance and sensitivity of tumor cells. *Cancer. Lett.*, 168–182.

Huncharek, M., & Muscat, J., (1998). Treatment of recurrent high-grade astrocytoma. Results of a systematic review of 1, 415 patients. *Anticancer Res.*, 1303–1312.

Huynh, G. H., Deen, D. F., & Szoka, F. C., (2006). Barriers to carrier-mediated drug and gene delivery to brain tumors. *J. Control. Release*, 236–259.

Ito, K., Suzuki, H., & Horie, T., (2005). Apical/basolateral surface expression of drug transporters and its role in vectorial drug transport. *Pharm. Res.*, 1559–1577.

Jahnke, K., Kraemer, D. F., & Knight, K. R., (2008). Intraarterial chemotherapy and osmotic blood-brain barrier disruption for patients with embryonal and germ cell tumors of the central nervous system. *Cancer*, 581–588.

Jain, R. K., Di Tomaso, E., Duda, D. G., et al., (2007). Angiogenesis in brain tumors. *Nat. Rev. Neurosci.*, 610–622.

Jones, A. R., & Shusta, E. V., (2007). Blood-brain barrier transport of therapeutics via receptor-mediation. *Pharm. Res.*, 1759–1771.

Juillerat-Jeanneret, L., (2008). The targeted delivery of cancer drugs across the blood-brain barrier: Chemical modification of drugs or drug-nanoparticles. *Drug. Discov. Today*, 1099–1106.

Karkan, D., Pfeifer, C., Vitalis, T. Z., et al., (2008). A unique carrier for delivery of therapeutic compounds beyond the blood-brain barrier. *PLoS. ONE*, *3*(6), e2469.

Kemper, E. M., Boogerd, W., Thuis, I., et al., (2004). Modulation of the blood-brain barrier in oncology: Therapeutic opportunities for the treatment of brain tumors. *Cancer. Treat. Rev.*, 415–423.

Kleihues, P., Burger, P. C., & Scheithauer, B. W., et al., (1995). The WHO Classification of Tumors of the Nervous System. *Journal of Neuropathology and Experimental Neurology.* Vol. 61, pp. 215-225.

Kruh, G. D., & Belinsky, M. G., (2003). The MRP family of drug efflux pumps. *Oncogene*, 7537–7552.

Kumar, V., Bhatt, P. C., Rahman, M., et al., (2017). Fabrication, optimization, and characterization of umbelliferone β-D-galactopyranoside-loaded PLGA nanoparticles in treatment of hepatocellular carcinoma: *In vitro* and *in vivo* studies. *Int. J. Nanomedicine.*, *12*, 6747–6758.

Lai, C. H., & Kuo, K. H., (2005). The critical component to establish in vitro BBB model: *Pericyte. Brain. Res. Rev.*, 258–265.

Laquintana, V., Trapani, A., Denora, N., et al., (2009). New strategies to deliver anticancer drugs to brain tumours. *Expert. Opin. Drug. Deliv. 6*(10): 1017–1032.

Levin, V. A., (1987). *Pharmacokinetics and Central Nervous System Chemotherapy.* Mcgraw-Hill, New York.

Levin, V. A., Patlak, C. S., & Landahl, H. D., (1980). Heuristic modeling of drug delivery to malignant brain tumors. *J. Pharmacokinet. Biopharm.*, 257–296.

Liebner, S., Fischmann, A., Rascher, G., et al., (2000). Claudin–1 and claudin–5 expression and tight junction morphology are altered in blood vessels of human glioblastoma multiforme. *Acta. Neuropathol.*, 323–331.

Loscher, W., & Potschka, H., (2005). Drug resistance in brain diseases and the role of drug efflux transporters. *Nat. Rev. Neurosci.*, 591–602.

Mathieu, D., & Fortin, D., (2007). Chemotherapy and delivery in the treatment of primary brain tumors. *Current. Clinical. Pharmacol.*, 197–211.

Medical Research Council Brain Tumor Working Party, (2001). Randomized trial of procarbazine, lomustine, and vincristine in the adjuvant treatment of high-grade astrocytomas, a medical research council trial. *J. Clin. Oncol.*, *19*, 509–518.

Mintz, A., Gibo, D. M., Mandhan, K. A. B., et al., (2008). Protein- and DNA-based active immunotherapy targeting interleukin–13 receptor alpha. *Cancer. Biother. Radiopharm.*, 581–589.

Moghimi, S. M., Hunter, A. C., & Murray, J. C., (2001). Long-circulating and target-specific nanoparticles: Theory to practice. *Pharmacol. Rev.*, 283–318.

Moghimi, S. M., Hunter, A. C., & Murray, J. C., (2005). Nanomedicine: Current status and future prospects. *FASEB. J.*, 311–330.

Muldoon, L. L., (2008). Chemotherapy delivery issues in central nervous system malignancy: A reality check. *J. Clin. Oncol.*, 2295–2305.

Musacchio, T., Laquintana, V., Latrofa, A., et al., (2009). PEG-PE micelles loaded with paclitaxel and surface-modified by a PBR-ligand: synergistic anticancer effect. *Mol. Pharm.*, 468–479.

Nappe, A., (2008). Drug discovery and development of innovative therapeutics -IBC's 13th annual world congress. *Approaches to Cancer Therapy, I Drugs.*, 705–709.

Neuwelt, E. A., (2004). Mechanisms of disease: The blood-brain barrier. *Neurosurgery*, 131–140.

Newton, H. B., (1994). Primary brain tumors: Review of etiology, diagnosis and treatment. *J. Am. Fam. Physician.*, *49*, 787–797.

Newton, H. B., (2004). Molecular neuro-oncology and development of targeted therapeutic strategies for brain tumors Part 2:PI3K/Akt/PTEN, mTOR, SHH/PTCH and angiogenesis. *Expert. Rev. Anticancer. Ther.*, 105–128.

Ogawara, K., Furumoto, K., Takakura, Y., et al., (2001). Surface hydrophobicity of particles is not necessarily the most important determinant in their *in vivo* disposition after intravenous administration in rats. *J. Control. Release*, 191–198.

Ohgaki, H., & Kleihues, P., (2005). Epidemiology and etiology of gliomas. *Acta. Neuropathol.*, 93–108.

Omuro, A. M., Faivre, S., & Raymond, E., (2007). Lessons learned in the development of targeted therapy for malignant gliomas. *Mol. Cancer. Ther.*, 1909–1919.

Omuro, A., (2008). Exploring multi-targeting strategies for the treatment of gliomas. *Curr. Opin. Investig. Drugs.*, 1287–1295.

Opel, D., Westhoff, M. A., Bender, A., et al., (2008). Phosphatidylinositol 3-kinase inhibition broadly sensitizes glioblastoma cells to death receptor and drug-induced apoptosis. *Cancer. Res.*, 6271–6280.

Osoba, D., Brada, M., Yung, W. K. A., et al., (2000). Health-related quality of life in patients with anaplastic astrocytoma during treatment with temozolomide. *Eur. J. Cancer.*, 1788–1795.

Parasramka, S., Talari, G., Rosenfeld, M., et al., (2017). Procarbazine, lomustine and vincristine for recurrent high-grade glioma. *Cochrane. Database. Syst. Rev.*, *7*, CD011773.

Pardridge, W. M., (1999a). Blood-brain barrier biology and methodology. *J. Neurovirol.*, 556–569.

Pardridge, W. M., (2003). Blood-brain barrier drug targeting: The future of brain drug development. *Mol. Interv.*, *51*, 90–105.

Pardridge, W., (1999b). Vector-mediated drug delivery to the brain. *Adv. Drug. Deliv. Rev.*, 299–321.

Pilkington, G. J., Parker, K., & Murray, S. A., (2008). Approaches to mitochondrially mediated cancer therapy. *Semin. Cancer. Ther.*, 226–235.

Poulsen, H. S., Grunnet, K., & Sorensen, M., (2009). Bevacizumab plus irinotecan in the treatment patients with progressive recurrent malignant brain tumors. *Acta. Oncol.*, 52–58.

Provenzale, J. M., Mukundan, S., & Dewhirst, M., (2005). The role of blood-brain barrier permeability in brain tumor imaging and therapeutics. *Am. J. Roentgenol.*, 763–767.

Rahman, M., Ahmad, M. Z., Kazmi, I., et al., (2012a). Emergence of nanomedicine as cancer-targeted magic bullets: Recent development and need to address the toxicity apprehension. *Curr. Drug. Discov. Technol.*, *9*(4), 319–329.

Rahman, M., Ahmad, M. Z., Kazmi, I., et al., (2012b). Advancement in multifunctional nanoparticles for the effective treatment of cancer. *Expert. Opin. Drug. Deliv.*, *9*(4), 367–381.

Rahman, M., Beg, S., Ahmed, A., et al., (2013). Emergence of functionalized nanomedicines in cancer chemotherapy: Recent advancements, current challenges and toxicity considerations. *Recent Patent on Nanomedicine*, *2*, 128–139.

Rahman, M., Kumar, V., Beg, S., et al., (2016). Emergence of liposome as targeted magic bullet for inflammatory disorders: Current state of the art. *Artif. Cells. Nanomed. Biotechnol.*, *44*(7), 1597–1608.

Reese, T. S., & Karnovsky, M. J., (1967). Fine structural localization of a blood-brain barrier to exogenous peroxidase. *J. Cell. Biol.*, 207–217.

Regina, A., Demeule, M., Che, C., et al., (2008). Antitumor activity of ANG1005, a conjugate between paclitaxel and the new brain delivery vector Angiopep–2. *Br. J. Pharmacol.*, 185–197.

Rittierodt, M., & Harada, K., (2003). Repetitive doxorubicin treatment of glioblastoma enhances the PGP expression-a special role for endothelial cells. *Exp. Toxicol. Pathol.*, 39–44.

Rousseau, J., Barth, R. F., Moeschberger, M., et al., (2009). Efficacy of intracerebral delivery of carboplatin in combination with photon irradiation for treatment of F98 glioma-bearing rats. *Int. J. Radiat. Oncol. Biol. Phys.*, 530–536.

Sampson, J. H., Archer, G. E., Mitchell, D. A., et al., (2008). Tumor-specific immunotherapy targeting the EGFRvIII mutation in patients with malignant glioma. *Semin. Immunol.*, *20*(5), 267–275.

Sarin, H., Kanevsky, A. S., Wu, H., et al., (2008). Effective transvascular delivery of nanoparticles across the blood-brain tumor barrier into malignant glioma cells. *J. Transl. Med.*, *6*, 80.

Sawada, T., Kato, Y., Kobayashi, M., et al., (2000). Immuno-histochemical study of tight junction-related protein in neovasculature in astrocytic tumor. *Brain. Tumor. Pathol.*, 1–6.

Schneider, T., Becker, A., Ringe, K., et al., (2008). Brain tumor therapy by combined vaccination and antisense oligonucleotide delivery with nanoparticles. *J. Neuroimmunol.*, 21–27.

Sekimata, K., Hatano, K., Ogawa, M., et al., (2008). Radiosynthesis and in vivo evaluation of N-[11C] methylated imidazopyridine-acetamides as PET tracers for peripheral benzodiazepine receptors. *Nucl. Med. Biol.*, *35*(3), 327–334.

Shai, R. M., Reichardt, J. K. V., & Chen, T. C., (2008). Pharmacogenomics of brain cancer and personalized medicine in malignant gliomas. *Future. Oncol.*, 525–534.

Sheleg, S. V., Korotkevich, E. A., Zhavrid, E. A., et al., (2002). Local chemotherapy with cisplatin-depot for glioblastoma multiforme. *J. Neurooncol.*, 53–59.

Sikic, B. I., Fisher, G. A., Lum, B. L., et al., (1997). Modulation and prevention of multidrug resistance by inhibitors of P-glycoprotein. *Cancer Chemother. Pharmacol.*, *40*, 13–19.

Soni, V., Kohli, D., & Jain, S. K., (2008). Transferrin-conjugated liposomal system for improved delivery of 5-fluorouracil to brain. *J. Drug. Target*, 73–78.

Stupp, R., Hegi, M. E., Gilbert, M. R., et al., (2007). Chemoradiotherapy in malignant glioma: Standard of care and future directions. *J. Clin. Oncol.*, 4127–4136.

Stupp, R., Mason, W., Van de Bent, M. J., et al., (2005). Radiotherapy plus concomitant and adjuvant temozolomide for glioblastoma. *N. Engl. J. Med.*, 987–996.

Tsutsui, Y., Tomizawa, K., Nagita, M., et al., (2007). Development of bionanocapsules targeting brain tumors. *J. Control. Release*, 159–164.

Van Landeghem, F. K. H., Maier-Hauff, K., Jordan, A., et al., (2009). Post-mortem studies in glioblastoma patients treated with thermotherapy using magnetic nanoparticles. *Biomaterials*, 52–57.

Vlodavsky, E., & Soustiel, J. F., (2007). Immunohistochemical expression of peripheral benzodiazepine receptors in human astrocytomas and its correlation with grade of malignancy, proliferation, apoptosis and serviva. *J. Neurooncol.*, 1–7.

Vukelja, S. J., Anthony, S. P., Arseneau, J. C., et al., (2007). Phase 1 study of escalating-dose OncoGel (ReGel/paclitaxel) depot injection, a controlled-release formulation of paclitaxel, for local management of superficial solid tumor lesions. *Anticancer Drugs.*, 283–289.

Vykhodtseva, N., McDannold, N., & Hynynen, K., (2008). Progress and problems in the application of focused ultrasound for blood-brain barrier disruption. *Ultrasonics*, 279–296.

Waldeck, W., Wiessler, M., Eheman, V., et al., (2008). TMZ-Bioshuttle: a reformulated temozolomide. *Int. J. Med. Sci.*, 273–284.

Westphal, M., Hilt, D. C., Bortey, E., et al., (2003). A phase 3 trial of local chemotherapy with biodegradable carmustine (BCNU) wafers (Gliadel wafers) in patients with primary malignant glioma. *Neuro. Oncol.*, 79–88.

Wong, H. L., Bendayan, R., Rauth, A. M., et al., (2007). Chemotherapy with anticancer drugs encapsulated in solid lipid nanoparticles. *Adv. Drug. Deliv. Rev.*, 491–504.

Yanagie, H., Ogata, A., Sugiyama, H., et al., (2008). Application of drug delivery system to boron neutron capture therapy for cancer. *Expert. Opin. Drug. Deliv.*, 427–443.

Yung, W. K. A., Prados, M. D., Yaya-Tur, R., et al., (1999). Multicenter phase II trial of temozolomide in patients with anaplastic astrocytoma or anaplastic oligoastrocytoma at first relapse. *J. Clin. Oncol.*, 2762–2771.

Zhou, O., & Gallo, J. M., (2009). Differential effect of sunitinib on the distribution of temozolomide in an orthotopic glioma model. *Neuro. Oncol.*, 301–310.

Zhou, Q., & Gallo, J. M., (2005). *In vivo* microdialysis for PK and PD studies of anticancer drugs. *AAPS. J.*, 659–667.

Zhou, Q., Guo, P., & Gallo, J. M., (2008). Impact of angiogenesis inhibition by sunitinib on tumor distribution of temozolomide. *Clin. Cancer. Res.*, 1540–1549.

Macrophage Targeting: A Promising Strategy for Delivery of Chemotherapeutics in Leishmaniasis and Other Visceral Diseases

JAYA GOPAL MEHER[1], PANKAJ K. SINGH[1], YUVRAJ SINGH[1], MOHINI CHAURASIA[2], ANITA SINGH[3], and MANISH K. CHOURASIA[1*]

[1]*Pharmaceutics and Pharmacokinetics Division, CSIR-Central Drug Research Institute, Sector–10, Jankipuram Extension, Lucknow–226031, U.P., India, E-mail: manish_chourasia@cdri.res.in*

[2]*Era College of Pharmacy, Era University, Lucknow-226003, India*

[3]*Department of Pharmaceutical Sciences, Bhimtal Campus, Kumaon University, Nainital–263136, India*

Corresponding author. E-mail: manish_chourasia@cdri.res.in

ABSTRACT

Visceral diseases, i.e., Condition which have something to do with deep-lying soft internal organs, have been a serious health concern for humans regardless of their geographical distribution, race, age or ethnicity. Amongst these, visceral leishmaniasis, tuberculosis, atherosclerosis, hepatic inflammation, and fibrosis are most common and are caused by deleterious pathogens as well as chemical and biological insults. Current therapeutic interventions for visceral diseases have been imperfect due two mutually perpetuating factors: (i) need for higher systemic doses to reach the deep-seated anatomical locations of the causative pathogens (a protective factor in itself), which leads to progressive alterations and adaptations in genetic configuration of pathogens; and (ii) development of resistance, which in turn raises the dose requirement even further, or may even render the applied approach ineffective. With increased insight into the pathophysiology of visceral diseases, a third confounding factor, immuno-compromisation of the host also seems to be highly consequential. Macrophages participate in physiological inflammatory processes, host-pathogen cross-talk and are often the first responders to an affront such as pathogenic infection. However, pathogens have evolved with host's immune response and developed pathways to compromise macrophage integrity, brazenly invading the toxic intramacrophage environment, making it their residence, and silently growing and proliferating inside the very defense mechanism which was intended to clear them. Therefore, going deep into macrophages' biology and studying its interactions with pathogens, seems to be the way forward for developing effective treatment options. Nano-/micro-technology-based delivery systems carrying drugs are promising approaches in targeting pathogens at the cellular level. These novel delivery tools exploit advantages of both passive and active targeting to ensure delivery of drugs at a target site in adequate quantity. Overexpression of various receptors on healthy/ infected macrophages has opened new avenues for active targeting that can be accomplished by decorating nanoformulations with appropriate ligands. Present chapter deals with macrophage targeting for treatment of leishmaniasis and other visceral diseases. It gives an overview of diseases, origin, and role of macrophages, phagocytosis, and receptor targeting as well as the latest updates on macrophage targeting.

13.1 INTRODUCTION

The rapid recent strides in immunology have opened up new echelons in understanding several previously incurable diseases, especially their inherent pathophysiology. Detailed molecular mechanisms involved in various diseases have come-up as a result of advanced investigative tools including proteomics and genomics (Cazzola and Skoda, 2000; Libby, 2007). Overall these explorations have revolutionized the medical intervention. Though still in the mainstream, classical drugs in conventional dosage forms used against diseases have issues of associated side effects, drug resistance as well as therapy relapse. Therefore, advanced therapeutic mediation such as personalized medicine, targeted drug delivery have been adapted (Lunshof et al., 2006). However, before any sort of therapy is initiated for treatment of an ailment, the first responder is always the body itself. Our own body is laden with defensive tools, honed over countless generations for tackling insults on both microscopic and macroscopic level. One such tool a group of cells known as macrophages, which form an important part of our immune system, being abundantly available in various organs viz. spleen, bone marrow, lungs (alveolar interstitial macrophages), brain, liver (kupffer cells), lymph nodes, gut, and connective tissue. Any pathogen that comes into systemic circulation is encountered by these macrophages and is usually, readily engulfed, by a process called as phagocytosis. Numerous mechanisms are involved in the process of phagocytosis, depending on the nature of pathogen and the pathway of immuno-activation. Phagocytosis appears to be a full proof method of ensuring hasty systemic clearance of pathogen before it can cause any harm. However, pathogens too have evolved with host's immune response and have developed evasive pathways to compromise macrophage integrity, blatantly invading the supposedly toxic (for that pathogen) intramacrophage environment, making it their residence, and silently proliferating inside the very defense mechanism which was intended to clear them. It has been proven beyond doubt that macrophages are in fact reservoirs for many pathogens that cause visceral or deep-lying diseases in human beings. Especially *Leishmania* spp. is known to reside and grow inside the host macrophage after corrupting its detection and clearance mechanisms. The conventional approach of treating such diseases, by systemic administration of high doses of therapeutics, yields mixed results. Firstly, the level of effective concentration attained at the site of infection is often inadequate, i.e., in vicinity of a causative organism residing in a macrophage, which itself might be situated in a deep visceral organ, shunned away from systemic circulation by sheaths of protective tissue like fat, muscle, endothelium, interstitial fluid, etc. Secondly, whatever frail fraction of an originally effective drug does reach the causative organism, may sometimes just stimulate the genetic and epigenetic faculties of that organism and induce rapid development of resistance (especially problematic in tuberculosis). Thirdly, the drug may itself be so overtly toxic (for instance nephrotoxicity associated with polyene antibiotics in the treatment of leishmaniasis) that its heavy systemic administration necessitated by the meager response attained due to the factors just discussed may be limiting in the successful completion of therapy. In many instances, clinical practice of conventional delivery is only followed because of unavailability of better or affordable alternatives. As a result, advancement in drug delivery technology is being aggressively pursued in various diseases like cancer, leishmaniasis, tuberculosis, diabetes, AIDS, inflammatory diseases as well as auto-immune disorders. The emphasis is on organ targeting, tissue targeting or the most advanced amongst these: cellular and intracellular targeting (Al-Jamal, 2013; Leucuta, 2014; Maity and Stepensky, 2015). In past few decades, pharmaceutical experts (scientists and industries) have offered novel drug delivery based therapeutics viz. AmBisome®, DOXIL®, Sandimmune®, Herceptin® as well as other sustained and controlled release based products, which are experimentally as well as clinically proven to be better than conventional medication. These novel drug delivery based products work by selectively reaching the desired site of action and exerting therapeutic action. Drug delivery via particulate and vesicular delivery viz. nano-/micro-particles, carbon nanotubes (CNTs), liposomes, micelles, noisome, dendrimers, SLNs, nanocrystals makes it a possibility to target any particular cell type such as cancer cells, immune cells, macrophages, etc. Special attention is given to receptor targeting in macrophages as it can ensure precise delivery of drugs required to kill pathogens, which were otherwise protected from systemic drug presence. Both infected or normal macrophages express a number of receptors on their surface, and a large number of molecules act as ligands for these receptors that can be used for targeting macrophages (Jain et al., 2013). Present chapter is focused on macrophage targeting of chemotherapeutics in leishmaniasis and other visceral diseases. It gives a brief overview of the diseases (leishmaniasis and other visceral diseases) and challenges encountered in their treatment. Figure 13.1 gives an artistic impression of visceral organs where macrophages are abundantly found and various therapeutic delivery strategies being researched for macrophage targeting of drug molecules.

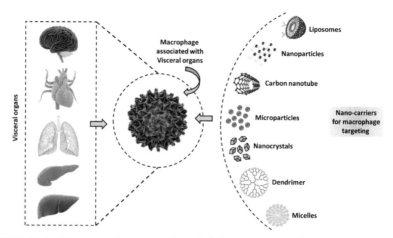

FIGURE 13.1 Diagrammatic presentation of vital organs, macrophage, and various nano-carriers for active targeting of macrophage. Macrophages are an important target site for drug delivery in different visceral organs as these are known to be residing sites for pathogens. Various novel drug delivery tools viz. liposomes, nanoparticles (polymeric/metallic), carbon nanotubes, microparticles, nanocrystals, dendrimers, micelles, etc. have been developed for macrophage targeting of chemotherapeutics.

13.2   A BRIEF OUTLINE OF LEISHMANIASIS AND SELECTED VISCERAL DISEASES

Leishmaniasis is a group of neglected vector-borne disease, mainly prevalent in the tropical and subtropical countries. Clinically there are three major forms of leishmaniasis (i) visceral, (ii) cutaneous, and (iii) mucocutaneous. It is estimated (report by World Health Organization (WHO)) that approximately 556- and 399-million people are affected by visceral and cutaneous leishmaniasis, respectively in twelve high-burden countries. Amongst the three major forms of leishmaniasis, visceral leishmaniasis is most dangerous and may cause death if left untreated. Although these diseases are predominant in countries like India, Nepal, Bangladesh, Brazil, and Sudan, in last few decades the disease has spread in other parts of the world as well. Emergence of drug resistance and poor preventive and curative measures are thought to be reasons for its prevalence. Leishmaniasis is caused by more than 20 species of protozoan parasites of genus *Leishmania,* family Trypanosomatidae and is spread by more than 30 species of sand fly. Parasites of this vector-borne disease have a digenetic life cycle (i) gut of sand fly in promastigote form and (ii) mammalian host in amastigote form. In mammalian host, i.e., inside macrophages instead of being killed, these parasites damage the inherent immuno-mechanisms of macrophages and reprogram them in their favor. Then they start fulfilling their nutritional and metabolic needs, proliferate inside the macrophages and continue to cause infection to host cells (2016; Lamotte et al., 2017). Initially, antimonials were used for treatment of leishmaniasis, and progressively other drugs viz. amphotericin B, miltefosine, paromomycin, and pentamidine started emerging. Liposomal formulation of amphotericin B is known to be highly effective in visceral leishmaniasis. However, since last few decades, leishmaniasis has developed resistance to several drugs, which has been attributed to inherent genome instability of the parasite, directly correlated to irrational usage of anti-leishmanial drugs. Recurrent gene and chromosome amplifications can lead to fitness gains by parasites. Even though numerous pre-clinical reports are available describing outstanding performance of novel drug delivery systems in treatment of experimental leishmanial models, only few have gained commercial footprint. However, it is anticipated that better understanding of host-pathogen interaction as well as identification of novel target sites in both host and pathogens might contribute in advancement of chemotherapy against leishmaniasis. As macrophages are reservoir of Leishmania, passive, and active targeting of these infected cells is an appropriate strategy to destroy the causative pathogens.

Pulmonary diseases, mainly tuberculosis is a major health concern over the globe with 8.3–9.0 million cases reported in 2012 by WHO. Women are more prone to tuberculosis infection, and high numbers are reported from Asia and Africa with 58% and 27% incidences respectively. India and China in the Asian countries are foremost contributors to tuberculosis with 2.0–2.4 million and 0.9–1.1 million cases as reported by WHO (2015). Although since last decade there has been a decrease in tuberculosis cases, but co-infection with HIV has become a serious issue. The most worrying matter with tuberculosis is the non-homogeneity of its prevalence amongst global population. It is known to be a poor man's disease, and hence in spite of availability of medication, it propagates among homeless people, chronic alcoholics, refugees, prisoners, and also HIV patients. As in case of leish-maniasis, macrophages play a significant role in initiation and progression of tuberculosis where the causative organism, *M. tuberculosis,* somehow masquerades into the alveolar macrophage and after reprogramming the host's immune-system starts proliferating inside. Currently used drugs for tuberculosis have several adverse effects and also lack efficient targeting ability due to the intracellular location of pathogen. Irrational use of drugs has further resulted in emergence of multi-drug resistant tuberculosis (MTR-TB). In 2012 alone, approximately 450,000 cases (an alarming) of MDR-TB were registered which included 3.6% of newly reported cases and 20.2% of previously reported cases. Another serious aspect of treating tuberculosis is patient incompliance. It requires long term therapy, that too in multiple daily dosages. Therefore, there is an urgent need for better therapeutic options that would cater to patient compliance issues as well as provide a high cure rate. In this regard inhalable liposomal and micro-particulate delivery systems are expected to provide breathing space in treatment and management of tuberculosis (Nuermberger et al., 2010; Sulis et al., 2014).

Liver is a vital organ of body that regulates many physiological functions and maintains homeostasis. Diseases in liver are caused by chemical or biological insults, pathogens, or even mechanical injury. In case of liver injury, activation of immune cells like kupffer cells occurs, leading to secretion of inflammatory mediators that increase cytotoxicity and chemotaxis. So kupffer cells, local residents in liver are a major player in hepatocellular destruction. At the same time these immune cells are also known for their protective action mediated via production of IL–10 and IL–18. Fatty liver is another major concern in liver-related complications which may be either alcoholic or non-alcoholic fatty liver. In both cases, progression leads to hepatic steatosis and steatohepatitis which results in cirrhosis and liver cancer. As these are chronic inflammatory diseases, hepatic macrophages are involved in their processing, and an increased level of these cells is detected in patients with fatty liver. Liver fibrosis is associated with physiological alterations in the anatomy and configuration of extracellular matrix of liver. Collagen type IV, proteoglycans, and laminin of subendothelial space are substituted by collagen type I and III leading to alteration in the hepatic architecture. It also exhibits a decreased production of matrix metalloproteinases as well as increased production of their inhibitors. Kupffer cells are known to participate in the above process and can be a target that can have significant role in management of this disease (Kolios et al., 2006).

Atherosclerotic cardiac disease is an inflammatory disease which occurs due to high plasma level of LDL, cholesterol, and progressive deposition of lipid on the arterial wall. It is observed on the muscular and elastic arteries of large and medium-sized hearts and is majorly of two different types, ischemic heart disease, and cerebrovascular disease. The atherosclerotic lesions may lead to infarct of heart or cardiac arrest causing death. These diseases accounted for 247.9 deaths per 100,000 persons in 2013, and were responsible for 84.5% of cardiac disease-related deaths. If inflammation does not subside, it causes an increase in recruitment of macrophages. These macrophages, also termed as foamy macrophages are rich in cholesterol, start infiltrating the arterial wall that ultimately leads to pathological intimal thickening lesion. This bulging lesion may convert into an atherosclerotic plaque which is vulnerable to rupturing and thrombosis (Barquera et al., 2015). Myocardial infarction is yet another leading cause of heart-related deaths. It is associated with occlusion of coronary vessel which leads to oxygen-deprived state of myocardium. All these events result in inflammation and myocyte necrosis. There is array of mechanisms involved in the origin of myocardial infarction which causes cardiac remodeling. Macrophages play an important role in the progression of such cardiac inflammatory conditions, hence targeting of these immune cells is understood to be one of the latest areas of therapy (Weinberger and Schulz, 2015).

13.3 CHALLENGES IN CONVENTIONAL CHEMOTHERAPY

Chemotherapy in any disease has its limitations and associated adverse effects. In case of leishmaniasis the existing chemotherapeutics have comparatively greater challenges: severe adverse effect, low cure rate as well as the most frightening, drug resistance. Whether it is antimonials, amphotericin B or orally active miltefosine, all reported antileishmanial drugs cause musculoskeletal pain, gastrointestinal disturbances, hypersensitivity, cardiotoxicity, renal failure etc. Long non-ambulatory administrative procedures pose additional discomfort to patients who are already distressed by unwanted adverse effects. Treatment failure is also sometimes encountered. In some parts of Asian countries like India, up to 65% of human treatment cases are found to be unresponsive to antimonials (Sundar, 2001). Investigational reports on such matters suggest a number of reasons viz. expression of multidrug resistance and efflux transporters, down-regulation of aquaporins, overproduction of thiol metabolizing enzyme. Recently, aberrant alteration in the sterol profiles of Leishmania parasite has been noticed where the cholesta–5,7,24-trien–3-ol is observed to be in membrane of parasite instead of ergosterol. This has significantly diminished selectivity of the gold standard drug amphotericin B. Miltefosine, is also on the verge of drug resistance, the reasons for which might be attributed to its long half-life as well as mutations in P-glycoprotein-LdRos3 and LdMT that are directly related to uptake of drug. Drug resistance is also reported for paromomycin, where parasites are observed to develop changes such as up-regulation of glycolytic enzymes and ATP-Binding cassette transporters. These modifications in parasites are acquired progressively with multiple stimulatory factors like inadequate treatment, partial or incomplete treatment as well as long term therapeutic interventions (Yasinzai et al., 2013). Infectious disease tuberculosis requires simultaneous treatment with multiple drugs viz. rifampicin, isoniazid, ethambutol, pyrazinamide, streptomycin, fluoroquinolones, kanamycin, capreomycin, amikacin, viomycinetc for long periods of time. All these drugs are reported to cause an array of adverse drug reactions. Apart from the side effects, the causative organism *Mycobacterium* has either acquired full or is on the verge of acquiring resistance to most of the listed drugs. Mutation in various genes is known to be responsible for drug resistance in *M. tuberculosis*. So chemotherapy with the current therapeutics is a challenge. Furthermore role of efflux pumps and porins are also caused drug resistance (Köser et al., 2015). In all of the infectious diseases, co-infection with HIV is reported to aggravate the case and treatment becomes more challenging. In such situations the duration of therapy increases multiple times and the socio-economic conditions of patients, their homelessness, poverty, non-compliance to therapeutic schedule remain the main reason for failure in therapy and development of drug resistance. Often the bioavailability of these drugs at the target site is also uncertain. Conventional therapy underperforms in other visceral diseases like hepatic diseases and cardiovascular diseases too (Acevedo, 2015).

13.4 MACROPHAGES AND THEIR ROLE IN PHYSIOLOGY AND DISEASE CONDITIONS

Macrophages were discovered by a Russian zoologist Ilya Ilyich Mechnikov in 1882, a ground-breaking exploration in immunology, which ultimately earned him a Nobel Prize. These cells play a role in various important biological phenomena viz. immune system, repair of tissues, homeostasis, removal of debris, as well as developmental processes. Macrophages are widely recognized as efficient immune cells capable of differentiating between the self and non-self, recognizing tissue injury, as well as sensing and fighting against invading pathogens. Being located at different sites for different functions, these cells have different transcriptional outlines. In normal conditions they are involved in the maintenance of physiological functions, whereas in diseased conditions their mode of action becomes pathophysiological. Hence these cells are part of an exceptionally specialized armory that responds to various natural as well as artificial challenges. Understanding complex biology that governs their diverse origins, phenotype switching on demand to homeostasis as well as their transcriptional profile are essential issues to address the novel concept of diseases treatment. As these cells are involved in various diseases and have specific identities, these could be striking therapeutic targets of future medicines. The following section will outline a brief overview of the origin and role of macrophages.

13.4.1 ORIGIN OF MACROPHAGES

Monocytes are considered to be the source of tissue-resident macrophages in living systems. These monocytes are derived specifically from blood and originate in the hematopoietic stem cells of bone marrow. After undergoing sequential differentiation, these cells become monocytes and circulating monocytes depending on the environmental demand are recruited to different locations in the body. After recruitment, monocytes differentiate into macrophages with a specialized function for that particular tissue. Apart from blood monocytes, macrophages are also derived from embryonic precursors, i.e., yolk sac (microglial cells; brain macrophages), and fetal liver (Langerhans cells; skin macrophages). The tissue-resident macrophages are from both prenatal and postnatal origins which are distributed (Figures 13.2A and 13.2B) as microglia; macrophage in brain, ocular macrophage, alveolar macrophage, cardiac macrophage, Langerhans cells; skin macrophage, splenic macrophage, kupffer cells; liver macrophage, intestinal macrophage, osteoclast; bone macrophages, and lymph node macrophage. Based on functional phenotype macrophages can be categorized into two types, i.e., M1 macrophages and M2 macrophages (Figure 13.2C). The classification is based on their activation status, where the M1 types are classically activated macrophages whilst the M2 are alternatively activated macrophages. Any external or cellular stimuli such as lipopolysaccharide (LPS), colony stimulating factor–2 (CSF–2), interferon-γ (IFN-γ), and tumor necrosis factor (TNF) could bring about the activation of M1 macrophages. These macrophages are known to secrete various degradative substances viz. NO, ROS as well as pro-inflammatory cytokines that help in killing pathogenic microorganisms. On the contrary, M2 type macrophages have involvement in the secretion of IL–4, IL–13, HIF–1α and TGF-β, IL–10 and cause immunosuppression. These macrophages have a critical role in angiogenesis, removal of debris as well as tissue remodeling (Pei and Yeo, 2016). Since few decades tumor-associated macrophages are in the limelight because of their significant involvement in the progression of cancer. Immuno-therapeutics concept is being employed for manipulation of balance between M1 and M2 types of macrophages (Singh et al., 2017).

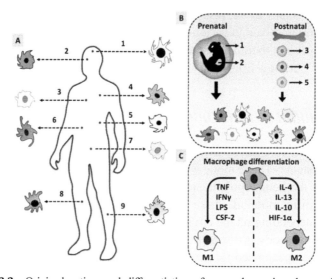

FIGURE 13.2 Origin, location, and differentiation of macrophages in a human being. (A) Diagrammatic representation of location of tissue-specific macrophages in human body; (1) microglia; macrophage in brain, (2) ocular macrophage, (3) alveolar macrophage, (4) cardiac macrophage, (5) splenic macrophage, (6) kupffer cells; liver macrophage, (7) intestinal macrophage, (8) osteoclast; bone macrophages, (9) lymph node macrophage. (B) Prenatal and postnatal origin of macrophages; (1) Yolk sac, (2) fetal liver, (3) hematopoietic stem cells, (4) common myeloid progenitor cells, (5) monocytes. (C) Differentiation of macrophage from monocyte to both M1 type pro-inflammatory macrophages and M2 type anti-inflammatory macrophages.

13.4.2   ROLE OF MACROPHAGES

On the basis of anatomical distribution, macrophages can be classified as per their role in the physiological system. Microglia, the local residents of brain act as a first-line defense system of the central nervous system and have a profound role in immune surveillance. They are actively involved in the maintenance of the brain, and play the role of scavengers on a routine basis by clearing deposited plaques, damaged neurons, as well as invading pathogens. Ocular macrophages are an integral part of the ocular immune system and play a pivotal role in the protection of eyes from microorganisms. Most of the macrophages in eyes remain in the uvea which is highly vascularized area devoid of lymph vessels. However, cornea which is the outer part of eye lacks vasculature and is in contact with environment which makes it vulnerable to microbial attack. Therefore, if a need for instigating an immunological response arises, macrophages from surrounding vascularized tissues take part in immuno-response. These macrophages work against any intra-ocular infection and participate in tissue repair and remodeling. Respiratory organ, lungs, are even more susceptible to microbial infection, owing to their massive surface area and the nature of their job, which requires them to be in direct contact with external environment throughout. Pathogens have easy entry via airways, and the moist oxygenated environment of lungs is just about right for them to flourish. Alveolar macrophages, the immune cells of lungs, therefore need to be especially alert to encounter invading microorganisms quickly. Kupffer cells are macrophages of liver, which have specialized functions such as scavenging waste cells and debris. Langerhans cells, gastrointestinal macrophages, osteoclasts, bone marrow macrophages, and red pulp macrophages are responsible for collaboration with T lymphocytes, mineral disruption, erythropoiesis support, maintenance of intestinal homeostasis, as well as removal of erythrocytes. Macrophages are also responsible for remodeling of tissues. They have crucial roles in tissue patterning as well as branching morphogenesis of different mammary gland. They take part in development of ductal structure and its branching. Their role as professional phagocytes is observed in ingestion of erythrocytes as well as neutrophils in liver and spleen. Angiogenesis is also regulated by macrophages, which has been seen in the development of ocular vasculature, circulatory system, and lymphangiogenesis. Macrophages dynamically peruse the metabolic activity of vital organs like liver and pancreas. In case of a bacterial infection, macrophages cause secretion of TNFα, IL6, and IL1β. These cytokines gradually enhance peripheral insulin resistance that ultimately causes decrease in storage of nutrients. Macrophages utilize glycolysis for acquiring fuel for their defense mechanism against pathogens, and the above-mentioned phenomena support in this regard. Similarly, in liver, kupffer cells expedite metabolic adaptation of hepatocytes to handle high caloric intake. The functions that have been discussed above are role of macrophages in the normal physiological conditions, but in diseased state, the functions, as well as characteristics of macrophages, change. In response to any injury circulatory monocytes differentiate into macrophages and quickly reach the site for action. Being of pro-inflammatory phenotype they secrete TNFα, IL1, and NO to degrade any pathogens which have gained entry. In the meantime, T_H1 and T_H17 cells are also produced which exhibit anti-microbial action. However, these cells also cause collateral damage of host tissue which can turn into chronic inflammatory state (Wynn et al., 2013).

13.4.3 MACROPHAGE: HOME TO PATHOGENS

It has been realized and experimentally observed that macrophages become the host for numerous pathogens. Moreover, pathogens because of their genome plasticity promptly achieve drug-resistant phenotype status. The genetic modulations are achieved by gene deletion, gene tandem duplication, as well as extra-chromosomal amplification. In case of leishmaniasis, the parasites repeatedly resort to chromosomal amplification that leads to modulation in the genetic configurations supporting their adaptability to change in environment. Leishmania parasites manage to manipulate the host cell signaling process by: employing over-expressed proteins and glycolipid effectors on their surface, alternating major histocompatibility complex (MHC)-peptide complex, or by destabilization of lipid rafts or even sequestration of antigens. This neutralizes the potency of macrophages by downplaying their ability to work via nitric oxide, reactive oxygen species, antigen-presentation or pro-inflammatory cytokine production. And allows the ingenious pathogen to occupy the macrophage, where they can grow, multiply, and invade other sites. Interfering with microRNAs, the regulators of cellular transcriptome involved in various immuno-modulatory activity is yet another strategy adopted by Leishmania to tame the macrophage. After being infected with Leishmania parasite, macrophages have quantifiably increased life spans. This is attributed to inhibition of host cell apoptosis as evident from the overexpression of anti-apoptotic proteins viz. Bcl2 in infected macrophages. In order to survive inside macrophages, pathogens also interfere in metabolic pathways, which ultimately facilitate their long and fruitful intracellular stay (Lamotte et al., 2017). *Mycobacterium tuberculosis* (MTB) is well known for its ability to resist the stress given by macrophage and to corrupt the host by adopting different mechanisms. Alike the Leishmania parasites, these pathogens are capable of residing in the host cells for a long period of time and to replicate as well as to grow there comfortably. After being inhaled, MTB gets engulfed by alveolar macrophages and undergoes phagocytosis. Thereafter it inhibits the maturation of phagosomes into phagolysosomes, which stops the whole process of cellular immunity. Reduction of proton ATPase in endosomes and removal of phosphatidylinositol 3-phosphate are the underlying mechanisms for such actions. The engulfed MTB in the meantime replicates and starts growing inside the macrophage. Furthermore, degradation of lysosomal enzymatic system takes place, and antigen-presenting efficiency of macrophage is also compromised (Schnappinger et al., 2003). Pathogens might interrupt or even stop fusion of early phagosomes with lysosomes, escape from early phagosomes/endosomes and reach cytosol, and induce adaptation of non-phagocytic pathways. In the last approach, instead of being phagocytized, microorganisms undergo receptor-mediated endocytosis which involves the formation of lipid rafts and clathrin-dependent internalization to avoid the classical lysosomal-lysis. Figure 13.3 gives an overview of the matter discussed in this section.

FIGURE 13.3 An arty presentation of how pathogens convert a macrophage from a purported death cell to a comfortable, safe house. (1) Mechanisms adopted by pathogens to corrupt the tools of host cells to degrade and eliminate microorganisms. This has also been seen to cause drug resistance in many pathogens. (2) Strategies implemented by host macrophages to kill pathogenic microorganisms inside the cells. Although these are robust systems, but pathogens have evolved to resist stress and survived as well as grow inside the macrophages.

13.5   MACROPHAGE AS A SITE FOR DRUG TARGETING

Previous sections explained how macrophages are important in various visceral diseases. Targeting macrophages have therefore become popular in drug delivery research as it can bring potential therapeutic benefit by targeting pathogens that reside inside the macrophage. Induction of cellular immune system is also an active field of *in vitro* and *in vivo* therapeutic research, which too requires some semblance of macrophage targeting. It has been reported that macrophages have greater tendency to uptake foreign substances in comparison to any other immune cells like dendritic cells and monocytes (Niu et al., 2014; Weissleder et al., 2014). Research shows that nanoparticles are readily cleared by the mononuclear phagocytic system and hence drug-loaded nanoparticles might be targeted to be intentionally engulfed by components of the mononuclear phagocytic system. It has also been shown that nanoparticles below 10 nm are excreted by the kidney and larger particles are taken-up by spleen, liver as well as bone marrow (Alexis et al., 2008). These indices turn macrophages into suitable targets for novel drug vectors. Alveolar macrophages are also known to uptake micro-/macro-sized particles; hence these are also suitable for targeting of anti-tubercular drugs. Although targeting these cells looks fascinating, but it is also expected that bio-materials used for formulation and development of nano-/micro-formulations may cause unwanted effects like activation of off-target immune response which could lead to tissue injury and severe adverse effects (Wolfram et al., 2015). In this context, it is necessary to evaluate the safety and specificity of formulations intended for macrophage targeting. Further, it is also desirable to understand the bio-distribution process of novel formulations. Some research deliberations in this aspect are pondered upon in the next portion of this chapter.

Ergen and team have tried to investigate the bio-distribution pattern of some novel drug delivery based formulations at cellular levels in different organs (Ergen et al., 2017). They have assumed that composition and size might be the determining factors in the uptake of delivery systems to various immune cells in the living system. Based on their hypothesis they selected three different formulations viz. (i) polymeric macromolecules (poly (N-(2-hydroxypropyl) methacrylamide, (ii) PEGylated liposomes and (iii) poly(butyl cyanoacrylate)-microbubbles. These formulations had different sizes, i.e., 10 nm, 100 nm, and 2 µ, respectively, and were composed of different materials. The delivery systems were administered in a mice model, and the *in vivo* fate of these was studied. Ergen et al. employed sophisticated multicolor flow cytometry to distinguish different immune cells in various organs and blood. Real-time two-photon laser scanning microscopy and computed tomography-fluorescence-mediated tomography were also used for analyzing up-takes in organs as well as intrahepatic bio-distribution. They reported that splenic red pulp macrophages and kupffer cells could quickly up-take microbubbles, whilst the liposomes could be easily taken-up in liver, lung, and kidney by dendritic cells. Contrarily polymeric delivery systems resided in the systemic circulation for longer time periods and were finally taken-up by endothelial cells, alveolar macrophages, liver, and neutrophils. Physicochemical properties of the delivery system are important in macrophage targeting. The surface of macrophages is negatively charged because of the presence of sialic acid residues, and hence, a positively charged drug delivery system is expected to be better attracted to macrophages and will have a greater uptake. Another factor which can be used to target macrophages potentially is their intracellular acidic pH (5 to 6), which is substantially distinct from prevailing metabolic pH. This opens up the prospects for utilizing smart delivery materials which can sense this subtle variation and respond by losing their integrity to release their drug content (Kunjachan et al., 2012).

13.5.1   PHAGOCYTOSIS BY MACROPHAGE

In 1880s, phagocytosis was discovered by Elie Metchnikoff as one of the central events in the innate immune system. In simple words, it is defined as the engulfment of larger entities, i.e., more than 0.5μ in a programmed manner which makes it different from other clearance mechanisms of body. Phagocytosis is a receptor-mediated activity which involves a cascade of cellular events. Intruders like bacteria, fungi, are recognized, engulfed, killed, and further cleared from the body as debris. However, a macrophage's work does not stop here, as the components of the phagocytosed entity are harvested, processed, and used to initiate an adaptive immune response. Macrophages also partake in phagocytic clearance of dead and decaying cells, cells which have undergone apoptosis, by additionally producing a selective cocktail of anti-inflammatory chemokines to ensure no harm is posed to the surrounding tissue. Therefore, broadly speaking, phagocytosis involves particle recognition, particle internalization, and phagosome maturation. As the catalog of foreign entities which a body encounters is truly gigantic, the adaptive response produced by the macrophages in the form of different surface receptors is also pretty diverse. Opsonic receptors are major players in the initial stages, which recognize foreign bodies (opsonin is an antibody or other substance which binds to foreign microorganisms or cells making them more susceptible to phagocytosis in blood/intestinal fluid). Phagocytes can then get attached to their prey and complete the initial process of recognition. The next step in phagocytosis is internalization which essentially includes binding of the ligand covered particles, clustering of the complex with signaling cascade and initiation of engulfment by an actin-mediated process. This is followed by formation of phagosome which is a membrane-bound vacuole. The membrane of this specialized vacuole resembles plasma membrane, and the inner fluid contents are like extra-cellular fluid. Endosomal compartments' interaction with the phagosomes makes these vacuoles acidic (pH; 4.5), extremely oxidative, and laden with hydrolytic enzymes. Here the phagosomes pass through various biochemical changes and traverse a journey from early to late phagosomes. They become microbicidal in due course of time acquiring the ability to digest lipids, proteins as well as carbohydrates. Thereafter phagosomes fuse with lyso-somes and become phagolysosomes which are a highly specialized lytic vacuole with pH 4.5–5.0, loaded with cathepsins (lysosomal hydrolases), oxidants, and cationic peptides. These components work together to degrade the prey, and the degraded materials are disposed by exocytosis (Handman and Bullen, 2002). Figure 13.4 gives an artistic view of phagocytosis as described in the text.

13.5.2 RECEPTOR TARGETING TO MACROPHAGE

Although being effective against the pathogens, lack of cellular internalization makes many chemotherapeutics intervention ineffective. Pathogens *viz. Leishmania spp., MTB, Mycobacterium leprae, Helicobacter pylori* are the most dangerous pathogens causing numerous deaths worldwide each year, and these are known to reside comfortably in the lap of macrophage (Ahsan et al., 2002). Infected macrophages are reported to overexpress a number of receptors which make them different from non-infected macrophages. Formulation scientist have been working on targeting such macrophages by taking advantage of specific-size of nanocarriers as well as their ability to be decorated with appropriate ligands. Figure 13.5 gives an artistic overview of receptor targeting to macrophages for effective delivery of chemotherapeutics. Several receptors have been reported which are expressed/over-expressed on the macrophage. Amongst them, scavenger receptors (SRs), complement receptors (CRs), integrin receptors, toll-like receptors (TLRs), Fc receptors, and mannose receptors (MRs) have been extensively studied for drug delivery of chemotherapeutics (Ahsan et al., 2002; Polando et al., 2013; Shannahan et al., 2015).

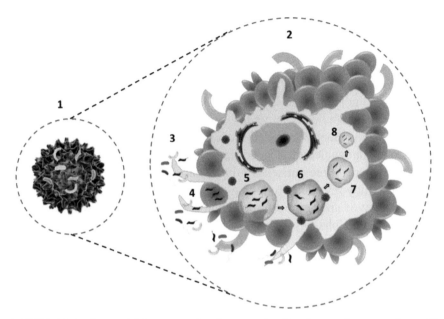

FIGURE 13.4 Sequential happening in phagocytosis of pathogens by macrophage. (1) Macrophage moving around the pathogens for phagocytosis. (2) Enlarged view of cascade of events happening during the phagocytosic process, (3) Pathogens are adjacent to the long pseudopod as the initial stage in phagocytosis, (4) pathogens are engulfed by the macrophage, (5) pathogens are confined within a lytic vacuole/vesicle, (6) lysosomes get attached to vacuole/vesicle and give enzymatic stress to the pathogen, (7) pathogen are killed inside the vesicle, (8) undigested debris remains as such after phagocytosis.

SRs are a superfamily of transmembrane receptors. SRs are basically of type A, and B/CD 36 types; however more (total eight types) have been identified. These proteins were previously known for binding and internalizing LDL only, but recent studies have demonstrated a large member of activities such as clearance of lipoproteins, removal of pathogens, and transportation of cargos to cells. The binding of modified LDL to SRsis responsible for initiating a cascade of events that lead to apoptosis. Studies have shown that these receptors are capable of recognizing and binding to lipoproteins, proteoglycans, phosphatidylserine, carbohydrates, poly-anionic ligands, negatively charged nano-carriers, chemically altered proteins as well as some un-opsonized entities (Shannahan et al., 2015). Tempone and co-workers have developed phosphatidylserine conjugated liposome entrapped with antimony that exclusively targeted *Leishmania (L.) chagasi* amastigotes via SRs of macrophage. The negatively charged device was found to be 16-fold more effective than plain drug solution. These findings improve the therapeutic potential of pentavalent antimony (Tempone et al., 2004). Another work on scavenger receptor targeting was reported where 17-per antisense phosphorothioateoligodeoxy-ribonucleotide was delivered to murine macrophages infected with *Leishmania mexicanaamazonensis.* Liposomes coated with maleylated bovine serum albumin was developed, and antisense mini-exon oligonucleotide was encapsulated into it. The surface decorated liposome could efficiently target macrophage, and after internalization, liposomes got conveniently degraded and released active molecule. The antisense mini-exon oligonucleotide loaded liposomes resulted in the killing of 90% parasites whereas free antisense mini-exon oligonucleotide could kill only 20%. The study revealed promising efficiency of scavenger-receptor mediated delivery of antisense phosphorothioate oligos in destroying pathogen hiding inside the phagolysosomes (Chaudhuri, 1997).

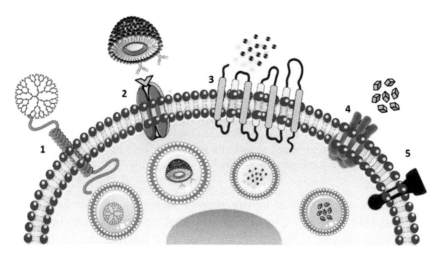

FIGURE 13.5   A graphic representing receptor targeting of macrophages. (1) Scavenger receptor, (2) Complement receptor, (3) Mannose receptor, (4), Toll-like receptor (5) Fc-receptor. Surface decorated novel particulate and vesicular carriers interact with respective receptors on surface of macrophage-based on attached ligand types and subsequently get internalized. After that, a series of events occurs leading to internalization of delivery vector and release of chemotherapeutics inside the targeted macrophage for therapeutic action.

MRs is a group of transmembrane receptors, are highly expressed on the surface of macrophage. These are from the CD 206 group and are mainly of C-type 1 and C-type 2. MRs have become a major area of macrophage drug targeting owing to the very specific feature of recognizing mannose and fucoseglyco conjugates that are broadly found on the surface of many pathogens. MRs are well known for their primary function of intracellular transport of *MTB* and also for antigen uptake and presentation. MRs has sialic acid residues on N-linked glycans which is very crucial for its binding to glycoproteins (sulfated and mannosylated). Pathogens like *Leishmania donovani* has glycan (with mannose terminal) on their surface and hence is recognizable to macrophages. Based on this information, mannose receptor targeting of chemotherapeutics has become a useful tool in treatment of visceral diseases (Irache et al., 2008). Our own group has developed and evaluated mannose grafted chitosan nanocapsules carrying amphotericin B for anti-leishmanial activity. The mannose decorated nanocapsules were observed to be highly internalized by infected macrophage when compared to non-decorated nanocapsule. Further studies revealed higher localization as well as accumulation of amphotericin B in the macrophage-rich visceral organs (liver and spleen) (Asthana et al., 2015). Zhu and co-workers (2011) developed mannose modified chitosan treated PLGA nanoparticles. They investigated the uptake ability of developed nanoparticles as a function of incubation time and concentration of nanoparticles. It could be concluded that the developed novel nanoparticles had high binding affinity and internalization capacity. Gao et al. (2013) have prepared PEGtidedendrons, and decorated them with alpha-d-mannopyranosyl-phenylisothiocyanate. They have reported a 12-fold higher uptake by surface coated dendrons than unmodified dendrons in murine macrophage.

CRs are an important class of membrane proteins that are expressed on various cells involved in immune system. These receptors bind to deposited C3-fragments of the complement system and help in opsonization process. Additionally, these receptors are involved in phagocytosis as well as extracellular killing of pathogens. The CRs are basically of two types, i.e., (i) SCR receptors that comprise CR1-CD35, and (ii) integrin CRs that includes CR3 and CR4 (CD18, CD11b/CD11c) (Leslie and Hansen, 2001). Abe and team have demonstrated role of complement component C3 and CRs type–3 in selective and rapid uptake of oligomannose-coated liposomes. They have shown a carbohydrate-dependent uptake of the decorated liposomes at both *in vitro* and *in vivo* level, where the delivery system was opsonized with CRs type–3, into peritoneal cavity and later to peritoneal macrophages (Abe et al., 2008). Fc receptors (FcRs) are another type of cell surface proteins found in a wide variety of cells including macrophages. When any microorganism or pathogens enter into the living system, it gets recognized and attached to specific antibodies. FcRs recognize those antibodies and binds to their Fc region, i.e., the tail part and hence tries to arrest the pathogens/microorganisms. Based on the specificity of antibody recognizing ability, the FcRs are mainly of three types viz. FcγR, FcαR, and FcεR for IgG, IgA, and IgE respectively. As the FcRs are a major player in the phagocytosis process, it has been employed for targeted delivery of drugs to macrophage for selective uptake and best therapeutic output (Masuda et al., 2009). Zubairi et al. have demonstrated increased granuloma maturation, CD4(+) T cell proliferation, as well as killing of *Leishmania* by employment of a combination of chimeric fusion protein (OX40L-Fc) and a monoclonal antibody that blocks CTLA–4 which is an inhibitory receptor on T cells. The authors also have validated that this combination can be used with conventional drugs used against leishmaniasis for better therapeutic outcome (Zubairi et al., 2004).

TLRs, a major class of proteins, also known as immune receptors are a very crucial and abundantly found receptors in immune cells including macrophages. There are 12 TLRs identified in mammals and amongst them TLRs 1–9 are conserved in human beings. Further, these TLRs are located at both extracellular (TLRs–1,–2,–4,–5,–6 and –11) and intracellular (TLRs–3,–7,–8,–9 and –13) regions of the cells and are able to recognize the pathogens at the outer cellular surface or even in the endosomal compartment. After recognizing pathogens, TLRs recruit adaptor proteins (MyD88 or TRIF) and consequently lead to a cascade of cellular events that cause phagocytosis. There are a number of ligands for the TLRs viz. triacyl lipopeptides, LPS, lipoproteins, profilin, alginate, chitosan, and many more that can be employed to target TLRs (Singh et al., 2012). The TLRs have been explored for vaccine delivery and immunotherapy against leishmaniasis. Raman et al. have developed an immunotherapeutic by employing the TLRs-synergy concept for improved immuno-response against cutaneous leishmaniasis. They have selected leishmania poly-protein vaccine (L110f) and combined with TLR4 agonist (monophosphoryl lipid A), and/or TLR9 agonist (CpG) and evaluated them against leishmaniasis. Researchers observed in their *in vivo* studies that a combination of three performed best against cutaneous leishmaniasis by boosting T cells, reducing parasite burden as well as lesions (Raman et al., 2010). Heuking et al. have worked on development of DNA vaccines for tuberculosis by using TLR2 agonist. They have employed chitosan polymer as a mucoadhesive material and complexed it with TLR2 agonist and plasmid DNA that can act in multiple way by imparting muco-adhesion for enhanced mucosal delivery as well as protection of DNA from degradative factors (Heuking et al., 2009).

13.5.3   FACTORS INFLUENCING MACROPHAGE TARGETING

There are certain factors that govern the magnitude of the maximum therapeutic benefit attainable when we employ the approach of macrophage targeting. These factors might be formulation related such as particle/ droplet size, charge (zeta potential), hydrophobicity, hydrophobicity, specific ligands or it may be physiological viz. pH of environment, types of interactions of host-carrier system, or pharmacokinetic such as duration of stay of drug-carrier in living system (Champion et al., 2008). It has been investigated and observed that particle below 10 nm are well engulfed by hepatic macrophages and also transported to lymph nodes after being phagocytized. However, particles above 10 nm are reported to be localized at the site of administration. Charge on the drug delivery systems also has an important role in the attraction as well as engulfment by macrophages. Experimentally it was found that phosphatidylglycerol and phosphatidylserine which are negatively charged entities, enhance the uptake of micelles composed of phosphatidylcholine by macrophage upon administration. There is still some controversy regarding the underlying mechanisms dictating this phenomenon. However, as a general concept, it is understood that presence of sialic acids, surface of macrophage is negatively charged. A particle, either polymeric or lipid-based in nature with positive charge might be easily attracted, attached as well as phagocytized by macrophages (Fröhlich, 2012). It has been found that hydrophobic core and surface of drug carrier systems are better recognized by macrophages. Hydrophobic surface of the drug carrier system hastens binding of blood proteins that ultimately facilitate the process of opsonization and subsequent phagocytosis. This is the primary reason for PEGylation of hydrophobic delivery systems intended for longer systemic circulation (anticancer therapy), in order to prevent their opsonization and systemic clearance (Ozcan et al., 2010). Although hydrophobicity is a required criteria, but it has also been seen that phagocytosis occurs at an optimum hydrophobicity level and not just at an aggregate manner. pH is another vital factor dictating the therapeutic effect of drug and carrier systems in macrophage targeting. It is well known that pH in the macrophages (infected with parasites) is acidic, hence selection of polymeric/lipid-based systems that could be degraded in a sustained manner at that pH greatly influences release of drug. Even mixtures of polymeric systems that could control release of drug at acidic pH are being investigated in macrophages, and it has been observed that such system could significantly alter the therapeutic outcome. pH inside and outside the macrophages, i.e., in cytosol, lysosomes or extracellular fluid, are different. Hence, it creates challenges to design such a carrier system that could perform desired activity at required sites (Kunjachan et al., 2012). The facets discussed till now fall in the purview of passive targeting, whereas factors such as ligand selection, ligand density on a carrier, play more specific role when macrophages are actively targeted. Table 13.1 in the subsequent section presents a compendium of ligand-targeted drug delivery carriers for treatment of leishmaniasis.

13.6 MACROPHAGE TARGETING IN LEISHMANIASIS

Anti-leishmanial drugs are known to become infective against the causative pathogens. So their delivery in a very specific and targeted manner is highly required to circumvent the current issues. Novel techniques for the delivery of anti-leishmanial drugs via liposomes, nanoparticles, and other carrier systems have been developed and are found to be suitable in treatment of leishmaniasis in cell lines and animal models. Macrophage targeting of these drugs are also investigated by researchers in order to enhance the thera-peutic potential. Amphotericin Bis the standard gold drug for treatment of visceral leishmaniasis and it is commercially available as both conventional and novel drug delivery based formulations. The novel formulations were claimed to reduce its well-known side effect: nephrotoxicity. However, there is controversy on these aspects whether there is a significant reduction in the toxicity associated with amphotericin B in its commercial formula-tions. Apart from the toxicity issues, the stability of formulations and their augmentation in efficacy has been challenged. Our research group have been working on novel drug delivery of anti-leishmanial drugs, and we have reported some promising delivery tools for treatment of leishmaniasis. We have developed a novel formulation of amphotericin B, i.e., chitosan polymer based nanocapsule by employing polymer deposition technique. The developed nanocapsule was stable, and chitosan was found to stabilize the formulation against any stress. *In vitro* toxicity study in J774A.1, cell line and erythrocytes showed the formulation to be comparatively less toxic than the marketed formulations. *In-vitro* anti-leishmanial activity study revealed very lower IC_{50} by the nanocapsules in macrophage-amastigote system. Our developed formulation exhibited higher *in vitro* internalization in macro-phages that the plain drug. Further nanocapsules containing amphotericin B were evaluated in hamsters infected with *L. donovani*, and a very encour-aging result ($86.1\% \pm 2.08\%$ parasite inhibition) was seen. To understand the molecular activities for such experimental findings, RT-PCR was employed,

TABLE 13.1 A Brief Account of Research Work Showing Receptor Targeted Drug Delivery Systems to Macrophage in Leishmaniasis

Drug	Formulation	Ligand	Receptor	Research outcomes and inference	Reference
Amphotericin B	Nanoparticles	1,2-diacyl-sn-glycero-3-phospho-l-serine	Scavenger, CD68, CD14, annexins, b2 glycoprotein I and GAS6	Ligand anchored formulation showed 1.20- and 1.69-fold higher *in vivo* efficacy in comparison to uncoated formulation and pure drug suspension.	Khatik et al., 2014
Amphotericin B	Lipo-polymersomes	Lectin	N-acetyl glucosamine) receptors	Ligand anchored formulation exhibited 1.18-, 1.26- and 1.39-fold higher *in vivo* efficacy in comparison to uncoated formulation, AmBisome, and Fungizone, respectively.	Gupta et al., 2014
Amphotericin B	Dendrimers	Mannose	Mannose receptor	Higher distribution of drug in liver and spleen. Ligand coated formulation resulted 1.31-and 1.84-fold higher *in vivo* efficacy in comparison to AmBisome and pure drug suspension, respectively.	Jain et al., 2015
Amphotericin B	Lipo-polymersomes	Glycol chitosan	Glucosamine carbohydrate receptors	Glycol chitosan conjugate formulation could bring 1.06- and 1.18- fold higher *in vivo* efficacy in comparison to AmBisome and Fungizone, respectively.	Gupta et al., 2014
Amphotericin B	Nanoparticles	Sodium Alginate	Fc and neuropilin–1 receptor	Sodium Alginate cross-linked formulation showed 1.32- fold higher *in vivo* efficacy in comparison to pure drug suspension.	Gupta et al., 2015
Amphotericin B	Nanoparticles	Mannose	Mannose receptor	Ligand anchored AmB formulation reduced 5.4-fold IC50 against infected J774.A.1 cells in comparison with pure drug.	Nahar and Jain, 2009

TABLE 13.1 *(Continued)*

Drug	Formulation	Ligand	Receptor	Research outcomes and inference	Reference
Amphotericin B	Nanoparticles	Mannose	Mannose receptor	Ligand anchored by PEG spacer formulation shoed 1.09-, 1.17-, 1.04- and 1.42-fold higher *in vitro* efficacy in comparison to ligand anchored formulation, uncoated formulation, AmBisome, and drug suspension, respectively.	Nahar et al., 2010
Amphotericin B	Carbon nanotubes	Mannose	Mannose receptor	Enhanced macrophage uptake as well as higher distribution of drug in liver and spleen.	Pruthi et al., 2012
Amphotericin B	Emulsomes	O-palmitoylmannan	Mannose receptor	Ligand anchored formulation exhibited 1.43- and 2.42-fold higher efficacy in comparison to uncoated formulation and Amphotericin B deoxycholate, respectively.	Gupta et al., 2007
Amphotericin B	Liposomes	Palmitoyl mannose and 4-SO4GalNAc	Mannose receptors	Enhanced macrophage uptake and higher distribution of drug in liver and spleen.	Singodia et al., 2012
Amphotericin B	Liposomes	p-aminophenyl-α-D-mannosid	Mannose receptors	Ligand anchored liposomal formulation resulted in 1.29- and 1.85-fold higher *in vivo* efficacy in comparison to liposomal Amphotericin B formulation and suspension.	Rathore et al., 2011
Amphotericin B	Liposomes	Tuftsin	Fc-receptor	Ligand anchored liposomal formulation showed 1.11- and 2.05-fold higher *in vivo* efficacy in comparison to liposomal formulation and drug suspension.	Agrawal et al., 2002
Amphotericin B	Nanoparticles	4-sulfated N-acetyl galactosamine	Mannose receptor	Ligand anchored chitosan nanoparticle exhibited 1.18- and 1.58-fold higher *in vivo* efficacy in comparison to chitosan nanoparticle and drug solution.	Tripathi et al., 2015

TABLE 13.1 *(Continued)*

Drug	Formulation	Ligand	Receptor	Research outcomes and inference	Reference
Doxorubicin	Nanocapsules	Chondroitin sulfate	Mannose receptor	Ligand coated formulation brought about 1.42-fold higher efficacy than pure drug.	Chaurasia et al., 2015
Doxorubicin	Nanocapsules	Sodium alginate	Fc and neuropilin–1 receptor	Ligand coated formulation exhibited 1.67-fold higher efficacy than pure drug.	Kansal et al., 2014
Doxorubicin	Nanocapsules	Phosphatidyl-serine	CD68, CD14, annexins, b2 glycoprotein I and GAS6	Ligand coated formulation could result in 1.99-fold higher efficacy than pure drug.	Kansal et al., 2012
Doxorubicin	Microparticles	Chitosan	Glucosamine carbo-hydrate receptors	Ligand coated formulation exhibited 2.35-fold higher efficacy than pure drug.	Kunjachan et al., 2011
Doxorubicin	Microparticles	Mannan	Mannose receptor	Mannan anchored formulation displayed 1.69- and 2.85-fold higher efficacy in comparison to the uncoated formulation and pure drug.	Sharma et al., 2011
Doxorubicin	Liposomes	Mannose	Mannose receptor	Mandoxosome exhibited 12-fold higher efficacy than pure drug.	Kole et al., 1999
Doxorubicin	-	Mannose-human serum albumin	Mannose receptor	Mannose-human serum albumin conjugated drug resulted in 95% inhibition.	Sett et al., 1992

and the quantitative mRNA analysis was performed. Results suggested that there was up-regulation of TNF-α, IL–12, and nitric oxide synthase whereas TGF-β, IL–10, and IL–4 were down-regulated. The overall research findings put forward that our developed amphotericin B bearing nano-capsules could perform well against leishmaniasis and can be further explored for other pre-clinical and clinical investigations (Asthana et al., 2013).

Leishmania pathogens are known to reside on the macrophages, and hence they are the obligate intracellular parasites. We further hypothesized that macrophages targeting by our previously developed nano-capsules could be improved by grafting mannose as a ligand. Additionally, we also investigated whether our developed system had any contributory effect in enhancing the immune response of host. Based on our assumptions we developed mannose grafted chitosan nanocapsules and entrapped amphotericin B into the same. After optimizing the nanocapsules, they were examined for various evaluation parameters like uptake in macrophages (J774A.1), immuno-modulatory activity, and anti-leishmanial activity in hamsters. Safety studies were also done in animal models and simultaneously bioavailability, and tissue localization was also estimated. Our optimized formulation exhibited particle size; 197.8±8.84 nm, PDI; 0.115±0.04, positive zeta potential; +31.7±1.03 mV. The encapsulation efficacy was good (97.5±1.13%) and was attributed as a factor responsible for higher therapeutic benefit. Fluorescence microscopy showed higher cellular uptake of mannosylated chitosan nanocapsules in comparison to chitosan nano-capsules. Tissue localization studies revealed higher accumulation of the formulation in liver and spleen rather than in kidney (Asthana et al., 2015).

Leishmaniasis has been treated for cure by various chemotherapeutic interventions, however, along with additional chemotherapy immunotherapy is also thought to be beneficial in potentiating antigen presenting cells to kill pathogens. In order to realize this concept of treatment in leishmaniasis, we hypothesized selection of immuno-modulatory agents that can exert desired immune response which can also be combined with any standard anti-leishmanial drug. It is well known that stearylamine pattern recognition receptors are overexpressed by macrophages infected with Leishmania parasites. By exploiting this information, we selected stearylamine as a ligand for our study that could not only target the stearylamine pattern recognition receptors but at the same time could also target the phosphatidylserine on the macrophage surface. Furthermore, stearylamine is reported to display biocompatibility, anti-protozoan, and immuno-modulatory activity. The whole idea was to develop a nanocarrier system with amphotericin B and stearylamine which would sensitize the antigen presenting cells and T-cells co-ordination to enhance production of TNF-α, nitric oxide, IL–12 as well as ROS. In order to devise a suitable and functional delivery system, we developed hybrid nanoparticles composed of lipid and polymer. The advantage of the developed system was that it possessed the characteristic features of both liposomes and nanoparticles which could together enhance the therapeutic efficiency of amphotericin B. It was observed that the developed nanoformulation formed a core-shell type assembly that could encapsulate higher amount of amphotericin B in its hydrophobic core. Developed nanoformulation showed higher cellular uptake in macrophages and consequently high therapeutic efficacy. Additionally, the delivery system was found to potentiate Th–1biased immune-alteration. Toxicity studies revealed low accumulation of drug in kidney tissues which was further supported by low level of BUN and creatinine (Asthana et al., 2015).

Mono-therapy in conventional dosage regimen against leishmaniasis has been found to be responsible for development of drug resistance. In order to circumvent this problem, researchers are also working on combination therapy. We have investigated combinations of amphotericin B with many suitable drugs that can enhance therapeutic efficacy. Doxorubicin, a well-known anticancer drug is reported to show anti-leishmanial activity, but it is also associated with undesired side effects viz. cardiomyopathy, typhlitis, and other common side effects of anti-cancer drugs. Therefore, we used low dose of doxorubicin along with amphotericin B for treatment of leishmaniasis. For the development of a novel carrier system chitosan, a linear polyamine was chosen as it is reported to target glucosamine-like receptors. From formulation development aspect also it is a suitable excipient which can be conveniently cross-linked and can be molded into particulate entities by affecting its desolvation in an anti-solvent, which in most instances is regular water. Chitosan was therefore employed as a decorative surface material for targeting macrophages. A brief compendium of research work for macrophage targeting in leishmaniasis is summarized in Table 13.1.

13.7 MACROPHAGE TARGETING TO HEPATIC DISEASES

Hepatic diseases are one of the growing, and serious issues as liver is the most functional and important organ in living human beings. Hepatic inflammation, hepato-carcinogenesis, fibrosis, and non-alcoholic fatty liver are amongst the more complicated liver diseases. Regardless of the region of liver affected, macrophages play a pivotal role in its progression as well as suppression (Tacke, 2017). As explained in the previous section, kupffer cells in liver could efficiently up-take various drug carriers. In a study by Ergen and fellow researchers, it was observed that 10–50% of newer formulations viz. microbubbles (2μ), liposomes (100 nm) and polymeric nanoformulations (10 nm) could be taken up by liver especially by kupffer cells (Ergen et al., 2017). Further decorating the nanoformulations with ligands specific to overexpressed receptors on kupffer cells has been found to be effective in targeting liver macrophages. Franssen et al. have worked on development of liver-targeted anti-inflammatory drugs by combining naproxen with human serum albumin and neoglycoproteins, galactose, and mannose terminated HSA (Franssen et al., 1993). They found SRs to be responsible for the uptake of developed drug-conjugates by liver endothelial cells. However, conjugation of naproxen to lactosaminated and mannosylated HAS, surprisingly shifted uptake to hepatocytes and Kupffer cells instead of endothelial cells.

Another study by Melgert et al. shows kupffer cells targeting of anti-inflammatory drug dexamethasone (Melgert et al., 2001). The researchers realized that kupffer cells are one of the vital players in pathogenesis of inflammatory liver diseases, which if not taken care of may lead to fibrosis. Accordingly, they have coupled mannosylated albumin with dexamethasone and examined its targeting efficiency to kupffer cells in liver employing organ cultures as well as fibrosis induced by bile duct ligation in rats. Their experimental findings suggested that the mannosylated albumin-dexamethasone conjugate could be selectively taken-up by kupffer cells in fibrotic and healthy rats. Significant reduction in intrahepatic ROS in animals and decrease in TNF-alpha production in *in-vitro* conditions supported better targeting of drug to target cells. So this combination could be treated as a suitable liver macrophage targeting alternative for drug delivery. Apart from inflammatory liver diseases, non-alcoholic fatty liver diseases are also one of the leading causes of death in human beings with chronic liver diseases. This has risen because of bad food habits such as consumption of high caloric carbohydrates which may also lead to non-alcoholic steatohepatitis. Researchers have shown that lipogenetic effects of fructose (high caloric intake) and activation of liver macrophages due to endotoxins are major reasons for development and progression of these liver diseases. In light of these findings, Svendsen et al. investigated kupffer cell targeting by using an anti-CD163-IgG-dexamethasone conjugate (Svendsen et al., 2017). They intended to target the hemoglobin scavenger receptor CD163 in Kupffer cells. A low dose of anti-CD163-IgG-dexamethasone was administered to rats on a high-fructose diet, and within few weeks a significant reduction in hepatocyte ballooning, glycogen deposition, inflammation as well as fibrosis was observed. Dexamethasone alone or dexamethasone conjugated to control-IgG could not produce such therapeutic effect in rats.

Macrophages involved in fibrotic disease including liver fibrosis are well known to overexpress Galectin–3 protein (gal–3 receptors) which has the ability to bind terminal galactose residues in glycoproteins. However, it is also reported that expression of gal–3 receptors is very low in normal liver. Taking this fact into consideration Traber and team hypothesized that targeting the gal–3 receptors in liver macrophages would be a suitable strategy in delivery drug to liver macrophages (Traber and Zomer, 2013). In continuation to their hypothesis, researchers employed GM-CT–01 and GR-MD–02 as inhibitors for treatment of fibrosis in murine model. As per their findings, these inhibitors could efficiently target liver macrophages and reduce gal–3 activated macrophage-related complications as well as hepatocyte ballooning, collagen deposition, inflammatory infiltration at intra-portal and intra-lobular level and fat accumulation. In various physiological conditions, activation of macrophages leads to release of pro-inflammatory cytokines like TNF which can aggravate progression of liver diseases. Under such conditions, anti-TNF antibody (e.g., infliximab) is given, but it suffers from lack of specificity and in many cases can immunocompromise the treated patient causing development of bacterial infections. So a target-specific strategy is needed, and delivery of siRNA against TNF has been found to be the best option. He et al. have developed mannose-modified trimethyl chitosan-cysteine (MTC) conjugate nanoparticles (He et al., 2013). They have employed TNF-α siRNA for gene silencing of TNF, and the nanoformulations were developed by ionic gelation technique. This unique delivery system was equipped with trimethyl, thiol, and mannose groups along with a chitosan backbone, which were altogether programmed in such a way that they could be activated at different time as well as environmental conditions ensuring the protection of TNF-α siRNA and delivery at target site. The researchers found that that the delivery system could efficiently deliver TNF-α siRNA to liver macrophages via a clathrin-independent endocytosis. Similar effects were also observed after oral administration in mice as they were protected against acute hepatic injury caused by inflammation-induced liver damage.

13.8 MACROPHAGE TARGETING TO PULMONARY DISEASES

Tuberculosis, aspergillosis, and pneumonia are the most common pulmonary diseases which are caused by a number of pathogens (bacteria and fungi) leading to intracellular infection. In many cases, the inhaled substances are cleared via phagocytosis. Yet, some pathogens are known to be smart and can survive the clearance mechanism adopted by alveolar macrophages in phagocytosis process. In such instances, the alveolar macrophages become home to pathogen, where not only they reside but also multiply and finally come out to cause further devastation. As a result, it has become very challenging to treat such intracellular infections by conventional chemotherapy. Microorganisms'viz. *M. tuberculosis, S. aureus* are mostly known to infect lung and reside inside the alveolar macrophages, which could protect them from chemotherapeutics. In recent years nanocarriers have been in boom for their ability to target macrophages and kill the pathogens effectively. They have added advantages of being passively targeted to macrophages owing to their nanometer size, and also their surface decoration with ligands have made them even more efficient in targeting specific macrophages. Rifampicin loaded PLGA microspheres have been developed by Diab et al. as inhalable formulation for better therapeutic effect. They have employed sucrose palmitate as chief excipient which is a biocompatible sucrose ester. The *ex vivo* uptake study in rat alveolar macrophages exhibited that the drug-loaded microparticles were efficiently internalized and accumulated in the perinuclear area. The uptake was 7-fold higher in case of rifampicin loaded PLGA microspheres than free drug (Diab et al., 2012). In a similar investigation Park and team formulated glutaraldehyde cross-linked chitosan microspheres and loaded ofloxacin as an anti-tubercular drug (Park et al., 2013). Researchers have employed water-in-oil emulsification method for microparticles preparation and could produce particles of 1–6 μm size. The uptake studies in alveolar macrophages (NR8383) revealed more than 3.5-fold higher uptake of developed microparticles than free drug after 24 h of incubation. The higher availability of inhalable microparticles containing drug is expected to exert better anti-tubercular action. An interesting study by van Noort et al. presented macrophage delivery of heat-shock protein alpha B-crystallin (an external protein) via porous PLGA microparticles for effective treatment of lung inflammation (van Noort et al., 2013). Researchers have shown that the protein loaded microparticles could activate macrophages through endosomal CD14 and Toll-like receptors, which was 100-fold more effective than free protein. *In vivo* study in mice with cigarette smoke-induced, COPD showed accelerated uptake of loaded PLGA microparticles to alveolar macrophages after intratracheal administration. They could conclude that heat-shock protein alpha B-crystallin via a porous PLGA microparticles might be an alternative for lung- inflammatory diseases. A novel liposomal formulation containing vancomycin for treatment of pneumonia was developed by Pomerantz and group (Pumerantz et al., 2011). They have targeted alveolar macrophage-specific delivery as conventional vancomycin in standard solution was found to be ineffective in curing pneumonia. Two different liposomal formulations were prepared, and the difference was surface modification by PEG coating. Researcher could find that the non-PEG coated liposome could effectively deliver vancomycin into the alveolar macrophages and kill the target pathogen Meticillin-resistant *Staphylococcus aureus*. However, the vancomycin in standard solution and PEGylated liposome bearing vancomycin was not effective. Table 13.2 gives a brief account of macrophage-targeted drug delivery systems in tuberculosis.

13.9 MACROPHAGE TARGETING TO CARDIOVASCULAR DISEASES

Macrophage targeting has been investigated in different cardiovascular diseases, especially the atherosclerotic cardiovascular disease which has assumed threatening proportions, as it grows without any prominent symptoms until serious blockade occurs in the vessels. Biological understanding of atherosclerosis reveals that as the atherosclerotic lesions grow, so does the accumulation of macrophages on arterial walls. SRs (CD 36) on surface of such macrophages are responsible for uptake of LDL which leads to progression of lesion (Bobryshev et al., 2016). Nie et al. have recognized CD 36 receptors to be an important target for detecting atherosclerotic lesions and inflammatory state. They have adopted a novel non-invasive technique for this purpose and developed CD36-targeted nanovesicles. Liposomes were formulated by taking soy phosphatidylcholine and then those vesicles were decorated with 1-(Palmitoyl)–2-(5-keto–6-octene-dioyl) phosphatidylcholine. *In vitro* examination in macrophages showed a very high selective uptake of nanovesicles. In vivo findings of the study revealed that nanovesicles were selectively taken up by aortic lesions. Authors claimed that this delivery system would be a suitable tool for early detection of atherosclerotic lesions via a non-invasive imaging technique (Nie et al., 2015).

For treatment of atherosclerotic plaques, Wennink and group have explored a novel strategy of photodynamic therapy. The idea was to remove the macrophages from the plaques that can help in development of smooth muscle on the vessel walls. Photodynamic therapy has basically three different components viz. the photosensitizer (non-aggregated), light source and tissue oxygen. When exposed to light source of suitable wavelength, the photosensitizer gets excited, and while returning to its ground state, it

TABLE 13.2 A Brief Compilation of Research Showing Receptor Targeted Drug Delivery to Macrophage Against Tuberculosis

Drug	Formulation	Ligand	Receptor	Research outcomes and inference	Reference
Isoniazid	Nanoparticles	Mannose	Mannose receptor	Developed ligand anchored formulation showed almost 2.5-fold reduction in bacterial counts in comparison to free drug.	Saraogi et al., 2011
Rifampin and Gallium (III)	Nanoparticles	Mannose and folic acid	Mannose and folate receptor	Ligand anchored formulation showed higher efficacy than pure drug.	Choi et al., 2017
Isoniazid and ciprofloxacin HCl	Liposomes	4-aminophenyl-α-D mannopyranoside	Mannose receptor	Higher drug accumulation achieved in lung by delivering ligand anchored liposomes as compared to conventional liposome.	Bhardvaj et al., 2016
Rifampicin	Micelles	Hyaluronic acid-tocopherol succinate	CD44 receptor	Ligand anchored micelles showed significantly higher uptake in comparison to free drug. Ligand anchored micelles induced higher concentration of Th1 cytokines than free drug, which ultimately helped in enhancing the anti-tubercular activity.	Gao et al., 2015
Rifampicin	Nanoparticles and Micelles	Hydrolyzed galacto-mannan (GalM-h)	Lectin-like receptors	Ligand anchored micelles showed significantly higher uptake in comparison to uncoated nanoparticles.	Moretton et al., 2013
Isoniazid	Nanoparticles	Mycolic acids	Scavenger receptors	Ligand anchored micelles showed significantly higher uptake in comparison to uncoated nanoparticles.	Lemmer et al., 2015
Isoniazid	Microspheres	Mannose	Mannose receptor	Ligand anchored formulations exhibited significantly higher uptake as compared to drug solution.	Tiwari et al., 2011
Rifampicin	Liposomes	Maleylated bovine serum albumin and o-steroyl amylopectin	Scavenger receptors	Ligand anchored formulations resulted in significantly higher distribution in lungs in comparison to free drug.	Vyas et al., 2004
Pyrazinamide	Nanoparticles	-	-	Ligand anchored formulations demonstrated significantly higher uptake as compared to drug solution.	Varma et al., 2015

TABLE 13.2 (Continued)

Drug	Formulation	Ligand	Receptor	Research outcomes and inference	Reference
Rifampicin	Microspheres	-	-	Microspheres of rifampin significantly reduced CFU compared to equivalent doses delivered as free drug.	Barrow et al., 1998
Isoniazid and rifampicin	Microparticles	-	-	Higher distribution of drug in targeting site in comparison to plasma.	Sharma et al., 2001
Isoniazid and rifampicin	Nanoparticles	-	-	Antimicrobial activity on day 1 and 5 with nanoparticles formulations was 3- and 5-fold greater than for native drug.	Edagwa et al., 2014
Rifabutin	Solid lipid nanoparticles	Mannose	Lectin-like receptors	Mannosylated formulation was found to be greatest in lungs and lower in liver and kidney.	Nimje et al., 2009
Rifampicin	Dendrimers	Mannose	Lectin-like receptors	Mannosylated dendrimer exhibited higher internalization in alveolar macrophages in comparison to free drug.	Kumar et al., 2006
Rifampicin, isoniazid, and pyrazinamide	Nanoparticles	-	-	No visible growth of *M. tuberculosis* occurred till day 25 in mice receiving free drugs (daily) and drug-loaded PLG-NP (every 10 days) with undiluted, 1:10 diluted or 1:100 diluted homogenates. However, control animals revealed a bacterial count of 4.72±0.05 and 4.73±0.03 cfu/ml in lung and spleen homogenates.	Pandey et al., 2003
Rifampicin, isoniazid, and pyrazinamide	Nanoparticles	-	-	No visible cfu growth appeared in the case of subcutaneous drug loaded formulation. However, the daily administration of oral free drugs for 5 weeks resulted in a 2.2 log10 cfu reduction ($P<0.001$ as compared with the untreated controls).	Pandey and Khuller, 2004
Rifabutin	Microparticles	β-glucan	Dectin–1 receptors	Ligand anchored formulations exhibited significantly higher uptake as compared to drug solution.	Upadhyay et al., 2017

TABLE 13.2 *(Continued)*

Drug	Formulation	Ligand	Receptor	Research outcomes and inference	Reference
Rifampicin and isoniazid	Microparticles	Mannose	Mannan receptors	Guar gum coated chitosan formulation showing 5 fold higher efficacy than control group.	Goyal et al., 2015
Rifampicin and isoniazid	Nanoparticles	Guar gum	Mannan receptors	The minimum inhibitory concentration of drug solution (135 ± 5.4 g/ml) was approximately 15–16 times higher than that of drug-loaded nanoparticles (chitosan nanoparticles 16.86 ± 0.9 g/ml and guar gum coated chitosan nanoparticles 8.53 ± 1.8 g/ml)	Goyal et al., 2016
Rifampicin and isoniazid	Nanoparticles	-	-	Preferential localized accumulation of developed formulation in lungs.	Garg et al., 2016
Rifampicin and isoniazid	Nano-aggregates	Guar gum	Mannan receptors	Higher distribution of formulation in lungs in comparison to pure drug.	Kaur et al., 2016

releases energy. The released energy interacts with the surrounding tissue oxygen and generates singlet oxygen that further causes cell death. This novel technique although effective in the treatment of atherosclerotic plaques, however, does produce some unwanted effects like activation at non-desirable sites due to aggregation of photosensitizer that may cause severe damage to host cells. In order to circumvent this problem nanocarriers have been employed which can take advantage of the enhanced permeability and retention effect to accumulate specifically in the target sites and exhibit a better effect. Wennink and team used m-tetra (hydroxyphenyl) chlorin, a photosensitizer which is currently used in the clinic for photodynamic cancer therapy and incorporated the same in polymeric micelles composed of benzyl-poly(ε-caprolactone)-b-methoxy poly(ethylene glycol) (Ben-PCL-mPEG). These materials have the special property of undergoing degradation only in an intra-macrophage environment, whereas it remains relatively intact in endothelial cells, hence the micelles could release actives at a faster rate and cause photo-cytotoxicity of specifically macrophages. *In vivo* studies also revealed higher accumulation of active medicament in the atherosclerotic lesions of mice aorta (Wennink et al., 2017).

Myocardial infarction is one of the main of cause of deaths in developed as well as developing countries, due to increasingly sedentary lifestyle, bad food habits, and lack of physical exercise as well as stressful life. Myocardial infarction is associated with cardiac remodeling where alterations in the cardiac structure and functions, as well as geometry, is observed. These alterations cause increase in wall stress that finally leads to heart failure. Macrophages have a role in cardiac remodeling, and they participate in the healing and repair of myocardial infarcts. Being major players in inflammation, these immune cells produce pro-/anti-inflammatory factors apart from participating in neo-angiogenesis and granulation tissue development. In case of acute myocardial infarction, the transition from pro-inflammatory macrophage to a reparative macrophage is impaired which results in hyperactivity of the earlier macrophage. The hyperactivation of pro-inflammatory macrophage is undesirable for cardiac patients and hence modulating the inflammatory macrophage to a reparative macrophage has been attempted (Gombozhapova et al., 2017). Ben-Mordechai et al. have hypothesized that the healing and repair of myocardium might be improved by activation of heme oxygenase–1 (HO–1) which is a known enzyme with anti-inflammatory and cytoprotective properties in macrophages. For this purpose, they have employed hemin, an iron-containing porphyrin and incorporated it in a lipid-based nano-carrier system. The results observed by researchers were interesting where the delivery system was found to inhibit secretion of TNF-α from the target macrophages and also it could polarize the macrophages of peritoneal and splenic origin into anti-inflammatory phenotype. Further in the *in vivo* set-up, authors found remarkable improvement in angiogenesis, reduction of scar expansion as well as positive development in infarct-related regional function (Ben-Mordechai et al., 2017).

Experimental autoimmune myocarditis is related to the inflammatory cardiomyopathy leading to heart failure. In autoimmune myocarditis, macrophages play a vital role. It has been verified that CD 68[+] macrophages are involved in this disorder and androgen receptors are found to be responsible for exerting pro-inflammatory reaction of macrophages. Considering the role of androgen receptors, Ma et al. have worked on macrophage targeting with an AR degradation enhancer ASC-J9 and they are *in vivo* findings suggested reduction in severity of autoimmune myocarditis, reduced macrophage infiltration as well as drop-in pro-inflammatory cytokines. Additionally, it was also observed that the treatment could restrict polarization of Raw264.7 cells to pro-inflammatory phenotype macrophages (Ma et al., 2017).

13.10   CONCLUSION AND FUTURE DIRECTIONS

Leishmaniasis and visceral diseases are one of the foremost cause of morbidity and mortality in human beings. Conventional chemotherapeutics after their initial promise in treatment of these diseases have been struggling. Their irrational use has resulted in development of drug resistance. The associated adverse effects had already been an alarming issue, which had contributed majorly to limited success of exercised therapies. Consequently, alternative new drugs, novel application of drug delivery techniques as a part of drug-discovery program have been started by healthcare sectors over the globe. Nano-/micro-formulations have been in focus for efficiently carrying drug molecules to the target site. Unbuckling of the fact that macrophages are central participants in the pathophysiology of diseases as well as a reservoir for multiple pathogens, has ushered in the aggressive approach of specifically targeting these immune cells only. Passive as well as active targeting of drugs by taking advantages of pathological conditions of diseases and pharmacokinetic peculiarities of nanoformulations has shown potential therapeutic benefit at pre-clinical levels. Ligand-based receptor targeting to macrophages is another emerging aspect in the treatment of leishmaniasis and other visceral diseases. Delivery of synthetic drugs, as well as proteins and peptides to infected macrophages, is also getting serious attention.

The future prospects of targeting macrophages solely rely on the cutting-edge understanding of host-pathogen interactions in a disease condition. Acquiring and validating the survival strategies adopted by pathogens at intracellular levels should be heavily investigated. Identification of novel proteins or other biological components that have a key role in the progression of diseases is another challenge for molecular biologists that can open avenues for the development of novel ligands as well as drugs. Apart from these, a detailed compendium of the toxic effects of nanomaterials in macrophage targeting should also be compiled. With changes in global climatic pattern, ongoing migration crisis instigated by geopolitical and climatic disturbances leishmaniasis has been found in places where it never used to occur previously. Open involvement of various private and government organizations via public-private partnership and adequate research funding as well as political will is consequently required for eradicating such diseases.

KEYWORDS

- **leishmaniasis**
- **macrophage**
- **phagocytosis**
- **receptor targeting**
- **visceral diseases**

REFERENCES

Abe, Y., et al., (2008). Contribution of complement component C3 and complement receptor type 3 to carbohydrate-dependent uptake of oligomannose-coated liposomes by peritoneal macrophages. *J. Biochem., 144*, 563–570.

Acevedo, J., (2015). Multiresistant bacterial infections in liver cirrhosis: Clinical impact and new empirical antibiotic treatment policies. *World J. Hepatol., 7*, 916–921.

Ahsan, F., et al., (2002). Targeting to macrophages: Role of physicochemical properties of particulate carriers-liposomes and microspheres-on the phagocytosis by macrophages. *J. Control. Release, 79*, 29–40.

Alexis, F., et al., (2008). Factors affecting the clearance and biodistribution of polymeric nanoparticles. *Mol. Pharm., 5*, 505–515.

Al-Jamal, K. T., (2013). Active drug targeting: Lessons learned and new things to consider. *Int. J. Pharm., 454*, 525–526.

Asthana, S., et al., (2013). Immunoadjuvant chemotherapy of visceral leishmaniasis in hamsters using amphotericin B-encapsulated nanoemulsion template-based chitosan nanocapsules. *Antimicrob. Agents Chemother.*, *57*, 1714–1722.

Asthana, S., et al., (2015a). Overexpressed macrophage mannose receptor targeted nanocapsules- mediated cargo delivery approach for eradication of resident parasite: *In vitro* and *in vivo* studies. *Pharm. Res.*, *32*, 2663–2677.

Asthana, S., et al., (2015b). Th–1 biased immunomodulation and synergistic antileishmanial activity of stable cationic lipid-polymer hybrid nanoparticle: Biodistribution and toxicity assessment of encapsulated amphotericin B. *Eur. J. Pharm. Biopharm.*, *89*, 62–73.

Barquera, S., et al., (2015). Global overview of the epidemiology of atherosclerotic cardiovascular disease. *Arch. Med. Res.*, *46*, 328–338.

Ben-Mordechai, T., et al., (2017). Targeting and modulating infarct macrophages with hemin formulated in designed lipid-based particles improves cardiac remodeling and function. *J. Control. Release*, *257*, 21–31.

Bobryshev, Y. V., et al., (2016). Macrophages and their role in atherosclerosis: Pathophysiology and transcriptome analysis. *BioMed. Res. Int.*, *2016*, 9582430.

Cazzola, M., & Skoda, R. C., (2000). Translational pathophysiology: A novel molecular mechanism of human disease. *Blood*, *95*, 3280–3288.

Champion, J. A., et al., (2008). Role of particle size in phagocytosis of polymeric microspheres. *Pharm. Res.*, *25*, 1815–1821.

Chaudhuri, G., (1997). Scavenger receptor-mediated delivery of antisense mini-exon phosphorothioate oligonucleotide to Leishmania-infected macrophages. *Biochem. Pharm.*, *53*, 385–391.

Diab, R., et al., (2012). Formulation and *in vitro* characterization of inhalable polyvinyl alcohol-free rifampicin-loaded PLGA microspheres prepared with sucrose palmitate as stabilizer: Efficiency for *ex vivo* alveolar macrophage targeting. *Int. J. Pharm.*, *436*, 833–839.

Ergen, C., et al., (2017). Targeting distinct myeloid cell populations *in vivo* using polymers, liposomes, and microbubbles. *Biomaterials*, *114*, 106–120.

Franssen, E. J., et al., (1993). Hepatic and intrahepatic targeting of an anti-inflammatory agent with human serum albumin and neoglycoproteins as carrier molecules. *Biochem. Pharmacol.*, *45*, 1215–1226.

Fröhlich, E., (2012). The role of surface charge in cellular uptake and cytotoxicity of medical nanoparticles. *Int. J. Nanomedicine.*, *7*, 5577–5591.

Gao, J., et al., (2013). Novel monodisperse PEGtide dendrons: Design, fabrication, and evaluation of mannose receptor-mediated macrophage targeting. *Bioconjug. Chem.*, *24*, 1332–1344.

Global Tuberculosis Report (2013). Geneva: WHO. See http://apps.who.int/iris/bitstream/ 10665/91355/1/9789241564656_eng. pdf) and also (http://www.who. int/tb/publications/ global_report/gtbr13_annex_4_key_indicators. pdf. 2015.

Gombozhapova, A., et al., (2017). Macrophage activation and polarization in post-infarction cardiac remodeling. *J. Biomed. Sci.*, *24*, 13.

Handman, E., & Bullen, D. V. R., (2002). Interaction of Leishmania with the host macrophage. *Trends Parasitol.*, *18*, 332–334.

He, C., et al., (2013). Multifunctional polymeric nanoparticles for oral delivery of TNF-alpha siRNA to macrophages. *Biomaterials*, *34*, 2843–2854.

Heuking, S., et al., (2009). Stimulation of human macrophages (THP–1) using Toll-like receptor–2 (TLR–2) agonist decorated nanocarriers. *J. Drug Target.*, *17*, 662–670.

Irache, J. M., et al., (2008). Mannose-targeted systems for the delivery of therapeutics. *Expert Opin. Drug. Deliv.*, *5*, 703–724.

Jain, N. K., et al., (2013). Targeted drug delivery to macrophages. *Expert Opin. Drug Deliv.*, *10*, 353–367.

Kolios, G., et al., (2006). Role of Kupffer cells in the pathogenesis of liver disease. *World J. Gastroenterol.*, *12*, 7413–7420.

Köser, C. U., et al., (2015). Drug-resistance mechanisms and tuberculosis drugs. *Lancet*, *385*, 305–307.

Kunjachan, S., et al., (2012). Physicochemical and biological aspects of macrophage-mediated drug targeting in anti-microbial therapy. *Fundam. Clin. Pharmacol.*, *26*, 63–71.

Lamotte, S., et al., (2017). The enemy within Targeting host-parasite interaction for antileishmanial drug discovery. *PLoS Negl. Trop. Dis.*, *11*, e0005480.

Leishmaniasis in high-burden countries: an epidemiological update based on data reported in 2014 (2016). *Wkly Epidemiol. Rec.*, *91*, 287–296.

Leslie, R. G. Q., & Hansen, S., (2009). *Complement Receptors.* eLS, John Wiley & Sons, Ltd. 10.1002/9780470015902.a0000512.pub2.

Leucuta, S. E., (2014). Subcellular drug targeting, pharmacokinetics, and bioavailability. *J. Drug Target*, *22*, 95–115.

Libby, P., (2007). Inflammatory mechanisms: The molecular basis of inflammation and disease. *Nutr. Rev.*, *65*, S140–S146.

Lunshof, J. E., et al., (2006). Personalized medicine: Decades away? *Pharmacogenomics*, *7*, 237–241.

Ma, W., et al., (2017). Targeting androgen receptor with ASC-J9 attenuates cardiac injury and dysfunction in experimental autoimmune myocarditis by reducing M1-like macrophage. *Biochem. Biophys. Res Commun.*, *485*, 746–752.

Maity, A. R., & Stepensky, D., (2015). Delivery of drugs to intracellular organelles using drug delivery systems: Analysis of research trends and targeting efficiencies. *Int. J. Pharm.*, *496*, 268–274.

Masuda, A., et al., (2009). Role of Fc receptors as a therapeutic target. *Inflamm. Allerg. Drug Target.*, *8*, 80–86.

Melgert, B. N., et al., (2001). Targeting dexamethasone to Kupffer cells: Effects on liver inflammation and fibrosis in rats. *Hepatology.*, *34*, 719–728.

Nie, S., et al., (2015). Detection of atherosclerotic lesions and intimal macrophages using CD36-targeted nanovesicles. *J. Control. Release*, *220*, 61–70.

Niu, M., et al., (2014). Biodistribution and *in vivo* activities of tumor-associated macrophage-targeting nanoparticles incorporated with doxorubicin. *Mol. Pharm.*, *11*, 4425–4436.

Nuermberger, E. L., et al., (2010). Current development and future prospects in chemotherapy of tuberculosis. *Respirology.*, *15*, 764–778.

Ozcan, I., et al., (2010). Pegylation of poly(gamma-benzyl-L-glutamate) nanoparticles is efficient for avoiding mononuclear phagocyte system capture in rats. *Int. J. Nanomedicine.*, *5*, 1103–1111.

Park, J. H., et al., (2013). Chitosan microspheres as an alveolar macrophage delivery system of ofloxacin via pulmonary inhalation. *Int. J. Pharm.*, *441*, 562–569.

Pei, Y., & Yeo, Y., (2016). Drug delivery to macrophages: Challenges and opportunities. *J. Control. Release.*, *240*, 202–211.

Polando, R., et al., (2013). The roles of complement receptor 3 and Fcgamma receptors during Leishmania phagosome maturation. *J. Leukoc. Biol.*, *93*, 921–932.

Pumerantz, A., et al., (2011). Preparation of liposomal vancomycin and intracellular killing of methicillin-resistant Staphylococcus aureus (MRSA). *Int. J. Antimicrob. Agents.*, *37*, 140–144.

Raman, V. S., et al., (2010). Applying TLR synergy in immunotherapy: Implications in cutaneous leishmaniasis. *J. Immunol.*, *185*, 1701–1710.

Schnappinger, D., et al., (2003). Transcriptional adaptation of Mycobacterium tuberculosis within macrophages: Insights into the phagosomal environment. *J. Exp. Med.*, *198*, 693–704.

Shannahan, J. H., et al., (2015). Implications of scavenger receptors in the safe development of nanotherapeutics. *Receptors Clin. Investig.*, *2*, e811.

Singh, R. K., et al., (2012). Toll-like receptor signaling: A perspective to develop vaccine against leishmaniasis. *Microbiol. Res.*, *167*, 445–451.

Singh, Y., et al., (2017). Targeting tumor-associated macrophages (TAMs) via nanocarriers. *J. Control. Release.*, *254*, 92–106.

Sulis, G., et al., (2014). Tuberculosis: Epidemiology and control. *Mediterr. J. Hematol. Infect. Dis.*, *6*, e2014070.

Sundar, S., (2001). Drug resistance in Indian visceral leishmaniasis. *Trop. Med. Int. Health.*, *6*, 849–854.

Svendsen, P., et al., (2017). Antibody-directed glucocorticoid targeting to CD163 in M2-type macrophages attenuates fructose-induced liver inflammatory changes. *Mol. Ther. Methods Clin. Dev.*, *4*, 50–61.

Tacke, F., (2017). Targeting hepatic macrophages to treat liver diseases. *J. Hepatol.*, *66*, 1300–1312.

Tempone, A. G., et al., (2004). Targeting Leishmania (L.) chagasi amastigotes through macrophage scavenger receptors: The use of drugs entrapped in liposomes containing phosphatidylserine. *J. Antimicrob. Chemother.*, *54*, 60–68.

Traber, P. G., & Zomer, E., (2013). Therapy of experimental NASH and fibrosis with galectin inhibitors. *PLoS ONE.*, *8*, e83481.

Van Noort, J. M., et al., (2013). Activation of an immune-regulatory macrophage response and inhibition of lung inflammation in a mouse model of COPD using heat-shock protein alpha B-crystallin-loaded PLGA microparticles. *Biomaterials*, *34*, 831–840.

Weinberger, T., & Schulz, C., (2015). Myocardial infarction: A critical role of macrophages in cardiac remodeling. *Front. Physiol.*, *6*, 107.

Weissleder, R., et al., (2014). Imaging macrophages with nanoparticles. *Nat. Mater.*, *13*, 125–138.

Wennink, J. W. H., et al., (2017). Macrophage selective photodynamic therapy by meta-tetra(hydroxyphenyl)chlorin loaded polymeric micelles: A possible treatment for cardiovascular diseases. *Eur. J. Pharm. Sci.*, *107*, 112–125.

Wolfram, J., et al., (2015). Safety of nanoparticles in medicine. *Curr. Drug Target.*, *16*, 1671–1681.

Wynn, T. A., et al., (2013). Origins and hallmarks of macrophages: Development, homeostasis, and disease. *Nature*, *496*, 445–455.

Yasinzai, M., et al., (2013). Drug resistance in leishmaniasis: Current drug-delivery systems and future perspectives. *Future Med. Chem.*, *5*, 1877–1888.

Zhu, L., et al., (2011). Preparation and evaluation of mannose receptor-mediated macrophage targeting delivery system. *J. Control. Release*, *152*, e190–e191.

Zubairi, S., et al., (2004). Immunotherapy with OX40L-Fc or anti-CTLA-4 enhances local tissue responses and killing of Leishmania donovani. *Eur. J. Immunol.*, *34*, 1433–1440.

CHAPTER 14

Nanomedicine for the Diagnosis and Treatment of Cancer

LAXMIKANT GAUTAM, ANAMIKA JAIN, NIKHAR VISHWAKARMA, RAJEEV SHARMA, NISHI MODY, SURBHI DUBEY, and S. P. VYAS*

Drug Delivery Research Laboratory, Department of Pharmaceutical Sciences, Dr. Harisingh Gour Vishwavidyalaya, Sagar, M.P., 470003, India, E-mails: vyas_sp@rediffmail.com, spvyas54@gmail.com

*Corresponding author. E-mail: vyas_sp@rediffmail.com, spvyas54@gmail.com

ABSTRACT

Cell annexation is an intrinsic cellular pathway whereby cells reciprocate to extracellular stimuli to drift and restrain the structure of their extracellular matrix (ECM) to evolve, reconstruct, and protect the body's tissues. In cancer cells, this process becomes peculiarly (varying from the usual) regulated and lead to cancer metastasis. So for the diagnosis and treatment (theranostic) of this unwanted cellular growth, researchers have worked on many potential molecular/therapeutic targets of cancer cell invasion. We aim to provide a perspective on how the advances in cancer biology and the field of nanomedicine can be combined to offer new solutions for treating cancer metastasis. Nanomedicine is an emerging technology which gives potential solution for approaching cancer metastasis by improving the specificity and potency of therapeutics delivered to invasive cancer cells. In this chapter, we describe different types of nanomedicines such as quantum dots (QDs), different nanoparticulate systems including silver nanoparticles, gold nanoparticles (GNPs), etc. For the diagnosis of cancer different types of testing procedures are applied to confirm the presence of disease and distinguish the correct tumor type, location, and stage. At present different types of cancer therapies are being used in the treatment of cancer like radiation therapy, chemotherapy, targeted therapy, hormone therapy, stem cell transplants, etc. It is further discussed how the field of nanomedicine can be applied to diagnose, monitor, and treat cancer cell annexation.

14.1   INTRODUCTION

Today science and technology have reached to an extent where almost every disease can be cured, but still, the rate of morbidity and mortality due to cancer is one of the major causes globally with nearly 8.2 million deaths in 2012. The annual number of new cases is projected to rise from 14.1 million in 2012 to 21.6 million in 2030 (Siegel, Naishadham, & Jemal, 2013). The foremost challenge encountered in the safe and effective treatment of cancer is difficulty in diagnosis as well as nonspecificity of the available chemotherapeutic agents towards cancer cells. The non-specificity of chemotherapeutic agents leads to the destruction of normal cells along with the cancer cells and thus leads to impairment of vital organs even at prescribed therapeutic dosages. If cancer can be diagnosed at the early stages, it is possible to treat cancer by chemotherapy. Presently, the research is largely based on the biological changes in the cellular system and the different pathological conditions of the cancer patients. The concept of theranostic nanotechnology is used to achieve a dual result, drug delivery and the monitoring of the disease (Elzoghby, Samy, & Elgindy, 2012). Theranostic systems implicate that it can be used for both diagnoses as well as in therapy of the disease which can facilitate the regimen of treatment with improved prognosis (Opoku-damoah, Wang, Zhou, & Ding, 2016). Nanotechnology-based systems have been already investigated for the targeted delivery of drug molecules (Rahman, 2016, 2017; Beg, 2017). Thus, nanosystem based theranostics enable us to deliver the therapeutic agents to a specific site as well as it will also promote the diagnostic and imaging functions (Figure 14.1) (Xie, Lee, and Chen, 2010). In addition, this approach also offers to provide an effective determination of adverse side effects in the starting phase of therapy and drug response monitoring (Theranostic Nanomedicine for Cancer, 2008). These theranostic nanosystems possess some advantages and disadvantages, which makes them suitable for use in the diagnosis and treatment of cancer. This review spotlights formulation and applications various nanosystem such as polymeric nanoparticles Gold nanoparticles (GNPs), Silver nanoparticles, Quantum dots (QDs), Carbon-Nanotube for theranostic, i.e., diagnostic, and therapeutic purposes (Sharma et al., 2017; Ahmad, 2015a, 2015b).

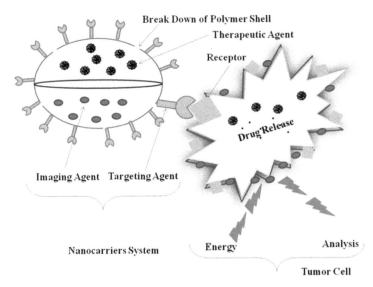

FIGURE 14.1 Carrier-mediated theranostic approaches for the cancer treatment.

Various uses of theranostic nanomedicines (Figure 14.2) include the study of pharmacokinetics, distribution of drug within the body, study of the amount of drug reaching the target site by non-invasive methods, prediction of fate of drug molecule and its effect on the body, determining therapeutic efficacy, etc. (Lammers et al., 2011).

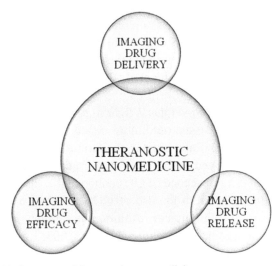

FIGURE 14.2 Various uses of theranostic nanomedicines.

Although theranostic nanomedicine has a great potential to improve the diagnosis as well as therapy, a large number of unanswered questions such as the dose of the diagnostic agent, therapeutic agents and the fate of the drug *in vivo* are yet to be answered (Table 14.1).

14.2   THERANOSTIC NANOMEDICINE IN CANCER THERAPEUTICS

14.2.1   QUANTUM DOTS (QDS)

In 1983, QDs were first characterized by the Louis E. Brus (1983). QDs also termed as colloidal semiconductor nanocrystals (NCs), contain an inorganic core of hundreds to thousands of atoms of semiconductor material, which is surrounded by an organic layer of surfactant molecules (ligands) (Reiss et al., 2009; Rahman, 2015a). They consist of spherical shape semiconductor particles with a diameter in a range of 2–10 nm. QDs show properties between bulk material and atom because its electronic structure is intermediate in-between to bulk and a single atom (Aneja, 2014). The energy gap between the HOMO and LUMO is a characteristic for QDs and semiconductors. HOMO and LUMO are the types of molecular orbital are which stand for Highest Occupied Molecular Orbital's (HOMO) and Lowest Unoccupied Molecular Orbital's (LUMO), respectively. Lowest energy state is referred to as a ground state while higher energy state is referred to as an excited state. Figure 14.3 represented the HOMO and LUMO concept QDs possess the semiconducting property and exhibit size-dependent fluorescence, which enables them to be used in the biological detection, optoelectronic devices and in the determination of size-dependent properties of nanomaterials (Smith et al., 2008; Parak et al., 2008; Rahman, 2015b; Ahmad, 2013).

The major advantage of the QDs is their use in the fluorescent and multi-modal imaging *in vivo*. Tissue autofluorescence is minimized due to high quantum yield (over 50%) and greater excitation/emission stokes shift (up to 300 nm). Timer resolved detection of fluorescence and high intensity of multiphoton excited fluorescence of QDs enhance the signal-to-background ratio in the living cells. With the use of QDs, it is easy to visualize cells and molecular targets, with greater resolution at significantly greater depth as compared to an organic fluorophore (Larson, 2003; Kosaka et al., 2010; Michalet, 2005; Rahman, 2012a) (Figure 14.4).

TABLE 14.1 Pharmaceutical Applications of Theranostic Nanomedicines in Cancer Chemotherapy

Diagnostic agents/ Contrast agents	Drug	Formulation	Cancer type	Route of administration	Outcomes	References
Cadmium telluride quantum dots (CdTe)	-	Micro-sphere	Breast cancer	-	Inhibition of autophagy by O-CdTe QDs shows effects on inducing cell cycle arrest at G2/M, leading to cell growth inhibition.	Fan, Lai, & Xie, 2016
CdS and ZnS quantum dots	-	Nano-conjugates	Common cancer	i.v.	Non-cytotoxic responses	Mansur et al., 2016
Near-infrared dye IR780	Gemcitabine	Nano-carriers	Pancreatic cancer	i.v.	The concentration of gemcitabine triphosphatewas significantly increased in tumor tissue, resulting from exhibiting superior tumor inhibition activity with minimal side effects	Han et al., 2017
Monodisperse quantum dot–antibody	-	-	Pancreatic cancer	-	Quantitative measurement for QDs analysis	Lee et al., 2012
CdSe/ZnS quantum dots	-	PAMAM dendrimers	Colon Cancer	i.v.	In vivo studies shows colloidal stability, solubility in physiological fluids, influence of the basic physiological parameters, and cytotoxic response	Bakalova et al., 2011
CdSe/ZnS quantum dots	Docetaxel	RGD-TPGS decorated theranostic liposomes	Brain cancer	i.v.	RGD-TPGS decorated theranostic li-posomes was 6.47- and 6.98-fold more effective than doceltm after 2 h and 4 h treatments	Sonali et al., 2016
Gadolinium-doped carbon quantum dots	Doxorubicin	Multi-walled carbon nanotubes	Common cancer	i.v.	100% tumor elimination was realized in irradiation at the laser power density of 2 W/cm2 and no recurrent was observed.	Zhang et al., 2017

TABLE 14.1 *(Continued)*

Diagnostic agents/ Contrast agents	Drug	Formulation	Cancer type	Route of administration	Outcomes	References
Near Infrared dye IR820	Doxorubicin	Pegylated Ormosil nanoparticles	Ovarian Cancer	i.v.	Encapsulation increased the in-vivo circulation time of IR820	Nagesetti & Mcgoron, 2017
Silver	-	Nanoparticles	Lung cancer	-	In vitro and ex vivo study performed, the dominant cytotoxic effect	Yeasmin, et al., 2017
Carbon	Doxorubicin and interleukin-6	Nanodots	Brain Cancer	i.v.	Inhibit the glioma growth and enhance the sensitivity of tumor cells towards chemotherapy	Wang et al., 2017
Fluorescent iron oxide	HER 2 Antibody	Nanospheres	Pancreatic Cancer	Intraperitoneal	Significant tumor regression and enhanced MRI in treatment groups	Jaidev et al., 2017
Gold & iron oxide	Peptide	Nanoparticles	Gastric cancer	-	In Vitro, results said multifunctional platform for MR/CT dual imaging of cancer cells	Kukreja et al., 2017
Gold	Heparin	Nanoparticles	Metastatic cancer	-	Heparin induces acute apoptosis and high potentials for optical imaging and apoptotic death of cancer cells	Lee, Lee, & Park, 2010
Superparamagnetic iron oxide	Docetaxel	Nanoparticles	Prostate cancer	i.v.	Promising multifunctional vesicles for simultaneous targeting imaging, drug delivery	Ling, Luo, & Zhong, 2011
Superparamagnetic iron oxide	Temozolomide	Nanoparticles	Malignant glioma	-	In vitro studied showed Good MRI negative enhance effect in C6 glioma -cells MR scanning	Ling, Zou, & Zhong, 2012
Super-paramagnetic iron oxide	Doxorubicin	Nanoparticle	Malignant Cancer	-	Chondroitin sulfate on receptor-mediated endocytosis and uptake of CS-capped SPIONs by selective receptors on cancer cells	Mallick et al., 2016

TABLE 14.1 (Continued)

Diagnostic agents/ Contrast agents	Drug	Formulation	Cancer type	Route of administration	Outcomes	References
Gold	-	Nanoparticles	-	-	3-mercapto–1-propane sulfonate and 1-thioglucose mixed thiols gold nanoparticles system is biocompatible with low or none cytotoxicity and potential users in well-improved radiotherapy and drug delivery applications.	Porcaro et al., 2016
Gold	-	Nanoparticles	Lung cancer	Pulmonary	Optimal biodegradation and release profiles enabled a sustained and controlled release of the embedded nanoparticles, with enhanced cellular uptake	Silva et al., 2017
Silver	-	Nanoparticles	Breast cancer	-	Green luminescence of the agents useful for diagnostic purpose while their anticancer activity and ability to disrupt bilirubin aggregates useful for therapeutic use	Mandal, Sarkar, & De, 2017
Superparamagnetic iron oxide	Hyaluronic acid	Nanoparticles	Squamous cell carcinoma	i.v.	Developed HA-SPIONs and HA-PEG10-SPIONs To compare both their T2 imaging contrast in MRI and their hyperthermia effects	Thomas et al., 2015

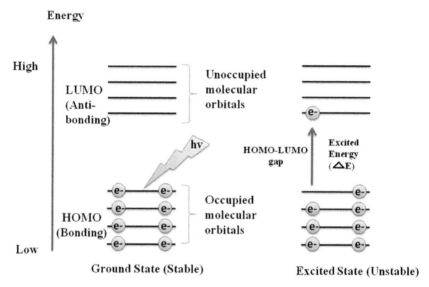

FIGURE 14.3   Representation of HOMO and LUMO state of the electron.

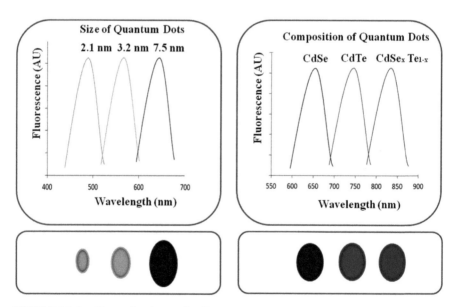

FIGURE 14.4   (See color insert.) Effects of the size and composition on the emission wavelength of QDs. (A) Change of the spectrum from 450 nm to 650 nm with a change of nanoparticle diameter from 2 nm to 7.5 nm. Below the spectrum size of the nanoparticles of the same composition is presented. (B) By changing the composition of QDs, while retain the constant size (5 nm diameter), emission vary be vary in the range of 610 nm to 800 nm (Bailey and Nie, 2003).

14.2.1.1 SYNTHESIS OF QUANTUM DOTS

Colloidal synthesis method was first introduced by Murray et al. by using organic solvents and organometallic precursors (Murray, Norris, & Bawendi, 1993) (Figure 14.5). QDs are produced by colloidal synthesis method, which is a chemically synthesized QDs suspended in a solution. Various methods are available for synthesis of specific semiconductor material. In general, an organic solvent containing the saturated solution of the semiconductor is prepared. A supersaturated solution is prepared by modulating the pH or temperature of the above solution, which forms small crystals after nucleation step (Rahman, 2012b). Size of the QDs can be modulated by varying the pH, temperature, and length of the reactions. Many of these steps can be processed in the lab by omitting the use of exotic reagents or equipment's; however precise temperature control is mandatory for producing larger quantity, therefore it is difficult to produce (Vachaspati, 2013; Ahmad, 2011).

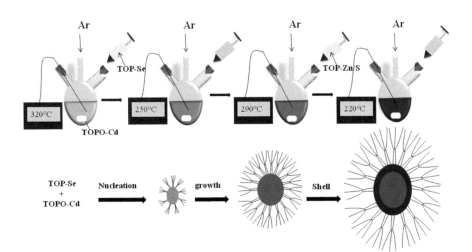

FIGURE 14.5 CdSe quantum dots formation by colloidal synthesis. In this reaction, a 3-necked flask is kept on a heating mantle and attaches a thermocouple and temperature controller with the flask. A cadmium precursor dissolved in an organic solvent, heated up to 320°C along with continuous stirring. At room temperature selenium precursor dissolved in another organic solvent is swiftly injected into the reaction mixture, causing the rapid nucleation which results in supersaturation of CdSe solution. As the temperature falls approx. 290°C, nucleation ceases rapidly, and growth of the existing crystals occurs. When the desired length of QDs has been synthesized, the reaction mixture is cooled to 220°C, to prevent further the growth of crystals. Further, QDs is coated with ZnS by injecting into the reaction mixture, to avoid any reaction with their environment. After cooling at room temperature, QDs are separated via precipitation method.

Although there is a major disadvantage associated with the colloidal synthesis, separation reaction should be developed for the individual semi-conductor material. Most of the reactions operate at high temperature and involving some inconvenient steps (Guzelian, Banin, Kadavanich, Peng, & Alivisatos, 1996; Akhter, 2011; Kumar, 2017; Pandey, 2018). Effect of the particle size to reaction parameters like pH and temperature should also be evaluated. The colloidal method also synthesizes wide size distribution of QDs. Reaction normally synthesizes size distribution of 10–15%, but other processes such as filtration and selection techniques reduce its size up to 5% (Murray et al., 2001). This colloidal synthesis method is well adapted for a greater quantity of QDs, like a solar cell.

14.2.1.2   MECHANISM OF QUANTUM DOTS

Different cells express numerous receptors and possess various internaliza-tion pathways, further mechanism of nanoparticles internalization and their behavior in endo-lysosomal vesicles still not clear. The possible mechanism may be cleaving of the nanoparticles ligands by protease enzyme in the endo-lysosomal vesicles. QDs (e.g., Cd, and Zn) induced cell cytotoxicity due to the release of Cd^{2+} or Zn^{2+} ions into the cytosol, which produces reactive oxygen species followed by lysosomal enlargement and further intracellular redistribution take place, which leads to the induction of apoptosis and death of the cell (Figure 14.6) (Mansur et al., 2016).

14.2.1.3   TYPES OF QUANTUM DOTS (QDS)

Properties of QDs are largely depended on their surface characteristic; however, they have a very high surface/volume ratio, which results in loss of their activity due to corrupted surfaces. Thus, preservation and protection of their surface is a prerequisite for their desired application. In general, passivation is carried out by a coating of a semiconductor precursor with the other substance, which have more bulk band gap as compared to core substance and both the electrons and holes exist within the core. It is called a type-I structures. The reverse arrangement is seen in the type-II structure in which the electrons and holes are spatially separated; one is present in the core material while another is located in the shell (Figure 14.7).

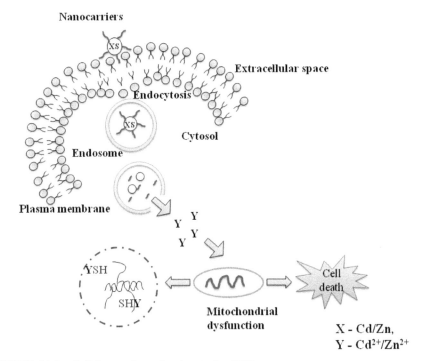

FIGURE 14.6   Cellular uptake and endocytosis of QDs.

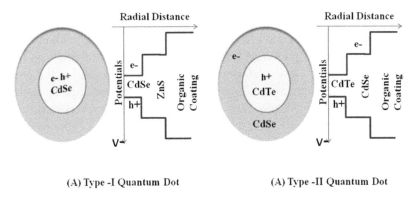

FIGURE 14.7   Types of quantum dots: (i) Type I quantum dots (ii) Type II quantum dots.

Type-I Core-Shell Structures: Quantum yield of the QDs is largely depended on their surface if the surface of QDs is perfectly passivated then the quantum yield is high, it means all the surface charge is saturated. However, these ligands are loosely bounded at the surface of QDs, which may be disturbed in the post-synthetic treatment of QDs. Thus, a more stable method is required to attain QDs with a high quantum yield which retains their property even after post-synthetic treatments. In order to overcome these obstacles and to increase quantum yield, core-shell structures were synthesized. The comparable increase of band gap fluorescence of CdS QDs was reported by Spanhel et al. In this process QDs prepared in water, treated with hydroxide ions in the presence of excessive Cd^{+2} ions. This passivated shell of Cd $(OH)_2$ throughout the QDs is responsible for the enhancement of quantum yield due to the blockage of free ions reaching the surface of QDs (Spanhel, Haase, Weller, & Hengleiir, 1987).

Nowadays there is much advancement in the core-shell QDs that are prepared in high boiling solvents, which is based on the CdE where E may be S, Se, Te as suggested by Murray et al. already discussed above. A lower degree of polydispersity has been recorded in this method as compared to the conventional water-based method. Guyot Sionnest et al. was first to successfully synthesize ZnS as a shell material onto the CdSe QDs in organic solvents (Hines & Guyot-Sionnest, 1996). Then after these types of QDs were subsequently explored by many researchers and reported. As compared to organic dye molecules type –1 core-shell structure represents a promising way for the labeling of biomolecules owing to the high photo and chemical stabilities as well as high fluorescence quantum yields (Bruns et al., 2009).

Type-II Core-Shell Structures: In type-II core-shell structures both the electrons and holes are spatially separated in different compartments. This arrangement minimizes the rate of recombination of charge carriers and enhances luminescence lifetime. In this structure staggered band alignment causes a spatially indirect transition, this phenomenon arises at lower energies as compared to the band gap of the precursor used. So, the type-II structure achieves higher wavelength, which cannot be attained by both the precursors alone. CdTe/CdSe core-shell QDs was the first reported type-II colloidal system. In these QDs, electrons are confined in a CdSe shell, and a hole is located in the core. These types of arrangement enable absorption and emission wavelength of the QDs to be a move towards lower energies. As a result emission wavelength of this type of structure can be obtained up to 1000 nm (Kim et al., 2003).

14.2.2   GOLD NANOPARTICLES (GNPS)

In recent years, GNPs -based research has attained prime attention in various fields such as molecular biology, bioengineering, and imaging. GNPs have also been considered to be of potential for use in the therapeutics for diagnosis & treatment of several diseases including cancer-based nanomedicine. GNPs are readily synthesized structures that absorb light strongly to generate thermal energy which in turn induces photothermal destruction of malignant tissue (Singh, Harris-Birtill, Markar, Hanna, & Elson, 2015). These nanoparticles can be readily synthesized in different geometrical forms like wires, cubes, spheres, cages, and rod, etc. for theranostic applications. The GNPs based therapy is based on the fact that these nanoparticles are capable of absorbing light and generates thermal energy. The thermal energy induces the photothermal degradation of malignant tumor (Mody, Agrawal, & Vyas, 2016). GNPs are biocompatible, bio-inert, modifiable, higher colloidal and photostability (Sharma et al., 2017; Ahmad, 2010). Due to the strong affinity of the thiol group with gold, surface modification is easily possible, and therefore thiolated species are broadly used in this new combination (Figure 14.8) (Daniel & Astruc, 2004).

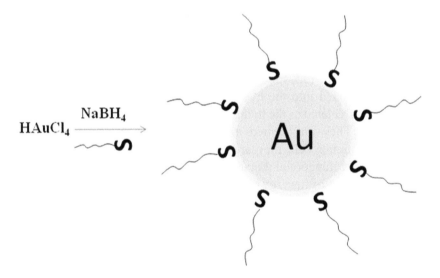

FIGURE 14.8 Formation of gold nanoparticles in the presence of thiol groups.

Ali and co-workers synthesized the AuNPs with three different ligands methoxy-polyethylene glycol thiol (enhanced the biocompatibility of AuNPs), RGD peptides (for attachment to the surface integrin of cancer cell and increased internalization) and nuclear localization signal (NLS) peptides (for nuclear targeting) in which aggrandize nuclear stiffness causing apprehension of cell motility upon GNPs treatment (Ali et al., 2017). Another formulation developed by Sun and co-worker synthesized elastin-like polypeptide (ELP) conjugated GNPs, and as given through i.v., the route for multimodal cancer imaging and photothermal therapy. The study reported that ELP-GNPs were novel intelligent theranostic agent stable at long-term storage condition, better passive diffusion ability into the cancer tissues, biocompatible (easy cleared from the body) and can be prepared with site-specific activity for use in a different type of cancer (Sun et al., 2017). Lin et al. developed in-situ formed pH-responsiveβ-cyclodextrin-{poly (lactide)-poly (2-(dimethylamino)ethyl-methacrylate)-poly[oligo(2-ethyl–2oxazoline)methacrylate]} doxorubicin loaded GNPs given by i.v., the route exhibited dual functions) for CT imaging and drug delivery for the effective treatment of cancer (Lin et al., 2017).

14.2.3   THERANOSTIC LIPOSOMES

Simultaneous diagnosis and therapy can be successfully assisted by theranostic liposomes in order to improve the treatment of the tumor. The nanosized imaging agents like QDs, GNPs, and iron oxide can be either entrapped inside the lipophilic core of liposomes or covalently linked to the surface of the theranostic liposomes while the chemotherapeutic agents can be encapsulated into the hydrophilic core of theranostic liposomes or embedded/ intercalated in the hydrophilic bilayer(s) of the vesicle (Rahman, 2013, 2016). They are termed as 'liposome-nanoparticle hybrids.' For targeting to a tumor cell, they are further conjugated with the molecular probe. Such multifunctional theranostic liposomes are retained in the blood for a longer period of time, evading the host defense system, and gradually release the chemotherapeutic agent to the targeted site and simultaneously facilitate in vivo and in vitro imaging (Rahman, 2012c, 2013). The combination of therapeutic agent and diagnosis agent improves the prognosis and monitoring of disease. Singh and coworkers prepared the targeted theranostic liposomes, which contain both QDs and docetaxel for imaging and therapy of brain cancer. The surface of these theranostic liposomes was decorated with transferrin and coated with D-alpha-tocopheryl polyethylene glycol 1000 succinate monoester (TPGS). The targeted and nontargeted theranostic liposomes were formulated and characterized for parameters such as drug encapsulation efficiency, polydispersity, size, morphology, and in-vitro release profile for brain theranostics. The encapsulation efficiency was found to be up to 71% with a mean diameter of liposomes below 200 nm. The drug release was sustained for 72 hr with 70% of drug released from targeted theranostic liposomes The in-vivo results demonstrated that transferring decorated theranostic liposomes prolonged the drug release and improved the brain targeting of docetaxel and QDs which may offer a promising drug carrier for brain theranostics and targeted chemotherapy (Singh et al., 2016). Muthu et al. prepared the theranostic liposomes for co-delivery of QDs and apomorphine for brain targeting. The liposomal formulation was characterized for *in vitro* brain endothelial cell uptake and *in vivo* bio-imaging. *In vitro* study indicated that receptor-mediated endocytosis was involved in cellular uptake of liposomes. The fluorescence derived from QDs was visualized during bio-imaging. Theranostic liposomes could improve the distribution and uptake in the brain as compared to free QDs. The free QDs were eliminated rapidly from the brain whereas the theranostic liposomes were retained and the fluorescence in the brain was observed for up to 1 h. These findings indicated that theranostic liposomes improved the bio-imaging in animal models for diagnosis as well as the treatment of cancer (Muthu & Feng, 2013). Jin et al. synthesized **liposome**-coated gadolinium-based on mesoporous silica nanoparticles which have improved colloidal stability, better biocompatibility, positive cellular uptake, and good MR imaging.

14.2.4 SUPER PARAMAGNETIC IRON OXIDE NANOPARTICLES (SPIONS)

Metallic nanoparticles have drawn the attention of many researchers as these could serve as target-oriented carriers and therefore reduce the side effects and improve the therapeutic index. The targeted action of SPIONs is due to their core, which is made up of iron oxides magnetically responsive to an external magnetic field (Rahman, 2012a, 2012b, 2012c). SPIONs are driven by the external magnetic field in a well-guided direction, ensuring the accumulation of SPIONs into the target site, enabling to trace and identify different cells and proteins *in vitro* (Schütt et al., 1999). Attractive features of SPIONs include superparamagnetism, extra anisotropic contribution, high saturation field and high field irreversibility (Kodama, 1999). Structure of SPIONs is consists of (i) a core made up of magnetic nanoparticles (magnetite, Fe_3O_4, or maghemite, γ-Fe_2O_3) which is coated with a biocompatible polymer; (ii) a porous biocompatible polymer (Hans & Lowman, 2002). The coating provides a confine to the magnetic particle and shields them from the surrounding environment. The coating also helps in carrier functionalization by linking with biotin, a carboxyl group, carbodiimide, and several other moieties. The attachment of target molecules at the surface of SPION allows personalized therapeutics to be used for the treatment of several diseases (Koneracká et al., 1999, 2002). Magnetic nanoparticles possess the exclusive properties such as (i) visualization through the magnetic resonance imaging (MRI); (ii) targeted action can be guided by an external magnetic field; (iii) provide hyperthermia for cancer treatment; (iv) formation of nontoxic iron ion *in vivo* (Weissleder et al., 1989). SPIONs-based therapy is affected or depended on various factors such as field strength, the gradients, and magnetic properties of particles. Targeted delivery of the drug through the SPIONs is due to the competition among forces acted on the particles by the blood compartment and force generated by the external magnetic field. Biocompatible drug/carrier complexes such as ferrofluid are injected into the body. Magnetic particles are localized at the target site by applying an external magnetic field; when the magnetic forces increase the blood flow rates in capillaries (0.05 cm/s) or arteries (10 cm/s). Drug releases at target site either via enzymatic activity or change in the osmolarity, pH, temperature (Alexiou et al., 2000) or uptake occurs by endothelial cells or tumor cells.

14.2.4.1   SUPER PARAMAGNETIC CORE

Alteration in the size of iron oxide from the macro-scale to nanoscale results in unique physical properties. One of the properties is superparamagnetic, in which individual particle has a single magnetic domain and thermal energy, which overcome the energy barrier of a magnetic flipping (usually, <20 nm). The body part which contains the SPIONs usually shows the darker image in MRI due to decreasing signal in the applied magnetic field because of the local homogeneities. This unique feature of SPIONs is adapted to direct the uptake of SPIONs to determine the efficacy of a treatment (Jain et al., 2009). Through the rapid alteration of magnetic field hyperthermia can be achieved (Jordan et al., 2006; Day, Morton, and West, 2009). Generation of heat is due to speedy rotation of SPIONs and the variation of the magnetic moment in the crystal lattice of SPION. The cells which contain SPIONs will heat up and die.

14.2.5   MAGNETO-LIPOSOME

Magneto-liposomes which were first introduced in 1988 have drawn the attention of researchers due to their multifunctional hybrid liposome/ nanoparticles assembly. Phospholipids are commonly used for the cellular delivery of drugs, DNA, RNA, and as a coating material for iron oxide core (Stock, 2017; Ito et al., 2007) or QDs (Al-Jamal et al., 2008). Liposome has many attractive features such as biocompatibility, size controllable, hydrophilic drugs can be incorporated into its hydrophilic core, and various modifications are possible such as attachment of antibodies for the targeted delivery (Al-Jamal and Kostarelos, 2007). Various strategies are adopted for the preparation of magnetoliposomes (MLs) such as (i) adsorption of a lipid on the large SPION, or (ii) encapsulation of a small SPION into the core of a liposome. Magnets are used to guide the MLs *in vivo* and thus increase the contrast for MRI imaging (Figure 14.9).

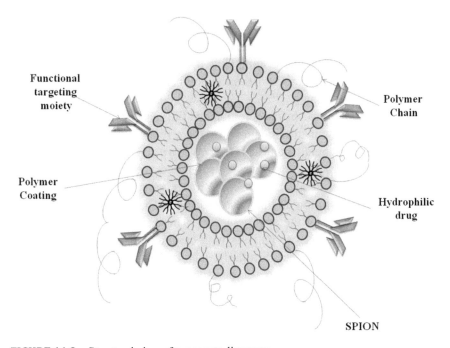

FIGURE 14.9   Structural view of a magneto-liposome.

14.2.6   *CARBON NANOTUBES*

Carbon nanotubes are nanosized structures seen for the first time in 1991 by a scientist Sumio Iijima et al. They are consisting of cylindrical, hexagonal nanosized structures of graphite and thus can be called as an allotrope of carbon with size in diameter ranging between 0.7 nm to few millimeters. Carbon nanotubes are known to have three different geometric structures based on how carbon tube is wrapped into a tube (Malik et al., 2013). The carbon nanotubes are broadly classified into six types:

1. single-walled;
2. multi-walled;
3. nanoflower;
4. nanobud;
5. fullerites; and
6. torus (Modi et al., 2011).

Carbon nanotubes with hollow structures inside can be used to accommodate a therapeutic moiety for theranostic purposes. CNTs are reported to be used for the photo-acoustic imaging, photo-thermal therapy, and delivery of genes. CNTs are safe for use when compared with other nanomaterials of inorganic origin. Thus it can be said and done that novel theranostic approaches based carbon nanotubes will be beneficial for drug delivery (Liu and Liang, 2012). Single-walled nanotubes are known to show absorption in near-infrared region and thus can be used for theranostic purposes after certain modification for their selective uptake. But multi-walled CNTs and metal filled CNTS are known to possess good physical, optical, and chemical properties as compared to that of single-walled nanotubes (Menon et al., 2013). Nanoflowers are nanostructures having flower-like configuration and possessed distinguished properties which can be used for delivery of therapeutic agents. These nanoflowers can be used for the delivery of drugs to patients with cardiac and ischemic disease for enhanced therapeutic effect. Similarly, nanobuds, fullerites, and torus are other morphologies where can also be used for theranostic purposes.

Multi-walled carbon nanotubes-magneto fluorescent carbon QDs/doxorubicin nanocomposites were prepared for dual modal imaging and photothermal effect on cancerous cells. These combinations have excellent water solubility and cell membrane permeability, which provides potentials for dual modal MR/fluorescence imaging *in vitro* and *in vivo*. The results showed the improved therapeutic effects by using NIR (near infrared) light and heat influenced release (Zhang et al., 2017). Another single-walled carbon nanotube was developed by Xie and co-workers. They developed multifunctional albumin/chlorin e6 loaded Evans blue carbon nanotube-based delivery system for theranostic application in the cancer chemotherapy. The resulting theranostic system showed a photothermal and photodynamic synergistic effect in tumor ablation and prevent metastasis (Xie et al., 2016). Wang and co-workers prepared carbon nanodots with interleukin–6 fragments for receptor targeting (pCDPI) and overcoming the biological barrier to imaging-guided site-specific delivery of drug to glioma. The *in vitro* and *in vivo* results have shown that pCDPI can overcome the BBB and preferentially penetrate and deeply accumulate into the orthotopic glioma in mice to inhibit interleukin 6 linked cell proliferation and achieve imaging-guided targeted drug delivery. Due to the small size, high water solubility and I6P8 peptide-mediated targeting, and the pCDPI nanodots can overcome the BBB and preferentially accumulate and deeply penetrate into the glioma. In the current work, they constructed a simple nanocarrier containing doxorubicin by covalently conjugating pCD with I_6P_8. The drug (DOX) induced inhibition of IL–6 cell proliferation is then visualized by FRET imaging (Wang et al., 2017). In another study, Xiaolong and his co-workers demonstrated that carbon nanoparticle/doxorubicin SiO2 nanocomposites can be effectively used as a platform for combined photothermal and chemotherapy. They prepared the nanocomposites by reverse microemulsion method with controlled size and possess high heat generating ability, pH-responsive drug delivery as well as heat-induced drug release. In vitro experiment showed that the combined photothermal and chemotherapy exhibited much higher toxicity to 4T1 cells and thus a single treatment with can effectively inhibit tumor growth as well prevent the cancer recurrence effectively (Tu et al., 2016) (Figure 14.10).

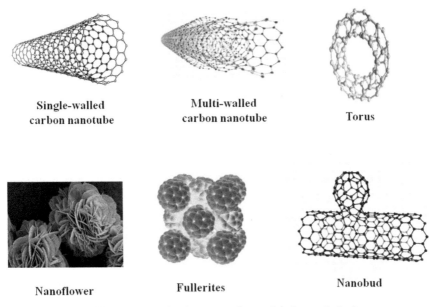

Single-walled carbon nanotube	Multi-walled carbon nanotube	Torus
Nanoflower	Fullerites	Nanobud

FIGURE 14.10 Different types of carbon nanotubes and their morphologies.

14.2.6.1   *PROPERTIES OF NANOTUBES*

These are the nanosized structures of diameter less than 100 nanometers and can be as thin as 1 or 2 nm. They can be modified physically and chemically for specific use in different ways. Some of their properties are as follows:

- They possess extraordinary electrical conductivity, heat conductivity, and mechanical properties.
- They are probably the best electron field-emitter known, largely due to their high length-to-diameter ratios.
- As pure carbon polymers, they can be manipulated using the well-known and the tremendously rich chemistry of that element.

14.2.6.2   *METHODS OF PREPARATION*

The methods used for the preparation of carbon nanotubes are given in the following subsections.

14.2.6.2.1   *Arc Discharge Technique*

It is one of the basic method used to prepare both single-walled carbon nano-tubes and multiwalled carbon nanotubes. The method involves the passage of current between two electrodes within an inert gas atmosphere which results into sublimation of carbon (Malik et al., 2013). One of the best qualities of carbon nanotubes can be prepared by using this technique (Figure 14.11).

14.2.6.2.2   *Laser Ablation Technique*

Preparation of carbon nanotube by laser ablation technique involves the vaporization and condensation of a target containing a mixture of graphite and a metal catalyst in an oven at a temperature of 1200°C using high-intensity laser beam (Figure 14.12) (Malik et al., 2013).

14.2.6.2.3   *Chemical Vapor Deposition Technique*

This method involves the preparation of CNTs using chemical reaction involving the deposition of solid using the mixture of gases (Kumar et al.,

2012). The method of preparation of CNTS using CVD technique includes two steps: The first and foremost step is deposition of a catalyst such as Ni, Fe or Co on the substrate followed by nucleation using chemical etching or thermal ammonia. In the second step carbon source such as methane, carbon mono-oxide or acetylene is kept in a reaction chamber consisting of a gas phase. The carbon molecules are then converted into anatomic from with the help of energy sources like plasma or heated coil which results into carbon diffusion towards the substrate, which has been coated with a catalyst followed by growth of nanotube onto this metal catalyst under the temperature range of 650–9000°C. The yield of CNTs formed using this technique comes around 30%.

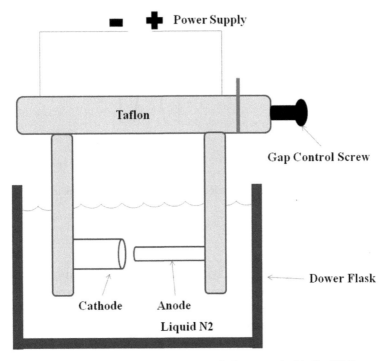

FIGURE 14.11   Schematic representation of Arc discharge method in liquid N2.

14.2.6.2.4   Ball Milling

CNTs can also be synthesized by Ball milling technique followed by annealing using carbon and boron nitride powder as a raw material. This technique involves the milling of graphite powder in the presence of argon in a stainless-steel container with four hardened steel balls at room temperature for up to 150 hours (Wilson et al., 2002). The milling process followed by annealing of graphite is performed under an inert gas flow at 1400°C for 6 hours and is used for the production of MWNTs.

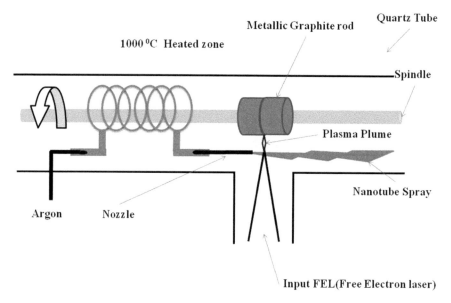

FIGURE 14.12   Schematic representation of Ultrafast laser evaporation method.

14.2.6.2.5   Other Methods

Other techniques used for the synthesis of CNTs are hydrothermal/sono-chemical methods, electrolysis using carbon rod as an electrode (cathode) at high temperature, synthesis using flame for the production of hydrocarbon gas at high temperature, solar energy and low temperature pyrolysis from metastable carbon-containing compounds at 1200°C to 1900°C to produces MWNTs of varying lengths of 0.1–1 μm and diameters ranging from 10 nm to 25 nm, etc. (Journet and Bernier, 1998).

14.3   CONCLUSION AND FUTURE PROSPECTS

Presently there are many theranostic nanomedicines which are under investigation for the targeted treatment of cancer. Theranostic nanoparticulate systems have a unique property either with the addition of QDs or differential coating technologies like gold coating, silver coating or introduction of iron oxide in different formulations, having an optical or magnetic property which helps in diagnosis and treatment of cancer. It has been shown that therapeutics of various forms, including those that are a small molecule, protein, and nucleotide-based, can be conveniently tethered onto nanoplatforms. The great capacity even allows the loading of a second or third functionality, a feature which encourages the formation of an all-in-one nanosystem with comprehensive features.

KEYWORDS

- **cancer**
- **carbon nanotubes**
- **liposomes**
- **nanomedicine**
- **nanoparticles**
- **quantum dots**
- **SPIONs**
- **theranostic**

REFERENCES

Ahmad, J., Akhter, S., Rizwanullah, M., et al., (2015a). Nanotechnology-based inhalation treatments for lung cancer: State of the art. *Nanotechnol. Sci. Appl., 19*(8), 55–66.

Ahmad, J., Amin, S., Rahman, M., et al., (2015b). Solid matrix based lipidic nanoparticles in oral cancer chemotherapy: Applications and pharmacokinetics. *Curr. Drug. Metab., 16*(8), 633–644.

Ahmad, M. Z., Akhter, S., Ahmad, I., et al., (2011). Development of polysaccharide-based colon targeted drug delivery system: Design and evaluation of Assam Bora rice starch-based matrix tablet. *Curr. Drug. Deliv., 8*(5), 575–581.

Ahmad, M. Z., Akhter, S., Anwar, M., et al., (2013). Colorectal cancer targeted Irinotecan-Assam Bora rice starch-based microspheres: A mechanistic, pharmacokinetic and biochemical investigation. *Drug. Dev. Ind. Pharm., 39*(12), 1936–1943.

Ahmad, M. Z., Akhter, S., Jain, G. K., et al., (2010). Metallic nanoparticles: Technology overview & drug delivery applications in oncology. *Expert. Opin. Drug. Deliv., 7*(8), 927–942.

Akhter, S., Ahmad, Z., Singh, A., et al., (2011). Cancer targeted metallic nanoparticle: Targeting overview, recent advancement, and toxicity concern. *Curr. Pharm. Des., 17*(18), 1834–1850.

Alexiou, C., et al., (2000). *Locoregional Cancer Treatment with Magnetic Drug Targeting Locoregional Cancer Treatment with Magnetic Drug Targeting, 1*(4), 6641–6648.

Ali, M. R. K., et al., (2017). Nuclear membrane-targeted gold nanoparticles inhibit cancer cell migration and invasion. *ACS Nano., 11*(4), 3716–3726.

Al-Jamal, W. T., & Kostarelos, K., (2007). Liposome–nanoparticle hybrids for multimodal diagnostic and therapeutic applications. *Nanomedicine, 2*(1), 85–98.

Al-Jamal, W. T., et al., (2008). Functionalized-quantum-dot-liposome hybrids as multimodal nanoparticles for cancer. *Small, 4*(9), 1406–1415.

Aneja, P., Rahman, M., Beg, S., et al., (2014). Cancer targeted magic bullets for effective treatment of cancer. *Recent. Pat. Antiinfect. Drug. Discov., 2*, 121–135.

Bailey, R. E., Smith, A. M., & Nie, S., (2004). Quantum dots in biology and medicine. *Physica E. Low-Dimensional Systems and Nanostructures, 25*(1), 1–12.

Bakalova, R., et al., (2011). Chemical nature and structure of organic coating of quantum dots is crucial for their application in imaging diagnostics. *International Journal of Nanomedicine, 6*, 1719–1732.

Beg, S., Rahman, M., Jain, A., et al., (2017). Nanoporous metal-organic frameworks as hybrid polymer-metal composites for drug delivery and biomedical applications. *Drug. Discov. Today., 22*(4), 625–637.

Bruns, O. T., et al., (2009). Real-time magnetic resonance imaging and quantification of lipoprotein metabolism *in vivo* using nanocrystals. *Nature Nanotechnology, 4*(3), 193–201.

Brus, L. E., (1983). A simple model for the ionization potential, electron affinity, and aqueous redox potentials of small semiconductor crystallites. *The Journal of Chemical Physics, 79*(11), 5566–5571.

Daniel, M. C. M., & Astruc, D., (2004). Gold nanoparticles: Assembly, supramolecular chemistry, quantum-size-related properties and applications toward biology, catalysis, and nanotechnology. *Chemical Reviews, 104*, 293–346.

Day, E. S., Morton, J. G., & West, J. L., (2009). Nanoparticles for thermal cancer therapy. *Journal of Biomechanical Engineering, 131*(7), 74001.

Elzoghby, A. O., Samy, W. M., & Elgindy, N. A., (2012). Protein-based nanocarriers as promising drug and gene delivery systems. *Journal of Controlled Release, 161*(1), 38–49.

Fan, J., et al., (2016). Inhibition of autophagy contributes to the toxicity of cadmium telluride quantum dots in Saccharomyces cerevisiae. *International Journal of Nanomedicine, 11*, 3371–3383.

Guzelian, A. A., et al., (1996). Colloidal chemical synthesis and characterization of InAs nanocrystal quantum dots. *Applied Physics Letters, 69*(10), 1432–1434.

Han, H., et al., (2017). Enzyme-sensitive gemcitabine conjugated albumin nanoparticles as a versatile theranostic nanoplatforms for pancreatic cancer treatment. *Journal of Colloid and Interface Science, 507*, 217–224.

Hans, M., & Lowman, A., (2002). Biodegradable nanoparticles for drug delivery and targeting. *Current Opinion in Solid State and Materials Science, 6*(4), 319–327.

Hines, M. A., & Guyot, S. P., (1996). Synthesis and characterization of strongly luminescing ZnS-capped CdSe nanocrystals. *The Journal of Physical Chemistry, 11, 100*(2), 468–471.

Ito, A., et al., (2007). 4-S-Cysteaminylphenol-loaded magnetite cationic liposomes for combination therapy of hyperthermia with chemotherapy against malignant melanoma. *Cancer Science, 98*(3), 424–430.

Jaidev, L. R., et al., (2017). Multi-functional nanoparticles as theranostic agents for the treatment & imaging of pancreatic cancer. *Acta Biomaterialia., 1*(49), 422–433.

Jain, T. K., et al., (2009). Magnetic resonance imaging of multifunctional pluronic stabilized iron-oxide nanoparticles in tumor-bearing mice. *Biomaterials, 1, 30*(35), 6748–6756.

Jordan, A., et al., (2006). The effect of thermotherapy using magnetic nanoparticles on rat malignant glioma. *Journal of Neuro-Oncology, 1, 78*(1), 7–14.

Kim, S., et al., (2003). Type-II quantum dots: CdTe/CdSe (core/shell) and CdSe/ZnTe (core/shell) heterostructures. *Journal of the American Chemical Society, 24, 125*(38), 11466–11467.

Kodama, R. H., (1999). *Magnetic Nanoparticles*, 200.

Koneracká, M. A., et al., (1999). Immobilization of proteins and enzymes to fine magnetic particles. *Journal of Magnetism and Magnetic Materials, 1, 201*(1–3), 427–430.

Koneracka, M., et al., (2002). Direct binding procedure of proteins and enzymes to fine magnetic particles. *Journal of Molecular Catalysis B: Enzymatic, 13, 18*(1–3), 13–18.

Kosaka, N., et al., (2010). Real-time optical imaging using quantum dot and related nanocrystals. *Nanomedicine, 5*(5), 765–776.

Kukreja, A., et al., (2017). Preparation of gold core-mesoporous iron-oxide shell nanoparticles and their application as dual MR/CT contrast agent in human gastric cancer cells. *Journal of Industrial and Engineering Chemistry, 25, 48*, 56–65.

Kumar, S. P., et al., (2012). Pharmaceutical application of carbon nanotube-mediated drug delivery system. *Int. J. Pharm. Sci. Nanotechnol., 5*, 1685–1696.

Kumar, V., Bhatt, P. C., Rahman, M., et al., (2017). Fabrication, optimization, and characterization of umbelliferone β-D-galactopyranoside-loaded PLGA nanoparticles in treatment of hepatocellular carcinoma: *In vitro* and *in vivo* studies. *Int. J. Nanomedicine, 12*, 6747–6758.

Lammers, T., et al., (2011). Theranostic nanomedicine. *Accounts of Chemical Research, 5, 44*(10), 1029–1038.

Larson, D. R., (2003). Water-soluble quantum dots for multiphoton fluorescence imaging *in vivo*. *Science, 30, 300*(5624), 1434–1436.

Lee, K. H., et al., (2012). Quantitative molecular profiling of biomarkers for pancreatic cancer with functionalized quantum dots. *Nanomedicine: Nanotechnology, Biology and Medicine, 1, 8*(7), 1043–1051.

Lee, K., et al., (2010). Heparin immobilized gold nanoparticles for targeted detection and apoptotic death of metastatic cancer cells. *Biomaterials, 31*(25), 6530–6536.

Lin, W., et al., (2017). pH-responsive unimolecular micelle-gold nanoparticles-drug nanohybrid system for cancer theranostics. *Acta Biomaterialia, 58*, 455–465.

Ling, Y., et al., (2011). Dual docetaxel/superparamagnetic iron oxide loaded nanoparticles for both targeting magnetic resonance imaging and cancer therapy. *Biomaterials, 32*(29), 7139–7150.

Ling, Y., et al., (2012). Temozolomide loaded PLGA-based superparamagnetic nanoparticles for magnetic resonance imaging and treatment of malignant glioma. *International Journal of Pharmaceutics, 430*(1–2), 266–275.

Liu, Z., & Liang, X., (2012). *Nano-Carbons as Theranostics*, 2(3), 10–12.

Malik, P., et al., (2013). *Carbon Nanotubes, Quantum Dots, and Dendrimers as Potential Nanodevices for Nanotechnology Drug Delivery Systems, 6*(3), 2113–2124.

Mallick, N., et al., (2016). Chondroitin sulfate-capped super-paramagnetic iron oxide nanoparticles as potential carriers of doxorubicin hydrochloride. *Carbohydrate Polymers, 151*, 546–556.

Mandal, R. P., et al., (2017). "Theranostic" role of bile salt-capped silver nanoparticles - gall stone/pigment stone disruption and anticancer activity. *Journal of Photochemistry and Photobiology B: Biology*, 269–281.

Mansur, A. A. P., et al., (2016). Surface biofunctionalized Cds and Zns quantum dot nanoconjugates for nanomedicine and oncology: To be or not to be nanotoxic? *International Journal of Nanomedicine, 11*, 4469–4490.

Menon, J. U., et al., (2013). *Theranostics Nanomaterials for Photo-Based Diagnostic and Therapeutic Applications, 3*(3).

Michalet, X., (2013). Quantum dots for live cells, *in vivo* imaging, and diagnostics. *Science, 307*(5709), 538–544.

Modi, C. D., et al., (2011). *Functionalization and Evaluation of PEGylated Carbon Nanotubes as Novel Drug Delivery for Methotrexate, 1*(5), 103–108.

Mody, N., et al., (2016). Chapter 5, Nanobiomaterials: Emerging platform in cancer theranostics. *Nanobiomaterials in Cancer Therapy* (7ᵗʰ edn., pp. 117–146). Elsevier Inc.

Murray, C. B., Norris, D. J., & Bawendi, M. G., (1993). Synthesis and characterization of nearly monodisperse CdE (E = S, Se, Te) semiconductor nanocrystallites. *Journal of the American Chemical Society, 115*(19), 8706–8715.

Murray, C. B., Sun, S., & Gaschler, W., (2001). Colloidal synthesis of nanocrystals and nanocrystal superlattices. *IBM Journal of Research and Development, 45*(1), 47–56.

Muthu, M. S., & Feng, S. S., (2013). Theranostic liposomes for cancer diagnosis and treatment: Current development and pre-clinical success. *Expert Opinion on Drug Delivery*, 151–155.

Nagesetti, A., Srinivasan, S., & McGoron, A. J., (2017). Polyethylene glycol modified ORMOSIL theranostic nanoparticles for triggered doxorubicin release and deep drug delivery into ovarian cancer spheroids. *Journal of Photochemistry and Photobiology B: Biology, 174*, 209–216.

Opoku-Damoah, Y., Wang, R., & Zhou, J., (2016). Versatile nanosystem-based cancer theranostics: Design inspiration and predetermined routing. *Theranostic., 6*(7), 986.

Pandey, P., Rahman, M., Bhatt, P. C., et al., (2018). Implication of nano-antioxidant therapy for treatment of hepatocellular carcinoma using PLGA nanoparticles of rutin. *Nanomedicine (Lond)*. doi: 10.2217/nnm–2017–0306.

Parak, W. J., Manna, L., & Nann, T., (2008). Fundamental principles of quantum dots. *Nanotechnology: Principles and Fundamentals, 1*, 73–90.

Porcaro, F., Battocchio, C., & Antoccia, A., (2016). Synthesis of functionalized gold nanoparticles capped with 3-mercapto–1-propane sulfonate and 1-thioglucose mixed thiols and "*in vitro*" response. *Colloids and Surfaces B: Biointerfaces, 142*, 408–416.

Qu, L., & Peng, X., (2016). *Control of Photoluminescence Properties of CdSe Nanocrystals in Growth, 124*(9), 2049–2055.

Rahman, M., Ahmad, M. Z., Ahmad, J., et al., (2015b). Role of graphene nano-composites in cancer therapy: Theranostic applications, metabolic fate, and toxicity issues. *Curr. Drug. Metab., 16*(5), 397–409.

Rahman, M., Ahmad, M. Z., Kazmi, I., et al., (2012a). Emergence of nanomedicine as cancer-targeted magic bullets: Recent development and need to address the toxicity apprehension. *Curr. Drug. Discov. Technol., 9*(4), 319–329.

Rahman, M., Ahmad, M. Z., Kazmi, I., et al., (2012b). Advancement in multifunctional nanoparticles for the effective treatment of cancer. *Expert. Opin. Drug. Deliv., 9*(4), 367–381.

Rahman, M., Ahmed, M. Z., Kazmi, I., et al., (2012c). Novel approach for the treatment of cancer: Theranostic nanomedicines. *Pharmacologia., 3*, 371–376.

Rahman, M., Akhter, S., Ahmad, M. Z., et al., (2015a). Emerging advances in cancer nano-theranostics with graphene nanocomposites: Opportunities and challenges. *Nanomedicine (Lond)., 10*(15), 2405–2422.

Rahman, M., Beg, S., Ahmed, A., et al., (2013). Emergence of functionalized nanomedicines in cancer chemotherapy: Recent advancements, current challenges, and toxicity considerations. *Recent Patent on Nanomedicine, 2,* 128–139.

Rahman, M., Beg, S., Verma, A., et al., (2017). Therapeutic applications of liposomal-based drug delivery and drug targeting for immune linked inflammatory maladies: A contemporary viewpoint. *Curr. Drug. Targets., 18*(13), 1558–1571.

Rahman, M., Kumar, V., Beg, S., et al., (2016). Emergence of liposome as targeted magic bullet for inflammatory disorders: Current state of the art. *Artif. Cells. Nanomed. Biotechnol., 44*(7), 1597–1608.

Reiss, P., Protière, M., & Li, L., (2009). Core/shell semiconductor nanocrystals. *Small, 5*(2), 154–168.

Schütt, W., Grüttner, C., & Teller, J., (1999). Biocompatible magnetic polymer carriers for *in vivo* radionuclide delivery. *Artificial Organs, 23*(1), 98–103.

Sharma, R., Mody, N., & Agrawal, U., (2017). Theranostic nanomedicine, a next-generation platform for cancer diagnosis and therapy. *Mini Reviews in Medicinal Chemistry, 17*(18), 1746–1757.

Siegel, R., Naishadham, D., & Jemal, A., (2013). *Cancer Statistics, 63*(1), 11–30.

Silva, A. S., Sousa, A. M., & Cabral, R. P., (2017). Aerosolizable gold nano-in-micro dry powder formulations for theragnosis and lung delivery. *International Journal of Pharmaceutics, 519*(1 & 2), 240–249.

Singh, M., Harris-Birtill, D. C. C., & Markar, S. R., (2015). Application of gold nanoparticles for gastrointestinal cancer theranostics: A systematic review. *Nanomedicine: Nanotechnology, Biology, and Medicine, 11*(8), 2083–2098.

Smith, A. M., Duan, H., & Mohs, A. M., (2008). Bioconjugated quantum dots for *in vivo* molecular and cellular imaging. *Advances Drug Delivery Reviews, 60*(11), 1226–1240.

Sonali, S., Singh, R. P., & Sharma, G., (2016a). RGD-TPGS decorated theranostic liposomes for brain-targeted delivery. *Colloids and Surfaces B: Biointerfaces, 147,* 129–141.

Sonali, S., Singh, R. P., & Sharma, N., (2016b). Transferrin liposomes of docetaxel for brain-targeted cancer applications: Formulation and brain theranostics. *Drug Delivery, 23*(4), 1261–1271.

Spanhel, Y., Haase, M., & Weller, H., (1987). Photochemistry of colloidal semiconductors. 20. Surface modification and stability of strong luminescing CdS particles. *Journal of the American Chemical Society, 109*(19), 5649–5655.

Stock, C. A., (2017). *Supporting Information © Pnas , 50,* 1–7.

Sun, M., Peng, D., & Hao, H., (2017). Thermally triggered *in situ* assembly of gold nanoparticles for cancer multimodal imaging and photothermal therapy. *ACS Applied Materials and Interfaces, 9*(12), 10453–10460.

Theranostic Nanomedicine for Cancer (2008), 137–140.

Thomas, R. G., Moon, M. J., & Lee, H., (2015). Hyaluronic acid conjugated superparamagnetic iron oxide nanoparticle for cancer diagnosis and hyperthermia therapy. *Carbohydrate Polymers, 131,* 439–446.

Tu, X., Wang, L., & Cao, Y., (2016). Efficient cancer ablation by combined photothermal and enhanced chemo-therapy based on carbon nanoparticles/doxorubicin@SiO2 nanocomposites. *Carbon, 97,* 35–44.

Vachaspati, P., (2013). Quantum dots : Theory, application, synthesis. *Massachusetts Institute of Technology, 2,* 1–6.

Wang, S., Li, C., & Qian, M., (2017). Augmented glioma-targeted theranostics using multifunctional polymer-coated carbon nanodots. *Biomaterials, 141,* 29–39.

Wehrenberg, B. L., Wang, C., & Guyot-Sionnest, P., (2002). Interband and intraband optical studies of PbSe colloidal quantum dots. *Journal of Physical Chemistry B., 106*(41), 10634–10640.

Weissleder, R., Stark, D. D., & Engelstad, B. L., (1989). Superparamagnetic pharmacokinetics iron oxide and toxicity. *American Journal of Roentgenology, 152,* 167–173.

Xie, J., Lee, S., & Chen, X., (2010). Nanoparticle-based theranostic agents. *Advanced Drug Delivery Reviews, 62*(11), 1064–1079.

Xie, L., Wang, G., & Zhou, H., (2016). Functional long circulating single-walled carbon nanotubes for fluorescent/photoacoustic imaging-guided enhanced phototherapy. *Biomaterials, 103,* 219–228.

Yeasmin, S., Datta, H. K., & Chaudhuri, S., (2017). *In-vitro* anti-cancer activity of shape-controlled silver nanoparticles (AgNPs) in various organ-specific cell lines. *Journal of Molecular Liquids, 242,* 757–766.

Zhang, M., Wang, W., & Wu, F., (2017). Magnetic and fluorescent carbon nanotubes for dual modal imaging and photothermal and chemotherapy of cancer cells in living mice. *Carbon, 123,* 70–83.

Zhong, X., Feng, Y., & Knoll W., (2003). Alloyed ZnxCd1-xS nanocrystals with highly narrow luminescence spectral width. Journal of the American Chemical Society, 125(44), 13559–13563.

Nanomedicine Therapeutic Approaches to Overcome Hypertension

MD. RIZWANULLAH[1], SADAF JAMAL GILANI[2], MOHD. AQIL[1], and SYED SARIM IMAM[3*]

[1]Department of Pharmaceutics, School of Pharmaceutical Education and Research, Jamia Hamdard, New Delhi–110062, India

[2]Department of Pharmaceutical Chemistry, College of Pharmacy, Jouf University, Aljouf, Sakaka 2014, Saudi Arabia

[3]Department of Pharmaceutics, Glocal School of Pharmacy, The Glocal University, Saharanpur, Uttar Pradesh–247121, India, Tel.: +91-9536572892, E-mail: sarimimam@gmail.com

*Corresponding author. E-mail: sarimimam@gmail.com

ABSTRACT

Hypertension (HT) is the most common cardiovascular disease, and their management requires long-term treatment. The conventional dosage forms may result in poor patient compliance with due to greater frequency of drug administration. The poor aqueous solubility and poor metabolic stability of drugs are major limitations in successful drug delivery systems. There are several nanotechnology-based drug delivery system approaches have shown the potential to overcome the challenges associated with the administration routes. It addresses the formulation and application of substances and devices on the nanometer scale. There is an availability of a plethora of nanotechnology-based delivery system of therapeutically effective antihypertensive molecules, and arguably presents an opportunity to deliver through a different route. The present chapter focuses on various nanoformulations available for different administration routes for improving solubility, dissolution, and consequently bioavailability of antihypertensive drugs.

15.1   INTRODUCTION

Hypertension (HT) is a long-term medical condition, also known as high blood pressure (HBP). It is usually defined by the presence of a chronic elevation of systemic arterial pressure above a certain threshold value. The blood pressure in the arteries is persistently elevated, and long-term problem lead to the various major risk factor (Lackland and Weber, 2015; Lau et al., 2017; Hernandorena et al., 2017). Worldwide prevalence estimates for HT may be as much as 1 billion individuals and approximately 7.1 million deaths per year may be attributable to HT (Alderman, 2007; Ahad et al., 2015). In 1980, about 600 million people were suffering from HT while this graph was raised significantly in the year 2008, where the data increased to 1 billion lead to big concern for dealing with this condition effectively (WHO, 2013). HT affects approximately 78 million individuals in the United States and approximately 1 billion worldwide (Go et al., 2014). As the population ages, the prevalence of HT will increase further unless broad and effective preventive measures are implemented. The report from the Framingham heart study suggest that individuals who are normotensive at the age of 55 years have a 90% lifetime risk for developing HT (Vasan et al., 2002).

The progression is strongly associated with functional and structural cardiac and vascular abnormalities that damage the heart, kidneys, brain, vasculature, and other organs and lead to premature morbidity and death (Giles et al., 2009) summarized in Table 15.1. It is as such not a disease in itself but is a risk factor for major cardiovascular events like heart stroke, ischemic heart disease, myocardial infarction, and heart enlargement. It is expressed by two measurements, the systolic and diastolic pressures, which are the maximum and minimum pressures, as shown in Table 15.2. At rest, the normal blood pressure in adults must be within the range of 100–130 millimeters mercury (mmHg) systolic and 60–80 mmHg diastolic and the HBP is present if the resting blood pressure is persistently at or above 130/90 or 140/90 mmHg. HBP is classified as either primary (essential) HBP or secondary HBP (Poulter et al., 2015). About 90–95% of cases are primary, defined as HBP due to nonspecific lifestyle and genetic factors. Average blood pressure may be higher in the winter than in the summer. Lifestyle factors that increase the risk include excess salt in the diet, excess body weight, smoking, and alcohol use.

Therefore, cost-effective approaches to optimally control blood pressure are very much needed (Selvam et al., 2010). There are many categories of antihypertensive agents, which lower blood pressure by different means; among the antihypertensive, most important and most widely used are the thiazide diuretics, β-blockers, the ACE inhibitors, calcium channel blockers and angiotensin II receptor antagonists. Most of the antihypertensive drug comes under BCS class II (low solubility and high permeability) which have low bioavailability as dissolution is the rate-limiting step.

TABLE 15.1 Affected Organ in Hypertension Condition

Complications of hypertension	
Organ	**Affected in high blood pressure**
Brain	• Reduced blood supply to the brain
	• Rapid loss of brain function
Eye	• Hypertensive retinopathy
	• Damage blood vessels in the retina
	• Loss of vision
Blood vessels	• Atherosclerosis- Narrowing of the artery
	• Heart attack and stroke.
Heart	• Difficulty in pumping the blood
	• Restricted blood flow
	• Heart muscle thick
Kidney	• Blood vessels damaged
	• Difficulty in filtering blood
	• Accumulation of waste and fluid
Bone	• High calcium content in urine.
	• Excess calcium loss leads to osteoporosis.

Several drugs in conventional dosage forms are available to treat HT, but the majority of the antihypertensive drugs is poorly water-soluble and therefore exhibits low bioavailability. These drugs are also substrate of P-gp and exhibit significant first-pass metabolism. The other challenges with these formulations are their short half-life and high dosing frequency. With the use of extended-release systems, these dosing frequencies can be reduced but as far as enhancement of bioavailability is concerned nanoparticles are a far better approach. The associated benefits with nanoparticle include their capability of circumventing the first-pass metabolism; P-gp mediated efflux and achieving targeting because entrapped drug in the carrier is directly taken into the systemic circulation (Alam et al., 2017).

This chapter takes into account challenges associated with conventional antihypertensive formulations and role of nanoparticulate drug delivery system in overcoming such hurdles and enhancing the treatment of HT. The present chapter covers a more recent and advanced technique for enhancing the efficacy of antihypertensive drugs through different strategies.

TABLE 15.2 Different Types of Hypertension Range with Treatment

Category	Systolic (mmHg)	Diastolic (mmHg)	Treatments	
			Compelling Indication	**Without Compelling Indication**
Hypotension	< 90	< 60		
Normal	90–119	60–79	None	None
	90–129	60–84		
Prehypertension (high normal, elevated)	120–129	60–79	Drug(s)	No antihypertensive drug
	130–139	85–89		
Hypertension (stage 1)	130–139	80–89	Diuretics, ARB, CCB, BB as needed	• Thiazide diuretics
	140–159	90–99		• ARB, CCB or BB
Hypertension (stage II)	>140	>90		Drug combination (Thiazide diuretics with ARB, CCB, BB)
	160–179	100–109		
Hypertensive crises	≥ 180	≥ 120		
Isolated systolic hypertension	≥ 160	< 90 to 110		

15.2 NANOMEDICINES FOR TREATMENT OF HYPERTENSION

Drugs with poor solubility possess difficulty in the formulation by applying conventional approaches as they present problems such as slow onset of action, poor oral bioavailability, lack of dose proportionality, failure to achieve steady-state plasma concentration, and undesirable side effects. The conventional dosage forms thus may result in over- or under medication and poor patient compliance (Patwekar and Baramade, 2012). These challenges can be overcome by applying novel drug delivery systems that offer benefits like reduction in dose frequency, lowering of dose size, site-specific targeting, enhanced permeability, and improvement in oral bioavailability (Akhter et al., 2018; Rizwanullah et al., 2018). Nanotechnology is a promising strategy in the development of drug delivery systems especially for those potent drugs whose clinical development failed due to their poor solubility, low permeability, inadequate bioavailability, and other poor biopharmaceutical properties (Imam et al., 2015; Ahmad et al., 2015; Baig et al., 2016; Imam et al., 2017). The application of nanotechnology to medicine and pharmaceutical formulations, generally referred to as nanomedicine. Different types of highly efficient nanomedicines have been used to enhance the physicochemical properties and efficacy of antihypertensive agents. Nanomedicines for treatment of HT can be broadly classified on the basis of formulation component into lipid-based and polymer-based nanomedicines. Various types of lipid and polymer-based nanomedicines have been extensively investigated for improvement of the bioavailability of antihypertensive drugs (Sharma et al., 2016) (Table 15.3).

15.2.1 LIPID-BASED NANOMEDICINES

Lipid-based nanomedicines are ideal candidate for drug delivery of antihypertensive drugs showing low solubility and high permeability. Lipid-based nanocarriers entrapped drug which is poorly soluble; the dissolution step is not needed as the drug is generally solubilized in lipid excipients (Porter et al., 2007; Khan et al., 2016; Rizwanullah et al., 2017). This solubilization is generally maintained throughout the gastrointestinal passage. The excipients used in preparing lipid-based nanomedicines include surfactants and co-surfactants apart from lipid which can promote permeability across intestinal wall. The mechanism underlying the enhancement of drug absorption includes increase in membrane fluidity, opening of tight junction, inhibition of P-glycoprotein efflux transporter, alteration of intestinal metabolism mediated by cytochrome P450, and lymphatic uptake thus bypassing hepatic first-pass metabolism (Lo et al., 2010; Allen and Cullis, 2013; ElKasabgy et al., 2014).

15.2.1.1 LIPOSOMES

Liposomes are vesicular structures having an aqueous core surrounded by a lipid bilayer shell. The basic method of their preparation comprises hydrating the mixture of natural or synthetic phospholipids, cholesterol, and tocopheryl acetate. Liposomes encapsulate drugs either in an aqueous compartment or in the lipid bilayer depending on their nature (Allen and Cullis, 2013). Liposomes have excellent potential for delivering various types of therapeutic drug for treatment of HT. In this context, ElKasabgy et

TABLE 15.3 Different Nanoformulations Used with Antihypertensive Drugs and Their Findings

Formulations	Drug	Class	Route	Findings	Reference
Niosomes	Lacidipine	CCB	T	In vivo antihypertensive study showed the developed formulation exhibited significantly higher reduction in blood pressure compared to oral drug suspension.	Qumber et al., 2017
Niosomes	Nifedipine	CCB	T	Niosomal formulations exhibited better permeability with good steady-state flux and enhancement ratio compared to control formulation.	Yasam et al., 2017
Invasomes	Isradipine	CCB	T	Isradipine invasomes formulation reduces 20% blood pressure by virtue of better permeation.	Qadri et al., 2017
Invasomes	olmesartan medoxomil	ARB	T	Transdermal nano-invasomes formulation showed 1.15 times improvement in bioavailability of olmesartan medoxomi compared to the control formulation.	Kamran et al., 2016
Liposomes	Lacidipine	CCB	O	The enhanced drug's oral bioavailability in human volunteers was achieved in comparison to the commercial product.	Elkasabgy et al., 2014
Liposomes	Valasartan	ARB	O	Liposomes exhibited significantly higher permeation of drug in Caco–2 monolayers compared to free drug suspension. Furthermore, Liposomes showed 202.36% improved relative oral bioavailability compared to free drug suspension.	Nekkanti et al., 2015
Transfersomes	Verapamil	CCB	IN	Chitosan composite transfersomal formulation exhibited absolute bioavailability of 81.83% compared to the oral solution which displayed only 13.04%.	Mouez et al., 2016
Transfersomes	Valsartan	ARB	T	Valsartan nanotransfersomes showed significantly enhanced flux with compared to rigid liposomes. The pharmacokinetic study showed that nanotransfersomes exhibited significantly enhanced C_{max}, AUC, and $T_{1/2}$.	Ahad et al., 2016b

TABLE 15.3 *(Continued)*

Formulations	Drug	Class	Route	Findings	Reference
Nanoemulsion	olmesartan medoxomil	ARB	T	A significant increase in the bioavailability (1.23 times) compared with oral formulation of olmesartan.	Aqil et al., 2016
Nanoemulsion	Mebudipine	CCB	O	Mebudipine nanoemulsion showed about 2.6-, 2.0- and 1.9-fold improved oral bioavailability compared to free drug suspension, ethyl oleate solution, and micellar solution respectively.	Khani et al., 2016
SLNs	Nisoldipine	CCB	O	SLNs exhibited 2.17-fold increase in oral bioavailability and significant reduction in the systolic blood pressure for a period of 36h when compared to free drug suspension.	Dudhipala et al., 2015
SLNs	Nitrendipine	CCB	ID	Bioavailability of nitrendipine was increased three- to four-fold after intraduodenal administration compared to that of nitrendipine suspension.	Kumar et al., 2007
SLNs	Ramipril	ACEI	O	Prolonged drug release, smaller particle size, and narrow particle size distribution was achieved.	Ekambram et al., 2011
NLCs	Carvedilol	BB	O	NLCs exhibited 3.95 fold improved oral bioavailability compared to free drug suspension.	Mishra et al., 2016
NLCs	Candesartan Cilexetil	ARB	T	Many folds increase in oral bioavailability than free drug suspension, which was further confirmed by antihypertensive activity in a murine model.	Paudel et al., 2017
NLCs	olmesartan medoxomil	ARB	O	In vivo performance showed that AUC_{total} and C_{max} of OM-NLC were found significantly higher as compared to the free drug.	Kathiwas et al., 2017
NLCs	Lercanidipine hydrochloride	CCB	O	NLCs released lercanidipine hydrochloride in a controlled manner for a prolonged period of time as compared to plain drug.	Ranpise et al., 2014

TABLE 15.3 *(Continued)*

Formulations	Drug	Class	Route	Findings	Reference
NC	Nitrendipine	CCB	O	Improvement in physical stability, *in vitro* drug release and bioavailability.	Alexa et al., 2012
NC	Nimodipine	CCB	O	The nanocrystals were taken up by enterocytes via macropinocytosis and caveolin-mediated endocytosis pathways.	Fu et al., 2013
NS	Valsartan	ARB	T	The solubility and dissolution efficiency of VAL in nanosuspension was significantly higher than pure form.	Vuppalapati et al., 2016
Dendrimer	Candesartan cilexetil	ARB	O	Improvement in the solubility of CC is achieved.	Erturk et al., 2017

Abbreviations: O: oral; T: transdermal; ID: intraduodenal; IN: intranasal; CCB: calcium channel blockers; ACEI: angiotensin-converting enzyme inhibitors; ARB: angiotensin receptor blockers; NC: Nanocrystal; NS: Nanosuspension; SLNs: solid lipid nanoparticles; NLC: nanostructured lipid carriers.

al. developed lacidipine (LAC) loaded liposomes to enhance oral bioavailability the drug (ElKasabgy et al., 2014). They reported in their study that the pharmacokinetic study in human volunteers showed significantly higher C_{max} compared to Tween 80 control formulation and marketed formulation. Also, the relative oral bioavailability of the developed liposomes was 5.4 as compared to Tween 80control preparation. The reason provided was that Tween 80 increases GI permeability, and lipid entrapped drug is circumvented by first-pass effect apart from its lymphatic uptake. In another study, Nekkanti et al. (2015) developed valsartan (VAL) loaded proliposomes (VAL-PL) for oral bioavailability improvement. The developed VAL-PL showed improved drug release in simulated gastric fluid compared to pure drug suspension. The averted rat intestinal perfusion study showed that *In-vitro* permeability using Caco–2 monolayers revealed significantly higher permeation of VAL-PL vis-a-vis free drug suspension. Further, In vivo pharmacokinetic study in male Sprague Dawley rats showed that the developed VAL-PL exhibited 202.36% improved relative oral bioavailability compared to free drug suspension.

15.2.1.2 NIOSOMES

Niosomes are self-assembling non-ionic vesicle systems also called as also called as non-ionic liposomes formed from nonionic surfactants in an aqueous environment. It has high potential to act as carriers for poorly soluble drugs. It is structurally similar to liposomes, but due to the better stability, they are ideal alternative to liposomes in drug delivery systems. They are biodegradable and biocompatible, relatively nontoxic and capable of loading both hydrophilic and lipophilic drugs (Lo et al., 2010). Niosomes have lamellar structures composed of amphiphilic molecules, known as surfactant; contain both hydrophobic (tails) and hydrophilic (heads) groups and leading self-assembling properties, aggregating into a variety of shapes like micelles or into a planar lamellar bilayer. Hydrophilic drugs are entrapped inside of vesicular aqueous core or adsorbed on bilayer surfaces while lipophilic substances are encapsulated by partitioning into the lipophilic domain of the bilayers (Moghassemi and Hadjizadeh, 2014). Qumbar et al. developed LAC loaded niosomes formulation for the management of HT after transdermal delivery (Qumbar et al., 2017). The optimized LAC loaded niosomal formulation showed vesicle size, entrapment efficiency and flux value of 676.98 ± 10.92 nm, 82.77 ± 4.34% and 38.43 ± 2.43 µg/cm^2/h respectively, with spherical morphology.

The comparative confocal laser scanning microscopic study revealed that optimized niosomal formulation showed significantly higher permeation (70.75 mm) as compared to control formulation (58.26 mm). In addition, the optimized LAC loaded niosomal gel exhibited 2.15 fold higher permeation enhancement compared to control gel. Furthermore, in vivo antihypertensive activity showed significantly prolonged reduction in blood pressure compared to oral suspension, which was maintained for 48 h. Similarly, Yasam et al. developed nifedipine loaded niosomes for transdermal treatment of HT (Yasam et al., 2017). The ex-vivo Franz diffusion studies using rat skin showed that the developed niosomal formulations exhibited better permeability with good steady-state flux and enhancement ratio compared to control formulation. Skin irritation studies for 7 days, showed that the developed nifedipine loaded niosomal gel formulation was non-irritant with no erythemia development compared to pure drug. Furthermore, the bio-distribution studies showed that the developed niosomal formulation maintained the nifedipine concentration over a prolonged period of time in the heart tissue.

15.2.1.3   ETHOSOMES

Ethosomes are novel lipid carriers developed by Touitou et al. in 1998, which are composed of phospholipid, ethanol, and water (Touitou et al., 2000). Ethosomes are promising nanocarriers for non-invasive transdermal delivery. Ethosomes were reported to enhance the skin permeation of drugs due to the interdigitation effect of ethanol on the lipid bilayer of liposomes and increasing fluidity of stratum corneum (SC) lipids (Song et al., 2012). The high flexibility of vesicular membranes from the added ethanol permits the elastic vesicles to squeeze themselves through the pores, which are much smaller than their diameters; thus, ethosomal systems are considerably more efficient in delivering substances to the skin in terms of quantity and depth than either conventional liposomes or hydro-alcoholic solution (Touitou et al., 2001; Verma and Pathak, 2012). Ahad et al. (2013) developed VAL loaded nanoethosomes for transdermal delivery. Their results indicated that the VAL loaded nanoethosomes of exhibited better flux, desirable vesicle size, reasonable entrapment efficiency, more effectiveness for transdermal delivery compared to rigid liposomes. VAL loaded nanoethosomes proved significantly superior in terms of, amount of drug permeated in the skin, with an enhancement ratio of 43.38 ± 1.37 when compared to rigid liposomes. Confocal laser scanning microscopy (CLSM) revealed an enhanced permeation of Rhodamine-Red loaded nanoethosomes to the deeper layers of the skin as compared to conventional liposomes. Furthermore, in vivo pharmacokinetic study showed that VAL loaded nanoethosomal formulation exhibited 3.03 fold higher bioavailability compared to oral suspension. In another study by the same research group, they evaluated the same nano-medicine, i.e., VAL loaded nanoethosomes for skin toxicity study and anti-hypertensive activity (Ahad et al., 2016a). Their results showed that VAL loaded nanoethosomes treated group showed significant and constant fall in blood pressure up to 48 h. The VAL ethosomal formulation was found to be effective, with a 34.11% reduction in blood pressure. Moreover, skin toxicity study revealed that the VAL ethosomal formulation was safe, less irritant, and well-tolerated for transdermal delivery.

15.2.1.4 INVASOMES

Invasomes are flexible, neutrally charged, phospholipid-based vesicular system containing a mixture of soy phosphatidylcholine, lysophosphatidyl-choline, terpenes, and ethanol. It has shown to improve skin penetration of hydrophilic and lipophilic drugs (Imam and Aqil, 2016; Dragicevic-Curic et al., 2009). Flexibility of the bilayer membrane is mainly due to the lysophos-phatidylcholine acting as an edge activator. Ethanol is a good penetration enhancer while terpenes have also shown potential to increase the penetra-tion of many drugs by disrupting the tight lipid packing of the SC (Qadri et al., 2017). Kamran et al. developed and evaluated the transdermal potential of nanoinvasomes, containing anti-hypertensive drug olmesartan (Kamran et al., 2016). The developed nanoinvasomes formulation showed vesicles size of 83.35 ± 3.25nm, entrapment efficiency of $65.21 \pm 2.25\%$ and transdermal flux of 32.78 ± 0.703 $\mu g/cm^2/h$. CLSM of rat skin showed that the developed formulation was eventually distributed and permeated deep into the skin. Pharmacokinetic study in Wistar rats showed that transdermal treatment of nanoinvasomes formulation exhibited 1.15-fold higher bioavailability compared to olmesartan oral suspension.

Similarly, Qadri et al. (2017) developed invasomes of isradipine for enhanced transdermal delivery against HT. The developed invasomes formulation showed vesicles size of 194 ± 18 nm, entrapment efficiency of 88.46%, and attained mean transdermal flux of 22.80 ± 2.10 $mg/cm^2/h$ through rat skin. CLSM revealed an enhanced permeation of Rhodamine-Red-loaded isradipine loaded invasomes to the deeper layers of the rat skin. During antihypertensive study, the treatment group showed a substantial and constant decrease in blood pressure, for up to 24 h. The isradipine loaded invasomes formulation was found to be effective, with a 20% reduction in blood pressure by virtue of better permeation through Wistar rat skin.

15.2.1.5   TRANSFERSOMES

The conventionally used liposomes remain confined to the upper surface, with little penetration into the SC. Hence, a new class of liposomes also called elastic liposomes, transfersomes, or ultra-deformable liposomes, have been developed. Several studies have reported that transfersomes were able to improve in vitro skin delivery of various drugs and to penetrate intact skin in vivo, transferring therapeutic amounts of drugs with efficiency comparable with subcutaneous administration. The key factors that confer ultradeformability to the liposomes have been considered to be edge activators (EAs), the special surfactants incorporated into Transfersomes (e.g., sodium cholate or sodium deoxycholate, SDC). Because Transfersomes are composed of surfactant, they have better rheology and hydration properties, which are responsible for their superior skin penetration ability (Benson, 2006). Ahad et al. developed VAL loaded nanotransfersomes for treatment of HT (Ahad et al., 2012). Their results showed that the developed VAL loaded nanotransfersomes showed significantly superior in terms of amount of drug permeated in the skin, with an enhancement ratio of 33.97 ± 1.25 when compared to rigid liposomes. This was further confirmed through a CLSM study. CLSM results revealed that the developed nanotransfersomes were fairly evenly distributed throughout the SC, viable epidermis, and dermis with high fluorescence intensity compared to rigid liposomes. Moreover, nanotransfersomes showed better antihypertensive activity in comparison to liposomes by virtue of better permeation through Wistar rat skin. In another study by the same researchers, VAL loaded nanotransfersomes showed significantly enhanced transdermal flux with an enhancement ratio of 26.91 compared to rigid liposomes (Ahad et al., 2016b). Skin irritation study revealed that the developed VAL loaded nanotransfersomes was safe, less irritant, and well-tolerated for transdermal delivery. Moreover, pharmacokinetic study showed that the developed formulation exhibited significantly enhanced C_{max}, AUC, and $T_{1/2}$. Furthermore, the formulation stored under a refrigerated condition showed greater stability, and results were found to be within the specification under storage conditions.

15.2.1.6 NANOEMULSION (NE)

Nanoemulsions (NE) are oil-in-water (O/W), water-in-oil (W/O) dispersion of two immiscible liquids stabilized using an appropriate surfactant. The mean droplet diameter attained is usually less than 500 nm. Small droplet size gives them a clear or hazy appearance which differs from milky white color associated with coarse emulsion (whose micron-sized droplets partake in multiple light scattering). Nanoemulsion is thermodynamically stable drug delivery system which can solubilize higher amount of drug. NE provide advantages of solubilization of hydrophobic molecules in the oily phase, modification of oil droplets with polymers to prolong circulation time, and targeting tumors passively and/or targeting ligands actively. Most commonly used methods to prepare NE are low-energy emulsification and high-energy emulsification (Singh et al., 2017). In a study, Gorain et al. developed olmesartan medoxomil (OM) loaded nanoemulsion (OM-NE) for improved oral bioavailability and extended antihypertensive activity in hypertensive rats. They developed OM loaded oil-in-water (o/w) NE using lipoid purified soybean oil 700, sefsol 218, and solutol HS 15. The developed OM-NE showed significantly enhanced permeability through the Caco–2cell monolayer compared to free drug solution. In vivo pharmacokinetic study showed that OM-NE exhibited 2.8 fold improved AUC upon oral administration compared to free drug suspension. Furthermore, in vivo antihypertensive studies with OM-NE demonstrated better and prolonged control of experimentally induced HT with 3-fold reduction in conventional dose compared to free drug suspension (Gorain et al., 2014). Chhabra et al. developed amlodipine besilate (AB) loaded nanoemulsion (AB-NE) for improved oral bioavailability (Chhabra et al., 2011). In vivo pharmacokinetics and biodistribution studies of the ^{99m}Tc-labeled, NE in male Swiss albino mice (p.o.) demonstrated a relative bioavailability of 475% against AB suspension. In almost all the tested organs, the uptake of AB from NE was significantly higher compared to AB suspension especially in heart with a drug-targeting index of 44.1%, also confirming the efficacy of NE at therapeutic site. Furthermore, AB-NE exhibited 3 fold higher overall residence time compared to AB suspension further signifies the advantage of NE as drug carriers for enhancing bioavailability of AB. In another study, Aqil et al. developed OM-NE for transdermal delivery (Aqil et al., 2016). Their results indicated that the developed OM-NE provides reasonable particle size (53.11 ± 3.13nm), polydispersity index (0.335 ± 0.008) and transdermal flux (12.65 ± 1.60 µg/cm^2/h). CLSM revealed an enhanced penetration of Rhodamine B loaded nanoemulsion carriers to the deeper layers of the skin.

Furthermore, in vivo pharmacokinetic study of optimized OM-NE showed a significant increase in the bioavailability (1.23 times) compared to oral formulation of OM by virtue of better permeation through rat skin.

15.2.1.7   NANOSUSPENSION

Nanosuspensions (NS) are biphasic, colloidal dispersions of drug particles which are stabilized by using surfactants. NS contain particles dispersed in an aqueous vehicle with the size of particle less than 1 μm. NS can overcome the problems related to the delivery of poorly water-soluble drugs due to their nanosize particle range. NS can be formulated with high solid content up to 40%, which reduces their dose size and improves patient compliance (Chavhan et al., 2011). In a study, Gora et al. developed VAL loaded nano-suspension (VAL-NS) for improved oral bioavailability and antihypertensive efficacy (Gora et al., 2016). In vivo pharmacokinetic study in Wistar rats showed that VAL-NS exhibited ~ 9.5 fold improved AUC compared to non-homogenized VAL suspension. Moreover, pharmacodynamic study showed that oral administration of VAL-NS significantly lowered blood pressure compared to non-homogenized VAL suspension in Wistar rat. In another study, Nagaraj et al. developed OM loaded nanoemulsion for improved oral bioavailability (Nagaraj et al., 2017). In vivo pharmacokinetic study in Wistar rats showed that OM loaded lyophilized nanosuspension exhibited significantly improved pharmacokinetic properties compared to free OM suspension and marketed formulation. OM loaded nanosuspension exhib-ited 2.45 and 2.25 folds higher oral bioavailability compared to marketed formulation and free OM suspension, respectively.

15.2.1.8   NANOCRYSTALS

Nanocrystals are comprised of aggregates of large number of atoms with size between 10nm and 400nm. The decrease in drug particle size to nanoscopic crystals results in an increased surface area to volume ratio (Sharma et al., 2016). The nanocrystals of nitrendipine was formulated, and result of the study revealed 15- fold and 10-fold increase in *in-vivo* study as compared to physical mixture and commercial tablet (Quan et al., 2011). Hecq et al. (2005) have investigated the nanocrystals of nifedipine for enhancement of solubility and dissolution rate. This study indicated that there was reten-tion of crystalline state upon particle size reduction and improvement in dissolution rate of nifedipine. In another study, chitosan coated nanocrystals containing nitrendipine was formulated and evaluated for bioavailability. The investigation results of disclosed modified nanocrystals showed remarkable improvement in bioavailability as compared to traditional dosage form. On the basis of experimental results, the surface modified nanocrystal with chitosan emerge as an efficient method for controlling *in vitro* and *in vivo* performance, therefore increasing the bioavailability of poorly water-soluble drugs (Quan et al., 2012).

15.2.1.9   SOLID LIPID NANOPARTICLES

Solid lipid nanoparticles (SLNs) have emerged as an efficient, nontoxic, and versatile colloidal drug carrier system that avoids some of the disadvantages of liposomes and polymeric nanoparticles. Because SLNs are composed of solid lipids (i.e., fatty acids; mono, di and triglycerides; phospholipids; etc.), they tend to exhibit high compatibility and biodegradability and have a lower risk of the acute and chronic toxicity that is often associated with other nanoparticles. Furthermore, because SLNs comprise a solid lipid core instead of an aqueous core, they offer better protection against chemical degradation of drugs than nanoemulsion, and also facilitate sustained release effect due to the zero-order degradation kinetics of the solid lipid matrix (Geszke-Moritz and Moritz, 2016; Thukral et al., 2014). Dudhipala and Veerabrahma (2016) developed candesartan cilexetil (CC) encapsulated solid lipid nanoparticles (CC-SLNs) for oral delivery for treatment of HT. Pharmacokinetic study showed that CC-SLNs exhibited 2.75 fold improved oral bioavailability compared to free drug suspension. Pharmacodynamic study of CC-SLNs in hypertensive rats showed a decrease in systolic blood pressure for 48 h, while free drug suspension showed a decrease in systolic blood pressure for only 2 h. The greater antihypertensive results are found due to significantly enhanced oral bioavailability coupled with sustained action of CC in SLN formulation. Similarly, the same researchers developed nisoldipine (ND) loaded solid lipid nanoparticles (ND-SLNs) for oral delivery (Dudhipala and Veerabrahma, 2015). Pharmacokinetic study showed that ND-SLNs exhibited 2.17 fold improved oral bioavailability compared to free drug suspension. Pharmacodynamic study of ND-SLNs in hypertensive rats showed a decrease in systolic blood pressure, which sustained for 36 h compared to free drug suspension. Zhang et al. developed CC loaded solid lipid nanoparticles (CC-SLNs) for enhancement of oral bioavailability (Zhang et al., 2012). Researchers also evaluated the mechanism of absorption of the developed nanoparticles. CLNs exhibited nanometer-sized spherical particles with high entrapment efficiency (91.33%). The absorption of CC-SLNs in the stomach was found to be only 2.8% of that in intestine. Absorption of CLNs in the GI tract mainly occurred in the intestine. Their results revealed that CC-SLNs could be internalized into the enterocytes by clathrin- and caveolae-mediated endocytosis pathway, and then transported into the systemic circulation via the portal circulation and the intestinal lymphatic pathway. After oral administration, CC-SLNs absorbed more rapidly than free drug suspension. Further, the pharmacokinetic results showed that CC-SLNs exhibited over 12 fold improved oral bioavailability compared to free drug suspension. In another study, OM loaded SLNs showed 2.32 fold improved oral bioavailability compared to free drug suspension (Nooli et al., 2017). From above results, it may be conclude that SLNs have great potential for increasing bioavailability and antihypertensive activity of different drugs.

15.2.1.10   NANOSTRUCTURED LIPID CARRIERS

SLNs has some limitation associated with it like expulsion of drug due to organization of solid lipid into more perfect crystal with time, which results in decrease in entrapment efficiency and loading capacity with time. This drawback associated with the SLNs led to the development of nanostructured lipid carriers (NLCs). NLCs are considered as second generation SLN composed of mixture of spatially different lipids; that is, solid lipid is blended with oil (liquid lipid) to overcome the limitations associated with SLNs. Liquid lipid is present within the solid lipid and does not undergo modification into stable structure; also solubility of drug in liquid lipid is higher than solid lipid, this results in enhancement of entrapment efficiency and loading capacity (Rizwanullah et al., 2016; Alam et al., 2016; Moghddam et al., 2017; Soni et al., 2018). Mishra et al. developed carvedilol (CAR) loaded nanostructured lipid carriers (CAR-NLCs) for the management of HT after oral delivery (Mishra et al., 2016). Their results indicated that the CAR-NLCs exhibited desirable particle size, reasonable entrapment efficiency, and drug loading. The ex vivo gut permeation study showed that CAR-NLCs exhibited about 2 fold improved permeation of drug compared to free drug suspension. Moreover, pharmacokinetic study in Wistar rats revealed that the developed CAR-NLCs exhibited 3.95 fold improved relative oral bioavailability compared to free drug suspension. Furthermore, In vivo antihypertensive study in Wistar rats showed significant reduction in mean systolic BP by CAR-NLCs compared to free drug suspension owing to the drug absorption through lymphatic pathways. Paudel et al. developed CC loaded nanostructured lipid carriers (CC-NLCs) for improved oral bioavailability and antihypertensive efficacy (Paudel et al., 2017). The results showed that the developed CC-NLCs showed reasonable particle size (183.5 ± 5.89 nm), PDI (0.228 ± 0.13), zeta potential (–28.2 ± 0.99mV), and significantly high entrapment efficiency (88.9 ± 3.69%). The developed CC–NLCs exhibited significantly enhanced gut permeation which was further confirmed by CLSM. In vivo Pharmacokinetic study in Wistar rats showed that CC-NLCs exhibited over 7 fold higher AUC compared to free CC suspension which was further confirmed by antihypertensive activity in a murine model. The developed CC-NLCs decrease the BP significantly into the normal range for longer time. Kaithwas et al. developed OM loaded nanostructured lipid carriers (OM-NLCs) for enhanced oral bioavailability (Kaithwas et al., 2017). The developed OM-NLCs was found to be stable in simulated gastric fluids as no significant changes were found in size, PDI and entrapment efficiency. In vitro release study showed extended release of OM from OM-NLCs. In vitro cellular uptake study revealed 5.2 folds higher uptake of OM-NLCs as compare to the free drug, when incubated with Caco–2 cells. Moreover, Pharmacokinetic study in Wistar rats showed that the developed OM-NLCs exhibited 4.84 and 5.01 fold improved AUC_{total} and C_{max} respectively compared to free dug. Similarly, Beg et al. developed OM-NLCs by Quality by Design (QbD) approach for improved oral delivery of OM (Beg et al., 2018). Their results showed that the developed and optimized OM-NLCs showed the particle size, zeta potential and entrapment efficiency of ~ 250 nm, < 25 mV, and > 75% respectively. In vitro drug release study revealed that OM-NLCs exhibited initial burst release (~ 80% within 4 h) followed by sustained release up to 24 h of study. Finally, researchers ratified that the developed formulation has high degree of formulation robustness and potential for improved therapeutic performance for the management of HT.

15.2.2 *POLYMER-BASED NANOMEDICINES*

Polymer-based nanomedicines, such as polymeric nanoparticles, polymeric micelles, and dendrimers for improvement in therapeutic efficacy of antihypertensive drugs have been extensively explored. The extensive scope for chemically modifying the polymer into novel nanocarriers leads to a versatile drug delivery system for antihypertensive therapeutics (Sharma et al., 2016; Alam et al., 2017).

15.2.2.1 POLYMERIC NANOPARTICLES

Polymeric nanoparticles are colloidal drug delivery carriers with particle size of 10 to 100 nm. The main advantages of nanoparticles are (i) enhanced bioavailability, (ii) increased specificity and targeting to desired site, and (iii) reduced toxicity and dose. All these benefits enable safe delivery of drugs especially to target sites without affecting normal tissue. These nanoparticles are prepared by employing biodegradable, synthetic, and natural polymers. The therapeutic effect of polymeric nanoparticles significantly depends on drug release and biodegradation of polymers. The drug release from nanoparticles follows diffusion mechanism, erosion mechanism, or both erosion and diffusion in combination (Mudshinge et al., 2011). Some of the most commonly employed polymers in the development of nanoparticles are poly(D,L-lactide-coglycolide), polycaprolactones, and poly(D,L-lactide) (Rao JP, Geckeler, 2011; Bamrungsap et al., 2012;). Shah et al. developed felodipine (FD) loaded PLGA nanoparticles (FD-PLGA-NPs) for improved absorption and antihypertensive efficacy after oral delivery (Shah et al., 2014). The in vitro and ex vivo release studies through stomach and intestine showed sustained drug release from the FD-PLGA-NPs. Pharmacodynamic studies showed control of blood pressure and normalization of ECG changes (ST segment elevation) by FD-PLGA-NPs for a period of 3 days in comparison to plain drug suspension. The researchers revealed that increase in pharmacodynamic response due to FD-PLGA-NPs in comparison to plain drug suspension could be attributed to bioavailability enhancement due to direct uptake by Peyer's patch of intestine. In another study, Mishra et al. developed cilnidipine loaded PLGA nanoparticles (CIL-PLGA-NPs) for improved oral bioavailability and antihypertensive efficacy (Mishra et al., 2017). The developed CIL-PLGA-NPs showed initial burst release for 2 h followed by sustained release over 24 hour and followed Korsmeyer-Peppas model for release kinetics indicating diffusion-erosion controlled drug release. In vivo pharmacokinetic study in Wistar rats showed that CIL-PLGA-NPs exhibited ~ 3 fold improved relative oral bioavailability compared to free drug suspension. Moreover, CIL-PLGA-NPs showed better antihypertensive effect on methylprednisolone-induced hypertensive rats compared to marketed formulation and free drug suspension. Similarly, Jana et al. developed and characterized an antihypertensive drug FD-PLGA-NPs and evaluated for acute toxicity in Albino Wistar rats (Jana et al., 2014). Their results showed that the developed FD-PLGA-NPs were in nanosize range with smooth and spherical morphology. The in vitro drug release study of PLGA nanoparticles showed longer duration of drug release with reduced burst release compared with pure FD. The in vivo toxicity study in Wistar rats showed no noticeable change in biochemical parameters and histopathology of organs.

15.2.2.2   POLYMERIC MICELLES

Micelles are a group of amphiphilic surfactant molecules that rapidly aggregate into a spherical vesicle in water and range from 10 to 100nm in size. Micelles arrange themselves in a spherical form upon contact with aqueous solutions. The center of the micelle is hydrophobic, and thus entrapment of lipophilic drug is easily possible. Polymeric micelles are formed of block copolymers consisting of hydrophilic and hydrophobic monomer units. Several polymers in combination are used in formation of micelles, for example, poly(ethylene oxide) (PEO) and poly(propylene oxide) (PPO), poly(lactic acid) (PLA), or other polyethers or polyesters. PEO–PPO–PEO structure with the trade names of Pluronic® and poloxamer is one of the commonly used polymers in micelles formulation. Polymer micelles have benefits like enhanced drug solubility, sustained circulation half-life, specificity at target sites, and reduced toxicity (Xu et al., 2013; Simões et al., 2015). Fares et al. developed LAC loaded pluronic P123/F127 mixed polymeric micelles for enhancement of dissolution and oral bioavailability (Fares et al., 2018). Their results showed that the developed LCDP micelles showed that particle size, PDI and entrapment efficiency of 21.8 nm, 0.11 and 99.23% respectively. LCDP loaded polymeric micelles showed saturation solubility approximately 450 times that of raw LCDP in addition to significantly enhanced dissolution rate. In vivo pharmacokinetic study in Albino rabbits showed that LCDP loaded exhibited 6.85-fold improved oral bioavailability compared to free LCDP suspension. In another study, Satturwar et al. (2007) developed CC loaded pH-responsive polymeric micelles of poly(ethylene glycol)-b-poly(alkyl(meth)acrylate-co-methacrylic acid) to improve loading capacity and drug release. Their results showed that the developed polymeric micelles showed particle size, entrapment efficiency, and drug loading 60–160 nm, 90% and 20% (w/w) respectively. In vitro drug release study revealed that the release of drug from pH-sensitive micelles was triggered upon an increase in pH from 1.2 to 7.2.

15.2.2.3   DENDRIMERS

Dendrimers are novel polymeric carrier having three-dimensional structure, nanometer size (1 and 100 nm) and controlled molecular structure. It is derived from a Greek word "Dendra," which means reminiscent of a tree (Hsu et al., 2017). Different types of dendrimers are available on the basis of different polymers such as polyamidoamines (PAMAMs), polyamines, polyamides (polypeptides), poly(aryl ethers), polyesters, carbohydrates, and DNA. PAMAM dendrimers are most commonly used. CC is a calcium channel blocker used in the treatment of HT. In a systematic study by Gautam and Verma can desartancilexetil loaded PAMAM dendrimers showed significant increase in water solubility of CC (Gautam and Verma, 2012). In another study, propranolol (a calcium channel blocker), a known substrate of the P-glycoprotein (P-gp) efflux transporter, has been conjugated to lauroyl-G3 dendrimers. The developed lauroyl-G3 based dendrimers exhibited significantly enhanced solubility of propranolol (D'Emanuele et al., 2004).

15.2.2.4   CARBON NANOTUBES

Carbon nanotubes (CNTs) are carbon allotropes, composed of one or more coaxial sheaths of graphite and exhibited physical, photochemical, and electrochemical properties. It can single-walled or multiwalled and also be used as drug transporters or as a basis for tissue regeneration (Kumar et al., 2015). Presently, CNTs are gaining tremendous attention as novel drug delivery carriers. It has the ability of high drug loading owing to their increased surface area. Several *in vivo* and *in vitro* studies on CNTs have proven them to be an effective drug delivery system. CAR loaded PAMAM functionalized multiwalled carbon nanotubes (MWNTs) were formulated and evaluated for drug loading and dissolution. The result of the study revealed that the developed formulation has shown marked increase in drug loading and solubility, so improves the dissolution of drugs (Zheng et al., 2013).

15.3   CONCLUSIONS AND FUTURE PERSPECTIVES

Nanotechnology-based delivery systems have shown great potential in effective delivery of antihypertensive drugs by improving solubility and oral bioavailability. It is promising approach in resolving several constraints of antihypertensive drugs. The targeted nanoparticle can effectively take antihypertensive drug to its site of action. These approaches have appeared as strategies to revitalize the development of new hydrophobic entities. The patient compliance can be achieved due to the low dose, colloidal size, drug targeting, reduced toxicity, and biocompatibility. There are significant advancements are achieved in nanotechnology based delivery system, but there are still some challenges to be encountered. Thus, nanotechnology-based delivery system offers opportunity for researcher to extend research and development to overcome the challenges related with current antihypertensive drugs, thereby improving the patient compliance and therapeutic efficacy.

KEYWORDS

- **blood pressure**
- **dendrimers**
- **hypertension**
- **nanoemulsion**
- **nanoparticle**
- **nanotechnology**

REFERENCES

Ahad, A., Aqil, M., et al., (2015). Systemic delivery of β-blockers via transdermal route for hypertension. *Saudi Pharm. J.*, *23*(6), 587–602.

Ahad, A., et al., (2012). Formulation and optimization of nanotransfersomes using experimental design technique for accentuated transdermal delivery of valsartan. *Nanomed.*, *8*(2), 237–249.

Ahad, A., et al., (2013). Enhanced transdermal delivery of an anti-hypertensive agent via nanoethosomes: statistical optimization, characterization, and pharmacokinetic assessment. *Int. J. Pharm.*, *443*(1 & 2), 26–38.

Ahad, A., et al., (2016a). Nano vesicular lipid carriers of angiotensin II receptor blocker: Anti-hypertensive and skin toxicity study in focus. *Artif. Cells Nanomed. Biotechnol.*, *44*(3), 1002–1007.

Ahad, A., et al., (2016b). The ameliorated longevity and pharmacokinetics of valsartan released from a gel system of ultradeformable vesicles. *Artif. Cells Nanomed. Biotechnol.*, *44*(6), 1457–1463.

Ahmad, J., et al., (2015). Nanotechnology-based inhalation treatments for lung cancer: State of the art. *Nanotechnol. Sci. Appl.*, *8*, 55–66.

Akhter, M. H., et al., (2018). Nanocarriers in advanced drug targeting: Setting novel paradigm in cancer therapeutics. *Artif. Cells Nanomed. Biotechnol., 2018, 46(5)*, 873-884.

Alam, S., et al., (2016). Nanostructured lipid carriers of pioglitazone for transdermal application: From experimental design to bioactivity detail. *Drug Deliv.*, *23*(2), 601–609.

Alam, T., et al., (2017). Nanocarriers as treatment modalities for hypertension. *Drug Deliv.*, *24*(1), 358–369.

Alderman, M. H., (2007). Hypertension control: Improved, but not enough! *Am. J. Hypertens.,* *20*(4), 347.

Allen, T. M., & Cullis, P. R., (2013). Liposomal drug delivery systems: From concept to clinical applications. *Adv. Drug. Deliv. Rev.,* *65*(1), 36–48.

Aqil, M., et al., (2016). Development of clove oil based nanoemulsion of olmesartan for transdermal delivery: Box- Behnken design optimization and pharmacokinetic evaluation, *J. Mol. Liq.,* *214,* 238–248.

Baig, M. S., et al., (2016). Application of Box-Behnken design for preparation of levofloxacin-loaded stearic acid solid lipid nanoparticles for ocular delivery: Optimization, *in vitro* release, ocular tolerance, and antibacterial activity. *Int. J. Biol. Macromol.,* *85,* 258–270.

Bamrungsap, S., et al., (2012). Nanotechnology in therapeutics: A focus on nanoparticles as a drug delivery system. *Nanomedicine,* *7*(8), 1253–1271.

Beg, S., et al., (2018). QbD-driven development and evaluation of nanostructured lipid carriers (NLCs) of olmesartan medoxomil employing multivariate statistical techniques. *Drug Dev. Ind. Pharm.,* *44*(3), 407–420.

Benson, H. A., (2006). Transfersomes for transdermal drug delivery. *Expert. Opin. Drug Deliv.,* *3*(6), 727–737.

Chavhan, S. S., Petkar, K. C., & Sawant, K. K., (2011). Nanosuspensions in drug delivery: Recent advances, patent scenarios, and commercialization aspects. *Crit. Rev. Ther. Drug. Carrier. Syst.,* *28*(5), 447–488.

Chhabra, G., et al., (2011). Design and development of nanoemulsion drug delivery system of amlodipine besylate for improvement of oral bioavailability. *Drug Dev. Ind. Pharm.,* *37*(8), 907–916.

D'Emanuele, A., et al., (2004). The use of a dendrimer-propranolol prodrug to bypass efflux transporters and enhance oral bioavailability. *J. Control Rel.,* *95*(3), 447–453.

Dragicevic-Curic, N., et al., (2008a). Topical application of temoporfin-loaded invasomes for photodynamic therapy of subcutaneously implanted tumors in mice: A pilot study. *J. Photochem. Photobiol. B.,* *91*(1), 41–50.

Dudhipala, N., & Veerabrahma, K., (2015). Pharmacokinetic and pharmacodynamic studies of nisoldipine-loaded solid lipid nanoparticles developed by central composite design. *Drug Dev. Ind. Pharm.,* *41*(12), 1968–1977.

Dudhipala, N., & Veerabrahma, K., (2016). Candesartan cilexetil loaded solid lipid nanoparticles for oral delivery: Characterization, pharmacokinetic and pharmacodynamic evaluation. *Drug Deliv.,* *23*(2), 395–404.

Ekambaram, P., & Abdul, H. S., (2011). Formulation and evaluation of solid lipid nanoparticles of ramipril. *J. Young Pharm.,* *3*(3), 216–220.

ElKasabgy, N. A., Elsayed, I., & Elshafeey, A. H., (2014). Design of liposomes as a novel dual functioning nanocarrier for bioavailability enhancement of lacidipine: *In-vitro* and *in-vivo* characterization. *Int. J. Pharm.,* *472*(1 & 2), 369–379.

Erturk, A. S., Gurbuz, M. U., & Tulu, M., (2017). The effect of PAMAM dendrimer concentration, generation size and surface functional group on the aqueous solubility of candesartan cilexetil. *Pharm. Dev. Technol.,* *22*(1), 111–121.

Fares, A. R., ElMeshad, A. N., & Kassem, M. A. A., (2018). Enhancement of dissolution and oral bioavailability of lacidipine via pluronic P123/F127 mixed polymeric micelles: Formulation, optimization using central composite design and in vivo bioavailability study. *Drug Deliv.,* *25*(1), 132–142.

Fu, Q., et al., (2013). Nimodipine nanocrystals for oral bioavailability improvement: Role of mesenteric lymph transport in the oral absorption. *Int. J. Pharm.,* *1,* *448*(1), 290–297.

Gautam, S. P., & Verma, A., (2012). PAMAM dendrimers: Novel polymeric nanoarchitectures for solubility enhancement of candesartan cilexetil. *Pharm. Sci.*, *1*, 1–4.

Geszke-Moritz, M., & Moritz, M., (2016). Solid lipid nanoparticles as attractive drug vehicles: Composition, properties, and therapeutic strategies. *Mater. Sci. Eng. C Mater. Biol. Appl.*, *68*, 982–994.

Giles, T. D., et al., (2005). Expanding the definition and classification of hypertension. *J. Clin. Hypertens.*, *7*(9), 505–512.

Go, A. S., et al., (2014). An effective approach to high blood pressure control: A science advisory from the American Heart Association, the American College of Cardiology, and the Centers for Disease Control and Prevention. *J. Am. Coll. Cardiol.*, *63*(12), 1230–1238.

Gora, S., et al., (2016). Nanosizing of valsartan by high-pressure homogenization to produce dissolution enhanced nanosuspension: Pharmacokinetics and pharmacodynamic study. *Drug Deliv.*, *23*(3), 940–950.

Gorain, B., et al., (2014). Nanoemulsion strategy for olmesartan medoxomil improves oral absorption and extended antihypertensive activity in hypertensive rats. *Coll. Surf. B Biointerfaces.*, *115*, 286–294.

Hecq, J., et al., (2005). Preparation and characterization of nanocrystals for solubility and dissolution rate enhancement of nifedipine. *Int. J. Pharm.*, *299*(1 & 2), 167–177.

Hernandorena, I., et al., (2017). Treatment options and considerations for hypertensive patients to prevent dementia. *Expert Opin. Pharmacother.*, *18*(10), 989–1000.

Hsu, H. J., et al., (2017). Dendrimer-based nanocarriers: A versatile platform for drug delivery. *Wiley Interdiscip. Rev. Nanomed. Nanobiotechnol.*, *9*(1), 1–21.

Imam, S. S., & Aqil, M., (2016). Penetration enhancement strategies for dermal and transdermal drug delivery: An overview of recent research studies and patents. In: Dragicevic, & Maibach, (eds.), *"Drug Penetration Into/Through the Skin: Methodology and General Considerations"* (pp. 337–350).

Imam, S. S., et al., (2015). Formulation by design-based proniosome for accentuated transdermal delivery of risperidone: *In vitro* characterization and *in vivo* pharmacokinetic study. *Drug Deliv.*, *22*(8), 1059–1070.

Imam, S. S., et al., (2017). Formulation by design based risperidone nanosoft lipid vesicle as a new strategy for enhanced transdermal drug delivery: *In-vitro* characterization, and *in-vivo* appraisal. *Mater. Sci. Eng. C Mater. Biol. Appl.*, *75*, 1198–1205.

Jana, U., et al., (2014). Felodipine loaded PLGA nanoparticles: Preparation, physicochemical characterization and in vivo toxicity study. *Nanoconvergence*, *1*, 31.

Kaithwas, V., et al., (2017). Nanostructured lipid carriers of olmesartan medoxomil with enhanced oral bioavailability. *Colloids. Surf. B. Biointerfaces.*, *154*, 10–20.

Kamran, M., et al., (2016). Design, formulation, and optimization of novel soft nanocarriers for transdermal olmesartan medoxomil delivery: *In vitro* characterization and *in vivo* pharmacokinetic assessment. *Int. J. Pharm.*, *505* (1–2), 147–158.

Khan, A., et al., (2016). Brain targeting of temozolomide via the intranasal route using lipid-based nanoparticles: Brain pharmacokinetic and scintigraphic analyses. *Mol. Pharm.*, *13*(11), 3773–3782.

Khani, S., Keyhanfar, F., & Amani, A., (2016). Design and evaluation of oral nanoemulsion drug delivery system of mebudipine. *Drug Deliv.*, *23*(6), 2035–2043.

Kumar, B. D., Kumar, K. K., & Bhatt, A. R., (2015). Single-walled and multi-walled carbon nanotubes based drug delivery system: Cancer therapy: A review. *Ind. J. of Cancer*, *52*(3), 262–264.

Kumar, V. V., et al., (2007). Development and evaluation of nitrendipine loaded solid lipid nanoparticles: Influence of wax and glyceride lipids on plasma pharmacokinetics. *Int. J. Pharm.*, *335*(1–2), 167–175.

Lackland, D. T., & Weber, M. A., (2015). Global burden of cardiovascular disease and stroke: Hypertension at the core. *Can. J. Cardiol.*, *31*(5), 569–571.

Lau, D. H., et al., (2017). Modifiable risk factors and atrial fibrillation. *Circulation*, *136*(6), 583–596.

Lo, C. T., et al., (2010). Controlled self-assembly of monodisperse niosomes by microfluidic hydrodynamic focusing. *Langmuir.*, *26*(11), 8559–8566.

Mishra, A., et al., (2016). Carvedilol nanolipid carriers: Formulation, characterization and *in-vivo* evaluation. *Drug Deliv.*, *23*(4), 1486–1494.

Mishra, R., Mir, S. R., & Amin, S., (2017). Polymeric nanoparticles for improved bioavailability of cilnidipine. *Int. J. Pharm. Pharm. Sci.*, *9*(4), 129–139.

Moghassemi, S., Hadjizadeh, A., (2014). Nano-niosomes as nanoscale drug delivery systems: An illustrated review. *J. Control. Rel.*, *185*, 22–36.

Moghddam, S. M., et al., (2017). Optimization of nanostructured lipid carriers for topical delivery of nimesulide using Box-Behnken design approach. *Artif. Cells Nanomed. Biotechnol.*, *45*(3), 617–624.

Mouez, M. A., et al., (2016). Composite chitosan-transfersomal vesicles for improved transnasal permeation and bioavailability of verapamil. *Int. J. Biol. Macromol.*, *93*(Pt A), 591–599.

Mudshinge, S. R., et al., (2011). Nanoparticles: Emerging carriers for drug delivery. *Saudi Pharm. J.*, *19*(3), 129–141.

Nagaraj, K., Narendar, D., & Kishan, V., (2017). Development of olmesartan medoxomil optimized nanosuspension using the Box-Behnken design to improve oral bioavailability. *Drug Dev. Ind. Pharm.*, *43*(7), 1186–1196.

Nekkanti, V., et al., (2015). Improved oral bioavailability of valsartan using proliposomes: Design, characterization and *in vivo* pharmacokinetics. *Drug Dev. Ind. Pharm.*, *41*(12), 2077–2088.

Nooli, M., et al., (2017). Solid lipid nanoparticles as vesicles for oral delivery of olmesartan medoxomil: Formulation, optimization and *in vivo* evaluation. *Drug Dev. Ind. Pharm.*, *43*(4), 611–617.

Patwekar, S. L., & Baramade, M. K., (2012). Controlled release approach to novel multiparticulate drug delivery system. *Int. J. Pharm. Pharm. Sci.*, *4*(3), 757–763.

Paudel, A., et al., (2017). Formulation and optimization of candesartan cilexetilnano lipid carrier: *In vitro* and *in vivo* evaluation. *Curr. Drug Deliv.*, *14*(7), 1005–1015.

Porter, C. J., Trevaskis, N. L., & Charman, W. N., (2007). Lipids and lipid-based formulations: Optimizing the oral delivery of lipophilic drugs. *Nat. Rev. Drug Discov.*, *6*(3), 231–248.

Poulter, N. R., Prabhakaran, D., & Caulfield, M., (2015). Hypertension. *Lancet*, *386*(9995), 801–812.

Qadri, G. R., et al., (2017). Invasomes of isradipine for enhanced transdermal delivery against hypertension: Formulation, characterization, and *in vivo* pharmacodynamic study. *Artif. Cells. Nanomed. Biotechnol.*, *45*(1), 139–145.

Quan, P., Shi, K., & Piao, H., (2012). A novel surface-modified nitrendipine nanocrystals with enhancement of bioavailability and stability. *Int. J. Pharm.*, *430* (1–2), 366–371.

Quan, P., Xia, D., & Piao, H., (2011). Nitrendipinenanocrystals: Its preparation, characterization, and *in vitro-in vivo* evaluation. *AAPS Pharm. Sci. Tech.*, *12*(4), 1136–1143.

Qumbar, M., et al., (2017). Formulation and optimization of lacidipine loaded niosomal gel for transdermal delivery: *In-vitro* characterization and *in-vivo* activity. *Biomed. Pharmacother.*, *93*, 255–266.

Ranpise, N. S., Korabu, S. S., & Ghodake, V. N., (2014). Second generation lipid nanoparticles (NLC) as an oral drug carrier for delivery of lercanidipine hydrochloride. *Colloids Surf. B Biointerfaces.*, *116*, 81–87.

Rao, J. P., & Geckeler, K. E., (2011). Polymer nanoparticles: Preparation techniques and size-control parameters. *Prog. Polym. Sci.*, *36*(7), 887–913.

Rizwanullah, M., Ahmad, J., & Amin, S., (2016). Nanostructured lipid carriers: A novel platform for chemotherapeutics. *Curr. Drug Deliv.*, *13*(1), 4–26.

Rizwanullah, M., Amin, S., & Ahmad, J., (2017). Improved pharmacokinetics and antihyperlipidemic efficacy of rosuvastatin-loaded nanostructured lipid carriers. *J. Drug Target.*, *25*(1), 58–74.

Rizwanullah, M., et al., (2018). Phytochemical based nanomedicines against cancer: Current status and future prospects. *J. Drug Target.*, 2018, 26(9),731–752.

Satturwar, P., et al., (2007). pH-responsive polymeric micelles of poly(ethylene glycol)-b-poly(alkyl(meth)acrylate-co-methacrylic acid): Influence of the copolymer composition on self-assembling properties and release of candesartan cilexetil. *Eur. J. Pharm. Biopharm.*, *65*(3), 379–387.

Selvam, R. P., Singh, A. K., & Sivakumar, T., (2010). Transdermal drug delivery systems for antihypertensive drugs–a review. *Int. J. Pharm. Biomed. Res.*, *1,* 1–8.

Shah, U., Joshi, G., & Sawant, K., (2014). Improvement in antihypertensive and antianginal effects of felodipine by enhanced absorption from PLGA nanoparticles optimized by factorial design. *Mater. Sci. Eng. C Mater. Biol. Appl.*, *35*, 153–163.

Sharma, M., Sharma, R., & Jain, D. K., (2016). Nanotechnology-based approaches for enhancing oral bioavailability of poorly water-soluble antihypertensive drugs. *Scientifica* (Cairo), 8525679.

Sica, D. A., (2006). Pharmacotherapy review: Calcium channel blockers. *J. Clinical. Hypertension.*, *8*(1), 53–56.

Simoes, S. M., et al., (2015). Polymeric micelles for oral drug administration enabling locoregional and systemic treatments. *Expert Opin. Drug Deliv.*, *12*(2), 297–318.

Singh, Y., et al., (2017). Nanoemulsion: Concepts, development, and applications in drug delivery. *J. Control. Rel.*, *252,* 28–49.

Song, C. K., et al., (2012). A novel vesicular carrier, transethosome, for enhanced skin delivery of voriconazole: Characterization and *in vitro/in vivo* evaluation. *Coll. Surf. B Biointer.*, *92,* 299–304.

Soni, K., Rizwanullah, M., & Kohli, K., (2017). Development and optimization of sulforaphane-loaded nanostructured lipid carriers by the Box-Behnken design for improved oral efficacy against cancer: *In vitro, ex vivo* and *in vivo* assessments. *Artif. Cells Nanomed. Biotechnol.*, 46: sup1, 15-31.

Thukral, D. K., Dumoga, S., & Mishra, A. K., (2014). Solid lipid nanoparticles: Promising therapeutic nanocarriers for drug delivery. *Curr. Drug Deliv.*, *11*(6), 771–91.

Touitou, E., et al., (2000). Ethosomes: novel vesicular carriers for enhanced delivery: Characterization and skin penetration properties. *J. Control Rel.*, *65*(3), 403–418.

Touitou, E., Piliponsky, A., Levi-Schaffer, F., et al., (2001). Intracellular delivery mediated by an ethosomal carrier. *Biomaterials*, *22,* 3053–3059.

Vasan, R. S., et al., (2002). Residual lifetime risk for developing hypertension in middle-aged women and men: The Framingham heart study. *JAMA, 287*(8), 1003–1010.

Verma, P., & Pathak, K., (2012). Nanosizedethanolic vesicles loaded with econazole nitrate for the treatment of deep fungal infections through topical gel formulation. *Nanomedicine, 8*(4), 489–496.

Vuppalapati, L., et al., (2016). Application of central composite design in optimization of valsartan nanosuspension to enhance its solubility and stability. *Curr. Drug Deliv., 13*(1), 143–157.

WHO, (2013). A global brief on hypertension [Online]. Available at: http://www.who.int/cardio-vascular_diseases/publications/global_brief_hypertension/en/ [last accessed 16 February 2018].

Xu, W., Ling, P., & Zhang, T., (2013). Polymeric micelles, a promising drug delivery system to enhance bioavailability of poorly water-soluble drugs. *J. Drug Deliv., 340315.*

Yasam, V. R., et al., (2016). A novel vesicular transdermal delivery of nifedipine: preparation, characterization and *in vitro/in-vivo* evaluation. *Drug Deliv., 23*(2), 619–630.

Zhang, Z., et al., (2012). Solid lipid nanoparticles loading Candesartan cilexitil enhance oral bioavailability: *In vitro* characteristics and absorption mechanism in rats. *Nanomedicine, 8,* 740–747.

Zheng, X., Wang, T., & Jiang, H., (2013). Incorporation of Carvedilol into PAMAM-functionalized MWNTs as a sustained drug delivery system for enhanced dissolution and drug-loading capacity. *Asian J. of Pharm. Sci., 8*(5), 278–286.

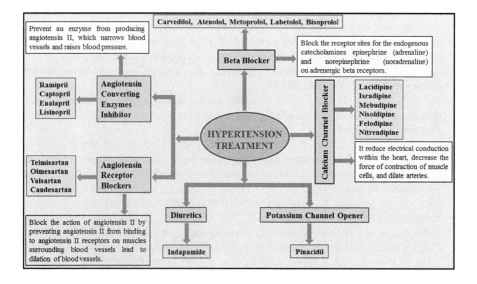

CHAPTER 16

Nanomedicine for the Treatment of Ocular Diseases

PRIYANKA PRABHU and SARITHA SHETTY*

Shobhaben Pratapbhai Patel School of Pharmacy & Technology Management, SVKM's NMIMS, Vile Parle West, Mumbai, India, E-mail: saritha.shetty@nmims.edu

Corresponding author. E-mail: sarithabhandary@gmail.com

ABSTRACT

Nanomedicine has indubitably revolutionized healthcare and the field of ocular delivery is no exception. The indisputable prowess of nanocarriers to enhance ocular delivery of therapeutics ranging from small molecular weight drugs to high molecular weight siRNA is exemplified by a plethora of research endeavors and few commercially available ocular nanoformulations. The innate benefits of nanocarriers include enhanced bioavailability, improved membrane permeation, ability to shield actives against degradation, capacity to provide controlled release, and amenability for surface functionalization. Mucoadhesive nanoparticles have also been designed to prolong ocular residence time and enhance ocular bioavailability. There is also a growing interest in nanocarriers as delivery vehicles in photodynamic therapy (PDT) for the treatment of age-related macular degeneration. Nevertheless, there are a few challenges which encumber ocular delivery. These include the inherent anatomy, physiology, and biochemistry of the eye, dearth of approved excipients, stringent requirements with respect to sterility, particulate matter, pyrogens, and scarcity of ocular toxicity studies on nanocarriers. The chapter highlights the incredible potential of nanotechnology for treatment of ocular diseases and focuses on utilization of a myriad of nanocarriers to deliver a variety of therapeutics to different regions within the eye at a desired rate of release. To conclude, nanotechnology will play a pivotal role in the treatment of ocular conditions by providing highly efficacious therapies with a concomitant decrease in dosing frequency and side effects.

16.1   INTRODUCTION

The eye is an important organ in the human body responsible for visual perception (Delplace et al., 2015). The grimacing statistics of visual disability highlight the pressing need to develop novel ocular therapeutic strategies (Saraiva et al., 2017). It is estimated that the statistics of visually impaired people will increase from 4.4 million to 10 million in 2050 (Joseph & Venkatraman, 2017). Aging and escalating cases of diabetes are responsible for increased episodes of visual disability (Hamdi et al., 2015).

Diseases affecting the eye can be categorized into anterior ocular diseases and posterior ocular diseases. Anterior ocular diseases include corneal wounds, keratitis, extracellular infections, glaucoma, uveitis, conjunctivitis, dry eye syndrome, and cataract (Vyas et al., 2015; Sunkara and Kompella, 2003; Wang et al., 2017). Posterior ocular diseases include glaucoma, age-related macular degeneration, diabetic retinopathy, choroidal neovascularization, uveitis, retinoblastoma, viral retinitis and bacterial/fungal endophthalmitis (Campos et al., 2017; Sunkara and Kompella, 2003; Wang et al., 2017; Qu et al., 2017). The anterior portion of the eye offers easy accessibility; however, the posterior portion of the eye is difficult to access owing to the elimination mechanisms and physical barricades (Hirani and Pathak, 2016).

Ocular therapies may be administered via various routes, the selection being dependent on the condition being treated, physicochemical characteristics of the therapeutic agent and the site of action of the drug within the eye. Commonly employed routes include topical, subconjunctival, and intravitreal. The former two are utilized for the treatment of diseases of the anterior segment of the eye, and the latter one is used for the treatment of diseases of the posterior segment of the eye (Dave and Bhansali, 2016).

Conventional ophthalmic products are unable to deliver the therapeutic agent to the eye in sufficient concentrations owing to the numerous elimination mechanisms and barrier properties of the ocular tissue (Bisht et al., 2017b). This necessitates the installation of large doses of the therapeutic agents at frequent intervals contributing to increased side effects and decreased patient compliance.

Nanocarriers have been explored in the field of ophthalmic drug delivery to overcome the problems of low bioavailability and poor residence time residence time in the ocular tissue. They have been utilized for enhancement of bioavailability, controlled release, gene delivery, photodynamic therapy (PDT), and so on for a variety of ailments ranging from inflammation, infections, glaucoma, retinitis, diabetic retinopathy, and age-related macular degeneration. Although nanocarriers have immense potential to improve delivery of ocular therapeutics, there are issues that need careful consideration before nanotechnologies can be made available in the clinic. There are constraints related to the design of drug delivery systems with respect to drug loading and issues of safety of components. Programming the drug release from controlled release ophthalmic systems is a huge challenge.

The chapter showcases the magnanimous potential of nanocarriers in ophthalmic delivery and recapitulates incessant research efforts which have explored nanocarriers for a myriad of ocular ailments.

16.2 ANATOMY OF THE EYE

The eye consists of the anterior and posterior segments (Figure 16.1). The anterior segment occupies about one-third of the eye and the posterior portion two-thirds of the eye (Patel et al., 2013). The anterior segment comprises the cornea, iris, aqueous humor, and the lens. The posterior segment includes the retina, macula, vitreous humor, optic nerve, sclera, choroid, etc. (Delplace et al., 2015). The anterior portion of the eye is easily accessible whereas corporeal barriers and elimination mechanisms preclude access to the posterior ocular segment (Hirani and Pathak, 2016). Table 16.1 enlists the main parts of the eye with their nature and function.

16.3 CHALLENGES IN OCULAR DELIVERY

The anatomy and physiology of the eye coupled with its biochemistry pose a significant barricade to effective delivery of ocular therapeutic agents (Macha et al., 2003). A very low percentage of the drug (approximately 5%) administered actually reaches the eye through the cornea owing to the impervious barrier posed by the anatomy of the eye, physiology, and biochemistry of the ocular tissue (Kim et al., 2015). The anterior and posterior part of the eye pose distinct challenges to drug delivery and place specific requirements on the respective dosage forms accordingly (Joseph & Venkatraman, 2017).

Delivery of large molecular weight hydrophilic agents to the eye such as nucleotides is a formidable task (Saraiva et al., 2017).

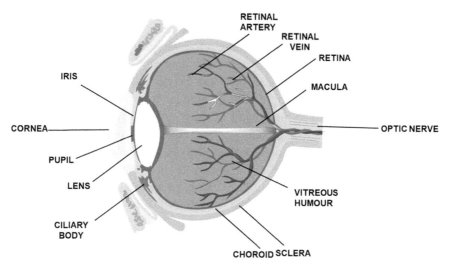

FIGURE 16.1   (See color insert.) Structure of the eye.

TABLE 16.1   Main Parts of the Eye

Structure	Nature	Function
Cornea	Avascular, contains three layers, the epithelium-stroma, and the endothelium (Dilnawaz and Sahoo, 2016)	Covers the front of the eye and acts as a protective barrier (Dilnawaz and Sahoo, 2016).
Iris	Shutter diaphragm of the eye (Malavade, 2016)	Has muscles responsible for constriction and dilation of pupil (Malavade, 2016)
Lens	Biconvex, avascular, transparent (Malavade, 2016)	Focuses the light rays which enter the eye via the pupil to get a sharp image on the retina (Malavade, 2016)
Retina	Comprises vascular cells, glial cells, and nerve fibers (Malavade, 2016)	Neurosensory part of the eye (Malavade, 2016)
Macula	Central part of the retina (Vyas et al., 2015)	Responsible for visual sharpness and color perception (Malavade, 2016)
Optic nerve	Around 7 cm long (Malavade, 2016)	Responsible for transmission of visual signal from the retina to the brain (Malavade, 2016)
Sclera	White of the eye (Dilnawaz and Sahoo, 2016).	Determines the shape and length of the eye (Dilnawaz and Sahoo, 2016).
Choroid	Vascular part of the eye (Malavade, 2016)	Provides nutrition to the retina (Sangave et al., 2016)

16.3.1 MEMBRANE BARRIERS

16.3.1.1 CORNEAL BARRIER

The cornea poses a powerful barrier to passage of drugs. Drug passage across the cornea involves transport through a series of strata endowed with different physicochemical characteristics. These layers include the epithelial layer, stroma, and the endothelial layer (Sangave et al., 2016). The tight junctions present in the corneal epithelium hinder the passage of drugs across the cornea (Kim et al., 2015). The middle layer called stroma which is made of collagen comprises around 90% of the cornea. It is hydrophilic and hinders the passage of lipophilic substances across the cornea (Macha et al., 2003). Corneal drug absorption is influenced by the molecular weight of the drug, its octanol-water partition coefficient and state of ionization. Formulation components such as viscosity builders, penetration enhancers also impact the penetration of drug through corneal membrane (Chastain, 2003).

16.3.1.2 BLOOD-AQUEOUS BARRIER

This exists in the anterior segment of the eye. It actually precludes entry of substances from plasma into the aqueous humor (Dave and Bhansali, 2016).

16.3.1.3 BLOOD-RETINAL BARRIER

This is comprised of retinal vascular endothelium (RVE) and retinal pigment epithelium (RPE). Both RVE and RPE contain tight junctions which obstruct the drug penetration into the intraocular compartment (Wang et al., 2017). Diabetic retinopathy and age-related macular degeneration are known to modify the RPE (Dave and Bhansali, 2016).

16.3.2 ELIMINATION MECHANISMS

16.3.2.1 NASOLACRIMAL DRAINAGE

This is an important factor which contributes to decreased ocular residence time (around 2 min) and low ocular bioavailability (5% corneal drug absorption) (Sharma and Taniguchi, 2017). This may also result in adverse effects owing to systemic absorption (Gallarate et al., 2013).

16.3.2.2   SYSTEMIC ABSORPTION

The noncorneal route of absorption includes transport of drugs across the sclera and conjunctiva into the intraocular tissues (Macha et al., 2003). The drugs absorbed by this pathway reach the aqueous humor and may enter the systemic circulation via local capillaries and show higher clearance (Dave and Bhansali, 2016).

16.3.2.3   EFFLUX PUMPS

Presence of efflux pumps (MRP1, MRP5, breast cancer resistance protein) on the corneal epithelium causes ejection of drugs out of the corneal epithelial cells (Kim et al., 2015).

16.3.2.4   DRUG METABOLISM

Enzymes such as esterases, lysozymes, oxidoreductases, and monoamine oxidases carry out drug metabolism (Sangave et al., 2016). Drugs administered via the topical route undergo metabolism by corneal enzymes and systemically and periocularly administered drugs are metabolized by the retinal enzymes (Dave and Bhansali, 2016).

16.3.2.5   BINDING OF DRUGS TO PROTEINS

Drug binding to melanin protein can occur in the anterior segment of the eye resulting in prolonged release of the drug from the drug-protein complex (Dave and Bhansali, 2016).

16.3.2.6   BLINKING REFLEXES

Blinking of the eyelids results in flow of the surplus amount of instilled drops into the lacrimal drainage system causing loss of drug from the ocular region within 2 min (Godse et al., 2016).

16.3.2.7   TEAR FILM DILUTION AND TEAR FILM TURNOVER

The eye is shielded from the ingress of harmful substances by continuous generation of a tear film which constantly washes the surface of the eye. Dilution of the instilled formulation with tear fluids results in significant reduction in the transcorneal flux of the drug. Certain factors such as irritant nature of drug and/ or excipients, paratonic nature of formulation, and pH may aggravate lacrimation resulting in further decrease in ocular bioavailability (Macha et al., 2003).

16.4 ROLE OF NANOCARRIERS

Nanotechnology which involves conception and application of materials at nanometer level has touched healthcare in manifold ways exemplified by its incessant contribution in a gamut of areas such as diagnosis, therapy, and prophylaxis (Al-Halafi, 2014). Nanocarriers possess immense potential to enhance the efficacy of ocular therapy through a multitude of mechanisms as discussed below.

1. Ability to Offer Sustained/Prolonged Drug Release

Nanocarriers can be formulated using various matrices to tune the rate of drug release. This feature of sustained release is beneficial to avoid frequent ocular administration/injection. Use of biodegradable nanosized carriers also eliminates the need for surgical insertion and removal of the carrier after the drug or gene has been released (Bisht et al., 2017b).

2. Amenable to Surface Modification/Functionalization

Nanocarriers can be designed for targeted ocular delivery. Nanocarriers such as dendrimers and liposomes are endowed with the ability to offer multivalent surface functionalization with ligands targeted to the desired site (Kambhampati and Kannan, 2013).

3. Enhanced Membrane Permeation

Nanocarriers are reported to enhance penetration of drugs (Vyas et al., 2015).

4. Increased Drug Loading

The huge surface area-to-mass ratio of nanocarriers enhances their drug loading capacity tremendously (Sharma and Taniguchi, 2017).

5. Augmented Solubility

Numerous nanocarriers such as nanosuspensions, micelles, microemulsions are known to enhance the solubility of poorly soluble drugs and enhance their bioavailability (Dave, 2016; Kaur and Kakkar, 2014).

6. Freedom from Ocular Irritation

Small size of nanocarriers precludes irritation to ocular tissue (Bisht et al., 2017b).

7. Easy Formulation in the Form of Drops

Nanocarriers encapsulating therapeutics can be developed into aqueous solutions which can be instilled into the eye (Bisht et al., 2017b)

8. Delivery of Genetic Materials

Nanocarriers help to preserve the stability of the entrapped/complexed genetic material (Bisht et al., 2017b). Nanocarriers having size of less than 25 nm are able to cross the pores of the nuclear membrane and aid in transfection (Kaur and Kakkar, 2014).

9. Shielding Sensitive Substances from Degradation

Encapsulation of drugs into the nanocarrier matrix can actually protect the drug against degradation (Reimondez-Troitiño et al., 2015).

10. Increasing Residence Time

Bioadhesive or mucoadhesive nanocarriers can be designed which prolong the ocular residence time by adhering to the mucosal layer (Reimondez-Troitiño et al., 2015). This feature enables reduction in dosing frequency which is beneficial from patient compliance point of view.

Figure 16.2 depicts the various nanocarriers explored for ocular delivery of therapeutics. Table 16.2 summarizes definition, benefits, and drawbacks of different nanocarriers.

16.5   NANOCARRIERS FOR TREATING DISEASES AFFECTING THE ANTERIOR SEGMENT OF THE EYE

Nanocarriers have been used extensively to treat conditions such as dry eye syndrome, bacterial, fungal, and viral infections of the eye. Tears Again Advanced Eyelid Spray® is a liposomal product which is helpful to treat dry eye syndrome (Gupta, 2016). Other liposomal products available for dry eye syndrome include TearMist®, Optix-Actimist®, and DryEyesMist®. Cationorm® and Lipimix™ are nanoemulsions which are available to treat dry eye syndrome (Hamdi et al., 2015).

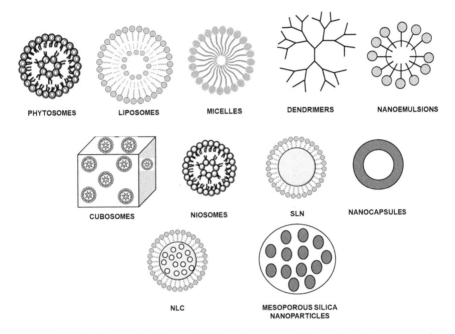

FIGURE 16.2 **(See color insert.)** Schematic representation of various nanocarriers explored for ocular delivery.

Li et al. developed micelles loaded with pirenzepine for treatment of myopia. Pirenzepine was complexed with sorbic acid through Coulombic interactions to form an ion pair. Sorbic acid is safe for ophthalmic use and enhances the lipophilicity of hydrophilic pirenzepine facilitating higher loading in the hydrophobic micellar core. Methoxypoly (ethylene glycol)-polylactide copolymer (mPEG-PDLLA), an amphiphilic block polymer was utilized to form micelles. The developed micellar system demonstrated 1.55 fold higher ocular bioavailability in rabbits than plain pirenzepine micelles and was non-irritant to rabbit ocular tissue. The developed micelles possess an edge over the existing treatment options for pirenzepine which include iontophoresis and intravitreal injections (Li et al., 2017).

Nanocarriers have been utilized for several purposes as follows.

16.5.1 ENHANCED PENETRATION

Kaur et al. have fabricated tropicamide containing carboxymethyl tamarind kernel polysaccharide (CMTKP) nanoparticles. The nanoparticles demonstrated

TABLE 16.2 Definition, Benefits, and Drawbacks of Different Nanocarriers

	Definition	Benefits	Drawbacks
Liposomes	Contain a lipid bilayer made of phospholipids and cholesterol enclosing an aqueous core in which therapeutic agents can be encapsulated (Kaur and Kakkar, 2014)	Biocompatible (Kaur and Kakkar, 2014), can entrap hydrophilic and hydrophobic moieties (Kaur and Kakkar, 2014), structure mimicking cellular membranes (Patel et al., 2013)	Limited shelf life, restricted drug loading, and problems in sterilization (Kaur and Kakkar, 2014), may cause vitreous clouding (Gupta, 2016)
Niosomes	Produced by self-assembly of nonionic surfactants (Gan et al., 2013)	Superior chemical stability, easy availability and low cost of excipients as compared to liposomes, biodegradable, biocompatible (Campos et al., 2017)	Low physical stability, possibility of drug leakage, fusion, aggregation (Gandhi et al., 2014)
Cubosome (liquid crystalline nanoparticles)	Contain a contorted continuous lipid bilayer having two congruent non-intersecting aqueous channels (Huang et al., 2017)	Enhanced surface area and higher physicochemical stability than liposomes, microstructure resembling biological membranes facilitating fusion with corneal epithelium (Huang et al., 2017)	Challenging scale up (Ramya Sri et al., 2017)
Phytosomes	Supramolecular complexes of phytoconstituent and phospholipid (Abdelkader et al., 2016)	Increased drug loading compared to liposomes (Abdelkader et al., 2016)	—
Solid lipid nanoparticles (SLN)	Contain a lipid core surrounded by a stabilizing corona of surfactants (Kaur and Kakkar, 2014)	Permit controlled drug release, possess long-term stability, can encapsulate both hydrophilic and hydrophobic compounds, physiologically compatible, amenability to autoclaving by sterilization, and simply scale up (Kaur and Kakkar, 2014), shielding actives from degradation (Vyas et al., 2015)	Poor drug loading and drug ejection on storage (Vyas et al., 2015)
Nanostructured lipid carriers (NLC)	Blend of solid lipids and liquid lipids stabilized by a layer of surfactants (Kaur and Kakkar, 2014)	Enhance drug loading than SLN owing to the imperfections and superior long-term stability (Kaur and Kakkar, 2014)	—

TABLE 16.2 *(Continued)*

	Definition	Benefits	Drawbacks
Polymeric Micelles	Produced by self-assembly of amphiphilic block copolymers above their critical micellar concentration with a hydrophobic core and hydrophilic shell (Kaur and Kakkar, 2014)	Augment solubility of poorly soluble drugs (Kaur and Kakkar, 2014), protect the drugs in inner core and outer core, allows surface functionalization for targeting (Shen et al., 2015); good kinetic and thermodynamic stability, narrow size distribution, enhanced corneal penetration owing to the nanosize (20–100 nm) (Li et al., 2015),	—
Polymeric nanoparticles	Polymer-based colloidal particles having size 10–1000 nm	Controlled release resulting in decreased dosing frequency (Kaur and Kakkar, 2014) Commonly used polymers include chitosan, polycaprolactone, Eudragit® RS100, hyaluronic acid, gelatin, carbopol, poly butyl cyanoacrylate, poly (lactic-co-glycolic acid) (PLGA) (Reimondez-Troitiño et al., 2015)	—
Dendrimers	Branched units resembling a tree endowed with an inner core and branches of different generations offering multiple sites for surface functionalization and binding (Saraiva et al., 2017)	Greater drug loading and possibility of surface functionalization for targeting due to multivalent nature (Shen et al., 2015), enhanced corneal residence time (Reimondez-Troitiño et al., 2015)	No clarity on in vivo safety (Honda et al., 2013)
Nanosuspensions	Colloidal dispersions containing submicron size drug particles stabilized by polymeric substances and/or surfactants (Patel et al., 2013)	Enhanced dissolution velocity and saturation solubility for poorly soluble drugs (Soltani et al., 2016); easy sterilization, non-irritation (Patel et al., 2013)	Thermodynamically unstable with propensity for crystal growth (Wang et al., 2013)

TABLE 16.2   *(Continued)*

	Definition	Benefits	Drawbacks
Nanocapsules	Composed of oily nanodroplets enclosed in a polymeric coat (Katzer et al., 2014)	Better entrapment than liposomes, useful for gene delivery owing to their property to protect encapsulated moiety against degradation (Patravale and Prabhu, 2014)	—
Nanoemulsions	Contain two immiscible liquids out of which one liquid is dispersed as droplets in another liquid stabilized by surfactants (Kaur and Kakkar, 2014)	Excellent permeation owing to small size and surfactant (Al-Halafi, 2014), possibility of sustained drug release (Yin et al., 2016) Biodegradable, simple sterilization, and enhanced drug solubilization (Gallarate et al., 2013), excellent stability against creaming, coalescence, and sediment formation (Gallarate et al., 2013)	—
Mesoporous silica nanoparticles	Silica nanoparticles with 2–6 nm mesopores (Vallet-Regi et al., 2017)	Biocompatible nature, amenability for surface modification, chemically stable, large surface area (Qu et al., 2017)	—

enhanced *ex vivo* permeation across isolated goat cornea compared to plain drug solution and showed good ocular tolerance. The nanocarriers also possessed mucoadhesive characteristic owing to the hydrogen bond formation between carboxylic groups of CMTKP and oligosaccharide chains of mucin (Kaur et al., 2012). Jain et al. have complexed natamycin with a cell-penetrating peptide Tat-dimer (Tat$_2$) via covalent bonding to treat fungal keratitis. Amide linkage was employed to form complexes between natamycin and Tat-dimer (Tat$_2$). Complexation resulted in an increase in the solubility of natamycin and augmented corneal penetration of the latter. The developed complexes showed 2 fold higher antifungal activity against clinical isolates of *Fusarium solani* than unconjugated natamycin (Jain et al., 2015). Gallarate et al. developed oil-in-water nanoemulsions for delivery of timolol maleate. The developed emulsions were stable and possessed desired osmolarity, viscosity, and surface tension for ocular instillation. Hydrophobic ion-pair formation with Bis-(2-ethylexyl)-sulfosuccinate resulted in enhanced corneal penetration. Inclusion of chitosan resulted in enhanced penetration owing to the cationic and mucoadhesive nature of chitosan (Gallarate et al., 2013). Ahuja et al. (2013) developed chitosan-based nanosuspension of itraconazole using the nanoprecipitation technique. Coprecipitation of chitosan and itraconazole was done by using aqueous sodium hydroxide solution since both chitosan and itraconazole exhibit pH-dependent solubility profile and show poor solubility at alkaline pH. The developed nanosuspension enhanced the saturation solubility of itraconazole by 12-fold and resulted in 97% dissolution of the drug at the end of 120 min as compared to 43% shown by the drug suspension. It also showed greater penetration across the isolated goat cornea than plain drug suspension (Ahuja et al., 2013). Mohanty et al. developed SLN of itraconazole using palmitic acid and stearic acid as solid lipids. Entrapment efficiency of drug was found to be more in SLN formulated using stearic acid than palmitic acid probably due to the lipophilicity of the former. The stearic acid containing SLN also demonstrated greater penetration across goat cornea as compared to the ones containing palmitic acid. The SLN also showed good antimicrobial activity against *Aspergillus flavus* (Mohanty et al., 2015).

16.5.2 SUSTAINED RELEASE

Katzer et al. developed prednisolone containing nanocapsules in the size range of 130–260 nm. The nanocapsules were formulated using castor oil and mineral oil, polysorbate 80 as surfactant and Eudragit® RS100 or poly ε-caprolactone as the polymeric coat. The nanocapsules exhibited a sustained release profile for a period of 5 h owing to the polymeric coat surrounding the oil nanodroplets. Hen's egg test Chorioallantoic membrane (HETCAM) test revealed the non-irritant nature of the formulation. This was further supported by the cytotoxicity test wherein they were found to be non-cytotoxic towards corneal epithelial cells (Katzer et al., 2014).

Mahor et al. developed positively charged gelatin nanoparticles containing moxifloxacin for controlled drug release. The nanoparticles showed a size of 175 nm and demonstrated initial burst release followed by sustained drug release up to 12 h. The nanoparticles exhibited superior *in vivo* antibacterial activity against *Staphylococcus aureus* as compared to marketed MoxiGram®. The nanoformulation was non-irritant to the rabbit eye. It showed symptomatic relief within 4 days when used twice daily as opposed to MoxiGram® which was given four times a day and failed to produce significant effect even at the end of six days (Mahor et al., 2016).

16.5.3   ENHANCED PENETRATION AND SUSTAINED DRUG RELEASE

Li et al. developed triblock copolymer micelles for enhanced drug penetration and sustained drug release. The micelles were formulated by self-assembly of the amphiphilic triblock poly(ethylene glycol) (PEG)-poly(ε-caprolactone) (PCL)-g-polyethyleneimine (PEI) copolymer. PEG and PEI constituted hydrophilic chains of the micelle and PCL the hydrophobic portion. The spherical micelles showed a size of 28 nm and were positively charged. Positively charged micelles enhance corneal penetration owing to the electrostatic interactions with the negatively charged mucin layer which prolongs precorneal residence time. The micelles demonstrated controlled release of fluorescein diacetate and showed enhanced corneal penetration (Li et al., 2015). Baig et al. developed levofloxacin loaded SLN for treatment of conjunctivitis. The SLN showed a globule size of 238 nm, demonstrated sustained drug release up to 12 h, and enhanced drug penetration across excised goat cornea. The SLN showed comparable efficacy as that of marketed eye drops against *Staphylococcus aureus* and *Escherichia coli*. HET-CAM test confirmed that the SLN were safe for ophthalmic application (Baig et al., 2016).

16.5.4   PROLONGED OCULAR RESIDENCE TIME

Nanocarriers employed for prolonging the ocular residence time commonly employ mucoadhesive polymers which help to increase the contact time between the formulation and the corneal surface to enhance drug absorption. Treatment of dry eye syndrome using conventional eye drop preparations necessitates frequent dosing owing to low ocular retention resulting in poor patient compliance. The developed mucoadhesive nanocarriers demonstrated ocular retention for over 24 h. Cyclosporine A loaded in these mucoadhesive nanoparticles was able to cause elimination of inflammation and promoted recovery of goblet cells. Administration of Restasis ® could only cause elimination of inflammation without promoting recovery of goblet cells. The developed mucoadhesive nanoplatform could treat dry eye syndrome at 50–100 fold lower dose of the drug as compared to Restasis® Liu et al., 2016. Shen et al. developed a thiolated non-ionic surfactant cysteine-polyethylene glycol stearate (Cys-PEG-SA) and fabricated NLC using the same for ocular delivery of cyclosporine A. Enhanced mucoadhesion was attributed to the formation of disulfide bonds between thiolated NLCs and cysteine-rich subdomains of mucus glycoproteins in mucus. The thiolated NLC were non-irritant to the rabbit ocular tissue. The thiolated NLC demonstrated increased precorneal retention time in rabbits than plain NLC (Shen et al., 2009).

16.5.5 ENHANCED PENETRATION AND PROLONGED OCULAR RESIDENCE TIME

Huang et al. fabricated cubosomes of timolol maleate from glycerol monooleate and Poloxamer 407 using the high-pressure homogenization method. The cubosomes demonstrated increased corneal penetration than timolol maleate eye drops. They also showed prolonged ocular retention and superior reduction in intraocular pressure than marketed eye drops in rabbits. The cubosomes were compatible with the corneal tissues. It is recommended that the cubosomes should possess a size of less than 200 nm for efficient corneal penetration (Huang et al., 2017). Conventional eye drops of gatifloxacin necessitate frequent administration of 4–6 times a day. Zubairu et al. fabricated chitosan coated niosomes to enhance the transcorneal penetration of the drug and also to prolong the ocular residence time. The developed niosomes exhibited more than double corneal penetration (87% in 24 h) as compared to plain gatifloxacin suspension (35% in 24 h). They also showed enhanced ocular residence time of more than 12 h. The developed niosomes showed superior antimicrobial activity against *Bacillus subtilis* than marketed formulation. Corneal hydration study revealed the non-irritant nature of the developed niosomes (Zubairu et al., 2015).

16.5.6   SUSTAINED DRUG RELEASE AND PROLONGED RESIDENCE TIME

Natamycin is used to treat fungal keratitis. However, its present dosage regimen involves administration of a 5% w/v suspension every 1 h or every 2 h for many days which is inconvenient for the patient. Chandasana et al. developed cationic chitosan functionalized polycaprolactone nanoparticles using the nanoprecipitation technique. The nanoparticles showed a size of 217 nm and a zeta potential of 43.5 mV. Chitosan conferred mucoadhesive properties to the nanoparticles plus the cationic charge helped to prolong the corneal contact time owing to electrostatic interaction with the negatively charge corneal mucin layer. Chitosan also has additional characteristics such as penetration enhancement, enhancement of paracellular transport, and fungicidal action. The nanoparticles showed prolonged release of natamycin up to 8h as compared to Natamet® which released the entire drug load within 2 h. The nanoparticles showed a higher zone of inhibition against *Candida albicans* and *Aspergillus fumigatus* in the disk diffusion assay as compared to Natamet®. *In vivo* pharmacokinetic studies in rabbits revealed 6-fold increased mean residence time, decreased clearance, and 6-fold enhanced bioavailability as compared to marketed preparation Natamet®. The nanoformulation resulted in a drastic decrease in the dosing frequency to once every 5h. The nanoformulation also reduced the dose of natamycin by five-fold (Chandasana et al., 2014).

16.5.7   ENHANCED PENETRATION, SUSTAINED DRUG RELEASE AND PROLONGED RESIDENCE TIME

Yin et al. developed palmatine loaded cationic lipid-based emulsions to treat ocular infections. The emulsions were fabricated with 1,2-dioleoyl–3-trimethylammonium-propane (DOTAP) as cationic modifier and had a size of 192 nm. The corneal surface is negatively charged as a result of which cationic nanocarriers can interact better through electrostatic interaction thus enhancing corneal residence time. The emulsions demonstrated increased corneal residence time. They also exhibited excellent internalization by human corneal epithelial cells. The emulsions demonstrated a sustained release pattern and showed enhanced antifungal activity in animal model of fungal keratitis (Yin et al., 2016). Ustundag-Okur et al. have developed ofloxacin-containing NLC modified with chitosan oligosaccharide lactate (COL) to treat bacterial keratitis. NLC were fabricated using oleic acid,

Compritol HD5 ATO, Tween 80, and ethanol by using microemulsion technique and high shear homogenization technique (HNLC). The NLC developed by the HNLC showed higher particle size (153.5 nm) than the one produced by microemulsion technique (MNLC) (8.6 nm). This reduced particle size of the MNLC resulted in a faster rate of drug release during the first 3 h as compared to HNLC. HNLC, MNLC, COL modified MNLC, and COL modified HNLC showed sustained drug release up to 12 h. The developed NLC demonstrated excellent penetration through excised rabbit corneas in comparison to marketed drug solution Exocin®. Modification of NLC with COL increased pre-corneal residence time in rabbits owing to the mucoadhesive nature of COL. The HNLC modified with 0.75% COL showed the greatest corneal retention time of 60 min whereas Exocin® could be retained for only 5 min. The COL modified HNLC showed highest drug concentration in the aqueous humor after installation in the rabbit eye. The Col modified NLC demonstrated good ocular safety as exemplified by the low irritation score as compared to Exocin® during the Draize test in rabbits (Üstündag˘-Okur et al., 2014).

Zhao et al. developed galactosylated chitosan nanoparticles containing timolol maleate. Chitosan possesses a plethora of properties which make it ideal for ophthalmic delivery. Firstly, being positively charged it interacts with the negatively charged mucus layer and prolongs corneal residence time. Secondly, its biodegradability and biocompatibility make it useful for ocular delivery. Galactosylated chitosan shows solubility at neutral pH unlike chitosan which is soluble only at acidic pH and also shows enhanced mucoadhesion and cell compatibility than chitosan. The developed nanocarriers showed sustained drug release up to 8 h as opposed to commercial timolol maleate eye drops. Estimation of ocular safety using the hen's egg chorioallantoic membrane assay revealed non-irritant characteristic of the formulation. The formulation also showed good corneal penetration and corneal retention in comparison to chitosan nanoparticles and timolol maleate eye drops. Confocal laser scanning microscopy study indicated the greater depth of penetration by the galactosylated chitosan nanocarriers which may be attributed to the combined effects of enhanced corneal retention due to chitosan, enhanced affinity of the developed nanocarriers for lipophilic corneal epithelium, nanosized dimensions (223 nm), and opening of tight junctions by chitosan. The formulation also showed superior lowering of intraocular pressure than timolol maleate eye drops (Zhao et al., 2017).

Liu et al. (2012) developed NLC of mangiferin for treatment of cataract. The NLC showed sustained release of mangiferin up to 12 h whereas the plain drug solution rapidly released the entire drug load within 3 h. They were stable for a period of 3 months and showed 4 fold higher penetration. The NLC showed an extended corneal residence time as compared to plain drug solution. Pharmacokinetic study revealed approximately 5 fold enhanced bioavailability compared to plain drug solution. This may be because of the following reasons. Incorporation of mangiferin into the lipophilic matrix of NLC enhances the flux across the cornea. Gelucire 44/14 used in the NLC acts as an absorption enhancer for hydrophobic molecules. Labrasol also acts as a penetration enhancer. Also, the nanosized dimensions of nanoparticles improve corneal transport. Fu et al. developed chitosan modified NLC for delivery of amphotericin B to treat fungal keratitis. The NLC showed a size of 185 nm and showed sustained *in vitro* release. The formulation was non-irritant in rabbits and enhanced corneal penetration. Ocular pharmacokinetics showed that the chitosan modified amphotericin NLC showed twice greater AUC than amphotericin NLC and 4 fold higher AUC than amphotericin B eye drops. The chitosan modified NLC showed prolonged residence time up to 24 h whereas the eye drops and amphotericin NLC failed to do so. The enhanced penetration and prolonged residence time may be attributed to chitosan which is known to act on the tight junctions present in the cornea and undergo strong mucoadhesion via electrostatic and hydrogen bonding (Fu et al., 2016).

Tan et al. developed timolol maleate loaded chitosan-coated liposomes. The liposomes had a particle size of 150.7 nm, demonstrated prolonged drug release and showed triple fold increase in the corneal permeation. Gamma scintigraphy studies showcased prolonged corneal retention of the liposomes than the commercial eye drops. The developed formulation was non-irritant to rabbit ocular tissue and also showed remarkable lowering of intraocular pressure in rabbits (Tan et al., 2017).

16.6 NANOCARRIERS FOR TREATING DISEASES AFFECTING THE POSTERIOR SEGMENT OF THE EYE

Posterior eye disorders may be treated using topical, scleral, intravitreal and subconjunctival routes, and intraocular dosage forms (Campos et al., 2017). However, the topical route is rarely useful owing to the barrier properties of the eye and the elimination mechanisms discussed before. Intraocular dosage forms include injections and implants. Therapeutics utilized for treatment of posterior eye disorders include drugs, genetic material, and photosensitizers for PDT.

16.6.1 DRUG DELIVERY

16.6.1.1 TARGETED THERAPIES FOR DELIVERY TO POSTERIOR SEGMENT

Epithelial cell adhesion molecule (EpCAM) receptor is overexpressed in retinoblastoma cells. Qu et al. fabricated EpCAM antibody-conjugated mesoporous silica nanoparticles (148 nm) loaded with carboplatin for treatment of retinoblastoma. Carboplatin was loaded onto mesoporous silica nanoparticles having carboxylic acid group on the external surface. This group was utilized for conjugation with amine group of EpCAM. The spherical nanoparticles demonstrated controlled drug release. The nanoparticles exhibited superior uptake in retinoblastoma cells than non-targeted counterparts and were found to localize in the lysosome. The developed nanoparticles showed a substantial decrease in IC_{50} value and double caspase—3 levels as compared to pure carboplatin (Qu et al., 2017).

Presently used therapy includes corneal or conjunctival administration of anti-VEGF antibody to treat corneal neovascularization which is invasive and excruciating. Epigallocatechin–3-gallate (EGCG) reportedly reduces angiogenesis and decreases VEGF-based corneal neovascularization. Topical administration of EGCG is unable to cross the corneal barricade to reach the posterior areas. Chang et al. developed gelatin nanoparticles loaded with EGCG and modified their surface with hyaluronic acid and RGD peptide. Amino group of RGD peptide was conjugated with carboxylic acid group of hyaluronic acid. The nanoparticles had a size of around 168 nm and demonstrated sustained release of drug (30% at 30 h). Twice-a-day application of the eye drops in mouse model for a week revealed reduction in number of blood vessels and their size (Chang et al., 2017).

16.6.1.2 SMART DRUG SYSTEMS AND IMPLANTS FOR POSTERIOR SEGMENT DISORDERS

The herculean task of delivery to the posterior segment of the eye continues to baffle the formulation scientists. Presently chronic therapy of posterior eye disorders necessitates frequent intravitreal injections of solutions because of the small half-life (Yasin et al., 2014). This leads to numerous issues such as endophthalmitis, infection, cataract formation, and increase in intraocular pressure (Bisht et al., 2017a). Implants which can deliver the drug at a constant rate are the solution to this problem. Examples of ocular implants include Ozurdex®, Vitrasert®, Retisert®, Iluvien®, etc. However, these implants release the drug at a preset rate. The need of the hour is to develop stimulus responsive or smart implants which can release the drug at different rates so as to match the disease conditions of the patient. Light-activated systems can be developed owing to transparent character of cornea and lens. Formulation of implants requires careful consideration of biocompatibility of polymers, tuning release kinetics, and cost of system (Yasin et al., 2014).

Bisht et al. developed photo-responsive in situ forming injectable implants made from methacrylate alginates loaded with PLGA nanoparticles for intravitreal peptide delivery. Photo-responsive in-situ forming injectable implants (ISFIs) are free-flowing liquids which get transformed to a gel on exposure to light. Peptide containing PLGA nanoparticles (149 nm) were formulated by nanoprecipitation. The formulation showed good syringe-ability and the formation of a transparent oval-shaped implant was seen in bovine vitreous after 24 h post intravitreal injection. Nanoparticles were loaded into the in situ forming implant to confer superior control over the sustained release profile of peptide and minimize burst release. Biodegrad-able property of the system was an added advantage in that there was no need to remove it after release of payload (Bisht et al., 2017a).

Huu et al. developed UV light responsive polymeric nanoparticles for delivery of nintedanib, an angiogenesis inhibitor. The polymer employed was made of o-nitrobenzyl moiety in monomer. The safety of the nanopar-ticles was confirmed using electroretinograms. The nanoparticles were able to preserve their payload and release drug in vitreous cavity up to 30 weeks post injection after irradiation. Intravitreal injection of the nanocarriers in Brown Norway rats resulted in greater inhibition of choroidal neovascular-ization as compared to free drug and drug-loaded PLGA nanoparticles (Huu et al., 2015).

Kakkar and Kaur developed elastic vesicles based on Span 60 and Tween 80 called "Spanlastics" for delivery of ketoconazole to the posterior segment of the eye. The vesicles showed twice greater corneal penetration than niosomes. The safety of the formulation was exemplified from results of Ames test, MTT assay, dermal irritation, and chronic eye irritation. Topical application in rabbits revealed presence of intact vesicles in interior of eye 2 h post application. Elastic nature of the vesicles enabled their deformation and squeezing through the pores of the corneal membrane. The "Spanlastics" system provided a non-invasive therapeutic platform for retinal delivery of drugs (Kakkar and Kaur, 2011).

16.6.1.3 SYNERGISTIC USE OF NANOCARRIERS WITH IONTOPHORESIS

Chopra et al. (2012) developed mixed micelles using sodium taurocholate and egg lecithin loaded with dexamethasone. These are known to undergo phase transition on dilution with water to form liposomes which can prolong drug release. They used these micelles in combination with transscleral iontophoresis. Transscleral iontophoresis using plain drug results in quick elimination from the eye necessitating frequent iontophoresis. Transscleral iontophoresis of the micelles provided sustained drug release. The system thus provided a non-invasive method for treating chronic inflammatory conditions of the posterior segment (Chopra et al., 2012).

16.6.2 GENE DELIVERY

Gene delivery involves the introduction of DNA or siRNA into the cell either to knock out any gene which exhibits improper function or to substitute for a mutated gene. Delivery of such material is hindered by rapid nuclease-based breakdown, compromised intracellular uptake, and inability for endosomal escape. Researchers have tried vectors of viral and non-viral origin to enable successful gene delivery to target sites (Prabhu and Patravale, 2012). Utilization of viral vectors for gene delivery may result in oncogenesis, immune responses, and accumulation of vectors in brain post intravitreal delivery. Puras et al. developed niosomes from cationic lipid 2,3-di(tetradecyloxy) propan–1-amine, squalene and polysorbate 80. The cationic lipid possesses inherent properties essential for gene transfection namely polar head, nonpolar hydrophobic chains, linker, and backbone. The niosome formed lipoplex with pCMS-EGFP plasmid. The nanocarriers resulted in good transfection in HEK–293 and ARPE–19 cells without any toxicity. The system showed good transfection in vivo in rats post intravitreal and subretinal injection (Puras et al., 2014).

Human retinal pigment epithelial cells are known to play a pivotal role in angiogenesis of choroid via increased expression of angiogenic factors. Delivery of siRNA to retinal pigment epithelial cells can be a potential dual preventive and therapeutic strategy for choroidal neovascularization in age-related macular degeneration. Formulation of siRNA is a formidable task owing to stability issues and low cellular penetration (Chen et al., 2011). Large size of siRNA coupled with its polyanionic nature precludes its passive cellular penetration (Chen et al., 2013). Integrin receptor $\alpha_v\beta_3$ is reportedly overexpressed in patients with age-related macular degeneration. Hence, Chen et al. fabricated integrin receptor targeted cationic liposomes loaded with siRNA and modified with arginine-glycine-aspartic acid (RGD) peptide. Tagging of fabricated nanocarriers with targeting ligands poses challenges in control over the number of targeting ligands attached. The lipid conjugate DSPE-PEG-RGD peptide was utilized to formulate liposomes which resulted in better control over targeting ligand. The liposomes showed a positive charge and size of 156 nm. The superior entrapment efficiency of the formulation could be attributed to the electrostatic interactions between dimethylaminoethane of the DC-cholesterol and the phosphate groups of the siRNA. The targeted liposomes showed 4-fold better siRNA delivery to retinal pigment epithelial cells than PEGylated liposomes (Chen et al., 2011). VEGF is a cardinal factor inducing angiogenesis. Anti-VEGF siRNA can be employed to silence VEGF expression for age-related macular degeneration. Chen et al. have also developed integrin peptide targeted vascular endothelial growth factor (VEGF)-siRNA loaded liposomes (123 nm). The liposomes showed superior delivery of siRNA and knockdown of VEGF expression in ARPE–19 cells (Chen et al., 2013).

Globally, numerous research efforts are directed towards enhancement of quality of life for people with diabetes. Diabetic retinopathy remains a major concern contributing to blindness. Increased VEGF signaling is also present in diabetic retinopathy. Hindering VEGF mediated angiogenesis can improve vision in diabetic retinopathy patients. Upregulation of human antigen R (HuR) protein in rats with diabetic retinopathy has been reported. Amodio et al. fabricated solid lipid nanoparticles and liposomes loaded with siRNA for HuR silencing. HuRsiRNA was bound to cationic vesicles made of ([2,3-dioleoxypropyl] N,N,trimethylamoniomethylphosphate) (DOTAP). Intraocular injection of liposomes and SLN drastically decreased retinal HuR and VEGF protein levels in diabetic rats (Amodio et al., 2016).

Hydroxyl-terminated PAMAM dendrimers possess greater safety than amine-terminated dendrimers. Mastorakos et al. developed hydroxyl-terminated polyamidoamine dendrimers modified with varying levels of amine groups for plasmid delivery. Triamcinolone acetonide was utilized to augment nuclear localization. The nanovector showed enhanced uptake and transfection in human retinal pigment epithelial cells (Mastorakos et al., 2015).

The ability of human beings to see light is contributed to a photochemical namely 11-cis-retinal. RPE protein 65 (Rpe65) regulates the levels of the chemical and deficiency of the Rpe 65 causes blindness. Rajala et al. developed liposome-protamine-DNA complex modified with TAT peptide and nuclear localization signaling peptide. The biocompatible composition of the system coupled with its enhanced capacity for gene transfection via cell penetrating function and nuclear targeting makes it a promising vector for gene delivery. The system showed persistent Rpe65 expression in mouse (Rajala et al., 2014). PEGylation of nanoparticles hinders their interaction with cellular membranes. Martens et al. (2015) developed hyaluronic acid coated polymeric pDNA gene complexes. Hyaluronic acid is favorable for ocular delivery as it is biocompatible, is a part of vitreous humor, and also binds to CD44 receptor expressed on RPE. The nanoparticles showed good uptake by and ARPE–19 cells and triggered GFP expression in the cells with least cytotoxicity. Hyaluronic acid coated complexes demonstrated excellent mobility in vitreous humor (Martens et al., 2015).

16.6.3   PHOTODYNAMIC THERAPY (PDT)

PDT represents a non-invasive and highly specific method to destroy unwanted tissues. It involves utilization of a photosensitizer, molecular oxygen, and visible or near infra-red radiation. Photosensitizers on exposure to radiation and molecular oxygen give rise to singlet oxygen and reactive oxygen species which are detrimental to target tissues (Benov, 2015). The first application of PDT in ophthalmic delivery was Visudyne®. It was a liposomal benzo-porphyrin derivative monoacid ring A (BPDMA) administered in the form of an intravenous infusion. It was intended for choroidal neovascularization in people with age-related macular degeneration. However, the advent of intravitreal Ranibizumab (anti-VEGF antibody) replaced the usage of PDT for choroidal neovascularization. PDT can be utilized for treatment of ocular surface conditions as well as intraocular tumors (Rishi and Agarwal, 2015). Ideta et al. developed negatively charged dendritic porphyrin and loaded the same into a polymeric micelle through electrostatic interaction for treatment of choroidal neovascularization lesions. The system demonstrated augmented accumulation and PDT effect in choroidal neovascularization lesions in rats (Ideta et al., 2005). Hypocrellin B is extremely lipophilic and has poor water solubility. Natesan et al. developed PLGA nanoparticles loaded with hypocrellin B and nanosilver (135 nm). Nanosilver reportedly acts as an anti-angiogenic. The nanoparticles showed sustained release of hypocrellin B. Evaluation of anti-angiogenic activity using chick chorioallantoic membrane assay revealed strong anti-angiogenic activity of the formulation (Natesan et al., 2016).

16.7   CONCLUSION

Nanocarriers are endowed with incredible potential to address major challenges in ocular delivery namely targeting to desired sites, controlled drug release, and improving drug penetration. Fabrication of mucoadhesive nanoparticles enhances binding to ocular mucosal tissue and prolongs the ocular residence time. Nanocarriers can be designed using specific matrix materials which will degrade at a specific rate and provide continuous release of the drug.

Development of novel nanomedicines for ocular conditions would be driven by persistent efforts from different disciplines of material science, formulation development, and pharmacology. Delivery of sufficient amounts of therapeutic agent to the posterior segments of the eye is a challenge in itself. Another challenge is delivery of genetic material for treatment of degenerative ocular problems. The labile nature of genetic material and its poor penetration are major hindrances to successful gene delivery. Although several research endeavors are being undertaken related to the application of nanocarriers for intravitreal delivery, which impede their journey from bench to bedside, include low volume that can be instilled into the vitreous cavity, loading ability of nanosystems, and intricate nature of industrial fabrication of nanocarriers. With respect to application of nanomedicine in ophthalmic delivery, there is certainly more to it than meets the eye. Optimization of nanocarrier size, selection of safe excipients, tuning release kinetics from nanomatrices, controlling elimination kinetics of nanocarriers from ocular tissue, sterilization of nanosystems, effects on visual acuity are issues which need to be adequately resolved. The dearth of *in vitro* or *ex vivo* models for evaluation of potential formulations further obstructs development of innovative dosage forms.

KEYWORDS

- **age-related macular degeneration**
- **anterior**
- **gene delivery**
- **liposomes**
- **micelles**
- **mucoadhesive**
- **photodynamic therapy**
- **posterior**

REFERENCES

Abdelkader, H., et al., (2016). Phytosome-hyaluronic acid systems for ocular delivery of L-carnosine. *Int. J. Nanomedicine, 11*, 2815–2827.

Ahuja, M., Verma, P., & Bhatia, M., (2015). Preparation and evaluation of chitosan–itraconazole co-precipitated nanosuspension for ocular delivery. *Journal of Experimental Nanoscience, 10*(3), 209–221.

Al-Halafi, A. M., (2014). Nanocarriers of nanotechnology in retinal diseases. *Saudi Journal of Ophthalmology, 28*(4), 304–309.

Amadio, M., et al., (2016). Nanosystems based on siRNA silencing HuR expression counteract diabetic retinopathy in rat. *Pharmacol Res., 111*, 713–720.

Baig, M. S., et al., (2016). Application of box–Behnken design for preparation of levofloxacin-loaded stearic acid solid lipid nanoparticles for ocular delivery: Optimization, *in vitro* release, ocular tolerance, and antibacterial activity. *Int. J. Biol. Macromol., 85*, 258–270.

Benov, L., (2015). Photodynamic therapy: Current status and future directions. *Med. Princ. Pract., 24*(1), 14–28.

Bisht, R., et al., (2017a). Preparation and evaluation of PLGA nanoparticle-loaded biodegradable light responsive injectable implants as a promising platform for intravitreal drug delivery. *Journal of Drug Delivery Science and Technology, 40*, 142–156.

Bisht, R., et al., (2017b). Nanocarrier mediated retinal drug delivery: Overcoming ocular barriers to treat posterior eye diseases. *Wiley Interdiscip. Rev. Nanomed. Nanobiotechnol.*, doi: 10.1002/wnan.1473.

Campos, E. J., et al., (2017). Opening eyes to nanomedicine: Where we are, challenges and expectations on nanotherapy for diabetic retinopathy. *Nanomedicine, 13*(6), 2101–2113.

Chandasana, H., et al., (2014). Corneal targeted nanoparticles for sustained natamycin delivery and their PK/PD indices: An approach to reduce dose and dosing frequency. *Int. J. Pharm., 477*(1 & 2), 317–325.

Chang, C. Y., et al., (2017). Preparation of arginine–glycine–aspartic acid-modified biopolymeric nanoparticles containing epigallocatechin–3-gallate for targeting vascular endothelial cells to inhibit corneal neovascularization. *Int. J. Nanomedicine, 12*, 279–294.

Chastain, J. E., (2003). General considerations in ocular drug delivery. In: Mitra, A. K., (ed). *Ophthalmic Drug Delivery Systems* (pp. 59–107). Marcel Dekker, Inc.: New York.

Chen, C. W., et al., (2011). Novel RGD-lipid conjugate-modified liposomes for enhancing siRNA delivery in human retinal pigment epithelial cells. *Int. J. Nanomedicine, 6*, 2567–2580.

Chen, C. W., et al., (2013). Efficient downregulation of VEGF in retinal pigment epithelial cells by integrin ligand-labeled liposome-mediated siRNA delivery. *Int. J. Nanomedicine, 8*, 2613–2627.

Chopra, P., Hao, J., & Li, S. K., (2012). Sustained release micellar carrier systems for iontophoretic transport of dexamethasone across human sclera. *J. Control. Release, 160*(1), 96–104.

Dave, V. S., & Bhansali, S. G., (2016). Pharmacokinetics and pharmacodynamics of ocular drugs. In: Pathak, Y., Sutaria, V., & Hirani, A. A., (eds.), *Nanobiomaterials for Ophthalmic Drug Delivery*, Springer: Switzerland, pp. 111–129.

Dave, V. S., (2016). Formulation approaches for ocular drug delivery. In: Pathak, Y., Sutaria, V., & Hirani, A. A., (eds.), *Nanobiomaterials for Ophthalmic Drug Delivery* (pp. 147–175). Springer: Switzerland.

Delplace, V., Payne, S., & Shoichet, M., (2015). Delivery strategies for treatment of age-related ocular diseases: From a biological understanding to biomaterial solutions. *J. Control Release, 219,* 652–668.

Dilnawaz, F., & Sahoo, S. K., (2016). Nanosystem in ocular bioenvironment. In: Pathak, Y., Sutaria, V., & Hirani, A. A., (eds.), *Nanobiomaterials for Ophthalmic Drug Delivery* (pp. 535–554). Springer: Switzerland.

Fu, T., et al., (2017). Ocular amphotericin B delivery by chitosan-modified nanostructured lipid carriers for fungal keratitis-targeted therapy. *J. Liposome Res., 27*(3), 228–233.

Gallarate, M., et al., (2013). Development of O/W nanoemulsions for ophthalmic administration of timolol. *Int. J. Pharm., 440*(2), 126–134.

Gan, L., et al., (2013). Recent advances in topical ophthalmic drug delivery with lipid-based nanocarriers. *Drug Discov. Today., 18*(5 & 6), 290–297.

Gandhi, M., et al., (2014). Niosomes: Novel drug delivery system. *Int. J. Pure App. Biosci., 2*(2), 267–274.

Gupta, D., (2016). Colloidal carriers in ophthalmic drug delivery. In: Pathak, Y., Sutaria, V., & Hirani, A. A., (eds.), *Nanobiomaterials for Ophthalmic Drug Delivery* (pp. 321–349). Springer: Switzerland.

Hamdi, Y., Lallemand, F., & Benita, S., (2015). Drug-loaded nanocarriers for back-of-the-eye diseases- formulation limitations. *Journal of Drug Delivery Science and Technology, 30(Part B),* 331–341.

Hirani, A., Pathak, Y., (2016). Introduction to nanotechnology with special reference to ophthalmic delivery. In: Pathak, Y., Sutaria, V., & Hirani, A. A., (eds.), *Nanobiomaterials for Ophthalmic Drug Delivery* (pp. 1–8). Springer: Switzerland.

Honda, M., et al., (2013). Liposomes and nanotechnology in drug development: Focus on ocular targets. *Int. J. Nanomedicine, 8,* 495–503.

Huang, J., et al., (2017). Ocular cubosome drug delivery system for timolol maleate: Preparation, characterization, cytotoxicity, *ex vivo,* and *in vivo* evaluation. *AAPS Pharm. Sci. Tech.,* doi: 10.1208/s12249-017-0763-8.

Huu, V. A., et al., (2015). Light-responsive nanoparticle depot to control release of a small molecule angiogenesis inhibitor in the posterior segment of the eye. *J. Control Release, 200,* 71–77.

Ideta, R., et al., (2005). Nanotechnology-based photodynamic therapy for neovascular disease using a supramolecular nanocarrier loaded with a dendritic photosensitizer. *Nano. Lett., 5*(12), 2426–2431.

Jain, A., Shah, S. G., & Chugh, A., (2015). Cell-penetrating peptides as efficient nanocarriers for delivery of antifungal compound, natamycin for the treatment of fungal keratitis. *Pharm. Res., 32*(6), 1920–1930.

Joseph, R. R., & Venkatraman, S. S., (2017). Drug delivery to the eye: What benefits do nanocarriers offer? *Nanomedicine (Lond)., 12*(6), 683–702.

Kakkar, S., & Kaur, I. P., (2011). Spanlastics—A novel nanovesicular carrier system for ocular delivery. *Int. J. Pharm., 413*(1 & 2), 202–210.

Kambhampati, S. P., & Kannan, R. M., (2013). Dendrimer nanoparticles for ocular drug delivery. *J. Ocul. Pharmacol. Ther., 29*(2), 151–165.

Katzer, T., et al., (2014). Prednisolone-loaded nanocapsules as ocular drug delivery system: Development, *in vitro* drug release and eye toxicity. *J. Microencapsul., 31*(6), 519–528.

Kaur, H., et al., (2012). Carboxymethyl tamarind kernel polysaccharide nanoparticles for ophthalmic drug delivery. *Int. J. Biol. Macromol., 50*(3), 833–839.

Kaur, I. P., & Kakkar, S., (2014). Nanotherapy for posterior eye diseases. *J. Control. Release.*, *193*, 100–112.

Kim, J., Schlesinger, E. B., & Desai, T. A., (2015). Nanostructured materials for ocular delivery: Nanodesign for enhanced bioadhesion, transepithelial permeability, and sustained delivery. *Ther. Deliv.*, *6*(12), 1365–1376.

Li, J., et al., (2015). Positively charged micelles based on a triblock copolymer demonstrate enhanced corneal penetration. *Int. J. Nanomedicine*, *10*, 6027–6037.

Li, Y., et al., (2017). Ion-paired pirenzepine-loaded micelles as an ophthalmic delivery system for the treatment of myopia. *Nanomedicine*, *13*(6), 2079–2089.

Liu, R., et al., (2012). Nanostructured lipid carriers as novel ophthalmic delivery system for mangiferin: Improving *In vivo* ocular bioavailability. *J. Pharm. Sci.*, *101*(10), 3833–3844.

Liu, S., et al., (2016). Prolonged ocular retention of mucoadhesive nanoparticle eye drop formulation enables treatment of eye diseases using significantly reduced dosage. *Mol. Pharm.*, *13*(9), 2897–2905.

Macha, S., Mitra, A. K., & Hughes, P. M., (2003). Overview of ocular drug delivery. In: Mitra, A. K., (ed.), *Ophthalmic Drug Delivery Systems* (pp. 1–12). Marcel Dekker Inc.: New York.

Mahor, A., et al., (2016). Moxifloxacin loaded gelatin nanoparticles for ocular delivery: Formulation and *in-vitro, in-vivo* evaluation. *J. Colloid Interface Sci.*, *483*, 132–138.

Malavade, S., (2016). Overview of the ophthalmic system. In: Pathak, Y., Sutaria, V., & Hirani, A. A., (eds.), *Nano-biomaterials for Ophthalmic Drug Delivery* (pp. 9–36). Springer: Switzerland.

Martens, T. F., et al., (2015). Coating nanocarriers with hyaluronic acid facilitate intravitreal drug delivery for retinal gene therapy. *J. Control Release*, *202*, 83–92.

Mastorakos, P., et al., (2015). Hydroxyl PAMAM dendrimer-based gene vectors for transgene delivery to human retinal pigment epithelial cells. *Nanoscale*, *7*(9), 3845–3856.

Mohanty, B., et al., (2015). Development and characterization of itraconazole-loaded solid lipid nanoparticles for ocular delivery. *Pharm. Dev. Technol.*, *20*(4), 458–464.

Natesan, S., et al., (2017). Hypocrellin B and nano silver loaded polymeric nanoparticles: Enhanced generation of singlet oxygen for improved photodynamic therapy. *Mater. Sci. Eng. C. Mater. Biol. Appl.*, *77*, 935–946.

Patel, A., et al., (2013). Ocular drug delivery systems: An overview. *World J. Pharmacol.*, *2*, 47–64.

Patravale, V., & Prabhu, P., (2014). Lipid based drug delivery systems. In: Udupa, N., & Mutalik, S., (ed.), *Recent Trends in Novel Drug Delivery* (p. 80). Prism Books Pvt. Ltd.: Bengaluru.

Prabhu, P., & Patravale, V., (2012). The upcoming field of theranostic nanomedicine: An overview. *Journal of Biomedical Nanotechnology*, *8*(6), 859–882.

Puras, G., et al., (2014). A novel cationic niosome formulation for gene delivery to the retina. *J. Control. Release*, *174*, 27–36.

Qu, W., et al., (2017). EpCAM antibody-conjugated mesoporous silica nanoparticles to enhance the anticancer efficacy of carboplatin in retinoblastoma. *Materials Science and Engineering C.*, *76*, 646–651.

Rajala, A., et al., (2014). Nanoparticle-assisted targeted delivery of eye-specific genes to eyes significantly improves the vision of blind mice *in vivo*. *Nano Lett.*, *14*(9), 5257–5263.

Ramya, S. V., et al., (2017). A review on Cubosomes drug delivery system. *Indian Journal of Drugs*, *5*(3), 104–108.

Reimondez-Troitiño, S., Csaba, N., Alonso, M. J., & De la Fuente, M., (2015). Nanotherapies for the treatment of ocular diseases. *Eur. J. Pharm. Biopharm.*, *95*, 279–293.

Rishi, P., & Agarwal, V., (2015). Current role of photodynamic therapy in ophthalmic practice. *Sci. J. Med. & Vis. Res. Foun.*, *XXXIII*(2), 97–99.

Sangave, N. A., Preuss, C., Pathak, Y., (2016). Pharmacological considerations in ophthalmic drug delivery. In: Pathak, Y., Sutaria, V., & Hirani, A. A., (eds.), *Nano-Biomaterials for Ophthalmic Drug Delivery* (pp. 37–56). Springer: Switzerland.

Saraiva, S. M., et al., (2017). Synthetic nanocarriers for the delivery of polynucleotides to the eye. *Eur. J. Pharm. Sci.*, *103*, 5–18.

Sharma, A., & Taniguchi, J., (2017). Emerging strategies for antimicrobial drug delivery to the ocular surface: Implications for infectious keratitis. *Ocul. Surf.*, S1542–0124(16)30115-X.

Shen, H. H., et al., (2015). Nanocarriers for treatment of ocular neovascularization in the back of the eye: New vehicles for ophthalmic drug delivery. *Nanomedicine (Lond).*, *10*, 2093–2107.

Shen, J., et al., (2009). Mucoadhesive effect of thiolated PEG stearate and its modified NLC for ocular drug delivery. *J. Control Release*, *137*(3), 217–223.

Soltani, S., et al., (2016). Comparison of different nanosuspensions as potential ophthalmic delivery systems for ketotifen fumarate. *Adv. Pharm. Bull.*, *6*(3), 345–352.

Sunkara, G., & Kompella, U. B., (2003). Membrane transport processes in the eye. In: Mitra, A. K., (ed.), *Ophthalmic Drug Delivery Systems* (pp. 13–58). Marcel Dekker Inc.: New York.

Tan, G., et al., (2017). Bioadhesive chitosan-loaded liposomes: A more efficient and higher permeable ocular delivery platform for timolol maleate. *Int. J. Biol. Macromol.*, *94*, 355–363.

Ustündağ-Okur, N., et al., (2014). Preparation and *in vitro–in vivo* evaluation of ofloxacin loaded ophthalmic nanostructured lipid carriers modified with chitosan oligosaccharide lactate for the treatment of bacterial keratitis. *Eur. J. Pharm. Sci.*, *63*, 204–215.

Vallet-Regi, M., et al., (2018). Mesoporous silica nanoparticles for drug delivery: Current insights. *Molecules*, *23*, 47.

Vyas, S., et al., (2015). Multidimensional ophthalmic nanosystems for molecular detection and therapy of eye disorders. *Curr. Pharm. Des.*, *21*(22), 3223–3238.

Wang, Y., et al., (2013). Stability of nanosuspensions in drug delivery. *J. Control Release*, *172*(3), 1126–1141.

Weng, Y., et al., (2017). Nanotechnology-based strategies for treatment of ocular disease. *Acta Pharm. Sin .B.*, *7*(3), 281–291.

Yasin, M. N., et al., (2014). Implants for drug delivery to the posterior segment of the eye: A focus on stimuli-responsive and tunable release systems. *J. Control Release.*, *196*, 208–221.

Yin, J., Xiang, C., & Lu, G., (2016). Cationic lipid emulsions as potential bioadhesive carriers for ophthalmic delivery of palmatine. *J. Microencapsul.*, *33*(8), 718–724.

Zhao, R., et al., (2017). Development of timolol-loaded galactosylated chitosan nanoparticles and evaluation of their potential for ocular drug delivery. *AAPS Pharm. Sci. Tech.*, *18*(4), 997–1008.

Zubairu, Y., et al., (2015). Design and development of novel bioadhesive niosomal formulation transcorneal delivery of anti-infective agent: *In-vitro* and *ex-vivo* investigations. *Asian Journal of Pharmaceutical Sciences*, *10*(4), 322–330.

CHAPTER 17

Nanomedicine for Radiation Therapy

FEIFEI YANG[1,3], YOULI XIA[2], and YU MI[1*]

[1]*Department of Radiation Oncology, University of North Carolina at Chapel Hill, Chapel Hill, North Carolina 27599, United States, E-mail: yumi1@med.unc.edu*

[2]*Department of Bioinformatics & Computational Biology Curriculum, University of North Carolina at Chapel Hill, Chapel Hill, North Carolina 27599, United States*

[3]*Institute of Medicinal Plant Development (IMPLAD), Chinese Academy of Medical Sciences & Peking Union Medical College, Haidian District, Beijing, P.R. China*

*Corresponding author. E-mail: yumi1@med.unc.edu

ABSTRACT

Radiation therapy (radiotherapy), which arose from and progressed through innovations in physics, engineering, and biology, is one of the most important modalities in cancer treatment. The limitation of radiotherapy is the damage to normal tissue during irradiation. A promising area that may overcome the limitation and improve radiotherapy is nanomedicine. Nanomaterials with unique size, shape and surface properties are ideally suitable for radiotherapy. As a nanodelivery system, they are able to deliver radioisotopes or radiosensitizers for enhanced therapeutic response and diagnosis by enhanced permeability and retention (EPR) effect. As nanotherapeutics, they are able to combine with radiation therapy for the improved therapeutic index by synergism. In this chapter, we depict the concept of nanomedicine for radiation therapy and summarize its applications in cancer treatment and diagnosis.

17.1   INTRODUCTION TO RADIATION THERAPY

17.1.1   HISTORY

Radiation therapy applies high-energy radiation to damage the DNA of cancer cells (see Figure 17.1). It is found that cancer cells dividing in an unregulated manner were more susceptible and more prone to radiation-induced DNA damage (Chen et al., 2014; Schaue, 2015). The history of radiation dated back to 1895 when Roentgen discovered X-ray and three years later when Marie Curie and Pierre Curie discovered radioactivity of radium (Connell, 2009). The first record for clinical use of ionizing radiation in cancer treatment was in late 19th century shortly after Roentgen's discovery (Sgantzos, 2014). From then on, the applications of radiation spread to many different types of cancers including head and neck, breast, cervix, prostate, eye, and thyroid cancer (Connell, 2009). Nowadays, radiation therapy is serving as curative, adjuvant, neoadjuvant or palliative therapy in cancer management and has become a common intervention for certain malignancies. More than 60% of the cancer patients receive radiotherapy in their course of treatment (Delaney, 2005; Durante, 2010).

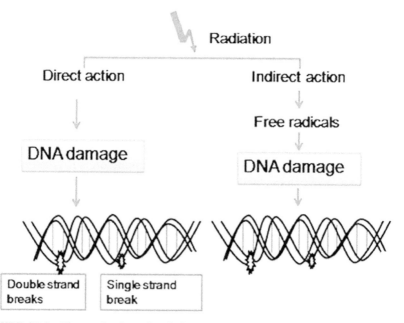

FIGURE 17.1   The mechanism of radiation therapy. (Source: Reprinted from permission via https://creativecommons.org/licenses/by/4.0 from Baskar et al., 2014.)

Basically, the high-energy radiation for radiation therapy can be obtained from X-rays, gamma rays, and charged particles (Barton et al., 2006; van de Bunt et al., 2006). External (electrons, protons, photons) and internal radiotherapy (brachytherapy) are utilized depending on the patient's clinical indications (Zhang et al., 2010; Guedea, 2014). For external therapy, the ionizing radiation such as X-ray or gamma ray comes from a source outside the patient's body and targets the tumor site. For internal therapy, a source of radiation such as ribbons or capsules containing radioisotopes is put inside the patient's body directly at the site of a cancerous tumor. The advantage of brachytherapy is decreasing the exposure of radiation to surrounding healthy tissues (Baskar et al., 2012; Mi et al., 2016).

With the development of imaging techniques such as computed tomography (CT), magnetic resonance imaging (MRI), ultrasound and X-ray imaging during the last decade, imaging-guided radiation therapy (IGRT) is utilized to improve the therapeutic efficacy of radiotherapy (Baskar et al., 2014). In IGRT, the radiation machine is equipped with imaging equipment, which allows the physician to localize the tumor before or during the irradiation by computer-assisted analysis. It is used to treat tumors on organs with continuous movements such as lung, liver and prostate gland, or the tumors located close to critical tissues. With the help of IGRT, patients are able to receive more precise radiation at the tumor and avoid damage to their healthy surrounding tissues.

17.1.2 CHALLENGES

The therapeutic response of treating cancer is dependent on the cancer's radio-sensitivity. Highly radiosensitive cancers, such as lymphomas and germ cell tumors, respond well to modest doses of radiation. However, many cancers such as pancreatic cancer and glioblastoma are relatively resistant to radiotherapy (Scaife et al., 2011). It is a challenge to improve the therapeutic index of radiotherapy in the less radio-responsive tumors. One strategy is combining radiation therapy with other therapeutic modalities such as chemotherapy and immunotherapy, which may also increase the toxicity for patients. For example, the mortality risk of chemo-radiotherapy in lung cancer is approximately 5%, which is higher than either chemotherapy or radiotherapy alone (Minami-Shimmyo, 2012).

Another challenge is the side effect of radiation to surrounding healthy tissues (Hubenak et al., 2014). External beam delivers the radiation to the tumor and its surrounding normal tissues due to the microscopic spread of cancer cells, as well as the movement of the tumor caused by breathing or unconscious body movement (Retif et al., 2015; Schaue, 2015). Therefore, radiation causes damage not only in cancer cells but also in normal cells near the tumor site. Generally, radiation therapy may lead to acute side effects such as skin irritation, hair loss, and urinary problems; and chronic side effects such as fibrosis, diarrhea, bleeding, memory loss, and infertility.

17.2   APPLICATION OF NANOMEDICINE IN RADIATION THERAPY

17.2.1   *APPLICATION OF NANOPARTICLES TO DELIVER RADIOISOTOPES*

Radioisotopes emit energy from the nucleus and generate ionized atoms or free radicals to induce single-strand cleavages in DNA. Radioisotopes used in clinical radiation therapy are alpha-emitters such as ^{225}Ac, ^{211}At, and ^{213}Bi; and beta-emitters such as ^{186}Re, ^{188}Re, ^{166}Ho, ^{89}Sr, ^{32}P, and ^{90}Y (Hamoudeh, 2008). Alpha-emitters have short half-life and strong damage to cells, but they are restricted by limited penetration around 50 to 80 μm. On the contrary, beta-emitters have relatively better tissue penetration ranging from 20 to 130 mm but low linear energy transfer.

Radioisotopes show limited effect due to their short circulating time and fast elimination from human body (Rogers, 1996). There are multiple mechanisms for how the human body eliminates radioisotopes. Many of the radioisotopes undergo rapid clearance by the kidney. In particular, renal clearance is size dependent. Those smaller than 5 nm will be excreted rapidly. Radioisotopes, as small molecules, suffer from short circulation time in blood and are unable to achieve enough therapeutic effect. Nanoparticle, such as micelle, liposome or polymeric complex, is an ideal tool to protect radioisotopes from rapid renal clearance. Generally, their size (in the range of 10 to 200 nm) and surface modification make them escape from renal elimination for a better pharmacokinetic profile. For example, through loading into nanoparticles, the half-life of ^{89}Sr increased from 47 h in plasma to 50.5 days (Brigger, 2002; Feng, 2007; Kim, 2010).

Another elimination for radioisotopes is the process of opsonization where radioisotopes are cleared by mononuclear phagocyte system. The opsonization can be prevented by introducing polyethylene glycol (PEG) on the surface of nanoparticles. The stereospecific blockade originated from PEG can impede the phagocytosis of opsonin, therefore decreasing the elimination (Yan, 2005). The PEGylation is widely used to increase the half-life of radioisotopes in blood circulation. For example, Wang et al. reported a PEGylated liposome containing ^{111}In and [177]Lu. The results demonstrated that the half-life time of ^{111}In and ^{177}Lu raised from less than 2 h to 10.2 and 11.5 h, respectively (Wang, 2006).

In addition to improving the pharmacokinetic profile, nanoformulation of radioisotopes increases tumor accumulation and decreases distribution in normal tissues. For example, Li and his colleagues developed beta-emitter ^{64}Cu-labeled nanoparticles for breast cancer radiation therapy (Zhou, 2015). They showed that over 90% of ^{64}Cu-labeled nanoparticles were retained in tumor with no distinct side effects. When combining with photodynamic therapy, the nanoparticles could not only delay the BT474 breast tumor growth, but also result in 7.6-fold longer survival time and significantly less metastasis to lungs, compared to the free radioisotopes. Vanpouille-Box et al. reported [188]Re-loaded nanocapsules (~50 nm) for internal glioblastoma radiotherapy. The biodistribution study showed that 86% of nanocapsules reached brain and 78% of initial dose resided in brain 96 h post injection (Vanpouille-Box, 2011). Their radioisotopes-loaded nanocapsules resulted in a better survival rate for rats bearing glioblastomas. In another study, the ^{131}I-loaded dendrimers were developed for single-photon emission computed tomography (SPECT) imaging and tumor therapy. The dendrimer showed significant improvement in efficacy and survival rate on rats bearing C6 xenografted tumor model, compared with free [131]I (DeSario, 2017).

17.2.2 APPLICATION OF NANOPARTICLES TO DELIVER RADIOSENSITIZERS

Radiosensitizer is a substance that improves the sensitivity of cancer cells to irradiation for more efficient radiotherapy. Example of radiosensitizers includes wortmannin, fluorouracil (5-FU, Adrucil) and cisplatin (Platinol). Radiosensitizers will accumulate in tumor more efficiently by being formulated with nanoparticles. For example, wortmannin is an inhibitor of phosphatidylinositol 3′kinases and phosphatidylinositol 3′ kinase-related kinases such as DNA-dependent protein kinases. It was confirmed by preclinical study as an effective radiosensitizer. However, its efficacy is limited by poor physicochemical properties including low solubility, low instability, and high toxicity. Wang et al. developed a core-shell nanoparticle with PLGA polymer core and DSPE-PEG-lipid shell for wortmannin delivery. It showed superiority in a KB cell-xenografted mice model with three to five times higher MTD, compared with free wortmannin (see Figure 17.2)

(Karve et al., 2012). Similar approaches have been applied in DNA double-strand repair inhibitors. DNA double-strand repair inhibitors such as histone deacetylase inhibitor (HDACI) show low efficiency in solid malignancies. This limitation can be overcome by loading them into nanoparticles. Nanoparticles can release HDACI in a sustained and controlled manner. It has been demonstrated that HDACI-loaded nanoparticles were more effective than free HDACIs in xenograft models of colorectal and prostate carcinomas (Wang, 2015).

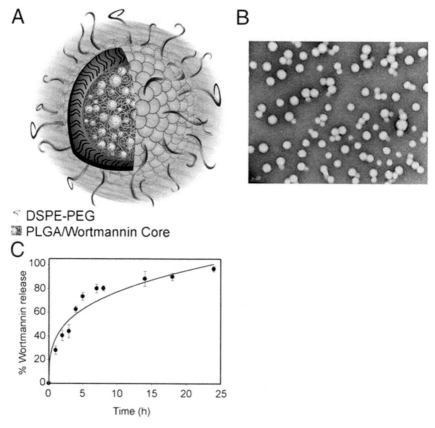

FIGURE 17.2 Characterization of NP Wtmn. (A) Cartoon of NP Wtmn depicting a PLGA core containing Wtmn surrounded by a lipid monolayer (green head groups) and a PEG shell. (B) TEM image of NP Wtmn. (C) Release profile of NP Wtmn in PBS at 37°C. Error bars correspond to SD of three separate sample preparations with duplicate samples per data point. (Source: Reprinted with permission from Karve et al., 2012. © PNAS. https://doi.org/10.1073/pnas.1120508109)

In addition to encapsulating radiosensitizers, some nanomaterials with high atomic numbers (Z) can work as radiosensitizers themselves. Gold nanoparticles are the most widely used nanoradiosensitizers. The dose absorbed by any tissue is related to the Z^2 of the material. If an agent can increase the overall effective Z of tumor without affecting the Z of nearby tissue, it can lead to increased radiotherapy dose to tumors and higher therapeutic efficacy. One example is the glutathione-coated gold nanocluster. Efficient uptake of gold nanocluster by tumors (~8.1% ID/g 24 h post injection) led to a significantly increased response of tumors to radiotherapy and reduced damage to normal tissues. In addition, there's no long-term side effect induced by nanocluster accumulation because they can be cleared by kidney (Zhang, 2015). Gadolinium (Z = 64) is another widely used radiosensitizer. Miladi et al. found DTPA modified gadolinium-based ultrasmall nanoparticles could increase nonrepairable DNA breaks by shortening the G2/M phase blockage *in vitro*. They also observed significant delay in tumor growth when combing with radiation *in vivo*. Shi et al. developed rattle nanoparticles composed of upconversion core as radiation amplifiers. Tirapazamine (a hypoxia-activated prodrug) was loaded in the nanoparticle as hypoxia-selective cytotoxin to overcome the oxygen dependency of radiotherapy. The nanoparticles showed high histocompatibility and low cytotoxicity *in vivo*. Moreover, hypoxic tumor cells were killed specifically when combining with low-dose radiotherapy (Liu, 2015). Other potential applications of inorganic nanoparticles in radiation therapy are still under investigation, such as superparamagnetic zinc ferrite spinel ($ZnFe_2O_4$) in treatment of human prostate cancer cells (Meidanchi, 2015), and Y_2O_3-based psoralen-functionalized nanoscintillators to kill cancer cells (Scaffidi, 2011).

17.2.3   APPLICATION OF NANOPARTICLES FOR IMAGE-GUIDED RADIATION THERAPY (IGRT)

IGRT utilizes high-resolution images, originated from CT, PET-CT, ultrasound or MRI, for disease diagnosis and treatment in radiotherapy (Mi et al., 2016). IGRT is particularly important in treating moving tumors by tracking the real-time location of the tumor (Sharp, 2004). The application of nanotechnology in IGRT improves the contrast between tumor and surrounding tissue, therefore increasing the precision of radiotherapy (Schaue, 2015).

Superparamagnetic iron oxide (SPIO) nanoparticles, such as ferumoxide, ferumoxtran–10 and ferucarbotran, have been approved clinically to enhance the T2 contrast of MRI. In one study, patients with cancer nodal metastases were successfully diagnosed by MRI using lymphotropic SPIO nanoparticles as T2 contrast. Its sensitivity increased from 35.4% to 90.5% with the help of SPIO nanoparticles (Harisinghani, 2003). SPIO nanoparticles can also assist to define the accurate location of lymphatic regions for radiation therapy (Vilarino-Varela, 2008; Ross, 2009; Meijer, 2012).

In addition to SPIO, gadolinium nanoparticles are also investigated for T1-enhanced MRI imaging. For example, AgulX nanoparticles are composed of polysiloxane network surrounded by gadolinium. When comparing with commercially used agents, it gave better T1-weighted imaging. Furthermore, the AgulX nanoparticle-guided radiotherapy showed an increased median survival (Le Duc, 2014).

Gold nanoparticles are widely used in CT-guided radiotherapy because they are able to offer three-fold more X-ray attenuation per unit weight than iodine (Dykman, 2012; Zhang, 2015). Andresen et al. designed a nanogel composition based on poly(N-isopropyl acrylamide) (PNIPAM)-coated gold nanoparticle as a novel liquid fiducial tissue marker for 2D X-ray image guidance. Excellent visibility of the nanogel was observed by micro-CT imaging after the injection of them in mice. The result was further validated in a cancer patient with a large spontaneous solid tumor, where the nanogel displayed high image contrast and was able to align radiation treatment for the patient (Jølck, 2015).

In addition to mono imaging-guided radiotherapy, nanoparticles can achieve co-encapsulation of agents for multimodal imaging. One example is the nanocomposite that contains both SPIO nanoparticles and gold nanoparticles. CT has the advantages in rapid scanning with high spatial resolution but suffers from poor soft-tissue contrast. Whereas MRI gives high soft-tissue contrast but suffers from long scanning time and sub-optimal geometrical accuracy. Multimodal imaging combing both CT and MRI is promising for better imaging-guided radiation therapy. Tsourkas et al. reported a PCL-PEG micelle system for co-delivery of SPIO and gold nanoparticles. It achieved improved MRI image with distinct tumor margins and significantly higher tumor accumulation. The survival rate increased from 25% in nanoparticle free group to 75% in nanoparticle-guided radiotherapy treatment group within 90 days (McQuade, 2015).

17.2.4 APPLICATION OF NANOPARTICLES FOR MONITORING IN RADIATION THERAPY

Nanoparticles can be utilized during and after radiation therapy for monitoring and evaluation. Real-time dose evaluation was one of the most important concerns during radiotherapy. A process called *in vivo* dosimetry assesses the delivery dose by measuring entry, exit or luminal dose. For example, Rege et al. reported plasmonic gold nanoparticles as nanosensor. The nanosensor is capable of determining the radiation dose from 0.5 Gy to 2 Gy in a linear fashion. Detection limit can rise to a dose of 5–37 Gy by adjusting the concentration and chemical property of the liquid surfactant. The nanosensor could quantitatively determine the radiation dose received by a organism simply with an absorbance spectrophotometer (Pushpavanam, 2015).

Another important issue is to monitor the side effect of radiation therapy. One of the side effects for patients after radiotherapy is vascular problem. For example, breast cancer patients have four times higher possibility to suffer cardiovascular events after radiotherapy (Baskar et al., 2012), which may further induce stroke, myocardial infarction, and atherosclerosis (Aleman, 2003). It was reported that cardiovascular issues induced by radiation are related to acute increase of proinflammatory cytokines and adhesion molecules at the endothelium of injured blood vessels (Halle, 2010). To monitor the side effect after radiation therapy, Wang et al. developed a basement membrane (BM)-targeting nanoparticle to detect radiotherapy-related vessel injury. In this study, the collagen IV-targeting peptide-conjugated PLGA-PEG nanoparticle was fabricated to bind the lesion of endothelium by targeting collagen IV-rich BM. A preclinical study on a murine model confirmed its ability to detect vessel injury at early-stage after high dose radiation (Au, 2015).

17.3 APPLICATION OF NANOMEDICINE FOR COMBINATION RADIATION THERAPY

17.3.1 COMBINATION WITH CHEMOTHERAPY

Combination of radiotherapy with chemotherapy is one of the most effective ways to improve patient's survival with locally advanced cancers. Chemotherapy can sensitize tumor cells to radiation-induced killing. In addition, concurrent therapy avoids the repopulation of cancer cells during the course of sequential treatment (Vaishnaw, 2010). However, increased toxicity is the main limitation of the strategy.

Nanotechnology enhances chemo-radiotherapy in two ways. One is to deliver chemotherapeutics by nanoparticles and combine with external irradiation. The chemotherapeutics, such as cisplatin, doxorubicin, and paclitaxel, can be radiosensitizers at the same time (Jung, 2012; Werner, 2013;

Xiong, 2015). The other is to co-deliver chemotherapeutics and radiosensitizers/radioisotopes by nanoparticles, which achieves precise ratio control of the agents at lesion. Both of the nanoapproaches decrease toxicity in normal tissues and promote accumulation of agents in tumors. For example, cisplatin is used as chemotherapeutic as well as radiosensitizer. Shi et al. reported to deliver cisplatin with a rattled-structured upconversion nanoparticle for chemo-radiotherapy. The nanoparticles co-delivered cisplatin and high-Z metal ions (Yb^{3+}, Gd^{3+}) and achieved enhanced treatment effect on mice bearing Hela xenograft tumors (Fan, 2013). In another study, the authors showed that combining docetaxel with wortmannin in PLGA nanoparticles maximized therapeutic efficacy due to the ability of nanoparticles to release the drugs in a desirable sequential fashion. Meanwhile, there was no increased toxicity *in vivo* (Au, 2015).

Targeted nanoparticles are also developed for chemo-radiotherapy. The cell penetration efficiency of nanoparticles can be boosted by surface modification of targeted ligand including folate, Arg-Gly-Asp (RGD) peptide or transferrin. In one study, docetaxel-loaded, folate-conjugated nanoparticles were developed as a nanoradiosensitizer. *In vivo* results revealed that targeted nanoparticles were more efficient, compared to the nanoparticles without targeting ligands. Interestingly, the radiosensitization efficacy was dependent on the timing of irradiation (Werner et al., 2011). Folate-targeted nanoparticles were utilized in co-delivery of paclitaxel and yttrium–90, which was demonstrated to be more efficient in a murine model with metastatic ovarian cancer (Werner et al., 2011). A similar design with aptamer as targeting ligand was developed for combining docetaxel with indium–111 and yttrium–90 (Wang et al., 2010).

Liposomal doxorubicin plus conventionally fractionated radiotherapy was the first example of nanochemo-radiotherapy in clinical trial for locally advanced non-small-cell lung cancer (NSCLC) and head and neck cancer (HNC). In a phase I trial, it achieved 40% complete response in NSCLC patients and 75% in HNC patients (Koukourakis, 1999). Liposomal cisplatin combining with radiotherapy was also reported to have 55% complete response at the primary tumor site for HNC patients (Rosenthal, 2002). There are also several clinical trials using albumin-stabilized nanoparticles containing paclitaxel (nab-paclitaxel) for chemo-radiotherapy. In a phase II trial, nab-paclitaxel and gemcitabine hydrochloride were used as chemotherapeutics followed by radiotherapy in treating patients with pancreatic cancer (NCT02427841). In addition, polymer-based nanoparticles, including polymer-drug conjugates or polymeric nanoparticles, are under evaluation in clinic. For example, a phase I trial was conducted to determine the maximal tolerated dose of poly(l-glutamic acid)-paclitaxel combining with radiation for patients bearing esophageal and gastric cancer (Dipetrillo et al., 2006).

17.3.2 COMBINATION WITH IMMUNOTHERAPY

Radiotherapy can trigger a phenomenon called abscopal effect where local tumor treatment induces systemic regression of metastatic lesions (Hiniker, 2012). With the development of cancer immunotherapy, evidences support that the abscopal effect after radiation therapy is caused by the activation of immune system. Radiation induces pro-inflammatory protein production for immune stimulation. It further increases the exposure of immune cells to cancer-specific antigens that are released following radiotherapy-induced cancer cell death (Rasaneh, 2015). In this process, nanoparticles are able to capture tumor-specific antigens and activate antigen-presenting cells (APCs) to improve the abscopal effect, thus enhance immunotherapy efficacy. For example, Wang et al. used PLGA nanoparticles to capture tumor-specific antigens after the exposure of tumor to irradiation. It showed a 20% cure rate using the B16F10 melanoma model, compared to 0% without the antigen-capture ring nanoparticles. It also showed increased CD4+/Treg and CD8+/Treg ratios (Min, 2017).

17.4 APPLICATION OF NANOMEDICINE TO OVERCOME RADIATION RESISTANCE

Radiation resistance is the main reason for failure of radiation therapy. It occurs through several mechanisms such as expression of anti-apoptotic proteins or DNA repair enzymes (Zhao, 2013; Al-Dimassi, 2014). Hypoxic tumor microenvironment, caused by abnormal vascular structure and rapid proliferation rates, plays an important role in radiation resistance. Hypoxia leads to the reduction of oxygen and reactive species such as ROS. It prevents the irreparable DNA damage. Cancer cells in the hypoxic environment are 2–3 fold more radio-resistant than that under normal oxygen supply (Willers, 2013; Barker, 2015). Patients with median oxygen tensions of more than 10 mmHg have disease-free survival rate of 78%; while patients with median oxygen tensions of less than 10 mmHg have the rate of 22% (Brizel, 1997).

Radiation resistance could be attenuated through down-regulating the expression of related genes with small interfering RNA (siRNA). However, the clinical application of siRNA is limited by high enzymatic degradation, fast clearance and low penetration for cancer cells (Zhao, 2015). Formulation of siRNA with nanoparticles is a practical way to overcome the limitations and maximize the function of siRNA. Co-delivery of radiotherapy agents and siRNA could effectively reduce radiation resistance and achieve synergistic effect. For example, Zhang et al. reported a PEGylated nanoparticle-based siRNA delivery system. SiApe1 was delivered by this system to increase the DNA damage after irradiation. The expression of Ape1 was knocked down over 75% in medulloblastoma cells and ependymoma cells. The down-regulation of Ape1 led to more than 3-fold reduction of LD50 after irradiation (Kievit, 2015). Kjems et al. delivered siTNFα by chitosan/siRNA complex and completely prevented the radiation-induced fibrosis in CDF1 mice after a single dose of 45 Gy (Nawroth et al., 2010). Gao et al. used PEG-PEI copolymer for siRNA delivery against sCLU protein. The viability of MCF–7cells was 38% at 0.5 Gy and 3% at 3 Gy for the combination group, compared with 93% at 0.5 Gy and 54% at 3 Gy for the radiotherapy only group (Sutton, 2006).

17.5 SUMMARY

In the past few decades, nanotechnology undergoes an explosive development, which benefits many therapeutic approaches including the traditional radiation therapy. Agents, failed on previous clinical trials due to delivery problem or high toxicity, get another chance to be reassessed for its clinical use. Radiosensitizers or radioisotopes, which often suffer from high toxicity in their molecular form, find a relatively safe way for translational application in medicine. However, it is also necessary to concern the acute systemic toxicity of nanodelivery system. While most molecular agents cause nephrotoxicity or neurotoxicity, the formulation of nanoparticles can cause hepatotoxicity and hypological toxicity.

Recent studies on tumor microenvironment may lead to other combinational strategies for radiation therapy. Radiation therapy causes increased exposure and presentation of tumor antigens, which triggers inflammatory cytokine signaling and immune cell recruitment. Cancer immunotherapy, such as checkpoint blockade inhibition or chimeric antigen receptor (CAR) T-cell therapy, shows promising results in clinic. Combination of radiotherapy with immunotherapy is promising to be more effective, and nanotechnology can help for this strategy in the future.

KEYWORDS

- computed tomography
- enhanced permeability and retention
- imaging-guided radiation therapy
- magnetic resonance imaging
- polyethylene glycol
- single-photon emission computed tomography

REFERENCE

Al-Dimassi, S. A. A., & El-Sibai, M. T., (2014). Cancer cell resistance mechanisms: a mini review. *Clin. Transl. Oncol., 16*, 511–516.

Aleman, B. M., Klokman, W. J., Van't Veer, M. B., Bartelink, H., & Van Leeuwen, F. E., (2003). Long-term cause-specific mortality of patients treated for Hodgkin's disease. *J. Clin. Oncol., 21*, 3431–3439.

Au, K. M. H., Wagner, S. N., Shi, C. K., Kim, Y. S., Caster, J. M., Tian, X., et al., (2015a). Direct observation of early-stage high-dose radiotherapy-induced vascular injury via basement membrane-targeting nanoparticles. *Small, 11*, 6404–6410.

Au, K. M. M., Tian, X. Y., Zhang, L., Perello, V., Caster, J. M., & Wang, A. Z., (2015b). Improving cancer chemoradiotherapy treatment by dual controlled release of wortmannin and docetaxel in polymeric nanoparticles. *ACS. Nano., 9*, 8976–8996.

Barker, H. E. P., Khan, J. T. A. A., & Harrington, K. J., (2015). The tumor microenvironment after radiotherapy: Mechanisms of resistance and recurrence. *Nat. Rev. Cancer., 15*, 409–425.

Barton, M. B., et al., (2006). Role of radiotherapy in cancer control in low-income and middle-income countries. *Lancet. Oncol., 7*, 584–595.

Baskar, R., et al., (2012). Cancer and radiation therapy: Current advances and future directions. *Int. J. Med. Sci., 9*, 193–199.

Baskar, R., et al., (2014). Biological response of cancer cells to radiation treatment. *Front. Mol. Biosci., 1*, 24.

Brigger, I. D. CCouvreur, P., (2002). Nanoparticles in cancer therapy and diagnosis. *Adv. Drug. Deliv. Rev., 54*, 631–651.

Brizel, D. M. S., Prosnitz, G. S. L. R., Scher, R. L., & Dewhirst, M. W., (1997). Tumor hypoxia adversely affects the prognosis of carcinoma of the head and neck. *Int. J. Radiat. Oncol. Biol. Phys., 38*, 285–289.

Chen, Y. H., et al., (2014). Radiation-induced VEGF-C expression and endothelial cell proliferation in lung cancer. *Strahlenther. Onkol., 190*, 1154–1162.

Connell, P. P. H. S., (2009). Advances in radiotherapy and implications for the next century: A historical perspective. *Cancer. Res., 69*, 383–392.

Delaney, G. J. S., Featherstone, C., & Barton, M., (2005). The role of radiotherapy in cancer treatment: Estimating optimal utilization from a review of evidence-based clinical guidelines. *Cancer, 104*, 1129–1137.

DeSario, P. A. P. J. J., Brintlinger, T. H., McEntee, M., Parker, J. F., Baturina, O., Stroud, R. M., & Rolison, D. R., (2017). Oxidation-stable plasmonic copper nanoparticles in photocatalytic TiO2 nanoarchitectures. *Nanoscale, 9*, 11720–11729.

Dipetrillo, T., et al., (2006). Paclitaxel poliglumex (PPX-Xyotax) and concurrent radiation for esophageal and gastric cancer: A phase I study. *Am. J. Clin. Oncol., 29*, 376–379.

Durante, M. L. J.S., (2010). Charged particles in radiation oncology. *Nat. Rev. Clin. Oncol., 7*, 37–43.

Dykman, L. K., N., (2012). Gold nanoparticles in biomedical applications: Recent advances and perspectives. *Chem. Soc. Rev., 41*, 2256–2282.

Fan, W. S. B., Bu, W., Chen, F., Zhao, K., Zhang, S., Zhou, L., et al., (2013). Rattle-structured multifunctional nanotheranostics for synergetic chemo-/radiotherapy and simultaneous magnetic/luminescent dual-mode imaging. *J. Am. Chem. Soc., 135*, 6494–6503.

Feng, S. S. Z. L. Y., Zhang, Z. P., Bhakta, G., Win, K. Y., Dong, Y. CChien, S., (2007). Chemotherapeutic engineering: Vitamin E TPGS-emulsified nanoparticles of biodegradable polymers realized sustainable paclitaxel chemotherapy for 168 h *in vivo. Chem. Eng. Sci., 62*, 6641–6648.

Guedea, F., (2014). Perspectives of brachytherapy: Patterns of care, new technologies, and "new biology." *Cancer Radiother., 18*, 434–436.

Halle, M. G. A., Paulsson-Berne, G., Gahm, C., Agardh, H. E., Farnebo, F., & Tornvall, P., (2010). Sustained inflammation due to nuclear factor-kappa B activation in irradiated human arteries. *J. Am. Coll. Cardiol., 55*, 1227–1236.

Hamoudeh, M. K. M. A., Diab, R., & Fessi, H., (2008). Radionuclides delivery systems for nuclear imaging and radiotherapy of cancer. *Adv. Drug. Deliv. Rev., 60*, 1329–1346.

Harisinghani, M. G. B. J., Hahn, P. F., Deserno, W. M., Tabatabaei, S., Van de Kaa, C. H., De la Rosette, J., & Weissleder, R., (2003). Noninvasive detection of clinically occult lymph-node metastases in prostate cancer. *N. Engl. J. Med., 348*, 2491–2499.

Hiniker, S. M. C. D. S., & Knox, S. J., (2012). Abscopal effect in a patient with melanoma. *N. Engl. J. Med., 366*, 2035.

Hubenak, J. R., et al., (2014). Mechanisms of injury to normal tissue after radiotherapy: A review. *Plast. Reconstr. Surg., 133*, 49e–56e.

Jølck, R. I. R., Christensen, A. N., Hansen, A. E., Bruun, L. M., Schaarup-Jensen, H., Von Wenck, A. S., et al., (2015). Injectable colloidal gold for use in intrafractional 2d image-guided radiation therapy. *Adv. Healthc. Mater., 4*, 856–863.

Jung, J. P., Chung, H. K., Kang, H. W., Lee, S. W., Seo, M. H., Park, H. J., et al., (2012). Polymeric nanoparticles containing taxanes enhance chemoradiotherapeutic efficacy in non-small cell lung cancer. *Int. J. Radiat. Oncol. Biol. Phys., 84*, 77–83.

Karve, S., et al., (2012). Revival of the abandoned therapeutic wortmannin by nanoparticle drug delivery. *Proc. Natl. Acad. Sci. USA., 109*, 8230–8235.

Kievit, F. M. S., Wang, K., Dayringer, C. J., Sham, J. G., Ellenbogen, R. G., Silber, J. RZhang, M., (2015). Nanoparticle-mediated silencing of DNA repair sensitizes pediatric brain tumor cells to gamma-irradiation. *Mol. Oncol., 9*, 1071–1080.

Kim, B. Y. R., & Chan, W.C., (2010). Nanomedicine. *N. Engl. J. Med., 363*, 2434–2443.

Koukourakis, M. I. K., Giatromanolaki, A., Archimandritis, S. C., Skarlatos, J., Beroukas, K., Bizakis, J. G., et al., (1999). Liposomal doxorubicin and conventionally fractionated radiotherapy in the treatment of locally advanced non–small-cell lung cancer and head and neck cancer. *J. Clin. Oncol., 17*, 3512–3521.

Le Duc, G. R., Paruta-Tuarez, A., Dufort, S., Brauer, E., Marais, A., Truillet, C., Sancey, L., Perriat, P., Lux, F., & Tillement, O., (2014). Advantages of gadolinium-based ultrasmall nanoparticles vs. molecular gadolinium chelates for radiotherapy guided by MRI for glioma treatment. *Cancer Nanotechnol.*, 5, 4.

Liu, J. N. B., & Shi, J. L., (2015). Silica-coated upconversion nanoparticles: A versatile platform for the development of efficient theranostics. *Acc. Chem. Res.*, 48, 1797–1805.

McQuade, C. A. Z., Desai, Y., Vido, M., Sakhuja, T., Cheng, Z., Hickey, R. J., Joh, D., Park, S. J., Kao, G., Dorsey, J. F., & Tsourkas, A., (2015). A multifunctional nanoplatforms for imaging, radiotherapy, and the prediction of therapeutic response. *Small*, 11, 834–843.

Meidanchi, A. A., Khoei, S., Shokri, A. A., Hajikarimi, Z., & Khansari, N., (2015). ZnFe2O4 nanoparticles as radiosensitizers in radiotherapy of human prostate cancer cells. *Mater. Sci. Eng. C. Mater. Biol. Appl.*, 46, 394–399.

Meijer, H. J. D., Kunze-Busch, M., Van Kollenburg, P., Leer, J. W., Witjes, J. A., Kaanders, J. H., Barentsz, J. O., & Van Lin, E. N., (2012). Magnetic resonance lymphography-guided selective high-dose lymph node irradiation in prostate cancer. *Int. J. Radiat. Oncol. Biol. Phys.*, 82, 175–183.

Mi, Y., et al., (2016). Application of nanotechnology to cancer radiotherapy. *Cancer Nanotechnol.*, 7, 11.

Min, Y. R., Tian, S., Eblan, M. J., McKinnon, K. P., Caster, J. M., Chai, S., et al., (2017). Antigen-capturing nanoparticles improve the abscopal effect and cancer immunotherapy. *Nat. Nanotechnol.*, 12, 877–882.

Minami-Shimmyo, Y. O., Yamamoto, S., Sumi, M., Nokihara, H., Horinouchi, H., Yamamoto, N., Sekine, I., Kubota, K., & Tamura, T., (2012). Risk factors for treatment-related death associated with chemotherapy and thoracic radiotherapy for lung cancer. *J. Thorac. Oncol.*, 7, 177–182.

Nawroth, I., et al., (2010). Intraperitoneal administration of chitosan/DsiRNA nanoparticles targeting TNFalpha prevents radiation-induced fibrosis. *Radiother. Oncol.*, 97, 143–148.

Pushpavanam, K. N., Chang, J., Sapareto, S., & Rege, K., (2015). A colorimetric plasmonic nanosensor for dosimetry of therapeutic levels of ionizing radiation. *ACS. Nano.*, 9, 11540–11550.

Rasaneh, S. R., & Johari, D. F., (2015). Activity estimation in radioimmunotherapy using magnetic nanoparticles. *Chin. J. Cancer. Res.*, 27, 203–208.

Retif, P., et al., (2015). Nanoparticles for radiation therapy enhancement: The key parameters. *Theranostics.*, 5, 1030–1044.

Rogers, B. E. A., Connett, J. M., Guo, L. W., Edwards, W. B., Sherman, E. L., Zinn, K. R., & Welch, M. J., (1996). Comparison of four bifunctional chelates for radiolabeling monoclonal antibodies with copper radioisotopes: Biodistribution and metabolism. *Bioconjug. Chem.*, 7, 511–522.

Rosenthal, D. I. Y., Liu, L., Machtay, M., Algazy, K., Weber, R. S., Weinstein, G. S., et al., (2002). A phase I study of SPI–077 (Stealth liposomal cisplatin) concurrent with radiation therapy for locally advanced head and neck cancer. *Invest New Drugs*, 20, 343–349.

Ross, R. W. Z., Xie, W., Coen, J. J., Dahl, D. M., Shipley, W. U., Kaufman, D. S., Islam, T., Guimaraes, A. R., Weissleder, R., & Harisinghani, M., (2009). Lymphotropic nanoparticle-enhanced magnetic resonance imaging (LNMRI) identifies occult lymph node metastases in prostate cancer patients prior to salvage radiation therapy. *Clin. Imaging*, 33, 301–305.

Scaffidi, J. P. G., Lauly, B., Zhang, Y., & Vo-Dinh, T., (2011). Activity of psoralen-functional-ized nanoscintillators against cancer cells upon X-ray excitation. *ACS. Nano.*, 5, 4679–4687.

Scaife, L., et al., (2011). Differential proteomics in the search for biomarkers of radiotherapy resistance. *Expert. Rev. Proteomics.*, 8, 535–552.

Schaue, D. M., (2015). Opportunities and challenges of radiotherapy for treating cancer. *Nat. Rev. Clin. Oncol.*, *12*, 527–540.

Sgantzos, M. T., Laios, K., & Androutsos, G., (2014). The physician who first applied radiotherapy victor Despeignes, on 1896. *Hell. J. Nucl. Med.*, *17*, 45–46.

Sharp, G. C. J., Shimizu, S., & Shirato, H., (2004). Prediction of respiratory tumor motion for real-time image-guided radiotherapy. *Phys. Med. Biol.*, *49*, 425–440.

Sutton, D. K., Shuai, X., Leskov, K., Marques, J. T., Williams, B. R., Boothman, D. A., & Gao, J., (2006). Efficient suppression of secretory cluster in levels by polymer-siRNA nanocomplexes enhances ionizing radiation lethality in human MCF–7 breast cancer cells in vitro. *Int. J. Nanomedicine*, *1*, 155–162.

Vaishnaw, A. K. G. J. G. V., Hutabarat, R., Sah, D., Meyers, R., De Fougerolles, T., & Maraganore, J., (2010). A status report on RNAi therapeutics. *Silence*, *1*, 14.

Van de Bunt, L., et al., (2006). Conventional, conformal, and intensity-modulated radiation therapy treatment planning of external beam radiotherapy for cervical cancer: The impact of tumor regression. *Int. J. Radiat. Oncol. Biol. Phys.*, *64*, 189–196.

Vanpouille-Box, C. L., Belloche, C., Lepareur, N., Lemaire, L., LeJeune, J. J., Benoit, J. P., Menei, P., Couturier, O. F., Garcion, E., & Hindre, F., (2011). Tumor eradication in rat glioma and bypass of immunosuppressive barriers using internal radiation with (188) Re-lipid nanocapsules. *Biomaterials.*, *32*, 6781–6790.

Vilarino-Varela, M. J. T., Rockall, A. G., Reznek, R. H., & Powell, M. E., (2008). A verification study of proposed pelvic lymph node localization guidelines using nanoparticle-enhanced magnetic resonance imaging. *Radiother. Oncol.*, *89*, 192–196.

Wang, A. Z., et al., (2010). ChemoRad nanoparticles: A novel multifunctional nanoparticle platform for targeted delivery of concurrent chemoradiation. *Nanomedicine (Lond).*, *5*, 361–368.

Wang, E. C. M., Palm, R. C., Fiordalisi, J. J., Wagner, K. T., Hyder, N., Cox, A. D., Caster, J. M., Tian, X., & Wang, A. Z., (2015). Nanoparticle formulations of histone deacetylase inhibitors for effective chemoradiotherapy in solid tumors. *Biomaterials.*, *51*, 208–215.

Wang, H. E. Y., Lu, Y. C., Heish, N. N., Tseng, Y. L., Huang, K. L., Chuang, K. T., Chen, C. H., Hwang, J. J., et al., (2006). Internal radiotherapy and dosimetric study for 111In/177Lu-pegylated liposomes conjugates in tumor-bearing mice. *Nucl. Instrum. Methods. Phys. Res. A.*, *569*, 533–537.

Werner, M. E. C., Sethi, M., Wang, E. C., Sukumar, R., Moore, D.T, & Wang, A. Z., (2013). Preclinical evaluation of Genexol-PM, a nanoparticle formulation of paclitaxel, as a novel radiosensitizer for the treatment of non-small cell lung cancer. *Int. J. Radiat. Oncol. Biol. Phys.*, *86*, 463–468.

Werner, M. E., et al., (2011a). Folate-targeted nanoparticle delivery of chemo- and radio-therapeutics for the treatment of ovarian cancer peritoneal metastasis. *Biomaterials.*, *32*, 8548–8554.

Werner, M. E., et al., (2011b). Folate-targeted polymeric nanoparticle formulation of docetaxel is an effective molecularly targeted radiosensitizer with efficacy dependent on the timing of radiotherapy. *ACS. Nano.*, *5*, 8990–8998.

Willers, H. A., Santivasi, W. L., & Xia, F., (2013). Basic mechanisms of therapeutic resistance to radiation and chemotherapy in lung cancer. *Cancer. J.*, *19*, 200–207.

Xiong, H. Z., Qi, Y., Zhang, Z., Xie, Z., Chen, X., Jing, X., Meng, F., & Huang, Y., (2015). Doxorubicin-loaded carborane-conjugated polymeric nanoparticles as delivery system for combination cancer therapy. *Biomacromolecules*, *16*, 3980–3988.

Yan, X. S., & Kamps, J. A., (2005). Liposome opsonization. *J. Liposome. Res.*, *15*, 109–139.

Zhang, L., et al., (2010). Delivery of therapeutic radioisotopes using nanoparticle platforms: Potential benefit in systemic radiation therapy. *Nanotechnol. Sci. Appl.*, *3*, 159–170.

Zhang, X. D. L., Chen, J., Song, S. S., Yuan, X., Shen, X., Wang, H., et al., (2015). Ultrasmall glutathione-protected gold nanoclusters as next-generation radiotherapy sensitizers with high tumor uptake and high renal clearance. *Sci. Rep.*, *5*, 8669.

Zhang, X., (2015). Gold nanoparticles: Recent advances in the biomedical applications. *Cell. Biochem. Biophys.*, *72*, 771–775.

Zhao, J. F., (2015). Nanocarriers for delivery of siRNA and co-delivery of siRNA and other therapeutic agents. *Nanomedicine (Lond)*, *10*, 2199–2228.

Zhao, J. M., & Feng, S. S., (2013). siRNA-based nanomedicine. *Nanomedicine (Lond)*, *8*, 859–862.

Zhou, M. Z., Tian, M., Song, S., Zhang, R., Gupta, S., Tan, D., Shen, H., Ferrari, M., & Li, C., (2015). Radio-photothermal therapy mediated by a single compartment nanoplatform depletes tumor-initiating cells and reduces lung metastasis in the orthotopic 4T1 breast tumor model. *Nanoscale*, *7*, 19438–19447.

Index

T - #0796 - 101024 - C562 - 234/156/25 - PB - 9781774634431 - Gloss Lamination